ENVIRONMENTAL SURFACES AND INTERFACES FROM THE NANOSCALE TO THE GLOBAL SCALE

ENVIRONMENTAL SURFACES AND INTERFACES FROM THE NANOSCALE TO THE GLOBAL SCALE

PATRICIA A. MAURICE

WILEY

A JOHN WILEY & SONS, INC., PUBLICATION

Published by John Wiley & Sons, Inc., Hoboken, New Jersey.
Published simultaneously in Canada.

Library of Congress Cataloging-in-Publication Data:

Maurice, Patricia A.
Environmental surfaces and interfaces from the nanoscale to the global scale / Patricia A. Maurice.
p. cm.
Includes index.
ISBN 978-0-470-40036-4 (cloth)
1. Surface chemistry. 2. Environmental chemistry. I. Title.
QD506.M346 2009
577′.14–dc22

2008045483

Printed in the United States of America

10 9 8 7 6 5 4 3 2 1

For Alexander

CONTENTS

PREFACE

This book came about because after teaching a water–rock interactions course for approximately 14 years at Kent State University and the University of Notre Dame, I felt that students would benefit from a single text that covered everything from the theory of charged particle surfaces to how minerals grow and dissolve to new frontiers such as nanoparticles, geomicrobiology, and climate change. As my course evolved, it took on a broader array of environmental surfaces and interfaces; hence, the title of this book is also more encompassing.

My classes have always been interdisciplinary, with students from such diverse fields as geology, biology, chemistry, physics, environmental engineering, chemical engineering, and electrical engineering. Over the years, I have been privileged to teach students ranging from freshmen undergraduates to Ph.D. students nearing completion of their degrees. I never know quite what to expect at the start of each semester, but I have always enjoyed the challenge of adapting course material to the particular backgrounds and interests of whoever walks through my door at the start of each semester. Teaching such a diversity of students has been an exciting experience; I have learned more from the students than they have learned from me. Many of my classes have involved team laboratory- and/or field-based research projects. Some of these projects have resulted in publications. Others have encountered show-stopping experimental or analytical problems. But we have learned from them, nonetheless.

This eclectic teaching experience is reflected in the content and style of this text. Because environmental surfaces and interfaces require interdisciplinary approaches and draw multidisciplinary interest, I have tried to assume very little about student backgrounds. Individual faculty may well choose to skip certain chapters depending on the makeup of students in their classes. I have also tried to provide basic understanding along with more complex subject matter and to encourage students to look beyond this text to ongoing research in the field. This is part of my fundamental belief that our greatest role as teachers is to inspire students to pursue life-long learning. Encourage your students to be critical of the material and treatments covered in this book. Nothing in this field is set in stone (pun intended); everything is evolving.

The content of any textbook ultimately comes from scientific research, and it is important for students to understand where material comes from, how it is produced and analyzed, and how scientific "knowledge" changes over time. Thus, I have endeavored to include several *case studies* highlighting papers or research approaches stretching back decades or hot-off-the press. Students should be encouraged to follow up on the case studies through library or web-based research. How did this research come about? What has it led to? What questions arose from the research? Have the approaches, interpretations, or conclusions been questioned?

Throughout the book, many terms of interest are italicized when first introduced or at times when students might want to pay particular attention to the terminology. Many, but not all, of the terms are defined in the Glossary. The italics are intended to assist the students in identifying terms that are particularly important to study and understand.

This book would not have been possible without the critical comments, suggestions, and enthusiasm of many friends and colleagues. Each and every person whom I contacted for help—from reviewing a section to requesting a needed figure—responded quickly and enthusiastically. Writing this textbook, therefore, has been a wonderful affirmation of the scientific and engineering communities' dedication to educating our next generation. A few people have been especially kind. Steve Cabaniss (University of New Mexico) encouraged me to write a textbook and provided many useful suggestions for Chapter 8. Bob Berner (Yale University), who is one of the great founders of this field, reminded me of the importance of plants and corrected some of the material on climate change. Mike Hochella (Virginia Tech) encouraged me to go forward when I was wondering whether I could ever complete the task. Domenic Grasso (University of Vermont) pointed out extended DLVO theory. Lesley Warren (McMaster University) offered feedback based on her considerable knowledge of acid-rock drainage and microbial associations. Steven Lower (Ohio State University) has set me straight on surface forces many times over the years and has been a constant reminder that our best students are our best teachers. Jerry Schnoor (University of Iowa) sent me a Thanksgiving gift of a pile of papers and figures related to phytoremediation. Alex Navrotsky (University of California, Davis) provided much useful insight on nanoparticles. Most authors whose work is described in "case studies" provided comments that helped to keep these sections focused. Frank Incropera (University of Notre Dame) encouraged me to increase the number and diversity of quantitative problems at the end of each chapter.

Of course, the responsibility for weaknesses and errors throughout the text lies firmly on my shoulders.

Editor Bob Esposito and the editorial staff at Wiley provided the professional assistance required to see this text through to publication.

I thank my husband Greg Madey and son Alexander for love, support, and friendship. And to Greg a big thanks for keeping my computers up and running and challenging me with dozens of questions. I thank the people of Playa Flamingo, Costa Rica, for their wonderful hospitality that kept me sane in the final stretch.

¡Pura Vida!

PATRICIA A. MAURICE

Playa Flamingo, Costa Rica
July 2008

CONSTANTS AND UNITS

Constant	SI units	non-SI units
Avogadro's number (N_A)	6.022×10^{23} mol^{-1}	
Boltzmann constant (k)	1.38×10^{-23} J K^{-1}	
Faraday constant (F)	96485 C mol^{-1}	
Planck constant (h)	6.626×10^{-34} J s^{-1}	
Planck constant over 2 pi ($\hbar$)	1.054×10^{-34} J s^{-1}	
Standard accel of gravity (g)	9.807 m s^{-2}	
Molar ("ideal") gas constant (R)	8.3145 J mol^{-1} K^{-1}	1.99 cal mol^{-1} deg^{-1}
Electronic charge (-e)	1.602×10^{-19} C	4.803×10^{-10} esu
Permittivity of free space ε_o	8.854×10^{-12} C^{2}J^{-1} m^{-1}	
Gravitational constant (G)	6.670×10^{-11} Nm^{2}kg^{-2}	6.607×10^{-8} g^{-1} cm^{3}s^{-2}

Units

1 Å = 10^{-1} nm = 10^{-8} cm = 10^{-10} m

1 L = 10^{-3}m^3

1 cal = 4.184 J

1 kcal mol^{-1} = 4.184 kJ mol^{-1}

1 dyne cm^{-1} = 1 erg cm^{-2} = 1 mNm^{-1} = 1 mJm^{-2}

1 atm = 1.013×10^6 dyne cm^{-2} = 1.013 bar = 1.013×10^5 Pa = 1.013×10^5N m^{-2}

1 N = 1 kg m s^{-2}

1 Bq = 1 Becquerel = the quantity of material that will undergo one nuclear decay in a second. 3.7×10^{10} Bq = 1 Ci (Curie)

1 C = 1 Coulomb = the amount of electrical charge in 6.241506×10^{18} electrons or other elementary charged particles.

1 1A	2 2A	3	4	5	6	7	8	9	10	11	12	13	14	15	16	17	18
1 H 1.008																	2 He 4.0026
3 Li 6.941	4 Be 9.0122																10 Ne 20.180
11 Na 22.990	12 Mg 24.305											13 Al 26.982	14 Si 28.086	15 P 30.974	16 S 32.065	17 Cl 35.453	18 Ar 39.948
19 K 39.098	20 Ca 40.078	21 Sc 44.956	22 Ti 47.867	23 V 50.942	24 Cr 51.996	25 Mn 54.938	26 Fe 55.845	27 Co 58.933	28 Ni 58.693	29 Cu 63.546	30 Zn 65.38	31 Ga 69.723	32 Ge 72.64	33 As 74.922	34 Se 78.96	35 Br 79.904	36 Kr 83.798
37 Rb 85.468	38 Sr 87.62	39 Y 88.906	40 Zr 91.224	41 Nb 92.906	42 Mo 95.96	43 Tc (98)	44 Ru 101.07	45 Rh 102.91	46 Pd 106.42	47 Ag 107.87	48 Cd 112.41	49 In 114.82	50 Sn 118.7	51 Sb 121.76	52 Te 127.60	53 I 126.90	54 Xe 131.29
55 Cs 132.91	56 Ba 137.33	57-71 Lanth.	72 Hf 178.49	73 Ta 180.95	74 W 183.84	75 Re 186.21	76 Os 190.23	77 Ir 192.22	78 Pt 195.08	79 Au 196.97	80 Hg 200.59	81 Tl 204.38	82 Pb 207.2	83 Bi 208.98	84 Po (209)	85 At (210)	86 Rn (222)
87 Fr (223)	88 Ra (226)	89- 103 Actin.	104 Rf (261)	105 Db (262)	106 Sg (266)	107 Bh (264)	108 Hs (270)	109 Mt (268)	110 Ds (281)	111 Rg (272)							

Lanthanide series	57 La 138.91	58 Ce 140.12	59 Pr 140.91	60 Nd 14.424	61 Pm (145)	62 Sm 150.36	63 Eu 151.96	64 Gd 157.25	65 Tb 158.93	66 Dy 162.50	67 Ho 164.93	68 Er 167.26	69 Tm 168.93	70 Yb 173.05	71 Lu 174.97
Actinide series	89 Ac (227)	90 Th 232.04	91 Pa 231.04	92 U 238.03	93 Np (237)	94 Pu (244)	95 Am (243)	96 Cm (247)	97 Bk (247)	98 Cf (251)	99 Es (252)	100 Fm (257)	101 Md (258)	102 No (259)	103 Lr (262)

1

SOME FUNDAMENTAL CHEMICAL THERMODYNAMIC AND KINETIC CONCEPTS

Although this text encompasses many disciplines—from aquatic chemistry to mineralogy to microbiology to hydrology—it is fundamentally concerned with surface and interface *chemistry*. Thus, a basic review of some key concepts, equations, terms, and approaches from environmental biogeochemistry is offered in this chapter. This background information is essential for understanding more complex concepts and equations presented throughout the remainder of the text. The chapter begins by introducing some commonly used chemical units. Next, the difference between thermodynamic and kinetic approaches is discussed. Finally, thermodynamic and kinetic considerations are addressed in greater detail.

CONCENTRATION UNITS

To perform calculations that involve the chemical compositions of natural waters, along with water–rock interactions, we must begin by establishing some units of concentration. Some of the most important units include (wt = weight):

$$Molality,\ m = \frac{\text{moles solute}}{1000\,\text{g water}} \quad \text{at } 25^{\circ}\text{C, 1 atm pressure (STP)} \tag{1.1}$$

$$= \frac{\text{wt solute in g} \times 1000}{\text{formula wt solute} \times \text{wt water in g}} \tag{1.2}$$

$$Molarity,\ M = \frac{\text{moles solute}}{\text{L solution}} \tag{1.3}$$

$$= \frac{\text{wt solute in g} \times 1000}{\text{formula wt solute} \times \text{volume solution in mL}} \tag{1.4}$$

For dilute aqueous solutions at STP, m generally $= M$.

Environmental Surfaces and Interfaces from the Nanoscale to the Global Scale, by Patricia A. Maurice

Equivalents per liter, eq L^{-1}, takes into consideration the charge on the ion. To convert from M to equivalents per liter, we simply multiply by the absolute value of the charge on the ion, z:

$$\text{eq L}^{-1} = M|z|; \tag{1.5}$$

z would be 1 for Na^+ and 2 for Ca^{2+}.

Mole fraction, N_i, is the ratio of the number of moles of the given constituent to the total moles of all constituents j,

$$N_1 = \frac{n_1}{n_1 + n_2 + n_3 + \cdots n_j}. \tag{1.6}$$

Parts per million (ppm) and parts per billion (ppb)

$$1\text{ ppm} = \frac{1\text{ g}}{10^6\text{ g}} = \text{concentration in } \mu\text{mol kg}^{-1} \times \text{formula wt} \tag{1.7}$$

$$1\text{ ppb} = \frac{1\text{ g}}{10^9\text{ g}} = \text{concentration in } \mu\text{mol kg}^{-1} \times \text{formula wt.} \tag{1.8}$$

Some Examples of Calculations

We want to make up a solution of 0.1 M in NaCl. How much NaCl in grams do we add to 1 L of water?

First, add the atomic weights, and then multiply by 0.1 mole L^{-1}:

$$\begin{aligned} \text{Na} &= 22.99\text{ g mole}^{-1} + \frac{\text{Cl} = 35.45\text{ g mole}^{-1}}{58.44\text{ g mole}^{-1}\text{ NaCl} \times 0.1\text{ mole L}^{-1}} \\ &= 5.84\text{ g NaCl added} \end{aligned}$$

What is the mole fraction of NaCl in a 0.1-m solution?

$$\frac{0.1\text{ moles NaCl}/1000\text{ g water}}{0.1\text{ moles NaCl}/1000\text{ g} + 55.51\text{ moles water per 1000 g water}}$$
$$= 0.0018\text{ dimensionless units.}$$

THERMODYAMIC VERSUS KINETIC APPROACHES

Consider a change in state of a system from state A to state B, as illustrated schematically in Figure 1.1. The system is closed or isolated from its environment. In this system, one state A changes or reacts to form a second state B according to the overall reaction

$$\text{A} \rightleftarrows \text{B}. \tag{1.9}$$

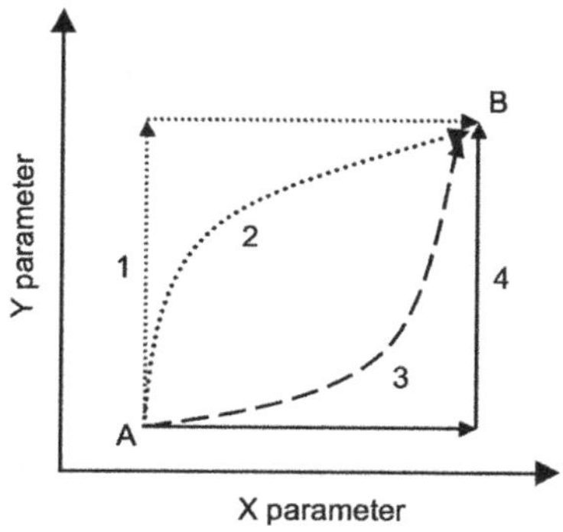

FIGURE 1.1 A schematic illustration of a change in a system from state A to state B. The x- and y-axes may represent changes in temperature and pressure, respectively, or some other parameters. For example, the state changes could be from a solid to a liquid or from a certain mineral assemblage to another.

Chemical thermodynamics is the study of energy and its transformations. Chemical thermodynamics describes the overall energy change that occurs when transitioning from state A to state B, which is independent of the pathway taken. *Chemical kinetics* is the study of rates and mechanisms of reactions. Chemical kinetics focuses on the detailed rates and mechanisms when transitioning from state A to state B along different potential pathways. Each pathway may consist of a different series of reactions or processes. Chemical thermodynamics tells us that the overall energy of the change from state A to state B is independent of the pathway taken. Kinetics focuses on the individual pathways, including the time dependency of reactions and the detailed reaction sequences.

INTRODUCTORY THERMODYNAMICS

In mineral–water interface studies, chemical thermodynamics is used to:

1. Describe all species present in water and at the mineral–water interface under equilibrium conditions. For example, how is Ca distributed in water? How much is Ca^{2+}? $CaHCO_3{}^+$? What is the *activity* or "effective concentration" (see below) of H^+ in solution or at the mineral–water interface?
2. Determine whether a given solution is in *equilibrium*, *supersaturated*, or *undersaturated* with respect to a given mineral or mineral assemblage, hence, whether any minerals should be precipitating or dissolving.
3. Predict the direction of spontaneous changes. If a system is not at equilibrium, then in what direction should it change? Based on the chemical analysis of a water sample, thermodynamic approaches allow a scientist to determine which minerals should dissolve or precipitate.
4. Determine whether a given system is an equilibrium assemblage. If it is not an equilibrium assemblage, then we need to ask ourselves whether something is wrong with the data or the thermodynamic speciation model used. If analyses are correct, then what process is preventing the system from attaining equilibrium?

5. Because rates of processes such as mineral growth and dissolution are ultimately related to saturation state relative to equilibrium, thermodynamics can provide insight into kinetics.

Gibbs Energy

Consider the system shown in Figure 1.1, in which state A transforms into state B. Each of these states will have some intrinsic energy associated with it. The measure of this energy at constant T and P is the *Gibbs energy* (*G*) (commonly known as *Gibbs free energy*). G is related to the heat content or *enthalpy H* and the *entropy* or randomness of the system *S*, by the equation:

$$G = H - TS, \quad (1.10)$$

where

G and H = SI units of Joules per kilogram, J kg^{-1} (or kilocalories per mole, kcal mol^{-1}),
S = SI units of J kg^{-1} K^{-1} (or cal mol^{-1} deg^{-1}), and
T = absolute temperature in Kelvin (273.15 + °C).

For changes to the system at constant pressure (P) and temperature (T),

$$\Delta G = \Delta H - T\Delta S. \quad (1.11)$$

For a spontaneous process, ΔG is negative; by definition at equilibrium, $\Delta G = 0$.

Chemical Potential and Activity

The *chemical potential* (μ) is a measure of how much the Gibbs energy changes with an infinitessimal addition of a particular component, and only that component, at constant temperature and pressure. For a component i,

$$\mu_i = (\delta G/\delta \mathrm{n}_i)_{\mathrm{T,P}}. \quad (1.12)$$

The chemical potential μ_i is an *intensive property*, meaning that its value is independent of the amount of a substance, whereas G is an *extensive property*, meaning that its value depends on the amount.

The chemical potential is encountered in the equation for relating *absolute concentrations* to *effective concentrations* of different species in solution. For example, in solution, there may be some shielding of reactive species by other ions in solution, and this shielding may decrease the reactivity.

The chemical potential of component i in a given state (μ_i) can be related to the chemical potential of the component in its standard state, at infinite dilution, (μ_i°)according to the equation

$$\mu_i = \mu_i^\circ + RT \ \ln \mathrm{a}_i, \quad (1.13)$$

where a_i is the activity or "effective concentration" of component i.

The activity of component i, a_i, can then be related to concentration through the activity coefficient γ_i, which accounts for nonideal behavior

$$a_i = \gamma_i \, m_i. \tag{1.14}$$

The activity coefficient is a dimensionless quantity. A variety of different equations are available for calculating γ_i for different ions in waters with different ionic strengths (see below and Langmuir, 1997). The value of $\gamma_i \rightarrow 1$ and the value of $a_i \rightarrow m_i$ as infinite dilution is approached.

One example of when activity rather than concentration is essential in environmental chemistry involves pH. A pH electrode does not measure the actual concentration of protons in solution but the effective concentration or activity

$$\mathrm{pH} = -\log a_{H^+} \tag{1.15}$$

$$= -\log \gamma_{H^+} \, m_{H^+}. \tag{1.16}$$

Equilibrium Constants

Consider a generalized equation for the reaction of b moles of reactant B with c moles of reactant C to form d moles of product D and e moles of product E

$$bB + cC \underset{k_{-1}}{\overset{k_1}{\rightleftarrows}} dD + eE. \tag{1.17}$$

At equilibrium, the rate of the forward reaction k_1 will be equal to the rate of the reverse reaction k_{-1}.

Also at equilibrium, the activities can be related by a thermodynamic *equilibrium constant* K_{eq}

$$K_{eq} = \frac{a_D^d \, a_E^e}{a_B^b \, a_C^c}. \tag{1.18}$$

Although many introductory texts relate equilibrium constants in terms of concentrations, the rigorous formulation is in terms of activities.

As another example, we can consider a mass-action equation for the dissolution of hematite, which is an Fe oxide mineral of importance in environmental surface chemistry (Chapter 3)

$$\underset{\text{hematite}}{Fe_2O_3(s)} + 6H^+(aq) \rightleftarrows 2Fe^{3+}(aq) + 3H_2O(l). \tag{1.19}$$

In Equation 1.19, the symbols in parentheses provide information on the phase: *s* for solid, *aq* for aqueous species, and *l* for liquid. The symbol *g* for gas (or vapor) is also used in this text.

The expression for the equilibrium constant is written as

$$K_{eq} = \frac{a_{Fe^{3+}}^{2} \, a_{H_2O}^{3}}{a_{Fe_2O_3} \, a_{H^+}^{6}}. \tag{1.20}$$

The activity of a solid in itself is 1, and the activity of water is also 1; hence, this will simplify to

$$K_{eq} = \frac{a_{Fe^{3+}}^{2}}{a_{H^+}^{6}}. \tag{1.21}$$

Reactions of environmental importance can also include gaseous components. An example is the formation of the carbonate mineral calcite during degassing of CO_2

$$Ca^{2+}(aq) + 2HCO_3^-(aq) \overset{\text{calcite}}{\rightleftarrows} CaCO_3(s) + CO_2(g) + H_2O(l). \tag{1.22}$$

This reaction occurs when groundwater in a limestone region enters a cave and degasses to form limestone formations (Chapters 10, 11, and 14).

For a reaction that occurs at a surface, such as the surface of a mineral or a bacterium, the reactive surface site is identified in this text using the symbol > coupled with S for "surface site" or "M" for "metal surface site" (such as Fe or Al). An example for deprotonation of a surface site (see Chapter 5) is

$$> SOH^o \rightleftarrows\ > SO^- + H^+(aq). \tag{1.23}$$

In the literature, the symbol ≡ is sometimes used as an alternative for > when indicating surface sites. The equilibrium constant for this sort of surface reaction cannot be written as easily as that involving ions in solution because of the complexities associated with the electrostatics of mineral surfaces (Chapter 6).

Calculating the Equilibrium Constant from Gibbs Energy Changes

The equilibrium constant for a reaction can be related to the Gibbs energy. If we return to the generalized reaction in Equation 1.17

$$bB + cC \rightleftarrows dD + eE, \tag{1.17}$$

at equilibrium, the sum of the free energies of products is equal to the sum of free energies of reactants, in each case taking into consideration their stoichiometric coefficients ν_i.

To calculate Gibbs energy changes, we must first obtain the *Gibbs energy of formation* (ΔG_f°) of each of the components. The *Gibbs energy of formation* is the energy of the reaction to form 1 mole of the substances in their standard states from the standard elements under standard conditions of temperature and pressure. We take by convention as zero the standard free energy of formation of the H^+ ion in aqueous solution. The values of ΔG_f° can be obtained from thermodynamic compilations such as at the National Institute of Standards and Technology (NIST) database.

The *standard free energy of a reaction* such as shown in Equation 1.17 is given by

$$\Delta G_r^\circ = \sum_i \nu_i \Delta G_f^\circ \text{ products} - \sum_i \nu_i \Delta G_f^\circ \text{ reactants}. \tag{1.24}$$

An Example Calculation

Let us take as an example the dissolution of calcite to release Ca^{2+} and CO_3^{2-} in aqueous solution at 25°C. We shall assume a simple system, neglecting formation of other species such as bicarbonate (HCO_3^-; see below).

$$\begin{aligned}
&\text{calcite} \\
CaCO_3(s) &\rightleftarrows Ca^{2+}(aq) + CO_3^{2-}(aq) \qquad (1.25)\\
\Delta G_f^\circ \text{ calcite} &= -1129.07\ \text{kJ mol}^{-1}\\
\Delta G_f^\circ\ Ca^{2+} &= -552.8\ \text{kJ mol}^{-1}\\
\Delta G_f^\circ\ CO_3^{2-} &= -527.9\ \text{kJ mol}^{-1}\\
\Delta G_r^\circ &= [(1)(-552.8) + (1)(-527.9)] - (1)(-1129.07)\\
&= 48.37\ \text{kJ mol}^{-1}
\end{aligned}$$

ΔG_r° for the reaction in Equation 1.17 is related to the equilibrium constant (at 1 atm, 25°C) by the expression:

$$\Delta G_r^\circ = -RT \ln K_{eq} \tag{1.26}$$

where K_{eq} is given by Equation 1.18.

We can then calculate the equilibrium constant $K_{eq} = K_{cal}$ at 25°C (298.15 K) for the reaction in Equation 1.25

$$\begin{aligned}
48.37\ \text{kJ mol}^{-1} &= -0.008314\ \text{kJ mol}^{-1}\ \text{K}^{-1} \times 298.15 \ln K_{cal}\\
K_{cal} &= 10^{-8.48}.
\end{aligned}$$

Equation 1.26 is the expression at equilibrium. In the more general case, the Gibbs energy of a reaction ΔG_r is related to the standard energy of reaction ΔG_r° and to the *reaction quotient* Q by the following relationship:

$$\Delta G_r = \Delta G_r^\circ + RT \ln\left(\frac{a_D^d a_E^e}{a_B^b a_C^c}\right), \tag{1.27}$$

$$\Delta G_r = \Delta G_r^\circ + RT \ln \text{Q} \tag{1.28}$$

Q, the reaction quotient, is distinguished from the equilibrium constant K because Q is based on the values of activities for a system whether or not it is at equilibrium.

For a system at equilibrium, $\Delta G_r = 0$, $Q = K_{eq}$, and

$$\Delta G_r^\circ = -RT \ln K_{eq}. \tag{1.29}$$

This fundamental equation is used to calculate the equilibrium constant from ΔG_r° at 25°C.

We shall see in Chapter 5 that when a new surface or interface is formed over the course of a reaction, the change in *surface free energy* also must be taken into consideration.

Temperature Effects on K_{eq}

If the temperature of a study system is close to 25°C, and if ΔH_r° is independent of temperature, then we can calculate the change in K_{eq} with the change in temperature from T_1 to T_2 from

$$\begin{aligned}\Delta G_r^\circ &= \Delta H_r^\circ - T\Delta S_r^\circ, \\ &= -RT \ln K_{eq}\end{aligned} \tag{1.30}$$

or

$$\ln K_{eq} = -\Delta H_r^\circ/RT + T\Delta S_r^\circ/RT. \tag{1.31}$$

This equation can be differentiated with respect to T to obtain

$$\mathrm{dln}\, K_{eq}/dT = \Delta H_r^\circ/RT^2 \tag{1.32}$$

And then upon partial integration from T_1 to T_1

$$\ln \frac{K_2}{K_1} = \int_{T_1}^{T_2} \Delta H_r^\circ RT^2\, dT \tag{1.33}$$

so that

$$\ln K_2 - \ln K_1 = \frac{\Delta H_r^\circ}{R}\left(\frac{1}{T_1} - \frac{1}{T_2}\right), \tag{1.34}$$

where K_1 is the equilibrium constant at T_1 and K_2 is the equilibrium constant at T_2.

An Example Calculation

Let us take as an example the solubility product ($K_{sp} = K_{eq} = K_{cal}$) of calcite in a system that is closed to the atmosphere so that CO_2 gas is not exchanging (Equation 1.25). We are given that $\Delta H_r^\circ = -9.61$ kJ mol^{-1}.

$$\underset{\text{calcite}}{CaCO_3(s)} \rightleftarrows Ca^{2+}(aq) + CO_3^{2-}(aq) \tag{1.25}$$

at 25°C 1 atm pressure,

$$K_{\text{cal}} = 10^{-8.48} = a_{Ca^{2+}} a_{CO_3^{2-}}. \tag{1.35}$$

How does the equilibrium constant change with temperature from 25°C to 35°C (298.15 to 308.15 K)?

$$\begin{aligned}
\Delta H_r^\circ &= -9.61\,\text{kJ mol}^{-1} \\
R &= 8.314\,\text{J mol}^{-1}\ \text{deg}^{-1} \\
\ln K_{308.15} - \ln(10^{-8.48}) &= \frac{-9610\,\text{J mol}^{-1}}{8.314\,\text{J mol}^{-1}\ \text{K}^{-1}}(1/298.15\,\text{K} - 1/308.15\,\text{K}) \\
\ln K_{308.15} &= -19.67 \\
K_{308.15} &= 10^{-8.54}.
\end{aligned}$$

What happens to the solubility of calcite with increasing temperature? It decreases. Conversely, as temperature decreases, calcite becomes more soluble. This dependence on temperature impacts C cycling and plays a role in determining how the distributions of carbonate-shelled organisms and carbonate deposition in the oceans changes with climate (Chapter 2). For most oxides, solubilities increase with increasing temperature. But, for calcite and some other minerals such as pyromorphite (Chapter 9), solubility decreases with increasing temperature.

Calculating Activities

Now, let us go back to the activity/molarity relationships. A method is needed to estimate the effective concentration from the actual concentration in solution. To do this, we first need to describe the *ionic strength* (I) of a solution

$$I = \frac{1}{2}\sum_i (m_i z_i^2), \tag{1.36}$$

where m_i is the concentration and z_i is the charge on each ion i.

If we take the simplest case of ionic interactions, where all interactions are considered to be *electrostatic*, all ions are taken as point charges, and the ions around any particular ion follow a *Boltzman* distribution, we get the simplified *Debye-Hückel equation*,

$$\log \gamma_i = -\text{A}z_i^2\,\text{I}^{1/2}, \tag{1.37}$$

which provides a good estimate for γ_i values to I ~ 10^{-3} M. At 25°C, 1 atm, A = 0.5085.

To I ~ 10^{-1} M, an extended form is used, which takes into account the finite size of ions (*Extended Debye-Hückel equation*):

$$\log \gamma_i = \frac{-\text{A}z_i^2\,\text{I}^{1/2}}{1 + \text{Ba}_\text{o}\,\text{I}^{1/2}}, \tag{1.38}$$

TABLE 1.1 Values of effective hydrated radius a_o of some ions of interest in environmental chemistry.

ion	a_o
H^+	9.0
K^+	3.5
HCO_3^-	5.4
Na^+	4.0
SO_4^{-2}	5.0
Ca^{2+}	5.0
Sr^{2+}	5.0
Mg^{2+}	5.5
Cl^-	3.5

Values are in Å.

where B at 25°C and 1 atm = 0.3281; a_o is the effective hydrated radius (Table 1.1).

For still higher ionic strengths (to I ~ 0.5 M), the *Güntelberg equation* is used

$$\log \gamma_i = \frac{-Az_i^2\, I^{1/2}}{1 + I^{1/2}} + bI, \qquad (1.38)$$

where b is an empirical constant, obtained from data tables. This is called the *Davies equation* when b is simply set equal to 0.3.

Other formulations for calculating activity coefficients for higher ionic strengths are described by Langmuir (1997). In Chapter 6, we shall see how calculating the activity coefficients for species at a surface is one of the primary challenges of interface chemistry.

An Example Calculation

What is the ionic strength of a solution of 0.05 M in NaCl?

$$1/2 \times [(0.05\,\text{M})(1)^2 + (0.05\,\text{M})(1)^2] = 0.05\,\text{M}.$$

What is the activity of Na in this solution, if a_o for Na = 4.0?

To calculate γ_{Na}, we can use the extended form of the Debye-Hückel equation

$$\log \gamma_{Na} = \frac{-0.5085(1)^2\,(0.05)^{1/2}}{1 + 0.3281 \times 4.0 \times (0.05)^{1/2}}$$

$$\log \gamma_{Na} = -0.08791$$

$$\gamma_{Na} = 0.82$$

$$a_{Na} = 0.82 \times 0.05\,\text{M} = 0.041\ \text{moles L}^{-1}.$$

We can see from this simple calculation that the effective concentration or activity is only about 80% of the concentration in moles L^{-1}. Consideration of activity is essential for many applications in aquatic and surface chemistry. Many calculations such as for determining species distributions and saturation state with respect to a given mineral can be incorrect by a significant amount if molarity (or molality) rather than activity is used. In many instances, an activity is raised to a power of 2 or 3 (or more), and the correction can become even more important.

Let us return to the calcium carbonate system as an example of a complex series of calculations involving activities. Assume for the current purposes, to simplify, that there is no complex ion formation in solution (e.g., no bicarbonate ion, HCO_3^-).

An Example Calculation

What are the activities of Ca^{2+} and CO_3^{2-} in equilibrium with calcite at 25°C in the system calcite-water (closed to the atmosphere, neglecting other species such as HCO_3^-; see below)?

$$K_{cal} = a_{Ca^{2+}}\, a_{CO_3^{2-}} = 10^{-8.48}. \quad (1.35)$$

To calculate the activities, we need first to come up with activity constants. To do this, an iterative approach is used in which we first approximate activity = concentration (here, concentration is symbolized by []), or $\gamma_i = 1$, and then use the concentrations to calculate I:

$$[Ca^{2+}] = [CO_3^{2-}] = (10^{-8.48})^{1/2} = 10^{-4.24}$$
$$I = 1/2[(10^{-4.24} \times (2)^2) + (10^{-4.24} \times (2)^2)]$$
$$I = 2.3 \times 10^{-4}$$

We can use the simple Debye-Hückel equation:

$$\log \gamma_i = -Az_i^2\, I^{1/2}, \text{ with } A = 0.5085$$
$$\log \gamma_{CO_3} = \log \gamma_{Ca} = -0.5085(2)^2\,(2.3 \times 10^{-4})^{1/2}$$
$$= -0.03$$
$$\gamma_{Ca} = \gamma_{CO_3} = 0.933$$

Using these provisional values, we can now recalculate Ca^{2+} and CO_3^{2-} concentrations and hence ionic strength

$$([Ca^{2+}] \times 0.933)([CO_3^{2-}] \times 0.933) = 10^{-8.48}$$
$$[Ca^{2+}] = [CO_3^{2-}] = 6.17 \times 10^{-5}$$
$$I = 1/2\,[(6.17 \times 10^{-5} \times (2)^2) + (6.17 \times 10^{-5} \times (2)^2)]$$
$$= 2.47 \times 10^{-4},$$

using the simple Debye-Hückel as above

$$\gamma_{Ca} = \gamma_{CO_3} = 0.929.$$

This is so close to the original value that we shall stop the iterations here and use the value 0.93.

The solubility of calcite in water, therefore, will be

$$(10^{-8.48}/\gamma_{Ca}\gamma_{CO_3})^{1/2} = (10^{-8.48}/0.93^2)^{1/2} = 6.19 \times 10^{-5}\ \text{moles L}^{-1}.$$

This concentration can be converted to g L^{-1} $CaCO_3$ by multiplying by the molecular weight of 100.08 g $mole^{-1}$ $CaCO_3$ to give 6.19×10^{-3} g L^{-1} $CaCO_3$.

If we had assumed that activity was equal to concentration, then the solubility would have been

$$(10^{-8.48})^{1/2} = 5.75 \times 10^{-5}\ \text{M} \times 100.08\ \text{g mole}^{-1} = 5.75 \times 10^{-3}\ \text{g L}^{-1}\ \text{CaCO}_3.$$

Thus, the calculated solubility is actually greater when the fact that activity deviates from concentration is taken into consideration.

Now, what happens if we add a nonreactive salt, such as 0.05 M of NaCl, to the system above? The effect of the salt will be to increase the ionic strength and, hence, to decrease the value of the activity coefficients. This will result in calcite being more soluble, which is the subject of one of the problems at the end of this chapter. This example provides insight into the surface and interface processes of mineral growth and dissolution discussed in Chapters 9 and 10.

Saturation Indices (SIs)

One of the most important ways in which thermodynamics is applied in environmental surface chemistry is to determine whether a given water sample is supersaturated, under-saturated, or in equilibrium with any mineral phases. This is of importance to studies of mineral growth and dissolution (Chapters 9 and Chapter 10). It is also important to sorption processes (Chapter 7 and 8) because the potential for dissolution of the mineral *sorbent* or precipitation of the *sorbate* could affect the experimental results (see Chapter 6).

Geochemists often want to know how close to equilibrium a system is with respect to a certain mineral phase. Commonly, saturation indices are used. For example, for dissolution of calcite, $CaCO_3$ (*s*),

$$\underset{\text{calcite}}{CaCO_3(s)} \rightleftarrows Ca^{2+}(aq) + CO_3^{2-}(aq) \tag{1.25}$$

the equilibrium constant, which in the case of mineral dissolution is known as the solubility product (K_{sp}), can be written as

$$K_{sp} = a_{Ca}\, a_{CO_3} = 10^{-8.48}. \tag{1.35}$$

The ion activity product (IAP) at equilibrium is given by

$$\text{IAP} = a_{Ca}\, a_{CO_3}. \tag{1.39}$$

The IAP is equal to the K_{sp} at equilibrium, but it will have another value if the solution is supersaturated or undersaturated with respect to calcite.

If the IAP is divided by the solubility product, the saturation ratio Ω is

$$\frac{\text{IAP}}{K_{sp}} = 1, \text{ at equilibrium; } >1, \text{ supersaturated; } <1, \text{ undersaturated.} \tag{1.40}$$

or

$$\frac{a_{Ca}\, a_{CO_3}}{10^{-8.48}} = 1, \text{ at equilibrium; } >1, \text{ supersaturated; } <1, \text{ undersaturated.}$$

More commonly, the saturation index is determined as a logarithm:

$$\log\frac{IAP}{K_{sp}} = 0, \text{ at equilibrium; } >0 \text{ in a system that is supersaturated; } <0 \text{ if undersaturated.} \tag{1.41}$$

In crystal growth, the *driving force*:

$$\sigma = \Delta\mu / k_B T = \ln(\text{IAP}/K_{sp}) \tag{1.42}$$

is related to the change in chemical potential $\Delta\mu$ and, hence, to supersaturation state (σ); k_B is the Boltzmann constant, and T is absolute temperature.

We shall encounter *driving force* again in discussing mineral nucleation, in Chapter 9.

Carbonate Equilibria in Open or Closed Systems

Thus far, we have made the assumption that the only carbonate species present in solution is the carbonate ion (CO_3^{2-}). However, in aqueous solutions, other species such as the doubly protonated carbonic acid (H_2CO_3) and singly protonated bicarbonate (HCO_3^-) will also exist (along with other species such as potentially sodium bicarbonate $NaHCO_3$, which we will ignore here), linked by the following equations:

$$H_2CO_3 \rightleftarrows H^+ + HCO_3^- \tag{1.43}$$

$$K_{H_2CO_3} = \frac{a_{HCO_3^-} a_{H^+}}{a_{H_2CO_3}} = 10^{-6.35} \tag{1.44}$$

and

$$HCO_3^- \rightleftarrows H^+ + CO_3^{2-} \tag{1.45}$$

$$K_{HCO_3} = \frac{a_{CO_3^{2-}} a_{H^+}}{a_{HCO_3^-}} = 10^{-10.33}. \tag{1.46}$$

Figure 1.2 shows a *Bjerrum plot* of the different species in the carbonate system as a function of pH at 25°C and with a total carbonate species concentration, Σcarbonate $= 10^{-2.5}$. This plot shows how the predominant species changes from doubly protonated carbonic acid to bicarbonate to carbonate as the pH of the water increases. Plots at different total carbonate concentrations would look similar, but they are shifted up or down accordingly.

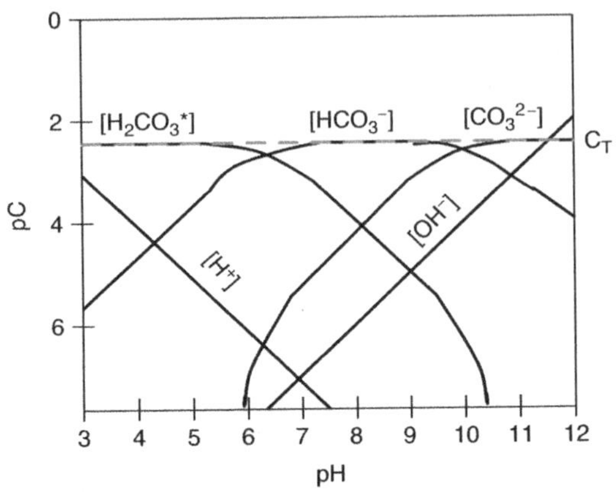

FIGURE 1.2 A Bjerrum plot for carbonate equilibria in a system closed to the atmosphere, at a total carbonate concentration of $10^{-2.5}$ M.

If a system is open to the atmosphere, H_2CO_3 can form through the association of water with CO_2:

$$CO_2(g) + H_2O(l) \rightleftarrows H_2CO_3(aq), \tag{1.47}$$

with the equilibrium constant written as

$$K_{CO_2} = \frac{a_{H_2CO_3}}{P_{CO_2}} = 10^{-1.47} \text{ at } 25°\text{C}. \tag{1.48}$$

H_2CO_3 can lose a proton to form HCO_3^-, and HCO_3^- can lose a proton to form CO_3^{2-}.

A Bjerrum plot for an open system with $P_{CO_2} = 10^{-3.5}$ (Figure 1.3) shows the transition from predominantly doubly protonated carbonic acid to bicarbonate to carbonate as pH increases. But because CO_2 is taken up more by solution at higher pH values, there is not a single total carbonate concentration.

Calcite Equilibria in a System Open to Atmospheric Carbon Dioxide

Example Calculation Let us consider a system containing an excess of the mineral calcite in equilibrium with water open to the atmosphere such that the atmosphere serves as an infinite reservoir of a fixed partial pressure of carbon dioxide, P_{CO_2}. What is the pH of this solution?

For this example, we shall set $P_{CO_2} = 10^{-3.50}$ and set activity equal to concentration to simplify the solution.

Because this problem includes solid calcite, we need to consider Equations 1.25 and 1.35:

$$\begin{array}{l}\text{calcite}\\ CaCO_3(s) \rightleftarrows Ca^{2+}(aq) + CO_3^{2-}(aq)\end{array} \tag{1.25}$$

$$K_{sp} = K_{CaCO_3} = a_{Ca}\, a_{CO_3} = 10^{-8.48}. \tag{1.35}$$

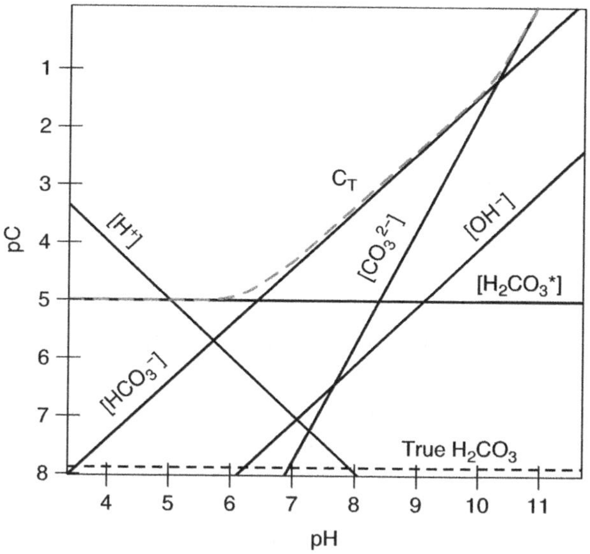

FIGURE 1.3 A Bjerrum plot for a system open to the atmosphere. After Stumm and Morgan, 1996 (with permission).

This problem can be solved in various ways. One of the most straightforward methods is to solve a series of simultaneous equations by first identifying the system unknowns and then coming up with several equations that are equal to the number of unknowns. We shall assume to simplify our calculations for the purpose of demonstration that concentration = activity.

Six unknown species concentrations are as follows: $[Ca^{2+}]$, $[CO_3{}^{2-}]$, $[HCO_3{}^{-}]$, $[OH^-]$, $[H^+]$, and $[H_2CO_3]$.

We thus need six equations. We can use the four above:

$$K_{CaCO_3} = [Ca^{2+}][CO_3^{2-}] = 10^{-8.48}$$

$$K_{H_2CO_3} = \frac{[HCO_3^-][H^+]}{H_2CO_3} = 10^{-6.35}$$

$$K_{HCO_3} = \frac{[CO_3^{2-}][H^+]}{HCO_3^-} = 10^{-10.33}$$

$$K_{CO_2} = \frac{[H_2CO_3]}{P_{CO_2}} = 10^{-1.47}$$

to which we can can add the expression for the dissociation of water:

$$H_2O(l) \rightleftarrows H^+(aq) + OH^-(aq) \tag{1.49}$$

$$K_w = [OH^-][H^+] = 10^{-14.0}. \tag{1.50}$$

We can also use the following equation for charge balance, as all systems must be charge balanced:

$$2[Ca^{2+}] + [H^+] = 2[CO_3^{2-}] + [HCO_3^-] + [OH^-]. \tag{1.51}$$

At this point, we work to put all of the unknowns in terms of the proton concentration and/or constants, so that we can solve the problem.

If we substitute $P_{CO_2} = 10^{-3.50}$ into Equation 1.48 and rearrange, we can calculate $[H_2CO_3]$

$$[H_2CO_3] = 10^{-1.47} \times 10^{-3.50} = 10^{-4.97}.$$

Substituting this value into Equation 1.44 and rearranging, we can obtain the bicarbonate concentration in terms of constants and the proton concentration

$$\begin{aligned}[HCO_3^-] &= 10^{-6.35}\ 10^{-4.97}/[H^+] \\ &= 10^{-11.32}/[H^+].\end{aligned}$$

This expression for $[HCO_3{}^-]$ can then be substituted into Equation 1.46 and rearranged to obtain

$$\begin{aligned}[CO_3^{2-}] &= 10^{-10.33}\ 10^{-11.32}/[H^+]^2 \\ &= 10^{-21.65}/[H^+]^2.\end{aligned}$$

Taking

$$[OH^-] = 10^{-14.00}/[H^+]$$

and substituting the expressions into Equation 1.36, we get

$$[Ca^{2+}] = 10^{-8.48}/[CO_3^{2-}] = 10^{13.17}[H^+]^2.$$

Because we now have expressions for $[Ca^{2+}]$, $[OH^-]$, $[CO_3{}^{2-}]$, and $[HCO_3{}^-]$ all in terms of $[H^+]$, we can substitute them into the charge balance equation (1.51) to obtain

$$2(10^{13.17}[H^+]^2) + [H^+] = 2(10^{-21.65}/[H^+]^2) + 10^{-11.32}/[H^+] + 10^{-14.00}/[H^+].$$

This equation in terms of $[H^+]$ and constants can be solved by a programmable calculator to obtain pH $\sim$ 8.3. An interesting exercise is to compare the pH in this system open to atmospheric CO_2 with that of calcite in water closed to the atmosphere, which is a problem to solve at the end of this chapter.

This value is close to the mean pH of surface ocean water (Chapter 2). The carbonate system exerts an important control on the ocean pH. The presence of other cations and anions does shift the pH of the ocean, but it does so primarily through effects on the charge balance equation. Note that to perform the calculation above rigorously, one should correct for activity.

Redox Reactions

Oxidation–reduction or "redox" reactions are encountered in many environments ranging from soils and sediments to acid mine drainage sites to the photic zone of surface waters and ocean waters. Chapters 12 and 14 contain examples of some environmentally important redox phenomena.

Redox reactions involve transfer of one or more electrons from an electron donor such as Fe(II) to an electron acceptor such as Fe(III). Examples of *redox couples* in nature include Fe(III)/Fe(II), O_2/H_2O, HS^-/SO_4^{2-}, NO_2^-/NO_3^-, NH_4^+/NO_3^-, NH_4^+/N_2, and CH_4/HCO_3^-. The tendency of a given chemical species to donate or accept electrons is defined in terms of the standard hydrogen electrode.

Let us consider the Fe system in terms of redox reactions. Fe is the fourth most abundant element in the Earth's crust, and it is an essential nutrient to most organisms. Its redox reactions have important implications for mineral growth and dissolution, Fe bioavailability, and other processes, such as organic matter degradation and aggregation phenomena. In addition to contributing to numerous abiotic geochemical processes, Fe has played an important role in microbial metabolism since Precambrian times. Ferrous iron, Fe(II), can function as an electron source for iron-oxidizing organisms under both anoxic and oxic conditions, and ferric iron, Fe(III), can serve as a terminal electron acceptor for Fe-reducing microorganisms (Weber et al., 2006).

A common redox reaction is the reduction of ferric iron to form ferrous iron

$$Fe^{3+}(aq) + e^- \rightleftarrows Fe^{2+}(aq), \tag{1.52}$$

which when coupled with the oxidation of hydrogen

$${}^{1}\!/_{2}H_2(g) \rightleftarrows H^+(aq) + e^- \tag{1.53}$$

can be written as:

$$Fe^{3+}(aq) + {}^{1}\!/_{2}H_2(g) \rightleftarrows Fe^{2+}(aq) + H^+(aq). \tag{1.54}$$

Using a thermodynamic approach, the potential difference relative to the standard hydrogen electrode can be calculated. This value is called the E_H of the solution, and it can be calculated using the *Nersnt equation*, which for Equation 1.54 is

$$E_H = E^\circ + 2.303(RT/nF)\log(a_{Fe^{3+}}/a_{Fe^{2+}}), \tag{1.55}$$

where E° is the electrode potential for the system at standard state and F is the Faraday constant.

For this reaction, the equilibrium activity of electrons in solution can be expressed either using the terminology E_H, which is in volts, or pε, which has units of activity,

$$p\varepsilon = -\log a_{e-}, \tag{1.56}$$

in analogy with pH (Stumm and Morgan, 1996), with

$$p\varepsilon = \frac{FE_H}{2.303\,RT}. \tag{1.57}$$

Although this thermodynamic approach can be valuable in predicting equilibrium behavior, it has problems in application to natural waters. First, measuring E_H can be tricky. Because oxidation/reduction involves transfer of electrons, both members of a redox couple (such as Fe and S or organic species) need to be measured. Many geochemically significant redox couples in nature do not have suitable electrodes for measurement. Second, no free electrons are in natural systems, so that a pε does not actually exist (Lovley et al., 1994). Finally, aqueous and sedimentary environments tend not to be at redox equilibrium. For nonequlibrium environments, it is important to measure the rates of electron transfer, and this approach is taken when possible in studying microbial redox reactions, such as for microbial fuel cells (Chapter 12).

In many instances in environmental chemistry, E_H is plotted against pH to give an E_H-pH diagram. In other instances, pε is plotted against pH, giving pε-pH diagrams that are more comparable with pC-pH diagrams discussed in Chapter 10. Figure 1.4 shows a pε-pH diagram for dissolved Fe(II) and Fe(III) species. Each field shows a pε-pH region where the marked species is predominant at equilibrium, meaning that it has highest concentration among all of the dissolved Fe(II) and Fe(III) species. The upper and lower limits of this

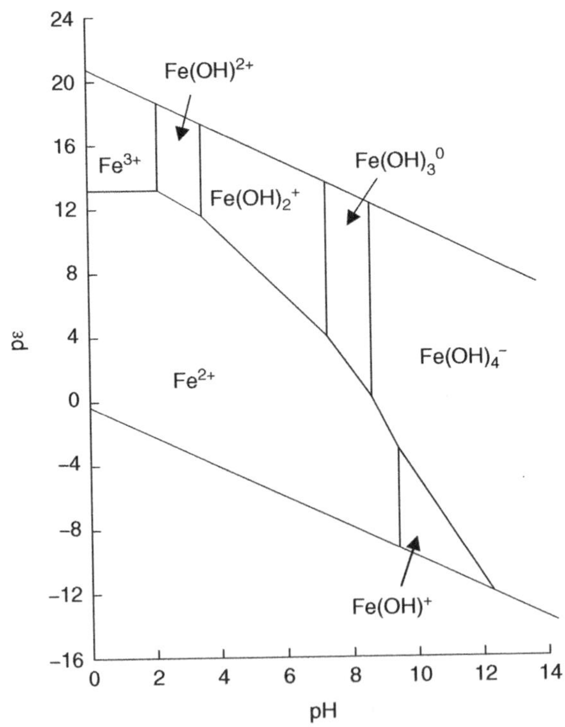

FIGURE 1.4 A pε-pH diagram for dissolved Fe(II) and Fe(III) species at 25°C and 1 atm pressure. After Nordstrom and Munoz, 1986 (with permission).

diagram (diagonal lines) represent the oxidation and reduction of water. In using a diagram such as this, it is essential to remember that it is based on thermodynamics, and that many systems are not at equilibrium. Nonetheless, it provides predictive capability regarding the direction in which a system should evolve toward equilibrium.

A typical redox reaction that involves minerals in aqueous solution is the oxidation of magnetite, Fe_3O_4, to form hematite, Fe_2O_3

$$\underset{\text{Magnetite}}{2Fe_3O_4(s)} + H_2O(l) \rightleftarrows \underset{\text{Hematite}}{3Fe_2O_3(s)} + 2e^- + 2H^+(aq). \tag{1.58}$$

Magnetite can also be formed by hematite reduction, for example in alkaline solutions in the presence of the antioxidant compound hydrazine

$$\underset{\text{Hematite}}{6Fe_2O_3(s)} + N_2H_4(aq) \rightarrow \underset{\text{Magnetite}}{4Fe_3O_4(s)} + 2H_2O\,(l) + N_2(g), \tag{1.59}$$

with electron transfer occurring between the hydrazine and the Fe(III) of the hematite (Cornell and Schwertmann, 2003).

Zerovalent Fe oxidizes to Fe(III) in the presence of oxygen, which is reduced. This forms Fe(III) oxide in the commonly observed process of rusting, generalized as

$$4Fe^{\circ} + 3O_2 + nH_2O \rightarrow 2Fe_2O_3{\cdot}nH_2O. \tag{1.60}$$

A great deal of effort has focused on microbially mediated oxidation–reduction processes that involve ferric and ferrous Fe and a wide range of Fe-bearing minerals. As an example, the organic compound formate produced by a variety of bacteria can reduce ferric Fe to ferrous Fe (e.g., Ehrlich, 2002) according to the equation

$$2Fe^{3+}(aq) + HCOOH(aq) \rightarrow 2Fe^{2+}(aq) + 2H^+(aq) + CO_2(g). \tag{1.61}$$

The redox conditions of aqueous environments exert important controls on the distributions not only of dissolved Fe species and of Fe-bearing minerals but also of microorganisms such as bacteria. For example, *iron bacteria*, which oxidize iron enzymatically, are limited to conditions that have relatively small reduction potential (fairly oxidizing) and range and are neither highly acidic (such as acid mine drainage streams with pH less than 2) nor highly basic (Ehrlich, 2002).

Metal Speciation Diagrams

Another common application of thermodynamics for environmental surface and interface chemistry is calculation of the distributions of metal species at equilibrium, which results in a plot of metal species concentrations versus pH. Which species are present for a given metal, such as Cu, Zn, Co, Cd, or Pb, can greatly impact phenomena such as metal adsorption to mineral surfaces (Chapters 7) and mineral solubility (Chapter 9 and 10). Determining the concentrations of metal species at equilibrium in a given solution requires

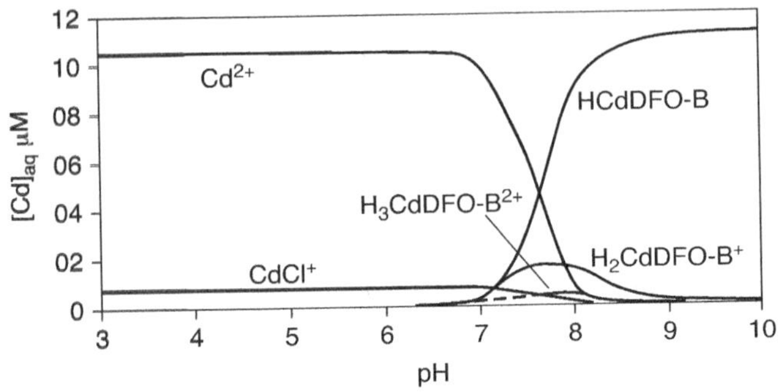

FIGURE 1.5 Speciation diagram for Cd (11.1 μM) in the presence of the organic microbial ligand DFO-B. In 0.1-M $NaClO_4$ to set ionic strength at 0.1 M and, at 22°C. After Hepinstall et al., 2005 (with permission).

a carefully reviewed and consistent thermodynamic compilation of stability constants and hand calculation or use of a themodynamic computer model such as PHREEQC (Parkhurst and Appelo, 1999).

An example of a speciation diagram for Cd(II) in the presence of the organic ligand desferrioxamine B (DFO-B) is shown in Figure 1.5.

In this diagram, the concentrations of the various Cd species are given on the y-axis (generally in moles L^{-1}) and pH is on the x-axis. At pH ≤ 7, the predominant Cd species is Cd^{2+}, with a small amount of $CdCl^+$. Other species present are at concentrations too low to see clearly on the plot. At pH > 7, Cd-DFO-B complexes HCdDFO-B, $H_2CdDFO\text{-}B^+$, and $H_3CdDFO\text{-}B^{2+}$ become important. The total concentrations of chemical constituents, the temperature, and the ionic strength play important roles in construction of speciation diagrams.

A BRIEF INTRODUCTION TO KINETICS

Whereas thermodynamics describes the overall energetics of a reaction and allows prediction of which way a reaction will go spontaneously, kinetics describes the rates and mechanisms of reactions. Many systems at and near the Earth's surface show strong kinetic influences, especially when processes are biologically mediated (Chapters 11 and 12). In kinetics, it is important to consider whether a reaction is *homogeneous*, in which only one phase is involved, versus *heterogeneous*, in which different phases are involved. We shall see examples of homogeneous versus heterogeneous reactions in discussing different modes of mineral nucleation in Chapter 9.

Overall Versus Elementary Reactions

Overall reactions are typically written, such as

$$A + B \rightarrow AB. \tag{1.62}$$

An example of an equation for an overall reaction is for dissolution of the clay mineral kaolinite to form gibbsite (Chapter 3)

$$\underset{\text{Kaolinite}}{Al_2Si_2O_5(OH)_4(s)} + 5H_2O \rightarrow \underset{\text{gibbsite}}{2Al(OH)_3(s)} + 2H_4SiO_4(aq). \tag{1.63}$$

This overall equation only describes the initial reactants and the products. In reality, reactions may take several or many steps, each of which is known as an *elementary reaction*. For example,

$$A + A \rightarrow A_2 \quad k_1 \tag{1.64}$$

$$A_2 + B \rightarrow AB + A \quad k_2, \tag{1.65}$$

where the k values are the rates of the different elementary reactions in the directions written. An elementary reaction is assumed to occur as a single step and to pass through a single transition state (see below).

It is impossible to tell from an equation for an overall reaction how many steps are involved. Most reactions occur by a sequence of elementary reactions, each of which generally involves only one or two molecules. An elementary reaction might be

$$H + Br_2 \rightarrow HBr + Br. \tag{1.66}$$

This equation for an elementary reaction signifies that a particular H atom attacks a particular Br_2 molecule to produce a molecule of HBr and a Br atom.

Molecularity and Reaction Order

The *molecularity* is the number of molecular entities coming together to react in the microscopic chemical event that is an elementary reaction. In a *unimolecular* reaction, a single molecule undergoes a transformation; a common example is *radioactive decay*. In a *bimolecular reaction*, a pair of molecules interacts. It is essential to distinguish between *molecularity* and *reaction order*. The *reaction order* is an empirical quantity, obtained from the experimental rate law. The *molecularity* refers to an elementary reaction only, where the elementary reaction has been postulated to be an individual step in some specific, detailed mechanism. The molecularity of an elementary reaction must be an integer because elementary reactions involve entire molecules and not fractions thereof. However, the reaction order may be noninteger or fractional if the reaction is not an elementary reaction.

Let us return to Equation 1.62 for the overall reaction

$$A + B \rightarrow AB. \tag{1.62}$$

This overall reaction is said to be *first order* with respect to A if the rate is proportional to [A], *zero order* if independent of A, *second order* if proportional to $[A]^2$, and so on. If proportional to [A] and [B], the reaction would be second order overall, but first order with respect to A.

The reaction mechanism cannot be inferred from the observed reaction order. The order of an overall reaction provides no evidence for *molecularity*. The reaction mechanism can only be inferred from reaction order if there is some independent line of evidence that the reaction is an elementary reaction. In environmental biogeochemistry, sometimes an investigator will find that a rate is proportional to some component(s) and a mechanism is proposed based on this observation. The problem is that thousands of different mechanisms may lead to the observed proportionality or dependency. Independent evidence may involve calculating activation energies, using microscopic and/or spectroscopic techniques, interrupting the reaction part way, or using poisoning or blocking agents that stop a particular step or component of a reaction, for example, blocking a specific reactive site on a mineral surface.

The rate law of a *unimolecular elementary reaction* that is *first order* in the reactant would look like

$$A \rightarrow \text{Products} \quad d[A]/dt = -k[A]. \tag{1.67}$$

Radioactive decay is an example of a first-order reaction; the change in the concentration of the decaying component M with respect to time is proportional to the amount of M present

$$-d[M]/dt = k[M], \tag{1.68}$$

where

M = amount (or concentration) of radioactive element
t = time
k = rate constant or decay constant.

Equation 1.68 integrates to give

$$[M] = [M]_0 e^{-kt}. \tag{1.69}$$

This equation can be linearized to:

$$\ln[M]_t = \ln[M]_0 - kt, \tag{1.70}$$

where the subscript 0 refers to time 0 and t refers to at some time, t.

When we compare the reactivities of chemicals that decay by first-order kinetics (Figure 1.6), it is customary to compare the reaction time for one half of the chemical to decay (or disappear)

$$[M] = \tfrac{1}{2}[M]_0, \tag{1.71}$$

and the half-life $t_{1/2}$ is given by

$$t_{1/2} = 0.693/k. \tag{1.72}$$

Many environmental particle/solution interface reactions have been described as being first order. Sometimes, there seems to be one or more "phases" to the reactions, for example,

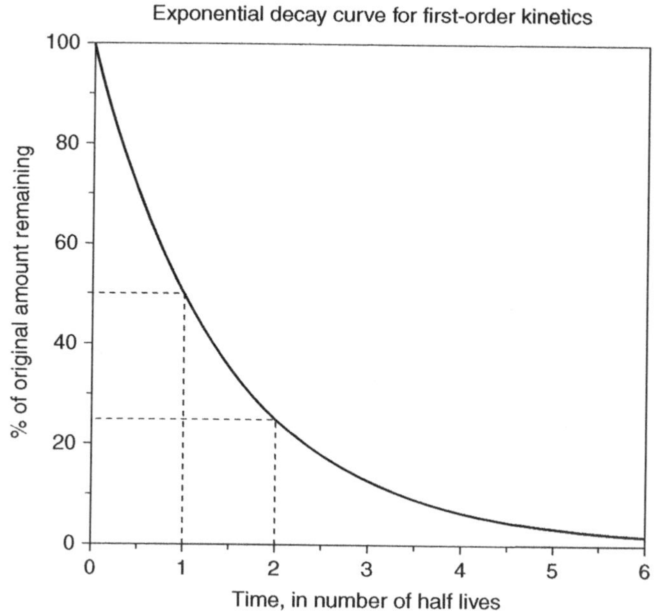

FIGURE 1.6 A typical exponential curve for first order kinetics, including radioactive decay.

an initial fast first-order phase followed by a latter, slower first-order phase (e.g., Sparks, 1998).

For a second-order reaction with respect to some chemical species A, the rate equation would be

$$-d(A)/dt = kA^2, \tag{1.73}$$

and for a reaction that is first order with respect to A and first order with respect to a reactant B, but second order overall, the equation is

$$\frac{-d[A]}{dt} = k[A][B]. \tag{1.74}$$

Most oxidation reactions

$$A + Ox \rightarrow \text{products} \tag{1.75}$$

are second-order overall (Larson and Weber, 1994), with the rate proportional to the concentrations of both the oxidizing agent (Ox) and the substrate (A)

$$-dA/dt = k[Ox][A]. \tag{1.76}$$

For reactions in water, there are few values of k available in the literature (Larson and Weber, 1994).

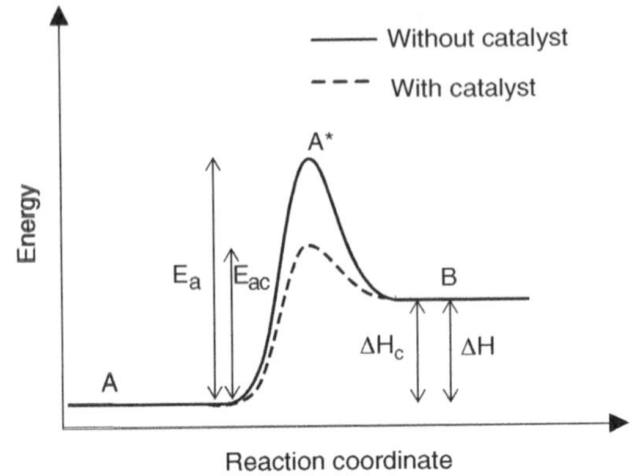

FIGURE 1.7 A schematic showing activated complex theory, including the effects of a catalyst. The activated complex A^* is unstable and intermediate between the Precursor A and the products B. A^* has the highest energy along the reaction coordinate. The presence of a catalyst decreases the activation energy from E_a to E_{ac}. The enthalpy is given by ΔH (without catalyst) and ΔH_c (with catalyst).

Transition State Theory and the Arrhenius Equation

Transition state theory states that there is a transitional state of higher Gibbs energy that occurs over the course of a reaction (Figure 1.7) between reactants and products. This transitional state has associated with it an energy barrier that must be overcome for the reaction to proceed. The energy barrier is known as the *activation energy* and is given the symbol E_a. The assembly of atoms at the transition state has been termed an *activated complex*.

A substance that increases the rate of a reaction without modifying the overall Gibbs energy change in the reaction is called a *catalyst*. A catalyst modifies the transition state to lower the activation energy, but it does not change the reactant energy and the product energy. A catalyst is not consumed over the course of the reaction. An *enzyme* is a biological catalyst. A substance that increases the rate of a reaction but is consumed is called an *activator*. An *inhibitor* decreases the rate of a reaction and may be consumed over the course of the reaction.

A well-known example of environmental catalysis is involved in destruction of stratospheric ozone. Anthropogenic chlorine-containing organic compounds known as chlorofluorocarbons (CFCs) are broken down by shortwave ultraviolet (UV) to produce Cl atoms:

$$CFCl_3 + h\nu \rightarrow {}^{\cdot}CFCl_2 + {}^{\cdot}Cl. \quad (1.77)$$

These Cl atoms then catalyze the conversion of stratospheric ozone (O_3) to molecular oxygen, in a process that also involves an oxygen atom

$$^{\cdot}Cl + O_3 \rightarrow {}^{\cdot}ClO + O_2 \quad (1.78)$$

$$^{\cdot}ClO + O \rightarrow {}^{\cdot}Cl + O_2, \quad (1.79)$$

yielding an overall reaction of

$$O_3 + O \rightarrow 2O_2. \tag{1.80}$$

Cl acts as a catalyst in the reaction sequence shown by Equations 1.78 and 1.79, speeding up the rate of ozone depletion while neither being created nor destroyed. Because stratospheric ozone is crucial for blocking harmful UVB radiation, CFC-promoted ozone destuction is of great environmental importance. Paul Crutzen, Mario Molina, and F. Sherwood Rowland were awarded the 1995 Nobel Prize in Chemistry for their work on stratospheric ozone (Rowland, 2007).

The rate constant (k) for most chemical reactions and processes can be described by the *Arrhenius equation*, which relates rate to temperature (T):

$$k = Ae^{-\frac{E_a}{RT}} \tag{1.81}$$

where

$A =$ *pre-exponential factor* can be calculated using statistical mechanics for simple systems or determined empirically for complex systems
$R =$ ideal gas constant
$E_a =$ activation energy.

For many processes in environmental surface chemistry, both A and E_a are themselves independent of temperature.

Michaelis-Menten Kinetics

One of the most widely used kinetic equations in the environment is the equation that describes the rate of many enzyme-catalyzed reactions, *Michaelis-Menten kinetics*, named for Leonor Michaelis and Maud Menten (1913). Because this approach has relevance for microbial processes (Chapter 12), it is described here. Assume a reaction mechanism in which a substrate (S) reacts with an enzyme (E) to form an enzyme-substrate complex (ES) in which S is bound to the reactive site of E. This initial part of the reaction sequence is reversible and in equilibrium. The product (P) is released and the free enzyme (E) recovered in an irreversible second part of the reaction sequence. Overall,

$$S + E \underset{k_{-1}}{\overset{k_1}{\rightleftarrows}} ES \overset{k_2}{\rightarrow} P + E. \tag{1.82}$$

The rates of reaction are given by k subscript values. An example of a substrate is H_2, and an example of an "enzyme" would be a respiring microorganism that uses H_2 in a process such as $Mn(IV)O_2$ or SO_4^{2-} reduction (simplifying the microorganism to represent its appropriate enzymatic behavior; Chapter 12).

When the following two common observations apply to a system such as this, Michaelis-Menten kinetics can be used (Vaccari et al., 2006):

1. If the substrate concentration is held constant, then the rate of the reaction at any particular substrate concentration is proportional to the total concentration of enzyme.
2. If the total enzyme concentration (E_T) is held constant, then at a low substrate concentration, the rate is proportional to the substrate concentration, but at a high substrate concentration, the rate levels off and is independent of concentration because of a saturation effect.

The Michaelis-Menten equation has the form

$$\nu = \frac{\nu_{max}[E_T][S]}{K_m + [S]}, \tag{1.83}$$

where ν is the rate of substrate uptake (in the example above, H_2 consumption in moles per unit time),

ν_{max} = maximum rate of substrate uptake when the uptake rate is not limited by substrate availability,
$[E_T]$ = total enzyme concentration, and
K_m = Michaelis-Menten rate constant = $(k_{-1} + k_2)/k_1$ and when $[S] = K_m$, $\nu = \frac{1}{2}\nu_{max}$.

The Elovich Equation for Chemisorption Kinetics

One equation commonly applied to the phenomenon of chemisorption (see Chapters 7 and 8) is the *Elovich equation*. Chemisorption is a means whereby a dissolved ion or molecule attaches to a surface forming a chemical bond. The *Elovich equation* is as follows:

$$dq/dt = ae^{-\alpha q} \tag{1.84}$$

(Roginsky and Zeldovich, 1934; Mclintock, 1967),
where

q = amount of sorbate (the material that is attaching to the surface) sorbed per unit mass of sorbent (the surface) at time t
a and α = empirical constants for a given experiment.

The integrated form of this equation is

$$q(t) = (2.303/\alpha)\log(1 + t/t_0). \tag{1.85}$$

Figure 1.8 shows a plot of the concentration of the organic ligand salicylate adsorbed to an alumina surface [Al:sal] against $\ln(1+t/t_0)$ from aqueous solution. The plot is linear, which indicates that the data can be described by the Elovich equation. These data are only for the first 30 s of the adsorption experiment. Data from longer reaction time periods did not fit well to the Elovich equation. Fits to the Elovich equation, or lack thereof, which include changes to fits over time, cannot necessarily be used to infer anything about the reaction mechanism; rather, more direct evidence is needed.

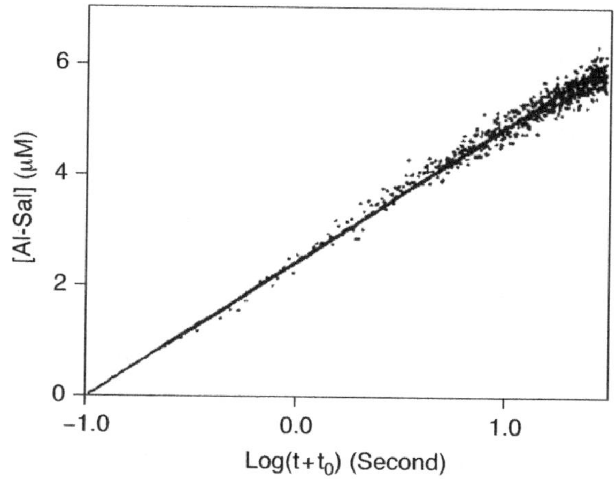

FIGURE 1.8 Elovich plot for salicylate adsorption kinetics to alumina at a reaction time of <30 s. After and using data from Wang et al., 2000 (with permission).

Simultaneous Versus Sequential Reaction Sequences

For a sequential reaction sequence in which one reaction step follows another, it is the step with the slowest rate that controls the overall rate of reaction. Thus, for sequential reactions

$$\text{Step 1} : A \rightarrow B \ \text{rate} : k_1 \tag{1.86}$$

$$\text{Step 2} : B \rightarrow C \ \text{rate} : k_2. \tag{1.87}$$

If $k_1 < k_2$, step 1 will be rate controlling and vice versa.

An example of sequential reactions is discussed below and in the section on organic ligand-promoted dissolution of Fe oxides in Chapter 10, in which the ligand first diffuses to the surface, then attaches to the surface and forms an activated complex, and finally, the ligand–metal complex removes from the solution and diffuses away. The slowest and thus rate-determining step is thought to be detachment of the ligand–metal complex from the surface.

At the mineral–water interface, reactions often occur simultaneously, for example, dissolution at microtopographic features such as step edges, kinks, and pits (Chapter 5). For simultaneous reactions, it is the fastest reaction that is rate controlling of the overall reaction unless the various reactions have similar rates; in which case, the overall rate is approximately the sum. Thus, although the rate of one of the individual mineral dissolution reactions at a step edge, kink, pit, or other feature may be controlled by the slow detachment rate, whichever site has the fastest rate would dominate in controlling the overall dissolution rate. Over the course of reaction, certain sites may become depleted and the rate control thus may change.

It is important to emphasize that we cannot tell what the reaction mechanism is from the overall reaction. Some Independent evidence must always be available. It is rare in environmental interface studies to have detailed, molecular-scale information on elementary reactions. Therefore, reaction kinetics often are spoken of in terms of reaction schemes or generalized reactions mechanisms. A combination of rate observations and spectroscopic techniques has been helpful for determining likely mechanisms.

Transport Versus Surface Control of Mineral Growth and Dissolution Rates

Three of the most important reactions that occur at the mineral–water interface are adsorption, mineral growth, and mineral dissolution. These processes are discussed in Chapters 7–10. The three processes include the following steps: (1) transport of reactant ions or molecules from the solution to the surface; (2) surface diffusion to a reactive site; (3) attachment to a site on the surface; and in the case of mineral dissolution, (4) detachment back to the solution (Lasaga, 1990). Other processes such as solvation and desolvation are also important, in many instances. Because (1)–(3)/(4) occur sequentially, it is the slowest step that is rate determining. As a result, the rates of mineral growth and dissolution are often considered in terms of transport versus surface control.

The rates of growth and dissolution of various minerals each may be controlled by *transport processes, surface processes,* or an intermediate case of both. Let us consider mineral growth in some detail, with dissolution being essentially (although often not exactly; Morse et al., 2007) the converse. *Transport-controlled crystal growth* is generally thought to be sensitive to flow (or stir) rate because this controls the rate of transport (by diffusion and advection) of reactants to the surface. This sensitivity to flow rate is only up to the point when flow rate is so fast that transport no longer limits the supply of reactants. *Surface-controlled crystal growth* is thought to be insensitive to flow rate (or stir rate) because the surface process is slow relative to transport. Surface-controlled growth is generally slower than transport-controlled growth and has greater temperature sensitivity associated with higher activation energy. Surface-controlled growth can be limited by surface diffusion or by attachment, depending on the particular conditions.

For surface-controlled growth, various imperfections, such as steps and kinks on a surface (Chapters 5, 9), can be important because sites associated with the imperfections have different coordinations and hence different energies. Therefore, different rates of reaction may be associated with different surface features. If growth occurs simultaneously at several different sites, then the site with the fastest growth rate generally determines the overall rate.

Figure 1.9 illustrates the concentration gradient of reactants that occurs around a growing crystal under static conditions and transport control versus surface control of rate. In transport-controlled growth, reactants become depleted from the vicinity of the surface. However, such a gradient in reactant concentration does not occur in surface-controlled growth. As the stir rate is increased, the gradient decreases and eventually disappears. An intermediate case has only a slight depletion near the surface.

Mineral dissolution also can be transport controlled or surface controlled. For transport-controlled dissolution, the rate-determining step is transport of the reaction product away from the surface. In transport-controlled dissolution systems, reaction products build up near the surface. The dissolution rate is then limited by the rate at which the dissolution products are transported (by diffusion and advection) away from the surface. The reactant also can be at a lower concentration near the surface than in the bulk solution, which can also affect the dissolution rate.

For surface-controlled dissolution, the rate-determining step is a reaction at the surface, probably most often the detachment from the surface of a metal or metal–ligand activated

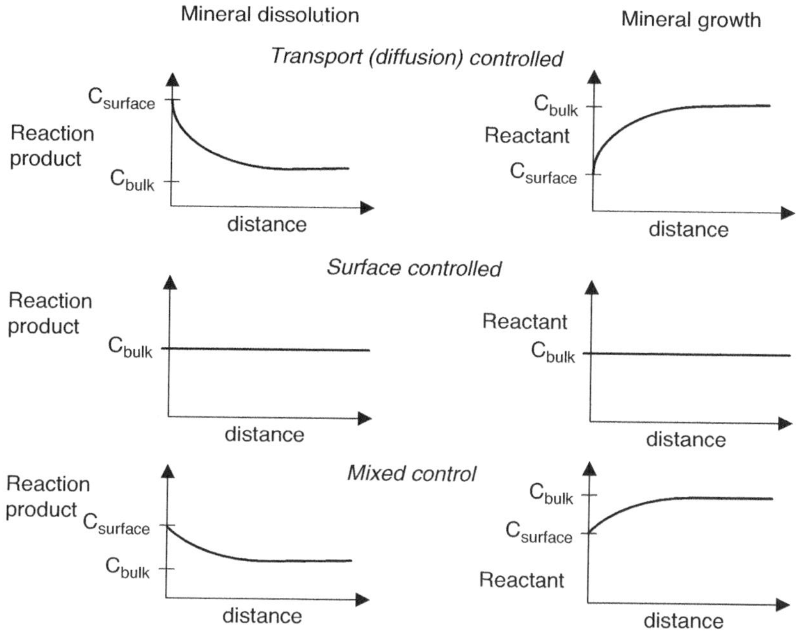

FIGURE 1.9 How the concentration of the reactant or product varies with distance from a surface for transport- and surface-limited growth and dissolution, along with the intermediate case of mixed controls.

complex. In this case, the concentrations of reaction products do not build up near the surface. At steady-state dissolution, the dissolution kinetics follow the rate law (rate of dissolution R_d in mol s^{-1}):

$$R_d = dC/dt = kA, \quad (1.88)$$

where

C = concentration of product
t = time
k = rate constant (in mol m^{-2} s^{-1})
A = surface area of the mineral (in m^2).

As for mineral growth, sometimes dissolution may be intermediate, with both controls contributing to the overall dissolution process.

When the dissolution rate is surface-reaction controlled, mineral surfaces tend to form structures such as pits and ledges because the dissolution process is fastest at certain high-energy sites on the surface (Chapter 10). However, when dissolution is transport controlled, the mineral surface tends to be more rounded without such distinct dissolution features. Minerals that dissolve faster, such as halite (NaCl) and gypsum ($CaSO_4 \cdot 2H_2O$), tend to do so by transport-controlled mechanisms. Minerals that dissolve more slowly such as the feldspars and hematite tend to do so by surface-controlled mechanisms (Berner, 1978). One of the most abundant and geochemically significant minerals, calcite ($CaCO_3$), can dissolve by either surface or transport control (e.g., Lasaga, 1990; Morse et al., 2007). Factors such as solution pH, saturation state, temperature, presence of impurities, and particle size of a mineral can influence whether a mineral dissolves by surface, transport, or combined or "complex" mechanisms in any particular setting.

Transport- and surface-limited growth and dissolution can be considered to be end-member situations, with many minerals that grow or dissolve at rates that are a more complex function of both transport and surface processes.

Rate Laws for Surface-Controlled Mineral Growth and Dissolution

Over the years, many different equations have been proposed for calculating rate laws for mineral growth and dissolution at the surface. Lasaga (1995) provided the following equation to encapsulate the controls on kinetic rate laws for surface-controlled mineral growth and dissolution:

$$\text{Rate} = k_0 A_{\text{min}} e^{-E_a/RT} X_{H^+,ads}^{n_{H^+,ads}} g(I) \prod_i X_{i,ads}^{n_i} f(\Delta G_r), \qquad (1.89)$$

where

k_0 = an intensive rate constant with units of moles $\text{cm}^{-2}\ \text{s}^{-1}$

A_{min} = *reactive* mineral surface area, or that part of the overall surface area that takes part in the reactions

E_a = apparent activation energy of the overall reaction,

R = ideal gas constant

T = temperature (K)

I = ionic strength of the solution

g(I) = indicates the potential importance of ionic strength, beyond the fact that it affects species reactivities

$X_{H^+,ads}^{n_{H^+,ads}}$ = mole fraction of H^+ at the surface (inner-sphere; see Chapter 6)

$X_{\text{i, ads}}$ = mole fraction of species at the surface

$\prod_i$ = multiplication of terms

ΔG_r = variation of rate with deviation from equilibrium, where $\Delta G_r = 0$ at equilibrium.

Sometimes, an isotherm equation (see Chapter 5) is used to describe the relation between adsorbed and solution H^+ and other species.

Much insight can be gained from this equation. Mineral precipitation and dissolution rates are sensitive to many factors, such as pH, which affects protonation of the surface, temperature, adsorbed species concentrations, ionic strength, and how far from the system is from equilibrium. The details of this equation also show that many fundamental and sometimes tricky factors need to be addressed. For example, how does the *reactive* surface area of a mineral differ from total surface area (BET or geometric; see Chapter 4)? How does one calculate the activity of H^+ at the surface? How does one calculate the mole fractions of adsorbed species? Rates and processes of mineral growth and dissolution are discussed in greater detail in Chapters 9 and 10, respectively.

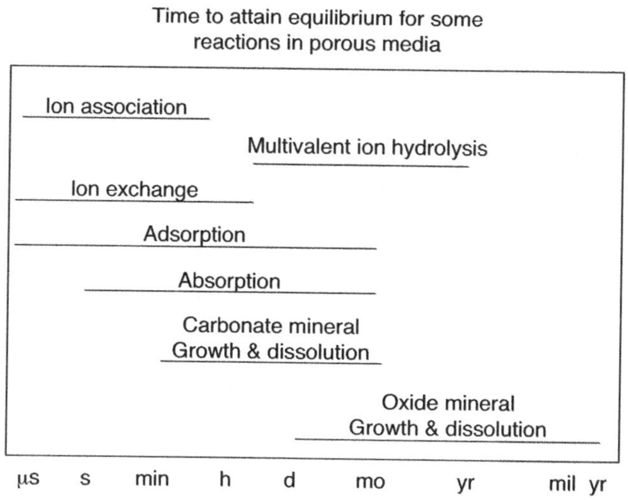

FIGURE 1.10 Time to obtain equilibrium for some reactions and processes in porous media. Substantially revised from Amacher, 1991.

Equilibration Time in Porous Media

Within the realm of environmental surface and interface chemistry, it is worthwhile to consider how long it typically takes for various reactions and processes to come to equilibrium in porous media such as soils and sediments, as shown in Figure 1.10. Figure 1.10 does not indicate reaction rate, as many reactions may occur prior to system equilibration. An example is the microbiologically mediated formation of the Fe(hydr)oxide mineral ferrihydrite (Chapters 3 and 12), which can occur on the order of seconds. However, this process most likely does not represent attainment of equilibrium because ferrihydrite is not the most stable phase in many environments.

QUESTIONS FOR FURTHER THOUGHT

1. Balance the following equations with correct stoichimetric coefficients. Be sure to include any electrons in redox reactions. Show your work.

   a) $\_NH_3(g) + \_O_2(g) \rightleftarrows \_NO(g) + \_H_2O(g)$

   b) $\_Ca_3(PO_4)_2(s) + \_ SiO_2(s) + \_ C(s) \rightleftarrows \_CaSiO_3 + \_ CO(g) + \_ P_2(g)$

   c) $\_OH^-(aq) + \_\_H_30^+(aq) \rightleftarrows \_H_20(l)$

   d) $\_FeS_2 + \_O_2 + \_H_20 \rightleftarrows \_Fe^{2+} + \_\_SO_4^{2-} + \_H^+$

   e) $\_Mg_4(CO_3)_3(OH)_2 \cdot 3H_2O(s) + \_H^+ \rightarrow \_ Mg^{2+} + CO_2(g) + \_H_2O$

2. You are given the following chemical analysis for a surface water sample. (a) Begin by converting the species' concentrations from parts per million to millimoles L^{-1} and milliequivalents L^{-1}. (Assume dilute solutions for these calculations). (b) Calculate

the ionic strength of the solution. (c) Calculate the activities of the various ions in solution, assuming that these are the only ions in solution.

Present your answers for (a) and (c) as two tables—one table for cations and the other for anions—giving molarity, equivalents per liter, and activity. (d) Does this solution seem to be charge balanced (to within 2–3%)? If not, then there is likely an error in the analysis or other species that need to be considered.

ion	ppm
K^+	399
HCO_3^-	145
Na^+	10760
SO_4^{-2}	2712
Ca^{2+}	412
Sr^{2+}	7.9
Mg^{2+}	1294
Cl^-	19350

3. You are asked to make up 1 L of solution containing 5 g of NaCl, 5 g of KCl, and 12 g of $CaCl_2$.
   a. What is the molar concentration (Molarity) of Cl^- in this solution?
   b. What is the activity of Cl^- in this solution?
   Show all work, and circle your final answers (concentration and activity).
4. You are asked to make up 1 L of each of the following solutions. How much reagent, in grams, would you add in each case? Assume that all the metal in solution is in the form indicated (e.g., no hydrolysis). Show all work, and circle the final answers.
   a. 0.5 M Na^+, using the reagent NaCl
   b. 0.06 eq L^{-1} Ag^+, using the reagent $AgNO_3$
   c. 0.057 M Ag^{3+}, using the reagent $AgCl_3$
5. Given the following thermodynamic data for 25°C and 1 atm pressure, (a) calculate the dissociation constant for water (Data from Langmuir, 1997), $K_w = [H^+][OH^-]$

Species	ΔH°_f (kcal/mol)	ΔS°(cal/mol K)
H^+	0	0
OH^-	−54.997	−2.560
$H_2O(l)$	−68.315	16.7

   (b) What is the pH of neutrality of water at 10°C? At 35°C?
6. (a) What is the pH of a solution in equilibrium with calcite but closed to the atmosphere such that no exchange of CO_2 occurs? Solve assuming activity = concentration to simplify. Take into consideration all three carbonate species. Hint: To obtain six equations for the six unknowns, replace the equation involving P_{CO_2} with the mass balance equation: $[Ca^{2+}] = [CO_3^{2-}] + [HCO_3^-] + [H_2CO_3]$, which is appropriate because the carbonate species in this closed system will come only from the dissolution of calcite. Compare the pH of a system in equilibrium with calcite closed versus open with atmospheric CO_2.

(b) If a beaker that contains otherwise pure water with calcite closed to the atmosphere were opened, what would happen to the calcite?
(c) What is the solubility of calcite present in excess in a beaker of otherwise pure water open to the atmosphere at 25°C and $P_{CO_2} = 10^{-3.5}$ versus $10^{-2.5}$?

7. In the landmark 1965 textbook *Solutions, Minerals, and Equilibria*, by Garrels and Christ, the equilibrium constant for the dissolution of calcite at standard temperature and pressure, according to the following reaction:

$CaCO_3(s) \leftrightarrow Ca^{2+}\ (aq) + CO_3{}^{2-}\ (aq)$ was calculated based on the following thermodynamic data:

Species	$\Delta G°_f$ in kcal/mole
Ca^{2+}	−132.18
$CO_3{}^{2-}$	−126.22
$CaCO_3(s)$	−269.78

A somewhat more recent data set is

Species	$\Delta G°_f$ in kcal/mole
Ca^{2+}	−132.30
$CO_3{}^{2-}$	−126.17
$CaCO_3(s)$	−270.04

a. Write the expression for the equilibrium constant for the reaction as written.
b. How much of a difference does the change in thermodynamic data make in the calculated equilibrium constant at 25°C?

8. The reaction rate v_1 for an enzymatic reaction is 0.2 mol L^{-1} min^{-1} of substrate consumed at a substrate concentration of $[S_1] = 0.01$ mol L^{-1}. When the substrate concentration is increased to $[S_2] = 0.02$ mol L^{-1}, the rate increases to $v_2 = 0.3$ mol L^{-1} min^{-1}. What are K_m and v_{max} for this enzymatic reaction? (Hint: plot 1/[S] on the x-axis and $1/v$ on the y-axis; the slope is K_m/v_{max}, and the y-intercept is $1/v_{max}$).
9. What is the pH of otherwise pure water in equilibrium with atmospheric carbon dioxide at 1 atm and 25°C, with $P_{CO_2} = 10^{-3.2}$? You may neglect activity corrections for this problem. Hint: Determine unknown species and set up a series of simultaneous equations. Solve for concentration of H^+ and use this to determine pH (neglecting activity correction). The system will be somewhat acidic.
10. Plot how activity coefficients γ_{Na} and γ_{Ca} vary with solution ionic strength from 0.0001 to 0.5. Use these plots to comment in a paragraph or two on how ionic strength may affect reactions in aqueous solutions.
11. (a) Calculate the solubility of calcite (in g L^{-1}) in a system closed to the atmosphere at 25°C, 1 atm pressure, when 0.1 M of NaCl is added to the solution? You may assume that the only carbonate species is $CO_3{}^{2-}$ to simplify, but you must correct for activities. The basis is the simple equation:

$$CaCO_3 \rightleftarrows Ca^{2+} + CO_3^{2-}$$

with $K_{cal} = 10^{-8.48}$.

Hint: Identify unknowns and solve a series of simultaneous equations, using an iterative approach in which concentration is used as an approximation for activity in the first iteration. (b) Quantitatively, how does the addition of 0.1 M of NaCl affect the solubility of calcite relative to a system without the added salt?

12. ^{14}C is a radioactive isotope of carbon with $t_{1/2} = 5730$ years. Willard Frank Libby was awarded the 1960 Nobel Prize in Chemistry for developing the method of radiocarbon C-14 dating. How useful is C-14 dating for samples of importance for human (e.g., archaeologic) versus geologic history? Explain your answer.
13. Plutonium-239 (^{239}Pu) is produced from uranium for use in nuclear weapons, according to

$$^{238}\text{Uranium} + \text{n} \rightarrow {}^{239}\text{Uranium} + \beta \rightarrow {}^{239}\text{Neptunium} + \beta \rightarrow {}^{239}\text{Plutonium},$$

where + n represents bombardment by a neutron and + β represents emission of a beta particle.

This "weapons grade" Pu has $t_{1/2} = 24{,}100$ years. During the Cold War, hundreds of metric tons of ^{239}Pu were manufactured and isolated for weapons systems. (a) If 1 g of ^{239}Pu were present at a contaminated site, how much would remain after 10, 100, 1000 and 10,000 years? (b) The 1980 *No Nukes* record album included a song titled, "Plutonium is Forever." Based on your calculations in (a), what are your opinions of this song title? Explain your reasoning.

14. At the U.S. Department of Energy's Hanford site, the chemical processing of fuel rods produced enormous amounts of radioactive waste. It has been calculated that about 4×10^{16} Bq of ^{137}Cs was leaked into the vadose zone, and sediment samples currently contain as much as 10^9 Bq g^{-1} ^{137}Cs (Serne et al., 1998). ^{137}Cs has $t_{1/2} =$ 30.17 years. (a) Hypothetically for the sake of calculation, if the total leak had occurred in one incident in 1950, then approximately how much of the original ^{137}Cs radiation should remain today (in Bq)? (b) If one considers the values of ^{137}Cs concentrations in sediments given above as representative of 1998, then approximately what would the soil concentrations in Bq g^{-1} have been in 1955 given the same scenario as in (a) and assuming no Cs mobility? (c) Approximately what should the ^{137}Cs concentrations (in Bq g^{-1}) be in the sediments in the year 2100, again assuming no mobility? (d) How many years will it take for the ^{137}Cs in the sediments to fall below 1 Bq g^{-1}? For comparison, natural granite samples often have radiation levels $\sim$ 1 Bq g^{-1}, because of the presence of radioactive U, Th, and K. (Note that in many real-world situations, including at the Hanford Site, Cs actually can be mobile.)

FURTHER READING

Drever, J.I. 1997. The Geochemistry of Natural Waters 3rd ed. Prentice Hall, Englewood Cliffs, NJ, 436 p.

Nordstrom, D.K., and Muñoz, J.L. 1986. Geochemical Thermodynamics. Blackwell Scientific Publications, Palo Alto, CA, 477 p.

Sparks, D.L. 1989. Kinetics of Soil Chemical Processes. Academic Press, San Diego, CA, 210 p.

Stumm, W., and Morgan, J. J. 1996. Aquatic Chemistry 2nd ed. John Wiley and Sons, New York. 1022 p.

2

THE HYDROLOGIC CYCLE AS CONTEXT FOR ENVIRONMENTAL SURFACES AND INTERFACES

This chapter begins by discussing the nature and properties of water and the average composition of the Earth's crust, both of which are essential components for understanding environmental surface and interface interactions. Next, the hydrologic cycle is introduced, as this is essential to understanding water–rock(–bios) interactions. Developing an in-depth understanding of water flow and its relation to (bio)chemical fluxes and interface reactions is beyond the scope of this book. Rather, for each of the major components—atmosphere, oceans, underground, and surface-water environments—some fundamental aspects of chemical composition are reviewed, and several key surface/interface reactions are highlighted.

THE STRUCTURE AND FUNDAMENTAL PROPERTIES OF WATER

Water is essential for all life as we know it, and it is an important component of most environmental interfaces. The term "water" is commonly used to refer to the liquid state, although water may also exist in solid (ice) or gaseous (water vapor) states. The water molecule consists of two hydrogen atoms and one oxygen atom, to give a formula of H_2O. Taking into consideration the relative abundances of H and O isotopes, water has a molecular mass of 18.0153 g mol^{-1}. The IUPAC name for water is *oxidane*, and water is also commonly known as *dihydrogen monoxide*.

The structure of the water molecule was determined in 1956 by infrared spectroscopy (Benedict et al., 1956). The H-O-H angle is 104.5° (Figure 2.1). The hydrogen atoms are symmetrical around the O atom with bond lengths of 0.9572 Å. The dipole moment points from the O apex toward the two H atoms with a relatively large value of 1.83×10^{-18} esu cm (Eisenberg and Kauzmann, 1969). This means that water is a very important polar solvent.

One of the main properties that makes water unique is the association of neighboring water molecules through hydrogen bonding. The energy of the hydrogen bond in

Environmental Surfaces and Interfaces from the Nanoscale to the Global Scale, by Patricia A. Maurice

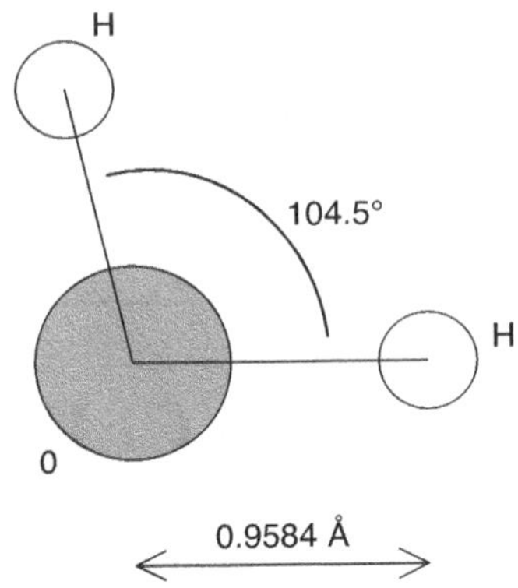

FIGURE 2.1 The water molecule.

water is ~20–40 kJ mol^{-1}, which is intermediate between van der Waals interactions (~1.3 kJ mol^{-1}) and covalent bonding (~400 kJ mol^{-1}) (Michot et al., 2002 and references therein). Each molecule of water in liquid water can form up to four hydrogen bonds with other water molecules; the average hydrogen bond length and number varies with temperature and pressure. The hydrogen bonding network in water tends to impart structure, whereas thermal motion tends to have a disordering effect. The behavior of the hydrogen-bonded network in water and the structure of liquid water are not fully understood and are the subjects of ongoing research. Michot et al. (2002) describe liquid water as "a gel structure where extensive connectivity of different regions is established by the hydrogen bonds." The hydrogen bonding network of water is responsible for the high boiling point, which in turn allows life as we know it to exist over a broad temperature range.

The *dielectric constant* of water (ε) is 87.9 at 0°C, 78.4 at 25°C, and 55.6 at 100°C (Murrell and Jenkins, 1994). Because of its relatively high dielectric constant and polarity, water tends to solvate ions in solution, with cations associated through the O atom and anions through the H atoms of the water molecule. Ions that are more polarizing (based on the ratio of charge to radius squared, $1/r^2$, known as the polarizing power; or the ratio of charge to radius, known as the ionic potential $I_P = z/r$) tend to associate more with water molecules. Ions such as Li^+, Na^+, Mg^{2+}, F^-, and OH^- have strong polarizing power, tend to break hydrogen bonds, and to be surrounded by multiple water layers. They are considered to be strongly hydrating and tend to dissolve easily from minerals. Ions such as K^+, Cs^+, and NO_3^- have medium polarizing power and can locally modify the hydrogen-bond network of water but do not have as many water molecule layers around them (Michot et al., 2002). They are considered to be weakly hydrating. Thus, they can lose their waters of hydration more easily to form bonds with surfaces. Ions that have still weaker polarizing power do not break hydrogen bonds and tend to be more "hydrophobic," which means that they tend not to associate with water.

Among organic compounds, nonpolar aliphatic hydrocarbons tend to have low aqueous solubilities. Aromatic hydrocarbons tend to be more soluble, most likely because of interactions between water and the π electron clouds (Larson and Weber, 1994). The solubilities of some example organic molecules are listed in Chapter 8 (Table 8.2). Within an aliphatic or aromatic homologous series, McAuliffe (1966) observed that solubility tends to decrease with increasing molar volume of the molecule.

The presence of a solid surface can perturb the hydrogen-bonding network of water and result in a different *dielectric constant* of water and a different structure of water at and near the interface than in the bulk solution. Water and ion distributions near surfaces are discussed in greater detail in Chapter 5.

Water is amphoteric, meaning that the water molecule can both donate and accept protons

$$H_2O + H_2O \rightleftarrows H_3O^+ + OH^-. \tag{2.1}$$

The pH (where $pH = -\log a_{H+} = -\log a_{H_3O^+}$) is a key parameter for describing many reactions and processes in and associated with water.

The density and dynamic viscosity of water change with temperature. The density of "pure"water at 0°C, 25°C, and 100°C = 999.8, 997.0, and 958.4 kg m^{-3}, respectively, and the dynamic viscosity at these temperatures is 1.79×10^{-3}, 8.91×10^{-4}, and 2.82×10^{-4} kg m^{-1} s^{-1}, respectively (CRC Handbook of Chemistry and Physics). At 1 atm, "pure" water freezes at 0°C (273.15 K, 32°F) and boils at 99.974°C (373.124 K, 211.95°F). At high elevations, water boils at lower temperatures, and in deep ocean environments such as at deep sea hydrothermal vents, water can remain liquid and not boil at up to several hundred degrees.

The *specific heat capacity* of (liquid) water at 15°C is 4.184 J g^{-1} K^{-1}. The value at 15°C is given because it is the approximate average temperature of the surface of the Earth. The high heat capacity of water plays an important role in regulating the temperature at the surface of the Earth. Every time that water changes state, for example evaporating from the oceans or condensing to form rain, heat is either absorbed or released by the water molecules. The heat energy associated with this exchange is called *latent heat*. The *latent heat of vaporization* of water at 15°C is 2467 kJ kg^{-1} water. This value is the highest latent heat of vaporization by far of common liquids. When water evaporates, about 600 calories (1 calorie = 4.1858 J) of energy are exchanged for every gram of water evaporated.

THE CHEMICAL COMPOSITION OF THE EARTH

The four most abundant elements on Earth: Fe, O, Si, and Mg, together account for > 99% of the Earth's mass. Other important elements of primary importance include Ni, S, Ca, and Al. Because of chemical zonation, the abundance of elements in the crust is different than the Earth as a whole, with O making up 46.6% of the crust by weight: Si 27.7%, Al 8.13%, Fe 5.00%, Ca 3.63%, Mg 2.09%, Na 2.83%, and K 2.59%. (Encyclopedia Britannica Online, 2008).

As determined by the "Father of Geochemistry" F.W. Clarke, most of the common minerals of the Earth's crust are oxides, with far lesser amounts of minerals based on S, Cl, and F. The feldspar minerals are the most abundant in the Earth's crust, and together with quartz, they compose ~63% by volume. Pyroxene and olivine together make up another 14%, hydrated silicates 14.6%, carbonates ~2%, and other minerals ~6% (Wyllie, 1971). The importance of different minerals with respect to the environmental interface interactions does not necessarily follow abundance; for example, Fe(III)(hydr)oxides play an inordinate role because they tend to be widely disseminated as small particles with high surface areas and to form coatings on other mineral grains. Some minerals of interest in environmental surface and interface interactions are discussed in Chapter 3.

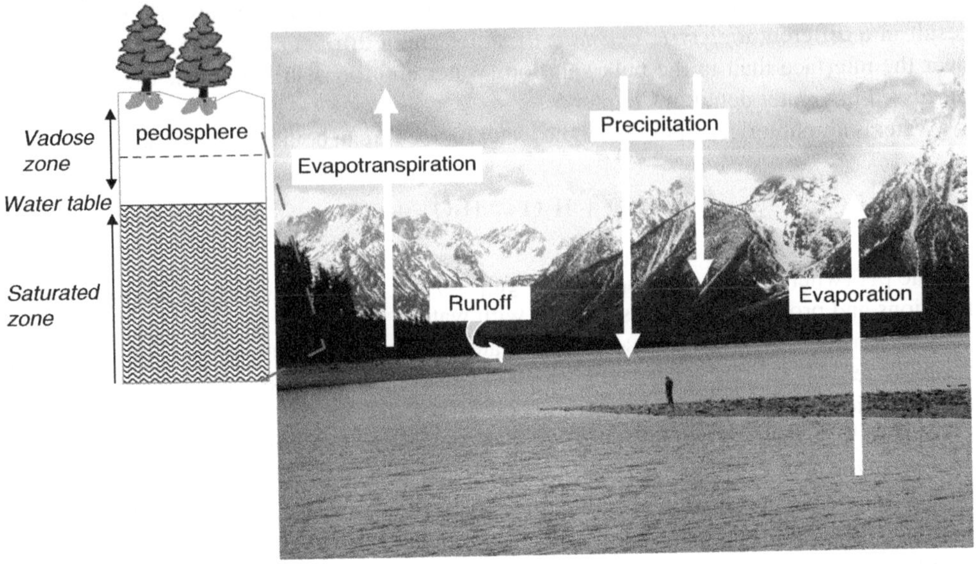

FIGURE 2.2 A schematic illustration of the critical zone showing some of the interactions in the Earth's near-surface environment.

THE CRITICAL ZONE

The "*critical zone*" (Figure 2.2) is the heterogeneous, near-surface Earth environment where the hydrosphere, biosphere, geosphere (specifically, the crust), and atmosphere interact with one another and with the human sphere in a complex manner that is of critical importance to life on Earth (e.g., NRC, 2001). The critical zone is a large-scale interface where numerous small-scale interfacial processes and reactions occur.

THE HYDROLOGIC CYCLE

Water covers two thirds of the Earth's surface and is essential for life on Earth. The complex system by which water circulates at and near the Earth's surface along *flow paths* that connect *reservoirs* is termed the *hydrologic cycle*. The hydrologic cycle is driven by and coupled to the Earth's energy cycle. Solar radiation and to a much lesser extent the Earth's internal energy, in combination with gravity, drive the circulation of water through the hydrologic cycle (Figure 2.3). This water circulation and *latent heat* exchange play an important role in the Earth's surface heat balance.

Ninety-seven percent of the Earth's water is in the oceans. The two other major reservoirs of the hydrologic cycle are water associated with land (~3%) and the atmosphere (0.001%). Most of the water associated with land is ice and snow, particularly in the polar ice caps. Additional reservoirs of water associated with land are underground water, which includes groundwater, soil porewater, and capillary fringe water; and surface waters, including rivers, lakes, and wetlands. A small percentage of water is present in the living material or *biota*. Surface water tends to be far more dilute than seawater, and atmospheric water is even more dilute (see below).

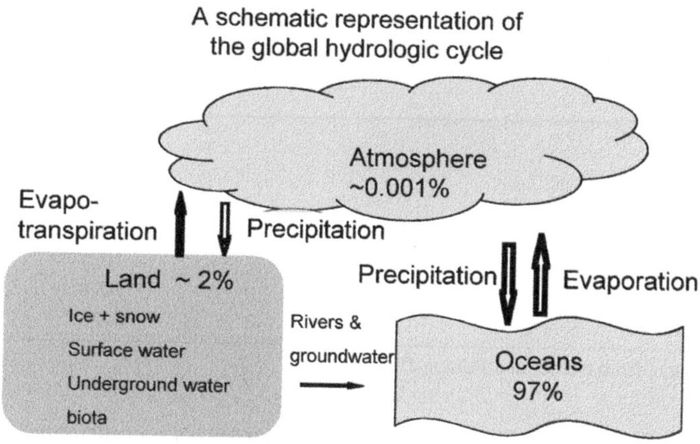

FIGURE 2.3 A schematic representation of the hydrologic cycle showing reservoirs and flow paths.

Average annual precipitation and evaporation are each ~ 1.0 m y^{-1} over the Earth as a whole. Over the land surfaces, average precipitation (P) is ~0.80 m y^{-1}, whereas over the oceans, the average is P ~1.1 m y^{-1}. Evaporation and evapotranspiration (E) over the land surfaces are ~0.5 m y^{-1}, whereas ocean evaporation is E ~1.2 m y^{-1}. P-E over land surfaces is ~0.3 m y^{-1}, whereas P and E are approximately balanced over the oceans (Brutsaert, 2005). The excess P-E over land surfaces leads to a constant transfer of water from the oceans to the land surfaces. Water is returned to the oceans by rivers, along with a small amount of direct discharge of groundwater to oceans.

The *mean residence time* (MRT) is the amount of time that the average drop of water spends between entering and leaving a given reservoir. Water in the atmosphere has the shortest MRT of only 2–3 days. Shallow groundwater has an MRT from days to weeks to years; deep groundwater MRT may be thousands or even millions of years. Water in the shallow oceans has a shorter MRT than water in the deeper oceans, which can be weeks to months versus thousands of years.

The *thermodynamics* of water–rock interactions were studied early on for relatively deep groundwaters systems (as groundwater-fed springs), which can have an MRT of thousands of years. Given the long MRT, *thermodynamics* often works reasonably well for deep groundwaters. However, microbiologic processes such as CO_2 production can prevent even deep groundwaters from attaining equilibrium. Shallow groundwaters, soils, and streams can have much shorter MRT values; hence, *kinetic* controls tend to be more important for these components of the critical zone. These examples are broad generalizations; for example, the mineralogy of a site can exert an important influence on attainment of equilibrium because different minerals tend to dissolve (or precipitate) more or less quickly.

OCEANS

Ocean water has a mean salinity of ~35 ppt (parts per thousand). Major dissolved constituents are Na and Cl along with Mg, SO_4, Ca, K, and HCO_3 (bicarbonate ion). Mean concentrations are shown in Table 2.1. Ocean waters are somewhat alkaline, with an average pH of ocean surface water ~8. The concentrations of minor constituents of ocean water, including nutrients, vary with depth and location. This variability, combined with

TABLE 2.1 Concentrations of major dissolved components of seawater with a salinity of 35 o/oo at 25°C.

Ion	mmol L^{-1}
Cl^-	558
Na^+	479
Mg^{2+}	54.3
SO_4^{2-}	28.9
Ca^{2+}	10.5
K^+	10.4
HCO_3^-	2.0

Data from Berner and Berner, 1987.

differences in temperature, light penetration, and other factors, contributes to pronounced spatiotemporal differences in ocean biological productivity.

Mineral–water, plant–water, and other interface reactions on land can help to control inputs of dissolved and particulate components to the oceans. Many mineral/biota-water interface reactions occur within the oceans, as well, and particularly along the margins of the continents where sediments are constantly being shed as the continents erode.

The saturation state of minerals helps to determine when and where calcareous versus silicic plankton can occur within the oceans. Changes in temperature, pH, and other parameters explain fluctuations in ocean planktonic life over geologic time. Most carbonate- and silicate-secreting organisms live in shallow surface waters (<200 m) where photosynthesis can occur where abundant food is present (Berner and Berner, 1987). The $CaCO_3$ minerals calcite and aragonite dissolve in the deeper oceans because of CO_2 that accumulates as respiration occurs in the absence of photosynthesis, such that

$$\underset{\text{Calcite or aragonite}}{CO_2(g) + H_2O(l) + CaCO_3(s)} \rightleftarrows Ca^2 + (aq) + 2HCO_3^-(aq) \tag{2.2}$$

and because the solubility of $CaCO_3$ increases with increasing pressure and decreasing temperature (Berner and Berner, 1987). Most of the $CaCO_3$ that passes into the deeper oceans as organisms die and shells sink is redissolved by these processes (Berner and Berner, 1987).

Ferromanganese nodules (often called manganese nodules) are microscopic to cm-scale concretions found in different parts of the ocean floor. They consist of concentric layers of Fe- and Mn-hydroxides around a core, and they can contain potentially economically viable concentrations of Mn, Ni, Cu, and Co. Other metals such as Pb and rare earth elements (REEs) such as Ce (cerium) can also be present. The mechanisms of these associations, such as whether primarily adsorbed or precipitated (Chapters 7, 9) or adsorbed to iron versus manganese components, is the subject of ongoing research in environmental surface chemistry (e.g., Takahashi et al., 2007).

ATMOSPHERE

Atmospheric water comes about mostly via evaporation and transpiration. When water evaporates, it leaves its dissolved consitutents behind so that atmospheric water tends to be dilute. The major sources of chemical components in the atmosphere include natural aerosols in the form of *sea salts*, terrestrially derived particles, pollutant particles, components released by forest and grassland burning, and other components. The source of ions in precipitation may be more marine or continental

$$Cl^- = Na^+ > Mg^{2+} > K^+ > Ca^{2+} > SO_4{}^{2-} > NO_3{}^- = NH_4{}^+, \tag{2.3}$$

where ions on the left have primarily a marine source, and those on the right have a more continental origin. Different storm patterns and/or seasonal differences can greatly affect the composition of precipitation. For example, along the east coast of the United States, hurricanes and tropical storms are large precipitation events with a strong marine signature (high Cl, Na). However, thunderstorms that travel eastward across the continent can bring more terrestrial-derived components, including SO_4 and NO_3 from anthropogenic sources (acid rain). Germer et al. (2007) showed marked changes in precipitation composition in the southwestern Amazon basin at the end of the dry season, mostly because of burning forests and pastures.

The pH of rainfall is controlled to a large extent by the uptake of CO_2 into water, to form carbonic acid and its various ionic species. The pH of otherwise pure water in equilibrium with atmospheric CO_2 is 5.65. "Natural" rainfall measured over the Amazon Basin is slightly more acidic because the trees of the Amazon Basin release a small amount of sulfur-containing compounds. In many parts of the world, *sulfur dioxide* (SO_2) and *nitrogen oxides* (NO_x) released by the burning of fossil fuels result in sulfuric acid and nitric acid in rainfall, which leads to *acid rain* or, more broadly stated (to include snow, sleet, etc.), *acid precipitation* (Chapter 14). In some parts of the world, alkaline particles in the atmosphere can drive the pH of rain higher.

When precipitation passes through the tree canopy, it can acquire organic matter and various ions associated with the plants, as well as components that were deposited on the vegetation surfaces as wet or dry deposition. In many instances, this *throughfall* acquires nutrients that are then carried to the forest floor. In regions where burning of forests occurs, organic material as well as K, Cl, NH_4, NO_3, and SO_4 can all be released to the atmosphere (Germer et al., 2007) and then deposited onto plant surfaces and eventually into throughfall.

Table 2.2 shows some values of solute concentrations in precipitation and throughfall in the Amazon Basin compared with the Atlantic Coastal Plain (New Jersey). First, note that the Cl concentrations are $\sim 10^4$ to 10^6 times more dilute than in mean ocean water. The Na and Cl concentrations in precipitation are considerably higher in the state of New Jersey and in the Central Amazonia region because of significant sea salt and/or marine-derived rainfall, but far less in Western Amazonia. The effects of acid rain are apparent in New Jersey rainfall as evidenced by relatively low pH and high SO_4 and NO_3 concentrations. Concentrations of many constituents are higher in throughfall than in precipitation at a given site (SW Amazonia, New Jersey). Throughfall composition can be influenced significantly by the type of canopy vegetation (e.g., Atlantic white cedar versus pine/oak forest, New Jersey).

Atmospheric particles range in size from 1 nm to 100 μm (Anastasio and Martin, 2001). They may come about from a wide variety of processes, such as windblown desert sands and soils, wave action, abrasion of plants, and volcanic activity. Very fine particles are often produced as primary emissions from combustion sources or are formed in the atmosphere via gas-to-particle conversions.

TABLE 2.2 Some volume weighted mean solute concentrations in precipitation and throughfall.

Site	pH	DOC	NH_4^+	Ca^{2+}	Mg^{2+}	Na^+	K^+	Cl^-	SO_4^{2-}	NO_3^-
Rainfall										
SW Amazonia[1]	6.3	591.8	24	25.4	4.3	7	13.1	11.4	4.5	4.1
C Amazonia[2]	4.8	—	1.1	1.3	2.2	24.5	1.1	27.3	7.9	3
W Amazonia[2]	5.7	---	---	4.6	0.8	1.7	1	3.4	1.9	—
New Jersey[3]	4.30	—	14	2.5	2.5	19	1	25	23	26.5
Throughfall										
SW Amazonia	6.3	1354	21.7	21.6	12	13	109.7	31.4	5.3	8.7
New Jersey										
Atl. White Cedar	3.9	1167	15	70	53.5	108.7	43.6	128.6	125	132
Pine/Oak	4.9	939	3.9	29.9	21.0	65.2	28.2	71.4	44.8	32.2

[1] Germer et al. 2007, based on repeated sampling in 2004–2005.
[2] Stallard and Edmond, 1981, as tabulated in Germer et al. 2007, based on single sampling.
[3] Johnsson and Barringer, 1993 and Lord et al., 1990, based on repeated sampling in 1983–1988.
All values except pH are in μmol L^{-1}. SW = southwest, C= central, CW = central west.

Atmospheric particles can play important roles in many environmental processes. They can influence climate change directly through radiative forcing (Chapter 14) or more indirectly through effects on properties and lifetimes of clouds. Interface reactions on atmospheric particles play a role in *ozone* depletion. Non–ozone-destroying chlorine forms such as $ClONO_2$ can be catalyzed by *polar stratospheric cloud* (PSC) particles into ozone-destroying chlorine compounds through reactions such as

$$ClONO_2 + H_2O \xrightarrow[\text{PSC particles}]{} HOCl + HNO_3 \tag{2.4}$$

and

$$ClONO_2 + HCl \xrightarrow[\text{PSC particles.}]{} Cl_2 + HNO_3 \tag{2.5}$$

The resulting HOCl and Cl_2 rapidly *photolyze,* or break apart through photo-reactions, to yield ozone-destroying Cl radicals (Molina et al. 1987). This ozone depletion allows more damaging ultraviolet (UV) radiation to penetrate the atmosphere.

Atmospheric particles must be considered in many portions of the critical zone, even when their importance might not be recognized at first glance. For example, aeolian (wind-blown) dust can be a major component of soils, riverine deposits, and marine sedimentation. Dust from the Sahara desert is transported to various parts of Africa and South America. Brimhall et al. (1988) showed that ore-grade aluminum enrichment in a Western Australian *bauxite* was primarily caused by input of aeolian dust from more highly weathered soils. It is essential to consider such inputs when investigating soil and riverine processes.

Atmospheric dust delivers Fe, which is an important and often limiting nutrient, to the oceans. Figure 2.4 shows dust blown off the Sahara Desert over the Atlantic Ocean. This dust carries nutrient Fe to the oceans.

Black carbon (BC) particles are produced by the incomplete combustion of fossil fuels and from fires. BC in the atmosphere can play a role in radiative forcing (Chapter 14), and BC deposited from the atmosphere onto snow in the Arctic can affect the albedo of the

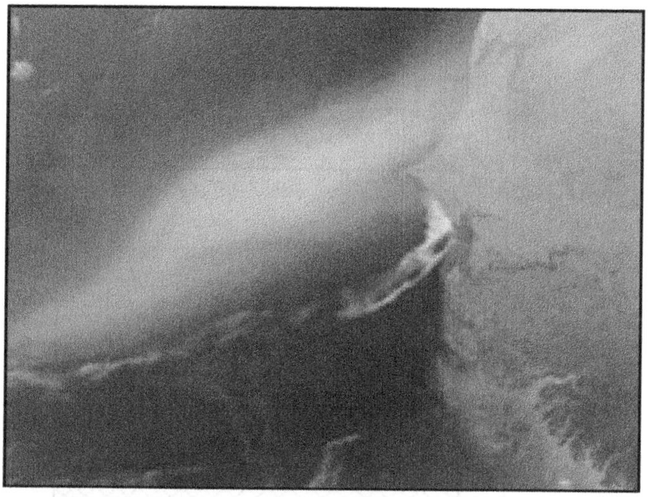

FIGURE 2.4 Windblown sand from the Sahara Desert is blown out over the Atlantic Ocean. A moderate resolution imaging spectroradiometer (MODIS) image captured by Terra satellite on April 30, 2003. Courtesy of Jacques Descloitres, NASA.

snow (the percent of sunlight reflected back versus absorbed) (McConnell et al., 2007). Both of these processes can play a role in climate change (Chapter 14).

UNDERGROUND WATER

The water beneath the land surface is collectively referred to as *underground water* or *subsurface water*. Much underground water (Figure 2.5) occurs in the *saturated zone*, which is the region in which pore spaces are completely filled with water; this water is known as *groundwater*. Above the saturated zone is the *unsaturated zone* or *vadose zone*, where the pores between mineral grains are only partly filled with water and the rest with soil gases. The top of the unsaturated zone is the *soil zone* or *pedosphere*, where a variety of processes such as physical and chemical erosion by plants and microbial activity work to transform rock into soil. The saturated zone is separated from the unsaturated zone by the *water table*, which is the top of the saturated zone. Directly above the saturated zone is the *capillary fringe*, which is the portion of the unsaturated zone at the upper boundary of the water table, where water is pulled upward against the force of gravity by capillary forces. *Capillarity* is an interfacial phenomenon in which surface tension plays a key role, as discussed in Chapter 5.

Two terms of importance to underground water environments are *porosity* and *permeability*. Porosity is defined as the fraction of the bulk volume of rock or sediment that is composed of voids that may be filled with water and/or air or soil gases. Permeability is a measure of the interconnectedness of the pores. A rock may be highly porous but not permeable, for example, if individual grains are cemented along some boundaries. Alternatively, a rock may have relatively low porosity but reasonably good permeability, for example along fractures.

When water passes into the underground environment, its composition can change dramatically. The concentrations of chemical constituents tend to be concentrated through

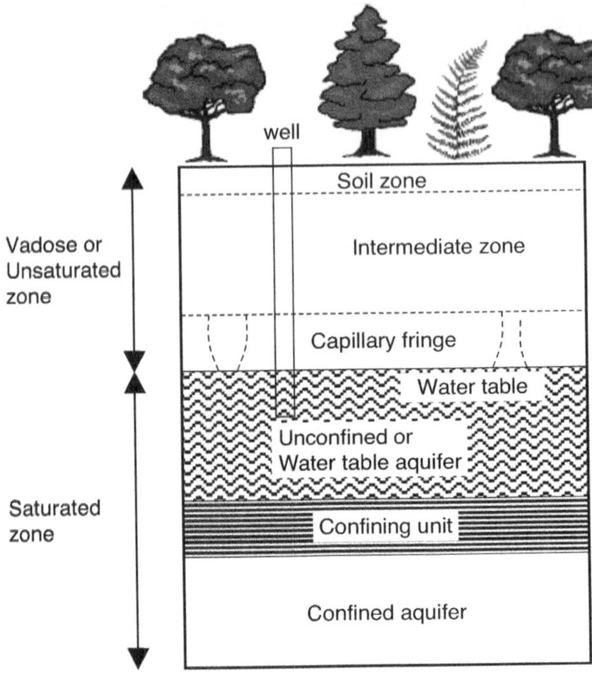

FIGURE 2.5 The components of underground water.

evaporation and evapotranspiration. Biological activity within the soil zone can contribute CO_2 and natural organic acids, which make the water more acidic and promote mineral dissolution, which releases additional chemical constituents such as Ca, Mg, and K from minerals into the water. Other processes such as ion exchange reactions, mineral precipitation, and a host of microbially mediated redox reactions also can impact the water composition or *water quality*.

Soils and Soil Water

Soils form as rock or sediments and are broken down by a combination of physical, chemical, and biological processes to form fine-grained materials. Such *weathering* processes are discussed in Chapter 10. Soil formation or *pedogenesis* is controlled by factors such as the parent-rock composition, climate, relief, time, and the presence and activity of organisms that range from plants to fungi and bacteria. Many different types of soils exist but one in which soil processes are highlighted as dramatically colored bands is the spodosols (Figure 2.6). *Spodosols* (sometimes termed podzols in earlier literature) are characterized by distinct horizons, which include the following from top to bottom:

1. *Zone of organic accumulation* (O horizon) that contains leaf litter and other dead organic matter, from which organic acids are carried into the lower soil zones
2. *Leached zone* (A or E horizon), which is often grayish in color and generally dominated by quartz
3. *Zone of accumulation* (B horizon) of Fe and aluminum oxides

FIGURE 2.6 A typical spodosol, photographed in the New Jersey Pinelands (USA). The underlying sediments show cross bedding.

4. Poorly weathered *regolith* (C horizon), which contains very little organic matter or living organisms

The change in chemical composition of water from throughfall into soil (spodosol) pore waters and then to a nearby shallow well in the New Jersey Pine Barrens (Atlantic Coastal Plain) is shown in Figure 2.7. The organic matter concentration increases in the organic horizon and then decreases into the lower soil and groundwater. Fe, which is primarily bound to organic matter, shows a similar profile. Al is also associated mostly with organic matter, although some inorganic Al remains mobile here and at other nearby sites. Sulfate remains mobile through this acid-rain–impacted spodosol and into the underlying groundwater. Si is released by weathering in soils.

From an interface standpoint, the fact that soils are by definition hydrologically unsaturated is significant. Not only mineral–water interface reactions, but also mineral–gas interface processes, or some combination, may occur. Soils undergo changes in saturation and may sometimes become fully saturated. Soils contain micropores as well as macropores. The saturation state of porewaters with respect to different minerals can be different in macropores versus micropores, with micropores often being closer to equilibrium because of longer contact times with the soil minerals. Soils also can have fluctuating redox boundaries. The area of soils that directly surrounds roots, and contains associated microorganisms, is known as the *rhizosome*, and it is the location of many complex mineral—water–biota-interface reactions as discussed in Chapters 11 and 12.

Groundwater

Water in the saturated zone is called *groundwater*. As part of the dynamic hydrologic cycle, groundwater is always moving, generally more quickly in shallow environments and more slowly in deep environments. The movement of groundwater is determined by *hydrologic head*, which can be thought of as the potential energy of the mass of groundwater at a given

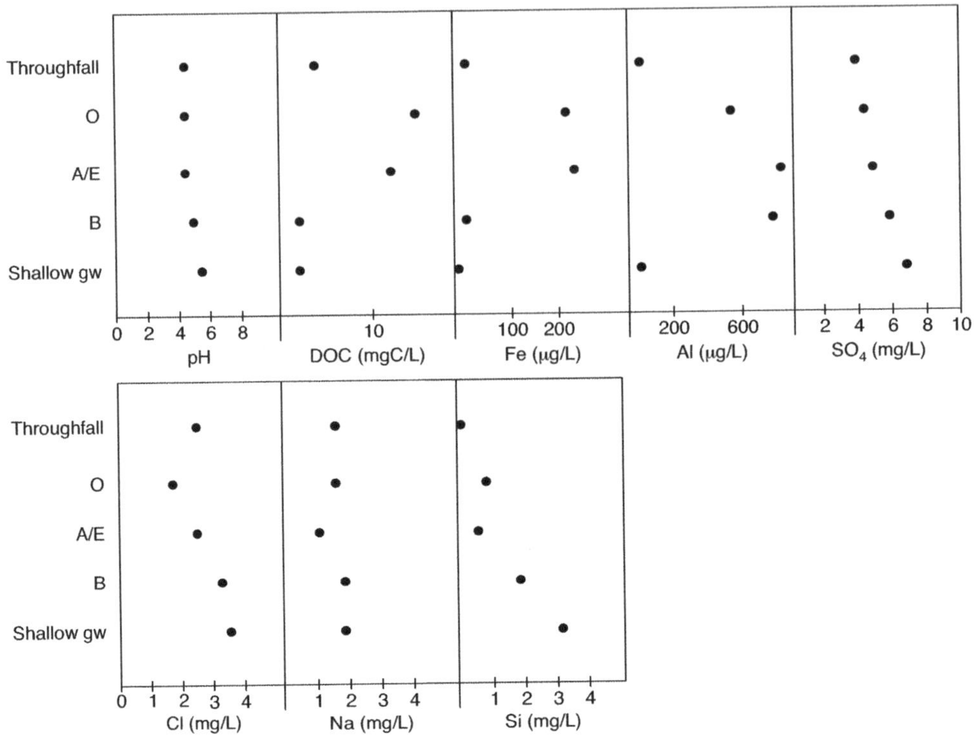

FIGURE 2.7 The change in chemical composition of water from throughfall into soil (spodosol) pore waters and then to a nearby shallow well in the New Jersey Pine Barrens (Atlantic Coastal Plain). Using data from Lord et al., 1990, and Johnsson and Barringer, 1993.

point relative to a specified base level, which is most often sea level. According to *Darcy's Law*, water always flows from locations of greater total head to locations of lesser total head

$$Q = K \times A \times I, \tag{2.6}$$

where

Q = flux, or the quantity of water moving through a given area over a given time
K = *hydraulic conductivity*, which depends on the flow properties of the geologic materials and of the water
A = cross-sectional area through which the water flows
I = *hydraulic gradient*, which is a measure of the change in head between two measurement points (e.g., wells) located at a given distance, dh/dl (where h is head and l is distance).

The relationship between groundwater head and flow is shown schematically in Figure 2.8.

Sometimes, water can actually flow upward geographically but still from higher head to lower head. This process occurs, for example, when a unit of groundwater underlies

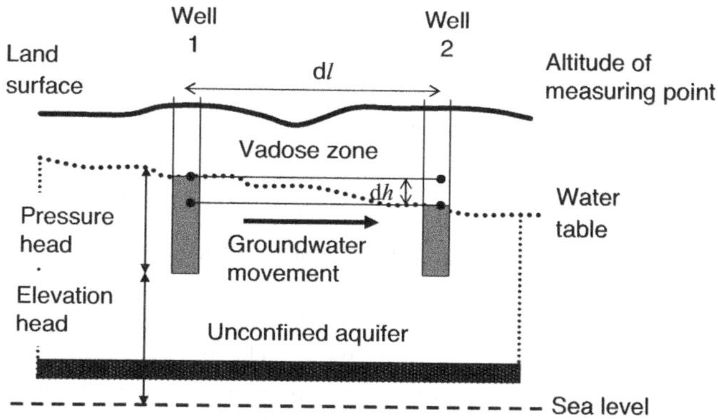

FIGURE 2.8 A schematic illustration of groundwater heads and flow, with symbols from Darcy's law equation. The value dl is the horizontal distance, and the value dh is the head difference between the wells. The level is the reference elevation. The direction of water flow is from Well 1, which has a higher head, to Well 2, which has a lower head.

an impermeable geologic unit such that pressure is put on the groundwater. If a well is placed through the impermeable unit, then the *confined* groundwater may flow upward toward the land surface without any need to pump. Such a free-flowing well is called an *artesian well.* A groundwater unit that contains a volume of water that can be supplied at reasonable cost to humans is known as an *aquifer*. An *unconfined aquifer* is not constrained by overlying impermeable units. Its surface is the *water table*, which is defined as the level in the saturated zone at which the hydraulic pressure is equal to atmospheric pressure. Aquifers that are bound by confining units are called *confined aquifers* with the confining units termed *aquicludes.*

Groundwater Chemical Composition. In groundwaters, seven major constituents, which are Ca, Mg, Na, K, SO_4, Cl, and HCO_3, tend to make up nearly 95% of the dissolved ions (Plummer et al., 2003). A variety of other trace constituents, which include natural and anthropogenic organic compounds, metals, inorganic anions, and isotopes, also can be present in various abundances.

Aquifer mineralogy can be broadly divided into the following two types: aquifers that contain relatively reactive minerals such as limestone, dolomite, gypsum, and halite, along with natural organic matter, and aquifers with relatively unreactive minerals, primarily quartz and highly weathered silicates (Plummer et al., 2003). Waters that pass through aquifers are characterized by reactive minerals can evolve significantly in composition along flow paths, which reflects a host of (bio)geochemical reactions. However, waters that pass through weathered silicates, particularly quartz sands, tend to retain more strongly the chemical characteristics that they had when first recharging the groundwater system, typically with less pronounced evolution over space and time.

Garrels and Christ (1965) and Garrels and MacKenzie (1967) developed an approach in which it was assumed that waters first entering an aquifer were far from equilibrium, but that they evolved toward equilibrium during reaction with aquifer minerals. If the mineralogy were known, then the evolution of the water could be considered in terms of mineral dissolution and precipitation. A mass-balance approach was developed, eventually also taking into consideration redox and isotopic information and accounting for the conservation of mass as water flows along a flow path

$$\text{initial water chemical composition} + \text{dissolving minerals} \\ -\text{precipitating minerals} = \text{final water chemical composition.} \qquad (2.7)$$

This approach has been expanded to account for other processes such as ion exchange on mineral surfaces.

Chapelle (2004) gives several examples of the application of the mass-balance approach to groundwater evolution along flow paths. The mass-balance approach must be applied carefully because different aquifers have more or less heterogeneous mineralogy and lithology (rock types), may or may not be at equilibrium, and may contain a variety of microbiologic processes that need to be identified.

The Snake River Aquifer is an example of a location that has fairly homogeneous (basalt) lithology. Rates of groundwater flow tend to be high at ~3 m per day, which results in a relatively low MRT of 200–250 years. Yet, the rates of mineral dissolution and precipitation tend to be relatively fast, so that no distinct trends in composition are observed along hydrologic flow paths. Olivine (a Mg/Fe silicate mineral), which represents a relatively small component of the mineralogy, accounts for much of the Mg and HCO_3 present in the water. Calcite precipitation exerts a control on the pH as well as on Ca and HCO_3 concentrations (Wood and Low, 1988, and Chapelle, 2004). Some examples of groundwater analyses from this aquifer are shown in Table 2.3.

The Upper Floridian Aquifer in Florida is a major source of freshwater, and it is formed primarily of limestone rocks (Ca and Mg carbonates). In contrast to the Snake River Aquifer, the lithology and mineralogy are more heterogeneous, and pronounced gradients are observed in groundwater composition along flow paths. As water flows from the topographically high recharge area near Polk City, southward to Ft. Meade and onward to Wauchula (Table 2.3), it evolves from relatively low to much higher total dissolved solids concentrations due to reactions with aquifer minerals (Plummer et al. 1983; Chapelle, 2004). Ca, Mg, and HCO_3 are released by weathering of carbonates. Near the coast, mixing with seawater that has penetrated the aquifer increases Na and Cl concentrations. Other

TABLE 2.3 Some examples of groundwater chemical analyses from the Snake River Aquifer and the Upper Floridian Aquifer.

Site	pH (units)	Ca	Mg	Na	K mmol L^{-1}	HCO_3	Cl	SO_4	SiO_2
Snake River Aquifer									
Snake River Aquifer 14	8.0	0.90	0.53	0.87	0.08	2.95	0.37	0.75	
Snake River Aquifer 10	7.5	1.6	1.2	1.6	0.14	4.59	1.55	0.65	
Upper Floridian Aquifer									
Polk City	8.0	0.89	0.23	0.14	0.01	2.03	0.13	0.20	
Ft. Meade	7.75	1.5	0.70	0.27	0.02	2.67	0.25	0.26	
Wauchula	7.69	1.7	1.2	0.36	0.05	2.75	0.28	0.30	

Based on data from compilation in Chapelle (2004); Snake River Aqufer data originally from Wood and Low (1988), and Upper Floridian Aquifer data originally from Plummer et al. (1983). Site locations given in Chapelle (2004).

minerals and processes also must be considered in applying a mass-balance approach, such as dissolution of gypsum ($CaSO_4 \cdot 2H_2O$) and iron hydroxide, precipitation of calcite ($CaCO_3$) and pyrite (FeS_2), and organic matter oxidation and CO_2 exchange. The realization that the Upper Floridian Aquifer is open to CO_2 exchange and microbial CO_2 production proved crucial for proper mass-balance modeling of chemical evolution along flow paths (Chapelle, 2004, and references therein).

Some Graphical Methods of Representing Groundwater Chemistry. A variety of approaches has been developed to represent graphically groundwater chemical composition. One common method is using bar graphs that represent the individual and total cation and anion concentrations in milliequivalents per liter (Figure 2.9). Diagrams of this type can be compiled for water from throughout the hydrologic cycle (e.g., precipitation and soil water) and have been used by researchers who investigate the effects of acid rain on watershed ecosystems. Data can be presented for specific study periods or (volume) averaged over time, but they are for one specific or average data set and cannot graphically present a large data set such as for a broad distribution of wells.

One graphical method frequently used to represent large data sets and to view trends in groundwater composition such as along flow paths is the *Piper diagram* (Figure 2.10). A Piper diagram (Piper, 1944) is actually a set of three graphs presenting concentrations of various constituents as percentages on an equivalent per liter basis, which includes a cation diagram, an anion diagram, and a combined diagram. To use a Piper diagram, the cation and anion concentrations in a water sample are converted to meq L^{-1} and then plotted as percentages of their respective totals in each of the two triangles. The cation and anion percentages in the two triangles are then projected onto the quadrilateral central diagram,

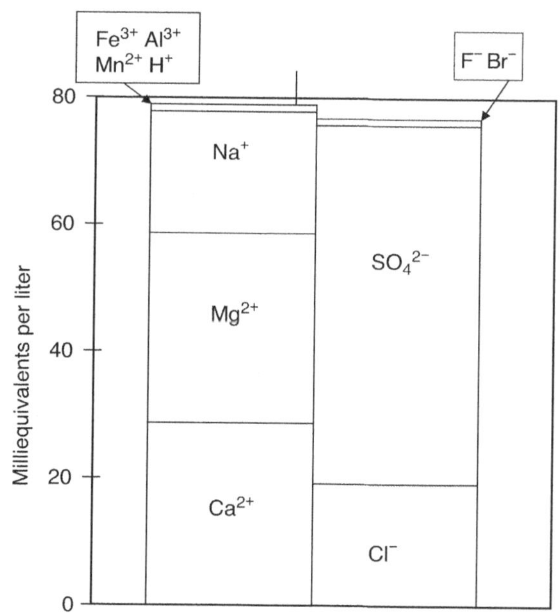

FIGURE 2.9 Bar graphs that show the concentrations in milliequivalents per liter of cations and anions at a location in the Florida Aquifer (based on data from USGS; Sprinkle, 1989).

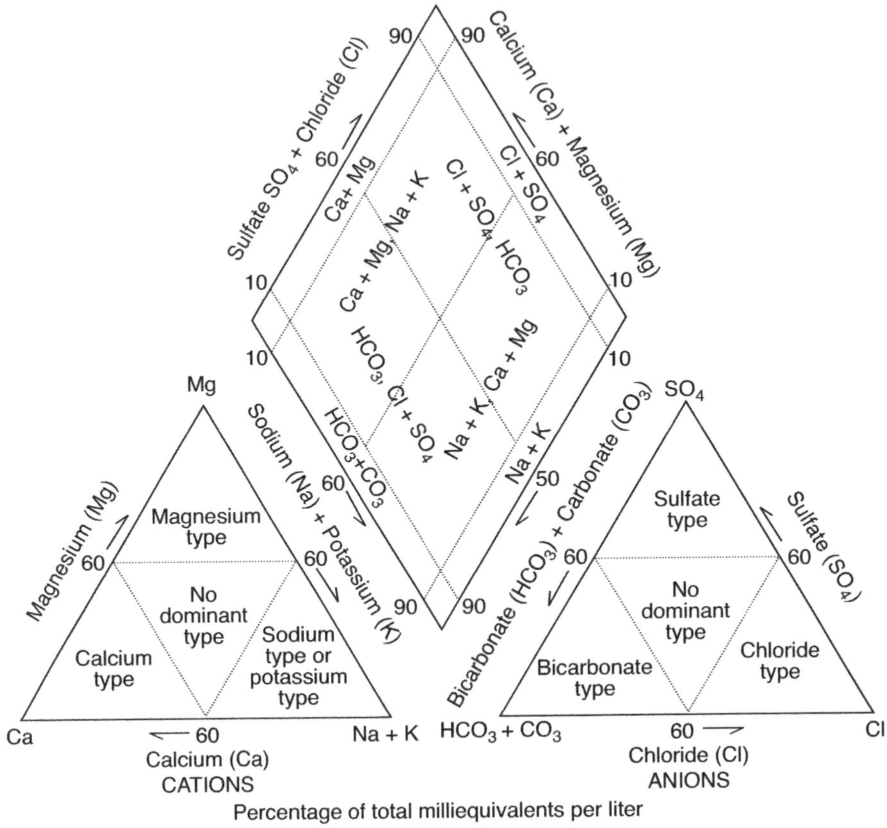

FIGURE 2.10 The regions of a Piper diagram, which are often termed *hydrogeochemical facies*. Based on Back, 1966; from Barton and Ogle, 2002, USGS publication.

which describes the water type or *hydrogeochemical facies*. The region in which a water plots on a combined diagram can serve as the basis for classifying the water, can be used to indicate "geochemical fingerprints," and can provide insight regarding the geologic media with which the water is interacting, the mixing of waters, and the hydrobiogeochemical evolution along a flow path in an aquifer. This last application must be performed with caution, as flow paths can be difficult to discern and the time component of flow needs to be considered.

The data in Piper diagrams need to be coupled with other factors, such as dissolved oxygen concentrations, concentrations of trace elements, and isotopic signatures, along with hydrologic analysis, to help determine factors such as likely sources of water, mixing of waters, and/or biogeochemical transformations.

Solute Transport. Solutes that are contained within groundwater are transported by the bulk motion of the fluid in a process known as *advection*. A solute moving through porous media will tend to spread out because of *hydrodynamic dispersion* brought about by mechanical mixing and molecular diffusion. The *dispersion* process causes the solute to be more dilute.

For the one-dimensional flow in a given direction (x) for a homogeneous medium

$$\text{transport by advection} = \bar{v}_x nCdA \quad (2.8)$$
$$\text{transport by dispersion} = nD_x(\partial C/\partial x)\,dA, \quad (2.9)$$

where

$\bar{v}_x$ = average linear velocity of the groundwater
n = porosity
C = mass of solute per unit volume solution
nC = mass of solute per unit volume of porous media
dA = elemental cross-sectional area
D_x = dispersion coefficient
x = distance in the x direction (Freeze and Cherry, 1979).

The processes of dispersion on a microscopic scale are shown in Figure 2.11.

As groundwater flows through an aquifer, it may undergo a wide variety of interface reactions that can affect the fate and transport of nutrients, pollutants, and other components. Minerals may dissolve or precipitate, which in turn can affect the porosity and permeability of aquifer materials. Contaminants may sorb or desorb from mineral surfaces. A host of aerobic or anaerobic microbially mediated processes may occur. Fine particles such as colloids, which includes nanoparticles, may migrate with the water. These fine particles may include viruses, bacteria, and abiotic amorphous and crystalline colloids such as Fe (hydr)oxides. Their interactions with aquifer mineral surfaces as related to surface and interfacial forces are discussed in Chapters 5 and 14.

Reactive transport models (Chapter 14) are used widely in attempts to describe pollutant transport in the subsurface and as part of risk assessment for contaminated sites. Many challenges are faced in reactive transport modeling, from flow in nonsaturated conditions

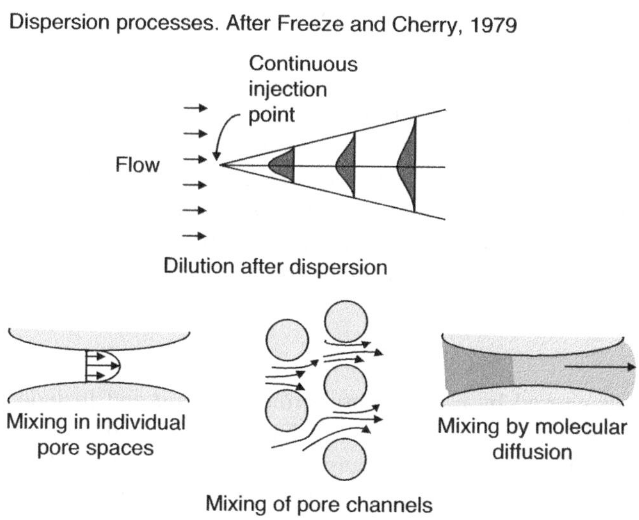

FIGURE 2.11 Schematic of the dispersion process overall and on a microscopic scale. Based on Freeze and Cherry, 1979.

or fractured flow, to developing sorption equations that work under a range of conditions (Chapter 14) but that can be integrated with flow, to dealing with complexities such as bacterial, nanoparticle, and colloid interactions.

Some Key Sites of Groundwater Investigations. Many contaminated sites in the United States have been the subjects of intensive investigation in groundwater hydrobiogeochemical processes. Several of these are discussed elsewhere in this text. Bemidji, Minnesota, was the site of a pipeline rupture in 1979 that contaminated a shallow aquifer with crude oil. This site has been particularly important for investigations of subsurface fate and transport of organic compounds and of microbial processes associated with redox boundaries and varying nutrient levels (e.g., Lovley et al., 1994; Delin et al., 1988; Bennett et al., 2001).

The U.S. Department of Energy's *Hanford Site* is a location along the Columbia River in southcentral Washington State that was used for Pu production and processing during the Manhattan Project and the Cold War. In addition to release of some Pu, a variety of other radioactive fission products were released, which include ^{137}Cs, ^{134}Cs, ^{135}Cs, ^{90}Sr, ^{131}I, ^{85}Kr, ^{144}Ce, ^{129}I, ^{3}H (tritium), ^{241}Am, and ^{99}Tc (Gephardt and Lundgren, 1998). The plume of contaminated groundwater beneath the Hanford site has been estimated to cover 150 square miles (Gephart and Lundgren, 1998). In particular, leaks of Cs, which was produced in substantial quantities and can be mobile in the subsurface, continue to be of major environmental concern and the focus of costly remediation.

The Department of Energy's Oak Ridge National Lab in Tennessee enriched uranium during the Manhattan Project and afterward and also has been the subject of a great deal of research on subsurface fate and transport of U, nitrate, and other contaminants. Research at these and other DOE sites has resulted in new knowledge of radionuclide sorption to minerals and interactions with microorganisms (e.g., Serne et al., 1998; McKinley et al., 2004; Um et al., 2007; Cardenas et al., 2008; Kelly et al., 2008).

SURFACE WATERS: FOCUS ON RIVERS

Surface waters include rivers, wetlands, and lakes, and all have a variety of interactions that involve environmental surfaces and interfaces. Here, we focus on rivers. Rivers transport solutes and sediments, ultimately from the land surfaces to the oceans. Many examples of water–rock interactions and other interface processes can occur within river systems, from colloid aggregation to photochemical reactions on particle surfaces (for example, photooxidation of organics and photoreduction of Fe(III) to Fe(II) in organic matter-Fe colloids) to *authigenic* formation of new minerals in streambeds.

Stream Load

The earth materials carried by a stream are termed its *load* and include *dissolved load, suspended load*, and *bed load* (Figure 2.12). The operational definition of *dissolved* is material that passes through a 0.45-μm pore-size filter. The *dissolved load* thus includes ions, molecules, as well as many small colloidal particles, including *nanoparticles*. The *suspended load* is the fine-grained sediment that is transported by the stream, such as silt, clay, and fine sand; the smallest particles overlap with the operationally defined dissolved load. The *bedload* consists of the heavier, coarser grained earth materials that travel along

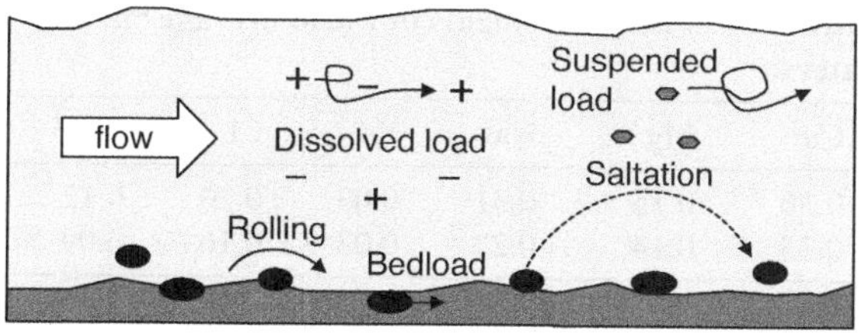

FIGURE 2.12 A schematic illustration of the components of stream load.

the bottom of the stream by sliding, rolling, or taking short skips or jumps in a process known as *saltation*.

The sources of the dissolved load to a stream are shown in Figure 2.13. Precipitation that falls on the stream's watershed may pass through the canopy vegetation to become throughfall. Water from precipitation may flow over the land surface as *overland flow*, through the soil zone directly to the stream as *interflow*, or pass through the deeper subsurface environment as groundwater *recharge* and eventual *discharge* to the stream. In each case, the chemical composition of the water may change along the flowpath. Precipitation may also fall on the stream directly. Once in the stream, water may pass in and out of the hyporheic zone, undergoing biogeochemical processes. Other sources of solutes to the stream include *nonpoint source pollution*, such as phosphate in agricultural runoff from widely distributed farms, and *point source pollution*, such as a heavy metal from a specific factory.

Table 2.4 shows the world average unpolluted and actual (including polluted and unpolluted) river water concentrations of major ions (Berner and Berner, 1987). Figure 2.14 describes the dominant sources, sinks, and controls of major ions in rivers and the global range of pristine river chemistry for medium-sized basins (Meybeck, 2004). The concentrations of dissolved constituents in rivers can vary greatly, sometimes by over three orders of magnitude in terms of both concentrations and yields (discharge weighted) (Meybeck,

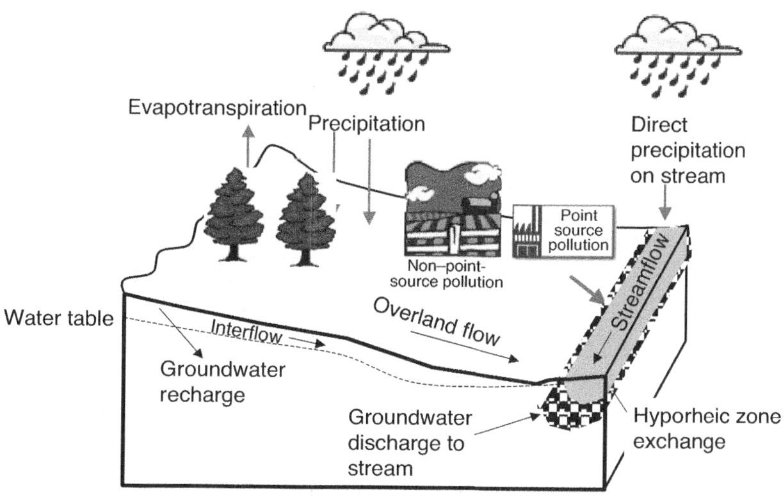

FIGURE 2.13 The sources of water and dissolved components to a stream.

TABLE 2.4 Concentrations of major elements in world average "actual" and "natural" (unpolluted) river waters.

	Ca^{2+}	Mg^{2+}	Na^{+}	K+	Cl^{-}	SO_4^{2-}	HCO_3^{-}	SiO_2
World actual	0.36	0.15	0.31	0.03	0.23	0.12	0.87	0.18
World unpolluted	0.33	0.14	0.22	0.03	0.16	0.09	0.85	0.17

TDS = total dissolved solids.
From data in Berner and Berner, 1987.

2004). The average river water concentration of Cl is about three orders of magnitude less than that of average ocean water, and one to several orders of magnitude greater than that of precipitation.

Gibbs Plots

In 1970, R. J. Gibbs proposed that the following three main mechanisms controlled the chemical composition of surface waters: (1) atmospheric precipitation, (2) rock weathering, and (3) evaporation-fractional crystallization. Freshwaters tend to have relatively higher Ca concentrations, and highly saline waters tend to have relatively higher Na concentrations. Based on this scenario, he further proposed that the relative contributions of each of these mechanisms could be determined by plotting the ratio [Na]/([Na] + [Ca]) on the x-axis of a diagram and the total dissolved salts in ppm on the y-axis (Figure 2.15).

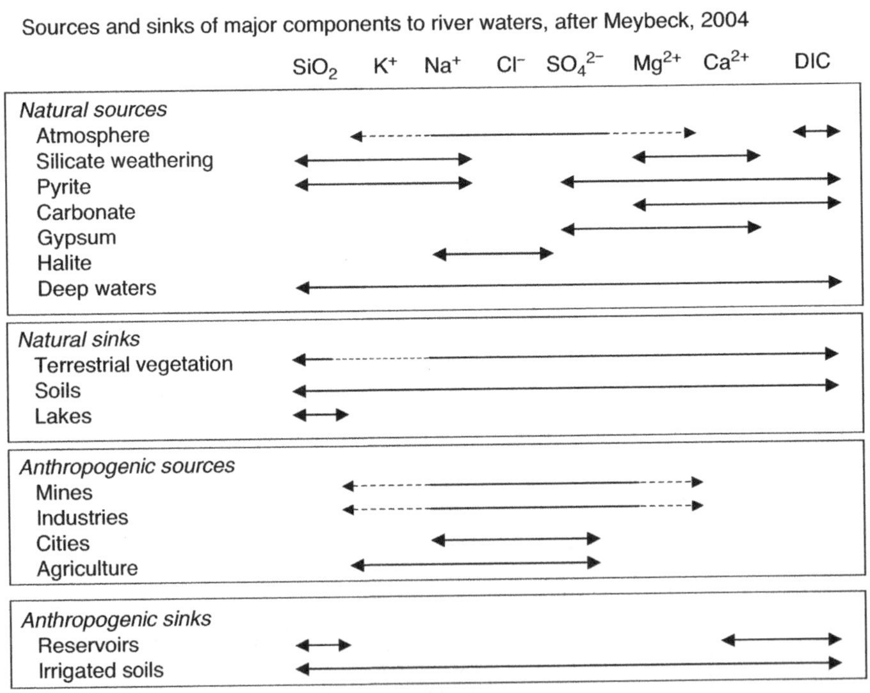

FIGURE 2.14 Natural and anthropogenic sources and sinks of major components of the world's river water. DIC = dissolved inorganic carbon. After Meybeck, 2004 (with permission).

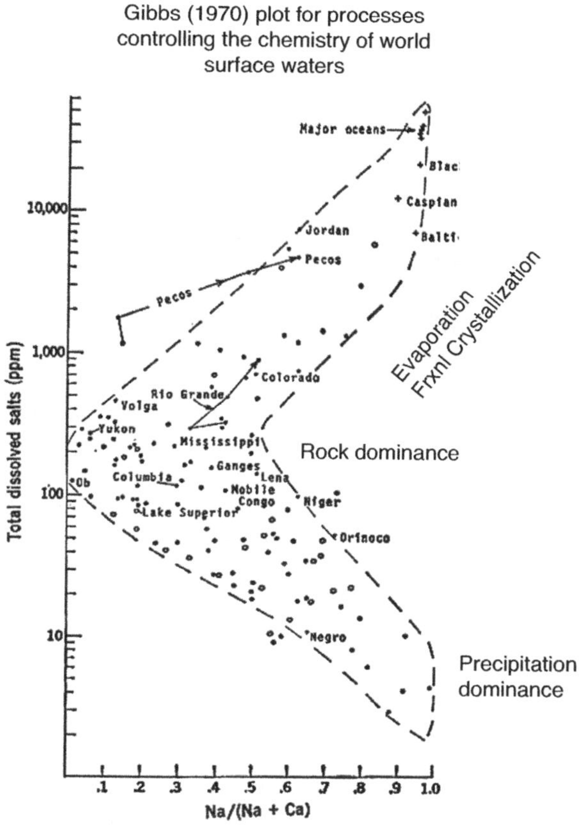

FIGURE 2.15 Gibbs plot for world rivers, using Gibbs' original 1970 data. The trends are shown to the right of the data in the figure. Ongoing research is refining this type of plot, especially as new data for river chemistry are obtained. Modified from Gibbs, 1970 (with permission).

When this is done, three trends occur, which form a boomerang-shaped curve as shown in Figure 2.15.

1. *Precipitation* (rainfall, snow, etc.) dominance is the lower arm. Because these surface waters are dominated by precipitation, which has relatively low dissolved solids, they plot in the lower portion of the graph. An example is the Rio Negro in South America, which is in an area with high rainfall and relatively little rock-weathering contribution.
2. *Rock dominance* is in the middle to the left. Many large tropical rivers plot in this region, where a good deal of rainfall occurs as well as a strong contribution from mineral weathering.
3. A combination of *evaporation/fractional crystallization* processes leads to the upper arm. The evaporation/fractional crystallization trend occurs because when evaporation occurs, salinity is increased, which causes $CaCO_3$ to precipitate out of solution and thereby increases the relative proportion of Na to Ca. Surface waters in this upper arm tend to be in hot and arid regions (Gibbs, 1970).

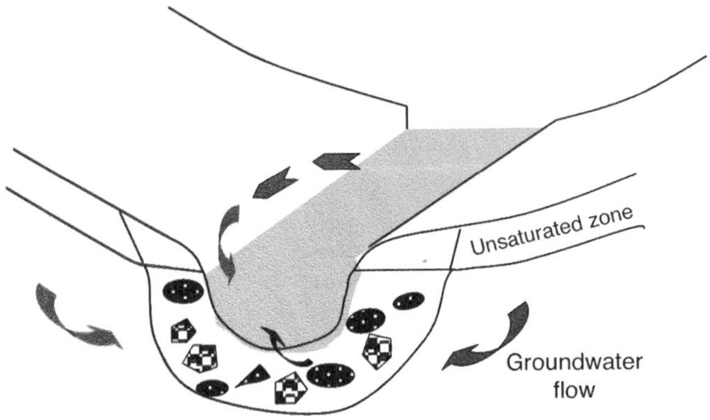

FIGURE 2.16 A schematic illustration of the hyporheic zone.

Gibbs plots have been useful for determining the controls on river chemistry, although they must be used with some caution. For example, Kilham (1990) observed that African waters did not follow the proposed trends because the evaporation-crystallization process produces sodium carbonate-bicarbonate lakes that contain little calcium.

The Hyporheic Zone

From the standpoint of interface interactions, one of the most important components of a stream system is the *hyporheic zone*. The hyporheic zone is the area of subsurface sediment and porous space adjacent to (and underlying) a stream through which the stream water readily exchanges. The hyporheic zone can be an area where surface water leaves the stream to recharge an aquifer, or where groundwater discharges into the stream. Recharge versus discharge can change downstream or even across a stream, and with season and flow conditions. In some cases, streamwater passes into and through the hyporheic zone, to reenter the stream further downstream. A model of the hyporheic zone is shown in Figure 2.16. The hyporheic zone is often the site of complex water–rock-biological interactions because of gradients in nutrient concentrations, redox conditions, temperature, and other parameters.

The OTIS Model and Solute Transport in Streams

The OTIS (one dimensional transport with inflow and storage; Runkel, 1998) model for solute transport takes into consideration advection, dispersion, lateral inflow, and transient storage, which includes both the stagnant water zones and the hyporheic zone. The OTIS model is governed by the following partial differential equations:

(in stream)

$$\frac{\partial C}{\partial t} = \frac{Q}{A}\frac{\partial C}{\partial x} + \frac{1}{A}\frac{\partial}{\partial x}\left(AD\frac{\partial C}{\partial x}\right) + \alpha\left(C_s - C\right) \tag{2.10}$$

(in storage zone)

$$\frac{\partial C_s}{\partial t} = \alpha\frac{A}{A_s}\left(C - C_s\right) + \lambda_s\left(C_s^* - C_s\right), \tag{2.11}$$

where

A = cross-sectional area of stream (m^2)
A_s = cross-sectional area of storage zone (m^2)
α = storage-zone exchange coefficient (s^{-1})
C = main channel solute (mM)
C_s = storage-zone solute concentration (mM)
$C_s{}^*$ = storage zone source (mM)
D = dispersion coefficient ($m^2\ s^{-1}$)
l_s = weathering rate coefficient (s^{-1})
x = distance downstream (m)
Q = stream flow rate ($m^3\ s^{-1}$).

An extension of OTIS that includes equilibrium chemistry—OTIS-OTEQ (OTIS—one-dimensional transport with chemical equilibrium; Runkel, 1995 and Runkel, 1998)—is also available from the U.S. Geological Survey web site. An application of the OTIS model to chemical weathering in the hyporheic zone of an Antarctic Dry Valley stream is discussed in Chapter 10.

Particle Transport and Sedimentation

The *suspended loads* of rivers are the dominant means of carrying chemical constituents on a worldwide basis. The total suspended load delivered by rivers to the oceans has been estimated at 13.5×10^9 t y^{-1} (Allen, 1997). The mean suspended sediment yield based on river suspended loads is ~116 t km^{-2} y^{-1}, which is ~3 1/2 times the mean annual solute load (Allen, 1997).

The suspended sediment load of rivers can have major environmental consequences. Many pollutant organic compounds and metals such as Cr, Ni, Pb, and Co as well as nutrients such as phosphate and organic matter are often associated with suspended clays and Fe oxides. Even as environmental legislation limits release of pollutants to rivers, old contaminated sediments may be remobilized during flood events. High suspended loads increase the turbidity of water, which decreases light penetration. During storms, peak sediment load generally is in advance of the peak surface water discharge. Sediment loads can be affected by human activities, such as deforestation, which leads to increased erosion.

The ability of streams to carry suspended sediments decreases as stream flow decreases, for example when a river enters a reservoir behind a dam or overflows its banks during a flood. During the Roman Empire, as today, engineers relied on sedimentation tanks to clarify water for municipal use. In modern-day construction sites in the United States, developers are often required to build stormwater trapping basins to capture sediments as a means of limiting surface-water pollution while also diminishing the potential for flood events. Understanding how suspended particles settle is an important component of water–rock interactions, from both basic and applied science perspectives.

To understand the settling of particles from the suspended relatively low-flow or quiescent load of surface waters, as may occur in a reservoir or trapping basin, we can consider the forces on a spherical particle of diameter d; the Gravitational Force (F_G), the Drag Force (F_D), and the Bouyancy Force (F_B) (Masters and Ela, 2008), taking as a simple case a system with only laminar (not turbulent) flow. These forces are in units of Pa m^2 or

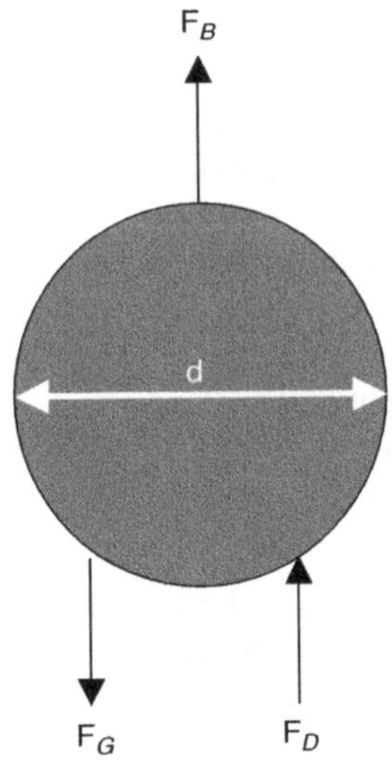

FIGURE 2.17 Forces on a particle. After Masters and Ela, 2008.

kg m s^{-2}. When the particle reaches its terminal or settling velocity (Figure 2.17), the forces are balanced

$$F_G = F_D + F_B. \tag{2.12}$$

Considering only laminar flow, which is appropriate for a settling basin scenario, the drag force is

$$F_D = 3\pi\mu v_s d_p, \tag{2.13}$$

and it is a function of the particle's *Reynolds number* (R_e)

$$R_e = \frac{r_w v_s d_p}{\mu}, \tag{2.14}$$

where

v_s = particle settling velocity (m s^{-1})
d_p = hydrodynamic diameter of the particle (m)
μ = absolute viscosity of the water (Pa s or kg m^{-1} s^{-1}).

The gravitational force on the particle is

$$F_G = mg = V_p \rho_p g, \tag{2.15}$$

and the buoyancy force is

$$F_B = m_w g = V_p \rho_w g, \tag{2.16}$$

where

g = acceleration due to gravity (9.807m s^{-2})
m = particle mass (g)
m_w = mass of water displaced by the particle (g)
V_p = particle volume = $\rho d^3 \pi/6$ (m^3)
ρ_w = density of water (kg m^{-3})
ρ_p = particle density (kg m^{-3}).

At the particle settling velocity, the sum of the three forces simplifies to *Stokes' Law*

$$\nu_s = g(\rho_p - \rho_w)d_p^2/18\mu, \tag{2.17}$$

which shows that the particle settling velocity (ν_s) increases with the particle size (d_p) and density (ρ_p). This equation is named after British mathematician and physicist George Gabriel Stokes who derived the equation for particles with small Reynolds number. This approach works well for particles less than 0.18 mm in diameter and under laminar flow conditions, but not for larger particles or under turbulent flow conditions (Pettijohn et al., 1987). Stoke's Law has important implications for particle transport and the distributions of reactive surfaces. As discussed in Chapter 13, when particle size decreases to the submicron scale, as for environmental colloids and nanoparticles, gravitational settling is no longer a factor, and Brownian motion becomes far more important in controlling mobility.

WATER BUDGETS AND CHEMICAL FLUXES IN TERRESTRIAL ECOSYSTEMS

Just as it is important for understanding the global hydrologic cycle to calculate global water budgets, it is essential for smaller scale studies of environmental hydrobiogeochemical processes to calculate *water budgets* for watersheds. A water budget for a forested ecosystem is generally calculated on a yearly basis, and then years are combined or compared by quantifying movement of water into and out of a basin. These data can then be coupled with measurements of dissolved chemical constituents as well as aerosols and particulates (Chapter 14). A yearly water budget can be represented by

$$P = SW + ET \pm GW \pm dST, \tag{2.18}$$

where

P = precipitation inputs to the basin
SW = surface–water discharge (outflow), with the basin chosen to include stream headwaters but not inflowing or throughflowing streams
GW = volume of groundwater discharge (+) or recharge (−)
dST = change of water storage in the basin, which includes surface water, soil water, and groundwater storage.

Biota can also store water and chemical constituents

ET = evapotranspiration, which consists of evaporation from the surfaces of plants and soils and transpiration from plants.

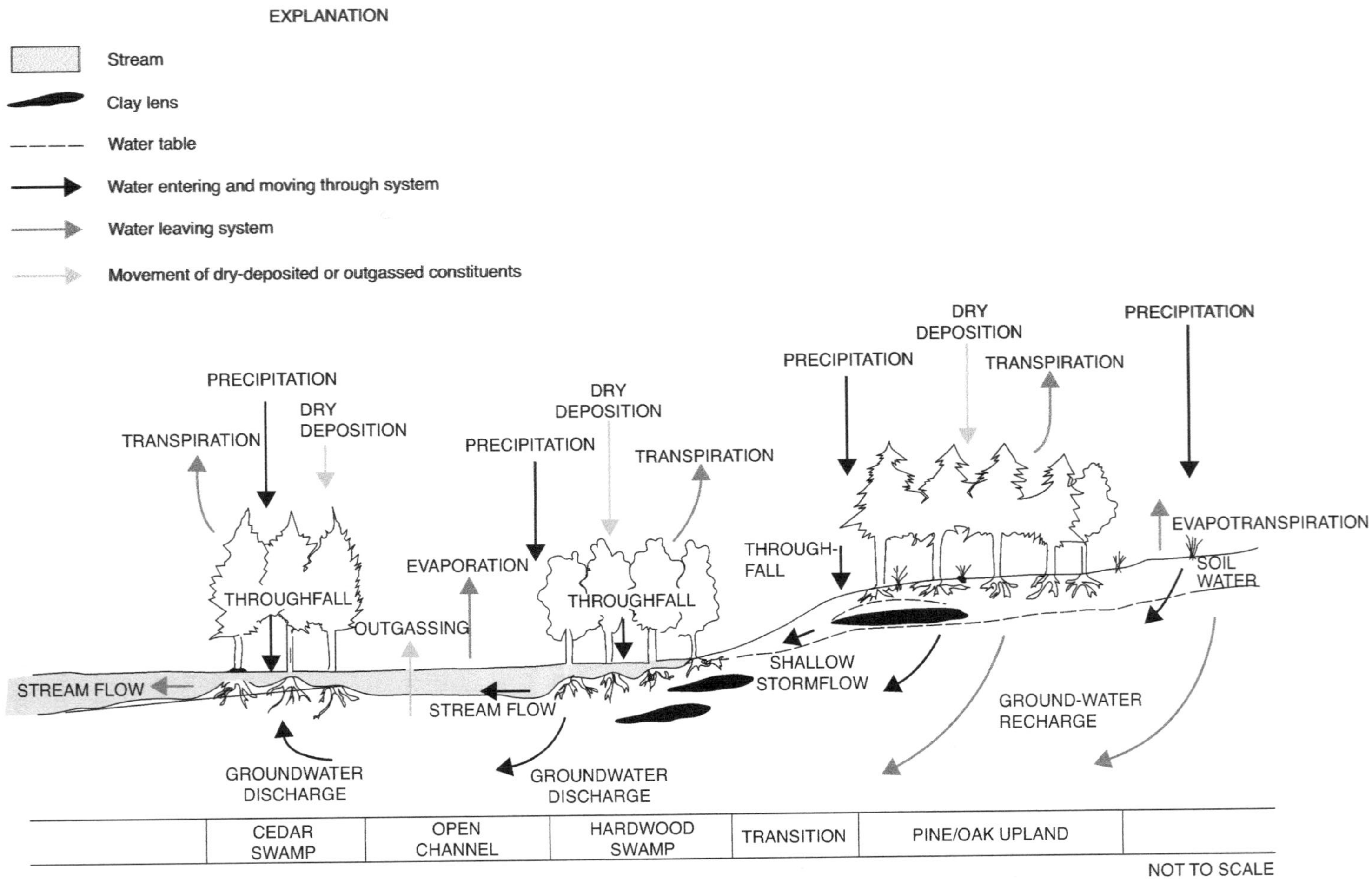

FIGURE 2.18 A conceptual model of hydrobiogeochemical processes at a small watershed, the McDonalds Branch basin in the New Jersey Pinebarrens. From Johnsson and Barringer, 1993.

The components of the water budget present challenges to accurate measurement from a hydrologic standpoint. Water input from snowfall can be challenging to monitor, especially in regions like the Sierra Nevada Mountains where snowfall can change dramatically with elevation. Evaporation and evapotranspiration can be some of the most difficult parameters to measure, especially because they can be highly variable spatially and temporally. Quantifying the chemical fluxes can be even more challenging. For example, both the dissolved and particulate loads of a stream tend to fluctuate over the course of storms. It can be difficult to measure these changing loads and to develop accurate models for the changes in load with flow. The chemical composition of snowfall can be difficult to determine, especially because particulates often accumulate between snowfalls.

Water-budget calculations coupled with geochemical measurements can help to identify potential processes that occur within terrestrial watersheds. By comparing the flux of chemical constituents, a *mass-balance approach* (above and Chapters 14) can be used to infer potential reactions, such as ion exchange or mineral dissolution (Chapter 6 and 10). A variety of models exists to help identify a potential biogeochemical reaction (e.g., NETPATH coupled with PHREEQC), based on water composition and flow data. However, a detailed understanding generally requires additional smaller scale and mechanistic approaches that focus on particular processes and reactions. It is essential to combine field measurements with laboratory experiments in an iterative fashion, using data and information gained at one scale to help understand and redesign experimental methods at other scales. The ultimate goal is to combine measurement of the fluxes of water, solutes, and particles with detailed hydrobiogeochemical investigations to understand the complex processes that occur within a watershed.

One example of a conceptual model that can be built from a watershed study is provided in Figures 2.18 and 2.19. This conceptual model was for the McDonalds Branch basin, which is a small watershed in the New Jersey Pinebarrens that contains a first-order stream. The stream begins in a hardwood swamp that is a groundwater recharge area, flows through

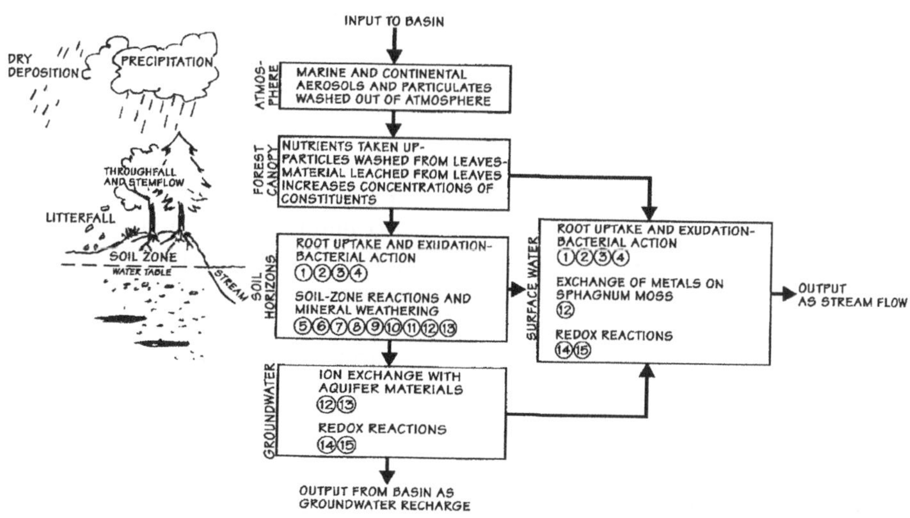

FIGURE 2.19 Some potential biogeochemical processes and reactions that may occur within a forested section of the McDonalds Branch basin. The numbers refer to Table 2.5. From Johnsson and Barringer, 1993.

TABLE 2.5 Example equations for the reactions shown in Figure 2.19.

Number on Fig. 2.19	Process	Example equation
1	Nitrification	$NH_4^+ + 2O_2 \rightarrow NO_3^- + H_2O + 2H^+$
2	Denitrification	$5CH_2O + 4NO_3^- + 4H^+ \rightarrow 5CO_2 + 2N_2 + 7H_2O$
3	Biological S cycling	$4R{\cdot}SH + 6\,H_2O + 7\,O_2 \rightarrow 4R{\cdot}OH + 4SO_4^{2-} + 12H^+$
4	Biological P cycling	$R{\cdot}H_2PO_4 + H_2O \rightarrow R{\cdot}OH + H_2PO_4^- + H^+$
5	Solubilization of organic C	$R^{\circ}{}_{OH} \rightarrow R_oO^- + H^+$
6	Metal complexation by DOC or hydroxide minerals	$HL + M^+ \rightarrow ML + H^+$
7 a.	Organic-metal-mineral ternary surface complexation (Chpt. 8)	Clay or oxide + ML $\rightarrow$ ternary surface complex
7 b.	Sulfate adsorption	$exch(OH) + SO_4^{2-} + H^+ \rightarrow exch(SO_4)^- + 2H_2O$
8	Microbial DOC decomposition	$CH_2O + O_2 \rightarrow CO_2 + H_2O \rightarrow HCO_3^- + H^+$
9	Reductive dissolution	Fe_2O_3 + DOC $\rightarrow$ Fe(II) + Organic C (oxidized)
10	Clay mineral weathering	$Al_2Si_2O_5(OH)_4 + 5H_2O \rightarrow 2Al(OH)_3 + 2Si(OH)_4$
11	Metal oxide dissolution	$\frac{n}{2}M\frac{2}{n}O + nH^+ \rightarrow M^{n+} + \frac{n}{2}H_2O$
12	Cation exchange on sphagnum	$M^+ + R^{\circ}{}_{OH} \rightarrow R^{\circ}{}_{OM} + H^+$
13	Cation exchange on clays	Ca^{2+} + Na-clay $\rightarrow Na^+ + {}^+$Ca-Clay
14	Iron redox	$Fe^{2+} + \frac{1}{4}O_2 + \frac{5}{2}H_2O \rightleftarrows Fe(OH)_3 + 2H^+$
15	Sulfur redox	$FeS + \frac{9}{2}O_2 + \frac{5}{2}H_2O \rightarrow Fe(OH)_3 + SO_4^{2-} + 2H^+$

After Johnsson and Barringer (1993), with revision. DOC = dissolved organic carbon.

an intermittently ponded open channel area, and exits the basin in an Atlantic White Cedar swamp that is a groundwater discharge area. The boundaries between recharge and discharge can change with different flow conditions, for example during drought or after a hurricane. The box model shown in Figure 2.19 identifies a variety of biogeochemical processes that may occur in a forested transect leading to the hardwood swamp section of the stream. Table 2.5 gives several generalized example reactions for the different processes. As our understanding of hydrobiogeochemical processes increases, particularly with respect to biologically driven or mediated processes, models such as this are becoming increasingly complex and approaching description of the "real world" ever more fully.

QUESTIONS FOR FURTHER THOUGHT

1. Shannon (1976) published values of unhydrated ionic radii in Å (for six-fold coordination) of Li^+ (0.76), Na^+ (1.02), K^+ (1.38), and Cs^+ (1.67). Calculate the I_P for

each of these ions. Which of these ions should be most hydrated? Which should be least hydrated? Do a similar comparison among the divalent cations Mg^{2+} (0.72), Ca^{2+} (1.00), Sr^{2+} (1.18), and Cd^{2+} (0.95). Remember this analysis for reference in later chapters on adsorption, absorption, cation exchange, and other phenomena.

2. Considering the average annual precipitation rate for the globe and the latent heat of vaporization at the mean Earth surface temperature (15°C), calculate the amount of energy that would be required to evaporate all the precipitation in a year (in $kJ\ y^{-1}$). The surface area of the Earth is 510,065,600 km^2. The average American household uses 10,650 KWh of energy per year, and there are ~100 million American households. (Note: 1 KWh = 3600 kJ or 1 W = 1 J s^{-1}). Compare the annual energy use of American households with energy required to evaporate all of the precipitation over the Earth's surface in 1 year. How does the amount of energy needed to evaporate all of the precipitation for an entire year (in kJ) compare with the energy released in the Hiroshima atomic bomb explosion (~10^{14} J) and the eruption of Mt. St. Helens (~5×10^{16} J)? The earthquake that resulted in the 2004 Indian Ocean Tsunami disaster has been estimated to have released 2×10^{18} J of energy. How many L of precipitation to the Earth's surface could have been evaporated by this same amount of energy (1 L = 1 kg water)?

3. The small watershed discussed above, the McDonalds Branch basin in the New Jersey Pinelands, had the the following components to the water budget for 2 years (Johnsson and Barringer, 1993):
   (a) Calculate the ET from these values.
   (b) Do the ET values seem reasonable?
   (c) How might you go about evaluating whether the value were reasonable (Hint: Look for values of evapotranspiration from the U.S. Geological Survey)? Values in inches. dST = change in total storage in the basin.

Year	P	SW	GW	dST
1	37.9 in	6.5 in	4.0 in	8.4 in
2	45.0 in	13.0 in	4.0 in	5.0 in

Note: Here, the value for GW is for water leaving the basin. The change in storage is an increase in storage for both years.

The volume-weighted mean concentrations of constituents in wet precipitation (rain, snow) to the basin and in shallow groundwater over several years in the mid-1980s are given below:

	Wet precipitation	Shallow groundwater
pH	4.4	4.8
Ca^{2+}	0.10 mg L^{-1}	1.8 mg L^{-1}
Mg^{2+}	0.06 mg L^{-1}	0.7 mg L^{-1}
Na^{+}	0.44 mg L^{-1}	1.9 mg L^{-1}
K^{+}	0.05 mg L^{-1}	0.29 mg L^{-1}
NH_4^{+}	0.25 $\mu g\ L^{-1}$	192 $\mu g\ L^{-1}$
SO_4^{2-}	2.22 mg L^{-1}	7.4 mg L^{-1}

Cl^-	0.89 mg L^{-1}	3.4 mg L^{-1}
F^-	0.03 mg L^{-1}	2.8 mg L^{-1}
NO_3^-	1.59 mg L^{-1}	below detection
HCO_3^- and CO_3^{2-}	below detection	below detection

(d) If Cl is conservative, meaning that it does not undergo any reactions in the basin, then how does the ET estimated by Cl concentrations vary from the ET based on the water budget data, above?

(e) If the various consistuents in rainwater are ratioed to Na and the ratios compared with those in ocean water, what does this tell you about the potential sources of various ions in the precipitation?

4. The river outflow from this same basin as in (3) has the following average chemical composition:

   (a) How does it compare with that of mean average river water as reported by Meybeck? (Note: You will need to convert to the same concentration units). Concentrations in mg L^{-1}: $Ca^{2+} = 0.5$, $Mg^{2+} = 0.5$, $Na^+ = 1.9$, $K^+ = 0.4$, $SO_4^{2-} = 7.6$, and $Cl^- = 3.6$.
   HCO_3^- and CO_3^{2-} are below detection by alkalinity measurements, $SiO_2 = 3.9$, and mean pH is 4.2 (Johnsson and Barringer, 1993).

   (b) Where do the mean river water and river outflow water plot on a Gibbs plot?

   (c) What does this tell you about the sources of ions to the river water at the McDonalds Branch basin?

5. In 1995, The Nobel Prize in Chemistry was awarded to Mario Molina, F. Sherwood Rowland, and Paul Crutzen for research elucidating the mechanisms of ozone destruction in the upper atmosphere. In 1987, Molina et al. published a paper that laid out the role of polar stratospheric clouds in catalysis to produce ozone-destroying Cl, as shown in the reactions above. Read this Science article. Write a brief description of the process as described by Molina et al. (1987) focusing especially on the role of environmental surfaces. If you have access to science citation indices, look for papers that have cited this work, and that also deal with environmental surfaces. Briefly describe several other environmental surface processes in the atmosphere, based on this glimpse of articles that have relied at least in some measure on Molina et al.'s (1987) work.

   Molina, M.J., et al. 1987. Antarctic stratospheric chemistry of chlorine nitrate, hydrogen-chloride and ice—Release of active chlorine. *Science* 238, 1253–1257.

6. Discuss the components considered in the OTIS model. Should any other processes be considered or included in modeling, to describe solute transport in a stream undergoing hyporheic exchange?

7. (a) Plot how the particle settling velocity v_s of a particle varies in "pure" water at STP = 25°C, 1 atm ($\rho_w = 998.23$ kg m^{-3}; μ for water at STP = 0.89×10^{-3} Pa s) for spherical quartz particles with diameter ranging from 2.0 um to 0.05 mm and density = 2.65 g cm^{-3}.

   (b) Compare v_s of 10.0-um diameter particles of hematite, quartz, feldspar, and magnetite (Note: You will have to do a small amount of research to answer this question).

   (c) What are some hydrologic and geologic consequences of the answers to (a) and (b)?

(d) By considering how settling velocity varies which particle size, what can you predict about the settling velocity of colloidal particles that are <1 um in size?

(e) Sediments are carried by rivers to the oceans, where they encounter salt water. Calculate the settling velocity of spherical quartz particles 2 um, 10 um, and 50 um in diameter in ocean water of salinity 35 ppt, temperature = 10°C, and assume that density of seawater at this salinity and temperature is 1027 kg m^{-3} and viscosity = 1.04×10^{-3} kg s^{-1} m^{-1}. For the purposes of this problem, assume laminar (nonturbulent) flow conditions.

8. For the waters of the McDonalds Branch basin described above (problems 3 and 4), plot meq L^{-1} as bar graphs for cations and anions for

(a) precipitation,

(b) surface water, and

(c) groundwater. To do this, you will need to convert mg L^{-1} to meq L^{-1} by multiplying concentration by the absolute value of the charge of the ion (see Chapter 1). For H^+, convert from pH and assume activity = concentration. The mean dissolved organic carbon (DOC) concentration for surface water is 5.1 mgC/L, the DOC of shallow groundwater is 1 mgC L^{-1}, and the (negative) charge on the DOC is 5 μeq mgC^{-1}.

(d) How well do the cations and anions balance for each type of water? Because of the requirement for neutrality, the cation/anion balance is often taken to indicate how reliable the chemical analysis is. A balance within a few percent is generally considered to be reasonable.

(e) What insight does comparison of the three types of waters provide for potential processes occurring within the basin?

9. Below are chemical composition data for 4 wells.

(a) Plot on a Piper diagram and classify the water types according to the hydrogeochemical facies approach. All values in mg L^{-1} except for pH.

(b) One well is in an area that has received saltwater intrusion. Can you make an educated guess as to which? Explain your reasoning.

(c) Two are in a dolomite [a carbonate mineral, $CaMg(CO_3)_2$] aquifer. Can you make an educated guess as to which? Explain your reasoning.

(d) What are some pitfalls involved in using Piper diagrams to make hydrogeochemical interpretations?

Well #	pH	Cl	SO_4	HCO_3	Na	Ca	K	Mg	TDS
1	7.5	2871	609	80	1459	247	41.0	203.0	5907
2	7.5	4557	1896	89	3200	541	69	268	10549
3	7.5	4	13.2	64	2	17	0.6	6.3	93
4	7.7	3	10.7	64	3	17	0.8	6.3	101

TDS = total dissolved substances.

10. (a) Plot the two sets of data from Table 2.3 on Piper Diagrams.

(b) What if anything do the plots suggest?

(c) What other types of data or information might be useful for discerning hydrobiogeochemical processes for the two aquifers?

FURTHER READING

Berner, E.K., and Berner, R.A., 1996. Global Environment: Water, Air, and Geochemical Cycles. Prentice Hall, Englewood Cliffs, NJ. 397 p.

Drever, J.I. ed. 2004. Surface and Groundwater, Weathering and Soils. Treatise on Geochemistry, vol. 5. Elsevier, LTD, New York, 626 p.

Freeze, R.A., and Cherry, J.A. 1979. Groundwater. Prentice-Hall, Englewood Cliffs, NJ, 604 p.

3

SOME MINERALS OF SPECIAL INTEREST TO ENVIRONMENTAL SURFACE CHEMISTRY

Water provides half of the context for mineral–water interactions; the other half is the mineral matter that is immersed in aqueous solutions. This chapter briefly reviews some minerals of particular interest to environmental surface chemistry. The focus is on mineral properties that one needs to understand before delving into interface interactions.

Terms such as Brunaver, Emmet, and Teller (BET) surface area, cation exchange capacity, and point of zero net proton charge are used that will become clearer in Chapter 6. Definitions of these terms are included in the Glossary. When pH of the point of zero net proton charge (p.z.n.p.c.) or BET surface area (surface area measured by gas adsorption) values are given, it should be understood that these values can vary from sample to sample, and the the values given are not absolute for any mineral. In the ball-and-stick mineral structure figures shown here, dark solids generally are metals, shaded larger balls are O, and small open balls are H.

GIBBSITE

The chemical makeup of gibbsite is γ-$Al(OH)_3$. Gibbsite is made up of aluminum octahedral units in which each aluminum atom is surrounded by six hydroxides in octahedral coordination (Figure 3.1). One third of the octahedral sites are vacant to allow for charge balance. Sheets are bonded together by hydrogen bonds between hydroxyl groups. Gibbsite is an important component of aluminum ore and is a common product of aluminosilicate weathering, particularly in the tropics. The gibbsite structure is also a key component of the structures of many clays. Gibbsite tends to be insoluble at circumneutral pH, but its solubility increases at either low or high pH (Chapter 10).

Yang et al. (2007) reported a pH p.z.n.p.c. of gibbsite = 8.1–8.3 and a BET surface area of a commercial gibbsite with a mean size of 226.8 nm = 9.97 m^2g^{-1}.

Environmental Surfaces and Interfaces from the Nanoscale to the Global Scale, by Patricia A. Maurice

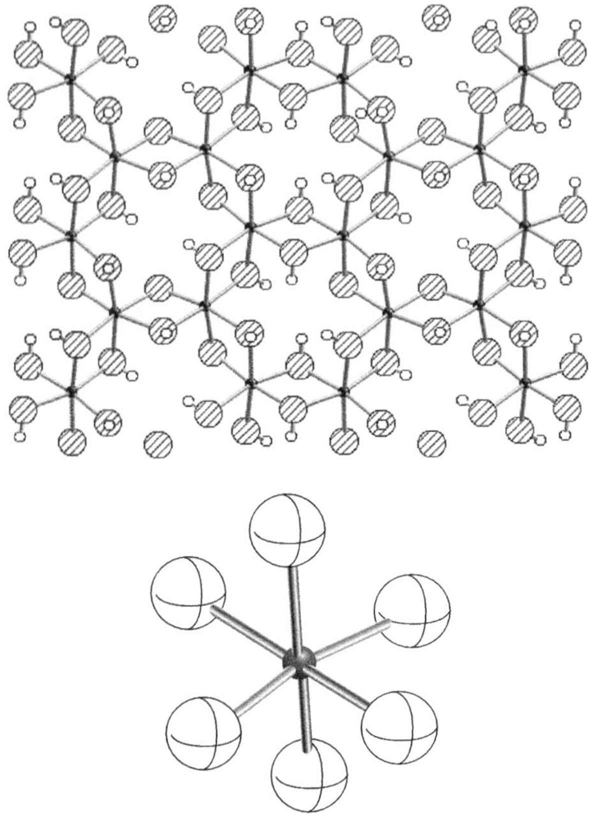

FIGURE 3.1 The gibbsite structure (left), consisting of Al octahedral units (right). In the ball-and-stick figure on the right, the crossed balls are OH groups.

QUARTZ

Quartz is crystalline SiO_2. Although a variety of SiO_2 minerals and amorphous solids is available, quartz is common and widely studied. Quartz is a framework silicate, which means that sharing of the tetrahedral oxygens around Si extends in three dimensions (tetrahedral unit shown in Figure 3.2). Quartz is important to mineral–water interface

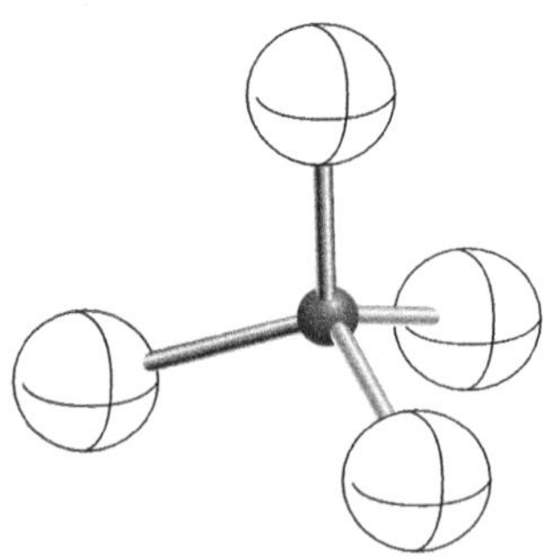

FIGURE 3.2 An SiO_4 tetrahedral unit. Here, the crossed balls are O atoms.

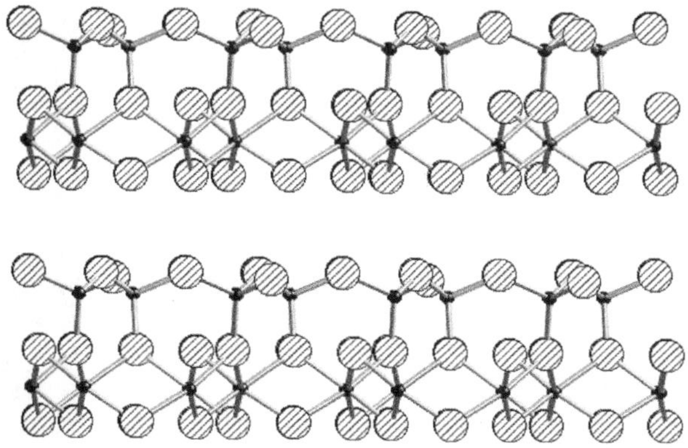

FIGURE 3.3 The kaolinite structure. A 1:1 nonexpanding clay.

geochemistry because of its widespread abundance and because it is often used to represent the Si end member of aluminosilicates. Quartz is highly stable in most environments, but its solubility increases at pH above ~10 (Chapter 10). The pH p.z.n.p.c. of quartz is ~1.8, and the BET surface area is often reported as $< 1\ m^2g^{-1}$. White et al. (1995) described quartz soil particles with a specific surface area = $0.1–0.2\ m^2g^{-1}$.

KAOLINITE

Kaolinite is a *clay mineral* with the chemical composition $Al_2Si_2O_5(OH)_4$. It is a layered silicate mineral, with one tetrahedral sheet linked through oxygen atoms to one octahedral sheet of alumina octahedral (Figure 3.3). Kaolinite often occurs in hexagonal or pseudohexagonal clay platelets (Figure 3.4), which may be aggregated from basal plane to basal plane to form *booksheets packeting*.

Because its structure consists of one tetrahedral unit and one octahedral unit, kaolinite is termed a 1:1 clay mineral. Hydrogen bonds form between protons exposed on the surface of one layer and oxygens on the next, which makes kaolinite a nonexpanding clay. The thickness of the two-layered unit of kaolinite is ~7 Å. The mineral tends to be close to stoichiometric, although small amounts of Fe may occur in the structure. Fe can also be adsorbed to kaolinite or present in an additional solid phase.

Different reactive sites for adsorption are discussed in detail in Chapter 7. Briefly, Si-O-Si sites on the basal plane surfaces are generally considered to be unreactive to ions, although they may adsorb nonpolar organics. A small amount of structural charge caused by isomorphic substitution can create diffuse negatively charged sites allowing at most a minor amount of cation exchange. Metal cations may associate with the ditrigonal cavities on the siloxane basal–plane surfaces. Sites on the basal plane where a hydroxide group is bridged between two Al atoms (Al-OH-Al) also can permit cation adsorption. The edge (Figure 3.5) sites tend to be far more reactive.

Figure 3.5 shows different types of reactive edge sites present on kaolinite: the aluminol sites, $Al(III){\cdot}H_2O$, and the silanol sites, Si(IV)OH. The H_2O on the aluminol sites can be replaced by hydroxide at relatively high pH (Sposito, 1984).

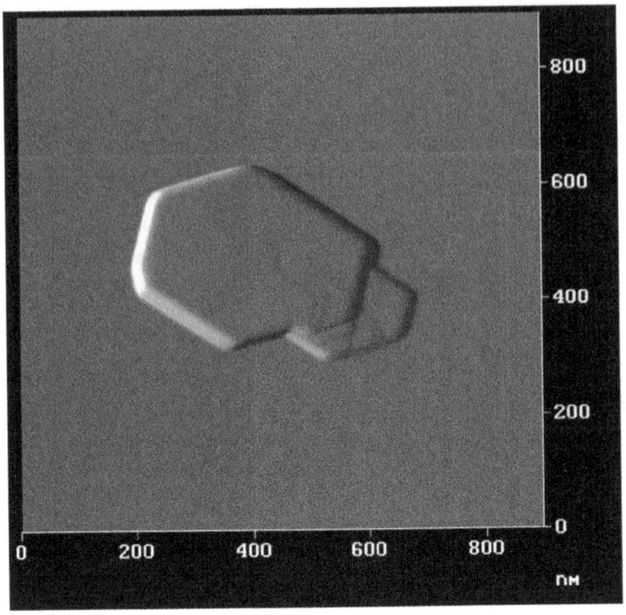

FIGURE 3.4 An atomic force microscopy image of kaolinite particles from the clay minerals society (CMS) source clay KGa-1b.

One of the main ways in which kaolinite forms is through the weathering of K-feldspar, by a variety of reactions. For example,

$$\underset{\text{K-feldspar}}{2KAlSi_3O_8(s)} + 2H^+(aq) + 9H_2O(l) \rightleftarrows \underset{\text{kaolinite}}{Al_2Si_2O_5(OH)_4(s)} + 2K^+(aq) + 4H_4SiO_4(aq). \quad (3.1)$$

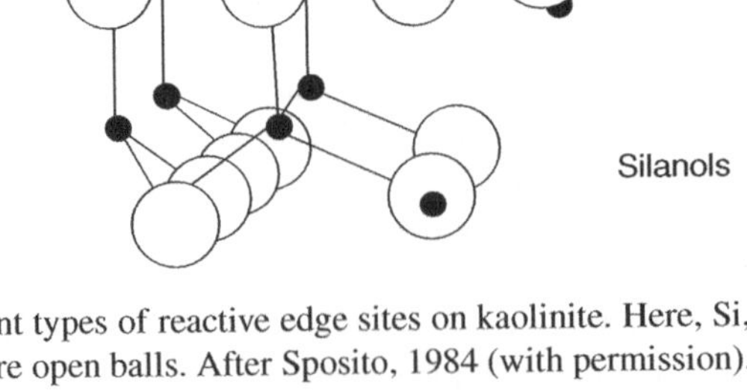

FIGURE 3.5 Different types of reactive edge sites on kaolinite. Here, Si, Al, and H are presented as solid balls, and Os are open balls. After Sposito, 1984 (with permission).

Thus, kaolinite is a *secondary weathering product* of importance to mineral–water interface reactions. Additional weathering of kaolinite can lead to gibbsite formation:

$$\underset{\text{Kaolinite}}{Al_2Si_2O_5(OH)4(s)} + 5H_2O(l) \rightleftarrows \underset{\text{gibbsite}}{2Al(OH)_3(s)} + 2H_4SiO_4(aq). \quad (3.2)$$

Dissolution of kaolinite tends to be congruent, which means that release to solution is in the same stoichiometry as the original mineral, far from equilibrium or in the presence of Al complexing ligands. However, incongruency can occur because of Al adsorption and/or precipitation of an Al-phase, such as amorphous Al hydroxide or gibbsite. (See Chapter 10.)

The CMS source clays Georgia kaolinites KGa-1 (and 1b for a sample taken from approximately the same locality) and KGa-2 are used in many adsorption and dissolution experiments. Schroth and Sposito (1997) measured a pH p.z.n.p.c. of Georgia kaolinite KGa-1 $= 4.99 \pm 0.03$ and of KGa-2 $= 5.36 \pm 0.02$, and pH point of zero net charge (p.z.n.c.) values of KGa-1 = 3.56 and KGa-2 = 3.52. The calculated values of the permanent structural charge density σ_0 were KGa-1 $= -6.3 \pm 0.1$ and KGa-2 $= -13.6 \pm 0.5$ mmol kg^{-1}. Sutheimer et al. (1999), using a similar methodology, measured a p.z.n.p.c for KGa-1b $= 5.1 \pm 0.2$ and for KGa-2 $= 4.9 \pm 0.2$. The cation exchange capacity (CEC) of KGa-1 has been reported to be ≤ 2 mEq $100g^{-1}$ (Brady et al., 1996). Sutheimer et al. (1999) reported a BET surface area of KGa-1b $= 12.5\ m^2g^{-1}$ and of KGa-2 $= 22.4\ m^2g^{-1}$.

SMECTITE: EXAMPLE MONTMORILLONITE

The term *smectite* refers to a group of clay minerals that includes montmorillonite as an important example. Smectites tend to swell when exposed to water. *Bentonites*, which are clays often used in well drilling, include minerals of the smectite group. A general equation for a smectite mineral is $(Al, Mg, Fe^{3+})_2((Si,Al)_4O_{10})(OH)_2$ (Krauskopf and Bird, 1995).

Smectite minerals tend to form when the acidity is not as high or the silica concentration as low as instances when kaolinite tends to form (Krauskopf and Bird, 1995). Smectite minerals are commonly observed as weathering products and secondary minerals in locations where *serpentinite* rocks are exposed (Wildman et al., 1968). For example, Lee et al. (2003) proposed the following reactions for formation of smectite from the minerals chlorite and serpentine in soils formed on serpentinite in the Klamath Mountains, California:

1. serpentine → low-charge Mg-rich, dioctahedral smectite. (3.3)
2. chlorite → interstratified chlorite/vermiculite → vermiculite → high-charge smectite. (3.4)

Montmorillonite is a more complex clay mineral than is kaolinite. Montmorillonite has the general formula $(Na,Ca)_{0.3}(Al,Mg)_2Si_4O_{10}(OH)_2{\cdot}n(H_2O)$. Montmorillonite has a 2:1 layer structure, which means that two silica tetrahedral layers are on either side of an aluminum octahedral layer (Figure 3.6). It is a common dioctahedral smectite clay. About one eighth of the octahedra contain Mg^{2+} substituted for Al^{3+}, which leads to a negative charge. Commonly, lesser amounts of substitution of Al^{3+} for Si^{4+} exist in the tetrahedral layers. The negative structural charge is neutralized by various weakly bound and

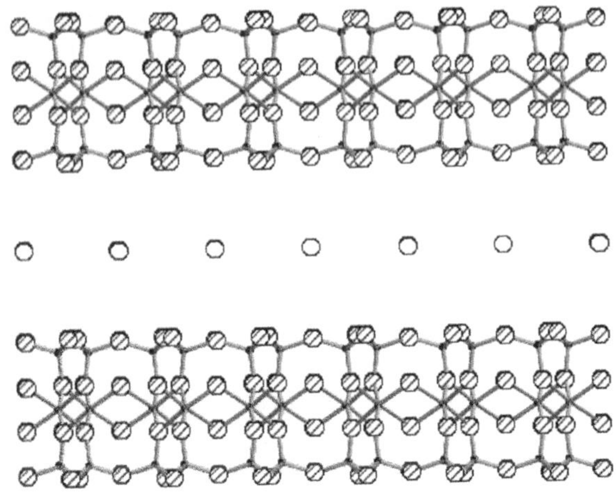

FIGURE 3.6 The montmorillonite structure, which is a 2:1 expanding clay. Here, the H atoms are not shown as part of OH groups, for clarity.

exchangeable hydrated cations that occur within the interlayers. Under humid conditions, water is drawn into the interlayer spaces so that the interlayer spacing expands and the clay swells. This means that X-ray diffraction measurements of interlayer thickness vary as a function of humidity. In the presence of ethylene glycol, the thickness of the 2:1 sheet, including the interlayer space, expands to 17 Å.

Many different types of surface and interface interactions can occur for montmorillonite. For example, the exchangeable hydrated cations in the interlayers can undergo cation exchange, and montmorillonite samples tend to have high cation exchange capacities. For montmorillonite in air, different relative humidities can lead to different interlayer spacings. McBride (1994) gives c-spacing distances for montmorillonite with K^+, N^+, or Mg^{2+} in the interlayer, as shown in Table 3.1.

Interlayer absorption of inorganic cations along with organic cations such as alkylammonium cations and neutral organic molecules can also occur (Chapters 7 and 8). Interlayer absorption can be site specific. Lateral interactions between hydrophobic components of absorbing organic molecules, such as alkylammonium or desferrioxamine B (Haack et al., 2008), can lead to bilayer structures within the interlayer. Absorption in the interlayer can affect interlayer spacing.

Cation-exchange capacity can be high for montmorillonite (~100 mEq $100g^{-1}$), and it is dominated by the interlayer cation exchange. The pH p.z.n.p.c. of montmorillonite has

TABLE 3.1 Measured c-axis spacings for montmorillonite saturated with exchangeable K^+, Na^+, or Mg^{2+}, and under different conditions.

Exchangeable cation	Air-dried	Wet	Glycerol Added
K^+	11.6 Å	≫ 20 Å	17 Å
Na^+	12.4 Å	≫ 20 Å	17 Å
Mg^{2+}	14.7 Å	19.5 Å	17 Å

Data from McBride, 1994.

been reported to be 8.5 ± 0.5. Van Olphen and Fripiat (1979) reported a BET surface area of 75.0 $m^2\ g^{-1}$, which includes internal surfaces.

FE(HYDR)OXIDES

Fe(III)(hydr)oxides have widespread abundance and tend to occur as small particles, often as coatings on other mineral grains. Thus, they can be important in controlling adsorption and other mineral–water interface interactions. There can be strong redox controls to their distributions, and the distributions can be biotically mediated. The iron oxides can have significant cationic substitution, which can alter their structure, morphology, and reactivity.

Fe(hydr)oxides tend to be insoluble in aerobic, circumneutral waters, but their solubility increases at higher and lower pH, under acidic anaerobic conditions, and in the presence of organic ligands; dissolution may also occur photochemically. Fe(hydr)oxides can serve as redox and photochemical catalysts; hence, experimental design needs to take into consideration factors such as dissolved oxygen content and exposure to light. The various Fe(hydr)oxides can undergo transformations from one to another, through processes such as thermal dehydration, dissolution/reprecipitation, oxidation, or reduction. These transformations can be complex and size dependent, particularly for nanoparticles (Chapter 13). All the Fe(hydr)oxides become competitive energetically in the nanosize range (Navrotsky, et al., 2008).

Fe is an essential element for most life forms. Because Fe(hydr)oxides tend to be highly insoluble in circumneutral, oxic waters, Fe often is a limiting nutrient. The bioavailability of Fe in oceanic waters can play an important role in controlling plankton photosynthesis and, hence, in the global C budget. Organisms that live in circumneutral, aerobic environments need to develop strategies for acquiring Fe, such as production and release of siderophores (Chapter 12).

Professor Udo Schwertmann (Technical University, Germany) has devoted his career to understanding and documenting the structure, composition, and reactivity of the Fe(hydr)oxides, and a comprehensive review is provided in *The Iron Oxides* by Cornell and Schwertmann (2003). This book is the basis for the structure, unit cell dimensions, and BET surface areas given here, unless otherwise referenced. Some examples of Fe(hydr)oxides are discussed below.

Hematite

Hematite (α-Fe_2O_3) has an hexagonal structure (Figure 3.7) (a = 0.5034 nm; c = 1.375 nm), which can also be indexed as rhombohedral. Hematite is a semiconductor. Magnetic properties change with temperature and may be antiferromagnetic, weakly ferromagnetic, or paramagnetic. Hematites synthesized from solution have a BET surface area = 10–90 m^2g^{-1}. Maurice et al. (1995) reported a BET surface area for hematite particles synthesized from $Fe(NO_3)_3$ solution = 40 m^2g^{-1}. Hematite pH p.z.n.p.c. ~8.4.

Goethite

Goethite (α-FeOOH), which is the most commonly occurring natural form of Fe(III)(hydr)oxide, is orthorhombic (Figure 3.8) (a = 0.9956 nm; b = 0.30215 nm; c = 0.4608 nm). The BET surface area = 8–200 m^2g^{-1}, and the pH p.z.n.p.c. = 7.7.

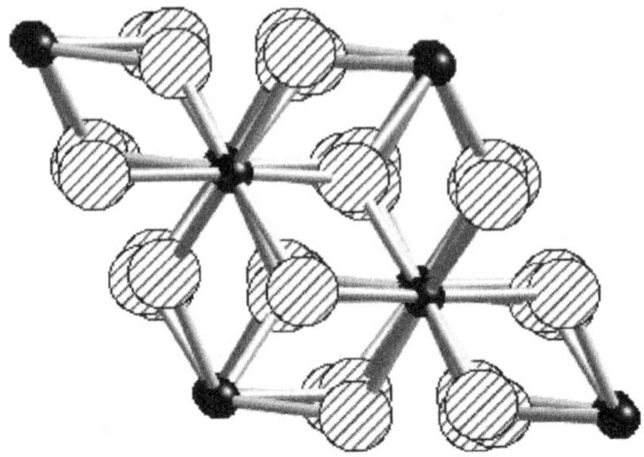

FIGURE 3.7 The hematite basal plane structure.

Impurities may effect the structure, shape, and reactivities of iron (hydr)oxides. Al^{3+} may substitute for Fe^{3+} in octahedrally coordinated sites, and the presence of Al impurities can affect the goethite structure and crystal morphology. Schulze and Schwertmann (1984) showed that Al substitution led to goethite needles becoming shorter and less multidomainic (i.e., fewer parallel subunits or domains). Muñoz (1988) observed that X-ray diffraction (XRD) line widths narrowed, suggesting more highly ordered structure, with increasing Al substitution. Maurice et al. (2000) measured changes to the BET surface area and particle length as well as the aspect ratio on aluminum substituted goethites grown in the method of Schwertmann and Cornell (1991). They observed that increasing Al substitution to a maximum of ~9% resulted in goethite particles with progressively decreasing mean particle length and aspect ratios (length:width), along with fewer domains (Table 3.2 and Figure 3.9). In agreement with previous results, changes to the XRD line widths (Figure 3.10) also suggested increased structural order with increased Al content, perhaps because goethite crystals grow more slowly in the presence of Al.

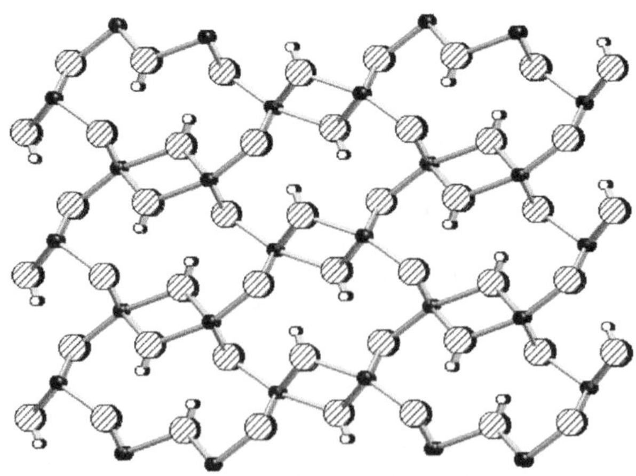

FIGURE 3.8 The goethite structure.

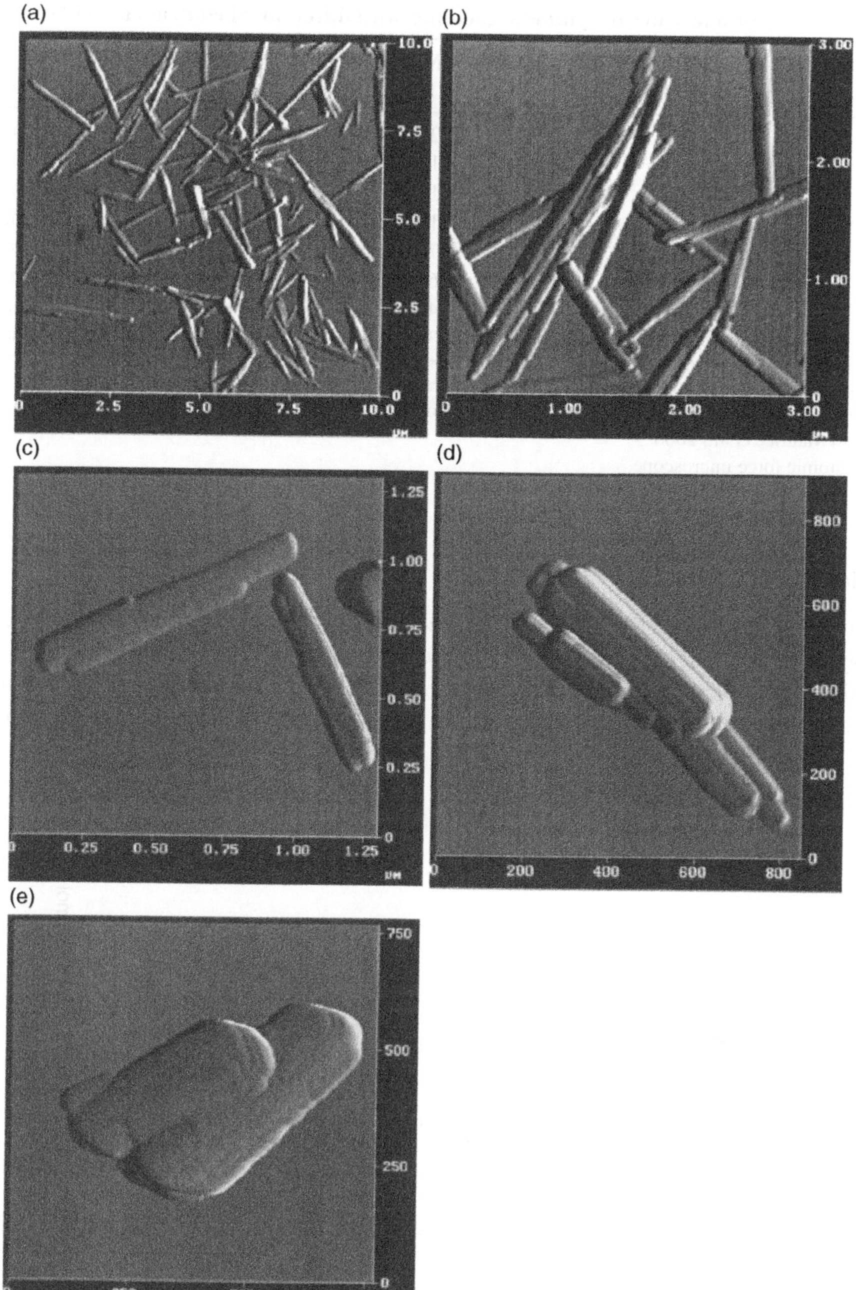

FIGURE 3.9 AFM images of goethite with different Al contents. (a, b) Pure, (c) 1.9%, (d) 3.5%, and (e) 8.8%. (a)–(c) in μm; (d) and (e) in nm. From Maurice et al., 2000 (with permission).

TABLE 3.2 Characteristics of synthetic goethites with different Al contents to ~9 %.

Al content (mol %)	BET surface area (m^2g^{-1})	AFM particles measured (no.)	Mean particle length (nm)[1]	Mean particle aspect ratio	Comments
0	33.4	23	2195(1116)	19:1(30:1)[2]	multidomainic
1.0	36.0	25	1812(961)	15:1	multidomainic
1.9	34.7	27	1161(663)	11:1	multidomainic
2.6	30.0	20	1125(546)	10:1	multidomainic
3.5	36.5	19	761(290)	8:1	multidomainic
4.2	35.6	21	616(218)	6:1	multidomainic
6.2	30.9	15	420(155)	5:1	few domains
8.8	26.9	20	379(81)	4:1	1–2 domains

[1]Length parallel to c-axis; standard deviation in parentheses.
[2]Ratio in parentheses represents aspect ratio for single-domain particles.
Data from Maurice et al., 2000.
AFM = atomic force microscope.

Lepidocrocite

Lepidocrocite (γ-FeOOH) consists of layered iron(III)(hydr)oxide octahedra bonded by hydrogen bonding via hydroxide layers. Lepidocrocite has an orthorhombic unit cell (a = 1.2520 nm; b = 0.3873 nm; c = 0.3071 nm). It occurs as a weathering product, and it is often contained in rust, for example, in aging steel pipes. The BET surface area = 15–260 m^2g^{-1}.

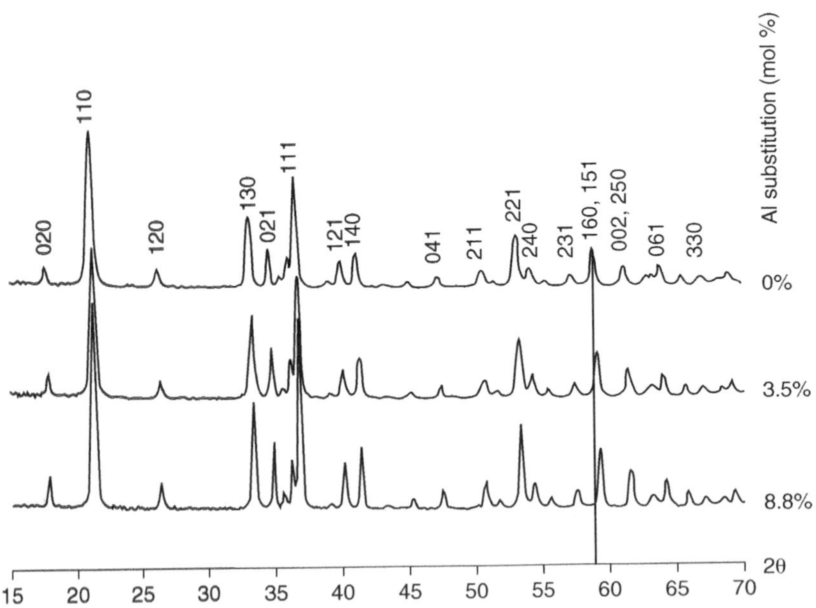

FIGURE 3.10 X-ray diffractograms of pure and Al substituted goethites. Revised from Maurice et al., 2000 (with permission).

Maghemite

Maghemite, which is known as γ-Fe_2O_3, has a cubic structure (a = 0.834 nm), with superlattice ordering that varies according to vacancies. Some Fe in maghemite can be Fe(II). Maghemite forms by weathering of magnetite and is ferromagnetic as well as a semiconductor. Commercial maghemite, which has a tetragonal unit cell, is important for magnetic recording. The BET surface area = 8–130 m^2g^{-1}.

Ferrihydrite

Ferrihydrite generally occurs as particles <10 nm; it is poorly ordered and can have a variable structural order. Based on XRD patterns, ferrihydrite is generally designated as two line or six line, depending on the number of reflections that appear as the order increases. Much research has focused on trying to decipher the ferrihydrite composition and structure. A recent structure proposed by Michel et al. (2007a, b) is discussed in Chapter 13.

Ferrihydrite is often called hydrous ferric oxide (HFO) in the literature, although this term is sometimes applied more broadly to the Fe (hydr)oxides in general or to amorphous forms. Ferrihydrite is a *nanomineral* that occurs as *nanoparticles* generally aggregated into micron-sized clusters (Chapter 13). Waychunas et al. (2005) report surface areas of 50–200 m^2g^{-1}. Cornell and Schwertmann (2003) report BET surface areas of 200–400 m^2g^{-1}, and they note that surface area can be difficult to measure because of particle aggregation.

Magnetite

Magnetite (Fe_3O_4) has a face-centered cubic cell (a = 0.839 nm = cell edge length) and a Fe(II)-to-Fe(III) ratio of 1:2. Magnetite is a semiconductor with nearly metallic conducting properties and is ferromagnetic.

Magnetotactic bacteria can take up complexed ferric iron and transform it into magnetite, which is deposited in membrane-bound structures known as magnetosomes (Ehrlich, 2002). The single-domain magnetic crystals allow the bacteria to align themselves with the Earth's magnetic field. Such bacterial–Fe(hydr)oxide interactions are discussed in greater detail in Chapter 12.

Surface areas of 4–100 m^2g^{-1} have been reported for magnetite.

MANGANESE OXIDES

More than 30 different Mn oxide minerals have been reported in natural environments, which range from soils and sediments to nodules at the bottom of lakes and the oceans (Post, 1999). Mn oxides are important to many industries; Mn is used in steel, batteries, and many other industrial applications. Like Fe oxides, Mn oxides tend to be fine grained with high surface areas, often forming coatings on other grains. This property makes Mn oxides important reactive surfaces that can affect the distributions of metals and radionuclides. Because of their redox properties, they can transform other constituents, for example, oxidizing As(III) into As(V). They can have cation exchange capacities exceeding that of montmorillonite (Post, 1999). Their compositions and distributions are controlled by redox conditions and are often biotically mediated. Mn tends to be mobile in acidic, organic-rich soils but less so in alkaline soils. Like Fe(hydr)oxides, Mn oxides can be important to a

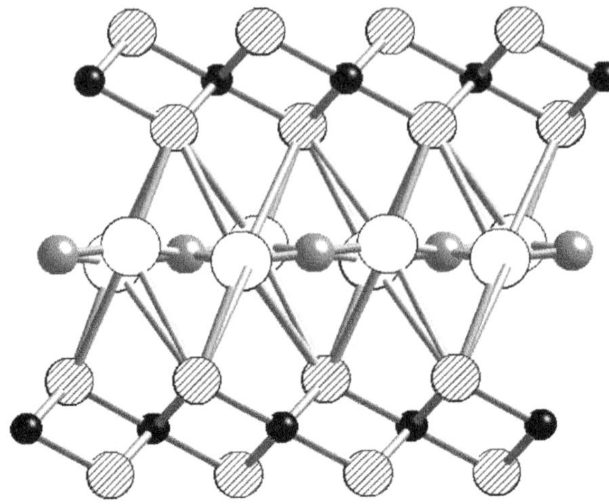

FIGURE 3.11 The birnessite structure.

wide range of biogeochemical processes. Both Mn-oxidizing and Mn-reducing bacteria and fungi exist and are widespread.

The principle Mn oxide minerals in soils are lithiophorite [$(Al,Li)MnO_2(OH)_2$], hollandite [$Ba(Mn^{4+}, Mn^{2+})_8O_{16}$], and birnessite [$(Na,Ca)_{0.5}(Mn^{4+}, Mn^{3+})_2O_4 \cdot 1.5H_2O$], along with poorly crystalline (or amorphous) forms. The birnessite structure is shown in Figure 3.11. A comprehensive review of manganese oxides minerals is provided by Post (1999). Manning et al. (2002) measured a BET surface area of synthetic birnessite = 32.0 m^2g^{-1}. The pH p.z.n.p.c. for birnessite ~2.

CALCITE

Calcium carbonate ($CaCO_3$) occurs as calcite (Figure 3.12); aragonite and vaterite have the same composition but different structures. Calcite is a widespread mineral that exerts a key control on the chemical compositions of many natural waters. Limestone [chiefly calcite and dolomite, $CaMg(CO_3)_2$] rocks are exposed at about 20% of the Earth's surface. Just a small amount of carbonate minerals present in an aquifer can dominate the water chemistry.

Calcite precipitation and dissolution have been studied extensively (e.g., Morse et al., 2007, and references therein). The reaction kinetics tend to be fast, and carbonate thermodynamic relationships and values have been reasonably well established for a long time. Calcite dissolves in the presence of carbon dioxide and water to yield calcium and bicarbonate ions in solution

$$\begin{array}{l}\text{calcite}\\ CaCO_3(s) + CO_2(g) + H_2O(l) \rightarrow Ca^{2+}(aq) + 2HCO_3^-(aq).\end{array} \quad (3.5)$$

Calcite mineral nucleation and growth are discussed in Chapters 9 and 10. Carbonate dissolution in the oceans is discussed briefly in Chapter 2. Calcite and aragonite (and, to a lesser extent, vaterite) often occur as biominerals, which are formed by the metabolic activity of organisms (Chapter 11 and 12).

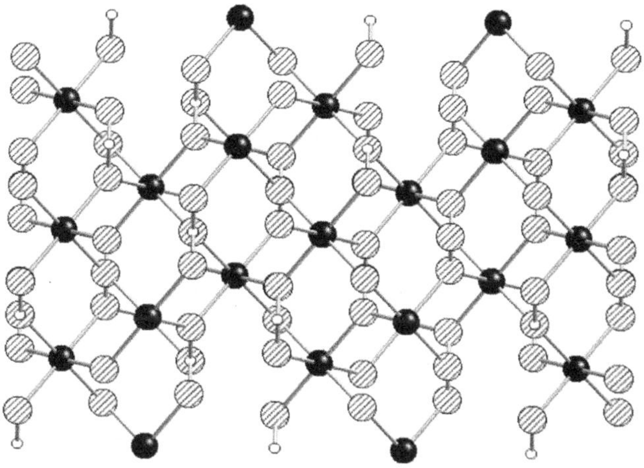

FIGURE 3.12 The calcite structure. Solid symbols are calcium, cross-hatched symbols are oxygen, and unfilled symbols are carbon.

The pH p.z.n.p.c. can be difficult to measure, especially because the mineral dissolves quickly during measurements. Calcite tends to be soluble in acidic waters and to precipitate in alkaline waters. In contrast to most oxides and aluminosilicates, calcite solubility decreases as temperature increases. Zelazny et al. (1996) give an isoelectric point (pH where the electrophoretic mobility drops to zero) of calcite $= 10 \pm 1$.

FELDSPARS

Taken together, the feldspar minerals are the most abundant minerals of the Earth's crust. They are important components of many igneous, metamorphic, and sedimentary rocks. Compositional endmembers based on cation identity include K-feldspar endmember $KAlSi_3O_8$, Albite endmember $NaAlSi_3O_8$, and Anorthite endmember $CaAl_2Si_2O_8$. Their dissolution rate tends to be greater at either high or low pH than at near-neutral pH. Feldspar dissolution and weathering have been studied extensively because of the vast importance of feldspars as rock-forming minerals. Figure 3.13 summarizes feldspar compositions, including solid solution.

Changes in the feldspar surface area after dissolution have been the subject of much study (Chapter 10). White et al. (1995) reported plagioclase from soils 3000 years old to have a specific surface area of 0.4 m^2g^{-1}, and of soils 6000 years old in a similar region to have a specific surface area of 1.5 m^2g^{-1}. Some feldspar samples can become porous because of weathering, which may lead to substantial increases in surface area (Chapter 10). Feldspar pH p.z.n.p.c. $\sim$2.

ZEOLITES

The word *zeolite* derives from Greek words that mean "boiling stones" (Dyer, 1995). They are a group of minerals that typically have unique structural arrangements of $(SiO_4)^{4-}$ and $(AlO_4)^{5-}$ tetrahedra joined at all four corners to produce three-dimensional frameworks

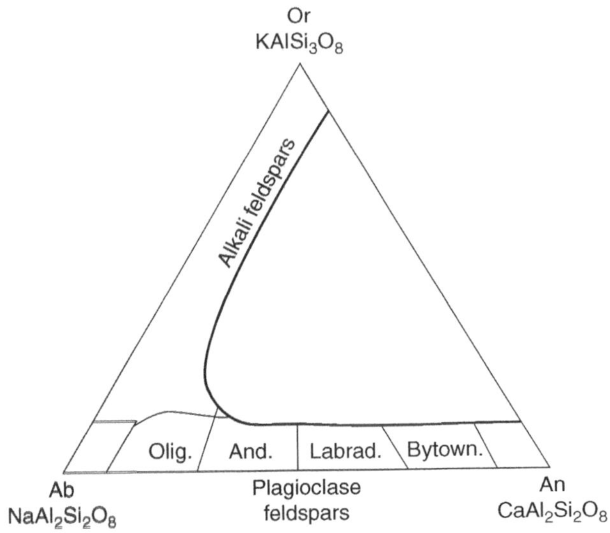

FIGURE 3.13 Ternary diagram of feldspar compositions. After Klein and Hurlbut, 1985. Olig. = oligoclase; And. = Andalusite; Labrad. = Labradorite; Bytown. = Bytownite.

that are often open to movement of water molecules and cations (Dyer, 1995). One common zeolite structure is the sodalite structure (Figure 3.14).

Zeolites can have important industrial applications, and synthetic zeolites are common, including nanoparticles (Chapter 13). Zeolites have been used as catalysts, as molecular

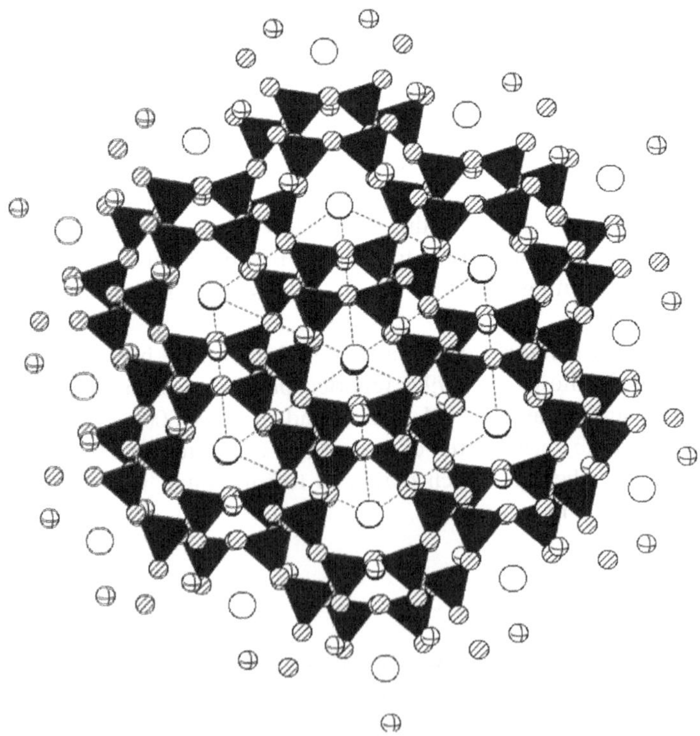

FIGURE 3.14 The sodalite structure.

sieves, in water treatment procedures, and in other applications. Zeolite BET surface areas are $\sim$100–1000 m^2g^{-1}, but these values are likely only estimates because the assumption of monolayer adsorption used in the measurements is unlikely for zeolites (Dyer, 1995).

QUESTIONS FOR FURTHER THOUGHT

1. There are hundreds of other minerals of interest in environmental surface chemistry. Pick a mineral or a mineral group, and write a brief description similar to the descriptions given above. If possible, try to include information on surface area and p.z.n.p.c. Also, try to include at least one structural model. Suggestions: muscovite, biotite, illite, nontronite, pyrite, galena, dolomite, and olivine.
2. Kaolinite has Lewis acid sites present at its surface. What is a Lewis acid? A Lewis base?
3. Find at least three papers that have used ferrihydrite for some aspect of research in mineral–water interface geochemistry. Write a short essay describing the following: What are some types of experiments in mineral–water interface geochemistry that this solid phase is used for? Why is it often chosen for use in mineral–water interface geochemistry projects? Describe how ferrihydrite is used in the paper, and (if possible) explain why this particular mineral was chosen for the research.
4. Suggest a plausible explanation for the increase in interlayer spacing for air-dried montmorillonite with exchangeable cation in the order: $K^+ < Na^+ < Mg^{2+}$

FURTHER READING

Cornell, R.M., and Schwertmann, U. 2003. The Iron Oxides: Structure, Properties, Reactions, Occurrences, and Uses 2nd ed. Wiley-VCH, Weinheim, Germany, 664 p.

Klein, C. 2001. Manual of Mineral Science, 22nd ed., Wiley, New York.

Post, J.E. 1999. Manganese oxide minerals: Crystal structures and economic and environmental significance. *Proceedings of the National Academy of Sciences of The United States of America* 96, 3447–3454.

4

SOME KEY TECHNIQUES FOR INVESTIGATING SURFACES AND INTERFACES

A wide variety of techniques has evolved for investigating surfaces and interfaces. Here, some of the most widely used techniques are discussed, some briefly and others in greater detail. This discussion provides a background for subsequent chapters that will rely on the knowledge of these techniques for explanation or understanding. Many of these techniques are not surface or interface techniques per se but are often used in studies of processes such as adsorption and dissolution.

A BRIEF OVERVIEW OF SOME COMMONLY USED TECHNIQUES

Many techniques are not reviewed in detail in this chapter, but they are used in studies of mineral–water interface processes; short descriptions are given below.

X-Ray Diffraction (XRD) XRD is a method for determining the arrangement of atoms within a crystalline solid based on the scattering of an X-ray beam by (primarily) the electrons associated with atoms within the solid. This method is commonly used to identify the mineral components of field samples and to verify the mineralogy and crystalline order of samples used in sorption, dissolution, and other experiments. XRD has been essential for understanding the complex structures of clays (Moore and Reynolds, 1997). It is also useful for determining how interlayer spacings in swelling clays change during different treatments, such as during organic absorption experiments. Analyses typically are performed on either single crystals or powdered samples. Grazing-incidence X-ray techniques can be used to record the atomic structure at the external surface of a mineral.

Inductively Coupled Plasma (ICP) Emission Spectrophotometry ICP emission spectrophotometry with either optical emission spectrophotometry (OES) or mass spectrometry

Environmental Surfaces and Interfaces from the Nanoscale to the Global Scale, by Patricia A. Maurice

(MS) detection: uses an inductively coupled plasma to produce excited atoms and ions that emit electromagnetic radition at wavelengths that are characteristic for a given element, with intensity indicative of the element's concentration within the sample. ICP is commonly used to measure concentrations of dissolved constituents, particularly metals, in field water samples and in sorption, precipitation, and dissolution studies. Rock samples may be dissolved and analyzed as well. Laser ablation may be applied to solid samples for chemical analysis.

Atomic Absorption (AA) Spectroscopy AA Spectroscopy relies on absorption of light at a particular wavelength by an element of interest and is used for measurement of dissolved metal concentrations. AA is particularly useful for analysis of trace metals and is complementary to ICP. AA may be conducted using flame or graphite furnace AA spectrometry (GFAAS) techniques. Generally, GFAAS is better than flame for trace metals and for samples that contain organic compounds such as natural organic matter or bacterial exudates.

Organic Carbon Analysis Organic carbon analysis is used for instruments that are often described as total organic carbon (TOC) analyzers. Depending on the instrument and setup, this technique may measure dissolved organic carbon (DOC) concentration or organic concentration of particulate samples [particulate organic carbon (POC)]. Both methods can be important for studies of organic matter adsorption to or dissolution of mineral surfaces. Schlautman (1992) measured organic carbon concentrations associated with particles using the particulate analyzer to quantify adsorption of organic matter. In many instances (e.g., Zhou et al., 2001), DOC is measured in solutions before and after adsorption, with the difference used to calculate the amount of organic matter adsorbed.

Fourier Transform Infrared Spectroscopy (FTIR) This vibrational spectroscopy, with Fourier transform analysis to account for the effects of water, can provide insight into molecular interactions, for example, between organic molecules and clay surfaces (e.g., Haack et al., 2008). The vibrational motions of atoms in solution or at a surface are highly sensitive to changes in their local environment. Hence, FTIR can provide detailed information on interactions of adsorbents with adsorbates. A variety of different infrared (IR) techniques exist that are applicable to dried samples, solutions, suspensions, surfaces, and so on. Such methods include, for example, diffuse reflectance (DR) and attenuated total reflectance (ATR). IR is generally not surface sensitive, but it can be used to probe organic sorbates when the organic compound is only present at the surface (or in interlayers). ATR works on a layer of metal oxide particles arrayed horizontally and in contact with mineral particles, which allows for analysis of the mineral–water interface. Reviews of IR spectroscopy as applied to determination of environmental surface species are provided by Johnston et al. (1996) and Johnston (1999).

Raman Spectroscopy This method is complementary to IR (Johnston, 1999). It is used to study vibrational, rotational, and other low-frequency modes of a given system. The technique relies on inelastic scattering, which is called Raman scattering, of monochromatic light generally from a laser in the visible, near IR, or near ultraviolet (UV) range. The shift in energy of the laser light during interaction with the sample gives information about the chemical bonding in the system. Some applications include analyzing molecules without a permanent dipole moment, which would not show up in an IR spectrum; determining bond lengths in non–polar molecules; and determining the identity of organic and inorganic species in solution as well as at surfaces. Pasteris et al. (2004) used laser Raman microprobe spectroscopy to characterize the degree of hydroxylation (OH) and the state of atomic order of apatite of biological (e.g., bone), geological, and synthetic origin (Chapter 9). An example of Raman application to selenate adsorption to goethite is provided in Chapter 7.

Nuclear Magnetic Resonance (NMR) Spectroscopy NMR spectroscopy relies on anisotropic spin interactions, such as magnetic dipole–dipole, chemical shift (shielding), and nuclear electric quadrupole. The application to surfaces and interfaces was reviewed by Brown et al. (1999). Both bulk and surface NMR techniques exist. Perhaps most notably, ^{1}H-NMR has been used to determine the nature and distributions of surface hydroxyl groups, with additional work on the nature of water at surfaces. ^{29}Si and ^{27}Al NMR have also been useful for studies of sites on silica and alumina surfaces.

Electron Spin Resonance (ESR) Spectroscopy ESR spectroscopy which is alternatively referred to as electron paramagnetic resonance (EPR) spectroscopy, is a nondestructive method that can be used to study chemical species with one or more unpaired electrons, such as organic free radicals and transition-metal ions with incompletely filled d-shells. ESR can provide local coordination environment information about an adsorbate, particularly certain oxidation states of paramagnetic minerals. Cheah et al. (1999) used ESR along with X-ray absorption fine structure (XAFS) (see below) in a study of copper(II)-2,2'-bipyridine sorption complexes on amorphous SiO_2 (Chapter 8). A review of ESR is provided by Senesi (1996).

Mössbauer Spectroscopy In Mössbauer absorption spectroscopy, a solid sample is exposed to a beam of gamma irradiation. Some gamma rays will be absorbed by the solid. Nuclear transitions occur, and a detector measures the intensity of the beam that passes through the sample. On environmental samples, ^{57}Fe is the most common elemental isotope studied using Mössbauer. For example, Stucki et al. (2007) used Mössbauer spectroscopy in a study of the effects of in situ biostimulation on Fe speciation in soil minerals. Zachara et al. (2004) applied Mössbauer to investigate the mineralogy of Fe minerals at the petroleum-contaminated *Bemidji* aquifer.

Auger Electron Spectroscopy (AES) AES is a technique for quantitative surface analysis of the top few monolayers of a surface; with high lateral resolution, it can permit surface composition mapping. The technique is based on the Auger effect, which occurs when an atom is ionized by removal of an electron from a core energy level, leaving a hole behind in one of the atomic-core levels. A second electron leaves a higher energy level shell to fill the core-level hole. This can be compensated for by the atom ejecting either a photon (X-ray) or a third electron known as the Auger electron. The energy level of the Auger electron is dependent not only on the original atom but also on the environment the atom is in. AES is often used in conjunction with ion sputtering to obtain depth profiles of elemental composition. Scanning Auger microscopy (SAM) can be performed using a rastering electron beam. Junta and Hochella (1994) used AES and SAM with other surface techniques to study Mn(II) oxidation at mineral surfaces. Eisenlohr et al. (1999) used AES, with sputtering for depth profiling, to study the effects of impurities on carbonate mineral dissolution. See Browning and Hochella (1999) for more discussion of the method and applications for environmental surfaces.

Low Energy Electron Diffraction (LEED) LEED is an ultrahigh vacuum (UHV) technique used to characterize the structure of a surface. Rather than relying on diffraction of X-rays, low-energy electrons are used that give surface sensitivity and allow for a diffraction pattern to be formed based on the crystalline structure of the surface. The surface sensitivity occurs because of the short mean free path of low-energy electrons, which is also why the technique requires UHV. LEED can experience difficulties for oxide surfaces that have insulating properties, such as feldspars (Brown et al., 1999). Stipp (1999) used LEED to analyze the surface relaxation and reconstruction (see Chapter 5) of calcite in UHV.

Secondary Ion Mass Spectrometry (SIMS) SIMS works by bombarding the surface of interest with a highly columnated beam of primary ions, which results in sputtering of the surface and emission of particles, some of which are ionized. These secondary ions are analyzed with a mass spectrometer to determine quantitatively the elemental, isotopic, or molecular composition of the area that has been sputtered. This technique can be used for depth profiling below surfaces. Nugent et al. (1998b) provide an example of using SIMS for depth profiling Al/Si ratios in a study of feldspar dissolution (Chapter 10). Detailed information on SIMS and application to geologic materials is presented by Metson (1999). Newly developed instruments are allowing for elemental and isotopic analysis down to 50-nm lateral resolution and 1-nm depth resolution (Herrmann et al., 2007).

Zeta Potential, ζ-Potential The $\zeta-$potential is a measure of electrokinetic potential for colloidal particles. It is not measured directly (even by so-called ζ-potential instruments), but it is calculated from modeling experimentally determined electrophoretic mobility or dynamic electrophoretic mobility data. Electrophoretic mobility can be measured in a variety of ways. A common technique is based on dynamic light scattering by particles in a fluid cell. The underlying basis for $\zeta-$potential calculations comes from the Henry equation, which for nonconducting particles has the form

$$\mu = \zeta\varepsilon/1.5\eta \; \mathrm{f}(\kappa \mathrm{a}) \tag{4.1}$$

(Shaw, 1992), where

μ = electrophoretic mobility, in $m^2s^{-1}V^{-1}$
ζ = zeta potential, in V
ε = permittivity of the liquid in Farads m^{-1}
η = viscosity in Pa s = kg m^{-1} s^{-1}.

The function $\mathrm{f}(\kappa \mathrm{a})$ takes on different values depending on the ratio of the radius of curvature of the particle surface to the double-layer thickness, $\kappa \mathrm{a}$. As $\kappa \mathrm{a}$ increases in size, $\mathrm{f}(\kappa \mathrm{a})$ varies from 1.0 to 1.5. When $\kappa \mathrm{a}$ is small, the charged particle can be treated as a point charge, but when $\kappa \mathrm{a}$ is large, the double layer can be treated as effectively flat. The resulting commonly used Smoluchowski equation for the larger $\kappa \mathrm{a}$ is

$$\mu = \zeta\varepsilon/\eta. \tag{4.2}$$

The basis for this equation is the assumption that the electrostatic driving force is countered by friction, ignoring other potential forces. This equation is used in the software of many commercial ζ-potential instruments for measuring electrophoretic mobility of colloidal particles. The Smoluchowski equation works fairly well for particles with $\kappa a > 100$, but it can lead to serious errors for particles with $\kappa a < 100$. Other approaches are sometimes necessary or preferred (e.g., based on work of O'Brien and White, 1978), especially for differently shaped particles or particles in particularly high- or low-density suspensions.

Mirau Vertical Scanning Interferometry (VSI) VSI is an optical microscopy technique that can provide microtopographic information with 1-nm vertical resolution in white light mode and 0.7 Å in green light mode. The application of this technique in environmental surface chemistry has been pioneered by A. Lüttge (Rice University; Lüttge et al., 1999 and Lasaga and Lüttge, 2001). The lateral resolution of VSI is 0.5–1.5 μm (Lüttge et al., 1999), although new techniques result in better resolution. VSI allows imaging of surfaces immersed in aqueous solution over the course of reactions such as mineral growth and dissolution or microbial attachment and is often used in conjunction with atomic force microscopy (AFM) and simulations (Chapters 9 and 12).

IN-DEPTH DESCRIPTIONS OF SOME KEY TECHNIQUES

Instrumentation continues to develop and evolve rapidly. In this portion of the chapter, several key techniques for surface and interface applications are discussed in some detail. The focus is on instruments likely to be encountered by students beginning to work in the field, rather than on the most state-of-the-art instruments that may be available within a relatively small number of laboratories. The reader is encouraged to search for ongoing advances within the scientific literature.

Scanning Electron Microscopy (SEM)

SEM is one of the most widely used techniques for imaging solid surfaces. The resolution is generally on the submicron scale, typically down to 50 nm, although some instruments can achieve nanometer-scale resolution on some samples. The images can appear to be three–dimensional (3-D), and have a characteristically high depth of field; an example is shown in Figure 4.1.

SEM imaging uses an electron beam, which typically has an energy that ranges from a few hundred eV to 100 keV. The electron beam is focused by one or two condenser lenses into a beam with a very fine focal spot sized 0.4 nm to 5 nm. This highly focused electron beam is rastered over a sample. The energy exchange between the electron beam and the sample results in the emission of electrons and electromagnetic radiation, which can be detected to produce an image.

Images may be produced in several ways. Secondary electrons of low energy that originate from beam interactions with the top few nanometers of the surface may be used to produce an image with a resolution potentially less than 1 nm. The images produced from secondary electrons tend to have three-dimensional appearance.

Backscattered electrons consist of high-energy electrons originating in the electron beam that are reflected or backscattered by the specimen. Because the brightness of the backscattered electron image tends to increase with an increasing atomic number of sample constituents, the contrast in backscattered images gives an indication of different average composition. Backscattered electrons can also be used to form an electron backscatter diffraction (EBSD) image that provides information on the crystallographic structure of the specimen.

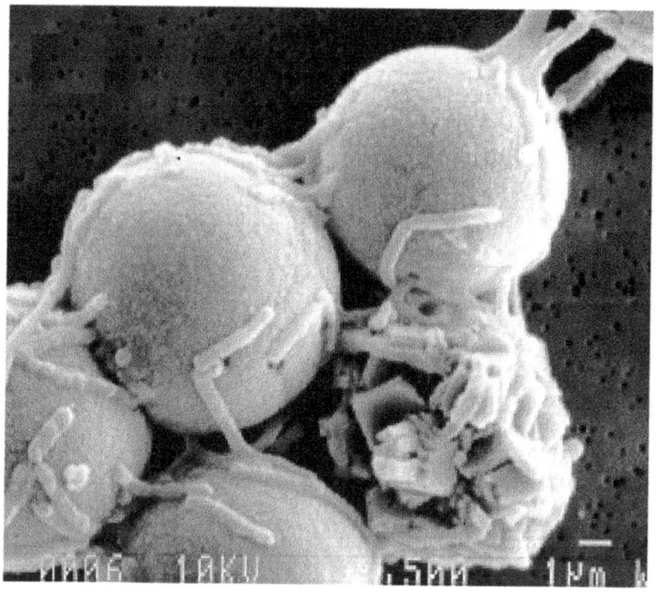

FIGURE 4.1 SEM image of *Bacillus pasteurei* on carbonate precipitates. From Warren et al., 2001 (with permission).

When the electron beam interacts with the sample, X-rays can be produced. Therefore, SEMs may be equipped with energy-dispersive X-ray spectroscopy (EDS) or wavelength-dispersive X-ray spectroscopy (WDS) to provide additional compositional information.

Typically, SEM is conducted in vacuum and, hence, on samples that are not in environmental conditions. For a typical SEM, many samples, especially nonconducting, are sputter coated with a conducting material such as gold or graphite carbon. This coating procedure increases contrast but can "destroy" a surface because most coatings cannot be removed. Coatings also have the potential to obscure some fine details of surface microtopography.

Environmental SEM (ESEM) allows samples to be observed in low-pressure gaseous environments (e.g., 1–50 Torr) and high relative humidity (up to 100%). This method was made possible by the development of a secondary-electron detector that can operate in the presence of water vapor. Also, ESEM instruments allow differential pumping of the vacuum regions around the gun and lenses and the sample chamber. Resolution can be 10–100s nm or μm scale depending on the sample and conditions. Nonconducting samples can be imaged without coating, which means that ESEM causes minimal sample disturbance. As for SEM, EDS compositional analysis is possible if an instrument is so equipped. An example of an ESEM image is shown in Figure 4.2.

Transmission Electron Microscopy (TEM)

The TEM operates on the same basic principles as the optical or light microscope, but electrons are used instead of light. The resolution of the light microscope is limited by the wavelength of light. By using electrons, TEMs typically can achieve Å-scale resolution. Because of this high resolution coupled with compositional analysis by EDS or electron energy loss spectroscopy (EELS), TEM has been an extremely powerful tool for studying

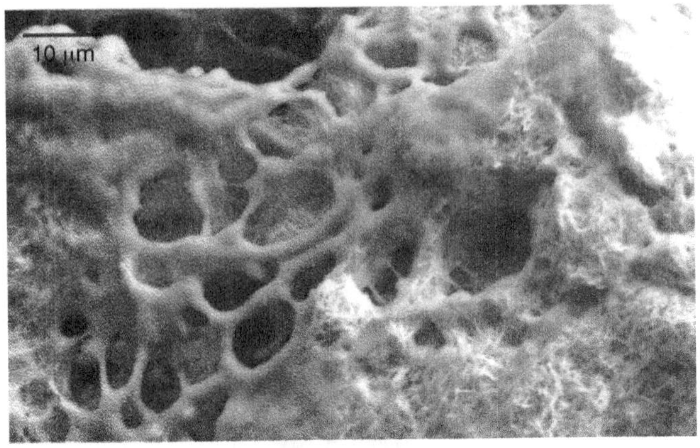

FIGURE 4.2 Environmental SEM image of biofilm formed by *Pseudomonas mendocina* on goethite. The sample is in a humid environment but not immersed in solution, and it shows some drying effects (likely film rupture).

minerals, bacteria, and other environmental samples. Although it is not generally a surface technique per se, surface decoration in which a thin gold layer is deposited on the surface can be used to obtain surface micromorphology. Samples can be sliced into extremely thin sections and then imaged by TEM; in this case, both the internal and the surface/near surface can be imaged and often analyzed spectroscopically.

A beam of electrons produced by an electron gun travels through vacuum from the top of the instrument through the column of the microscope toward the specimen to be imaged (Figure 4.3). These electrons are focused by electromagnetic lenses into a very thin beam.

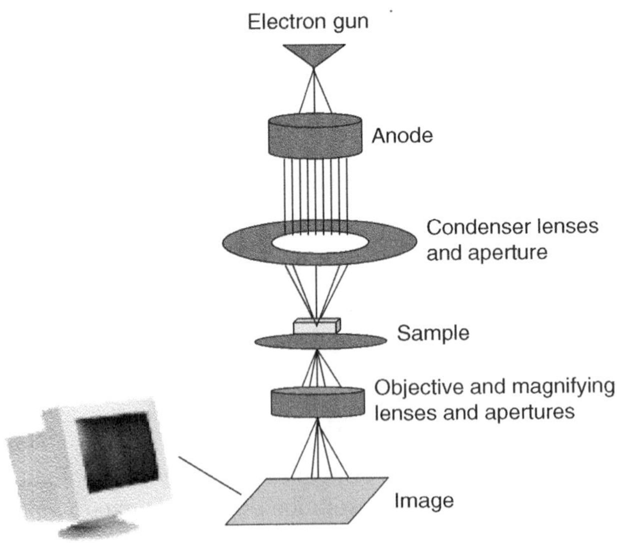

FIGURE 4.3 Schematic of a TEM.

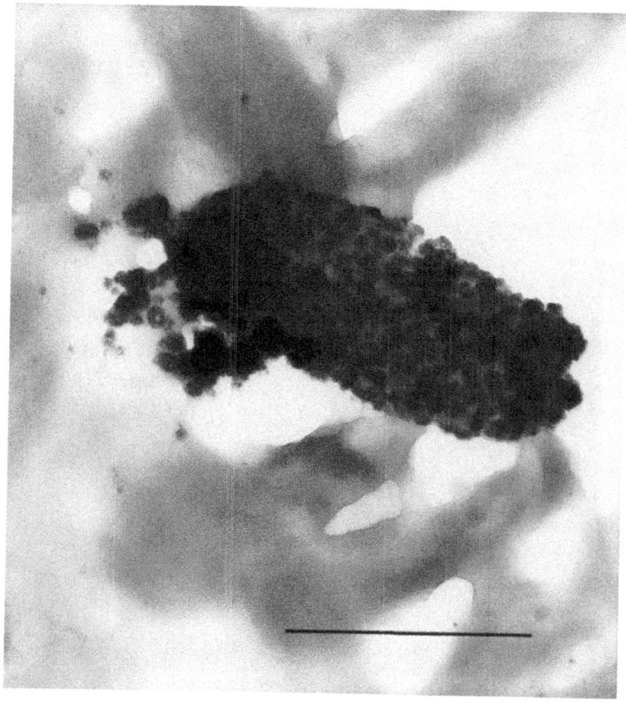

FIGURE 4.4 TEM image of *P. mendocina* bacteria with mineral precipitates containing Fe, Cd, and phosphate. Scale bar 1 μm. From Schramm, Haack, and Maurice, unpublished data.

The electron beam then travels through the specimen. The most dense materials scatter electrons, so that these electrons do not pass through to produce an image. The less-dense materials allow more electrons to pass through. After passing through the specimen, the electrons that have not been scattered hit a fluorescent screen, producing a shadow image with darkness related to the density of the sample material. The image can be magnified and captured on film, digitized to form a computerized image, or just viewed by the operator. Scanning transmission electron microscopy (STEM) is also widely used in studies that involve environmental particles. Like TEM, electrons pass through the specimen; but like SEM, the electron optics focus the beam into a narrow spot that is scanned over the sample in a raster pattern. With the use of a high-angle detector, atomic-scale images can be obtained in which the contrast is related directly to atomic number.

TEM can provide Å to nm-scale resolution and can be used on biotic and abiotic samples (Figure 4.4). Some state-of-the art (S)TEM instruments now have the ability to provide sub-Å resolution (Wiggington et al., 2007). With EDS, the composition of an area a few nanometers in diameter can be quantified. EELS and elemental mapping can provide more analytical capabilities. At high resolution, TEM can provide crystallographic information, including the arrangement of atoms and the degree of order (Figure 4.5). Atomic-scale defects may be imaged. Samples may be sectioned to allow for imaging of internal structures, and staining may be used to highlight different materials such as biofilm exopolysaccharides. Given that the technique requires high vacuum, imaging and analysis does not occur under "environmental" conditions.

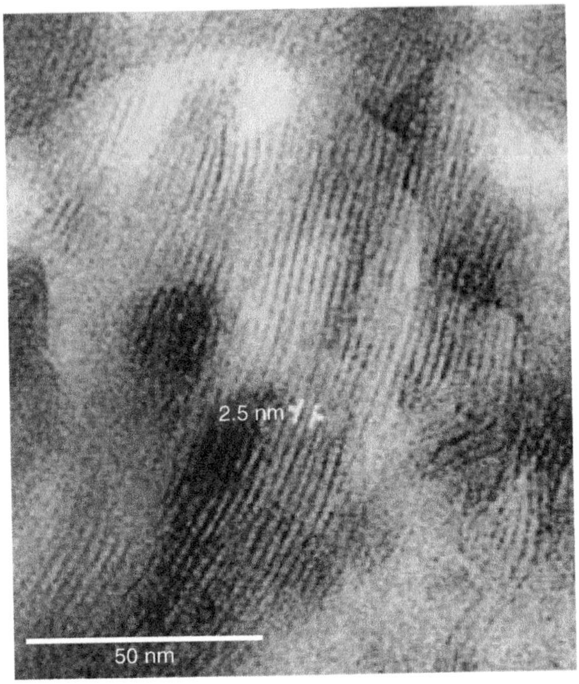

FIGURE 4.5 TEM lattice fringe image of smectite in a soil formed on serpentinite rocks in the Klamath Mountains, CA. From Lee et al., 2003 (with permission).

Scanning Tunneling Microscopy (STM)

STM was developed in the early 1980s, and it is described in papers by inventors Binnig and Rohrer, who quickly received the Nobel Prize in 1986. STM allows for imaging of conducting and semiconducting surfaces in air, solution, vacuum, or under oil down to the atomic level.

STM (Figure 4.6) works by using a sharp metallic tip that is rastered over the surface of interest with a piezoelectric translator that can be controlled to within 0.1 Å in x-, y-, and z-directions. When the tip is brought to within several Å of the surface, a small bias mV voltage is applied. Electrons can then "tunnel" across the gap between the tip and the surface.

The electron tunneling is very sensitive to the tip-sample separation distance, s (Å), as described by the equation for tunneling current, I_T (in nA)

$$I_T = e^{[(-4s\pi/h)(2m\phi)1/2]}, \tag{4.3}$$

where

ϕ = tunneling barrier height, which is proportional to the sample and tip electronic work functions, and which can be variable across a surface (typically 3 to 5 eV)

m = electron rest mass (9.110×10^{-13} kg)

h = Planck's constant (6.626×10^{-34} J s = 4.136×10^{-15} eV s) (Hochella, 1990).

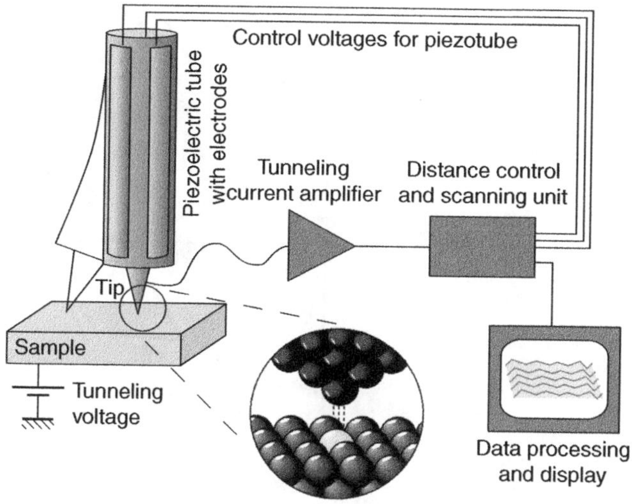

FIGURE 4.6 Schematic of STM from Michael Schmid, Technical University, Wien, Germany.

Equation 4.3 is a simplification, but it demonstrates the surface sensitivity of STM because of the (negative) exponential relationship between tip-sample distance and tunneling current. As the tip is rastered over the surface, a 3-D image is formed with very high resolution and sensitivity to surface microtopography and to bonding at high resolution. Atomic-scale images can be produced.

Case Study: Imaging Parameters and High-Resolution Imaging of Hematite

Hematite is an n-type semiconductor, which means that it conducts by motion of negative charge carriers (electrons) through the conduction band. Impurities present in natural samples can provide charge carriers (generally electrons in the conduction band) that make hematite sufficiently conductive to be imaged by STM. Both Fe and O atoms are present on the hematite (001) surface (Chapter 3). Eggleston and Hochella (1992) imaged the hematite (001) surface at different bias voltages and concluded that images at negative bias showed the oxygen sublattice, whereas images at positive bias showed the Fe sublattice.

Eggleston (1999) systematically varied bias voltage and tunneling currents while conducting high-resolution imaging. The images in Figure 4.7 were collected at different tunneling currents, all at the same positive bias voltage where higher current corresponded to smaller tip-sample distance. When the tip was relatively far from the surface, periodicities of both 0.3 and 0.5 nm were observed. As the setpoint current was increased, the 0.5-nm periodicity became the only periodicity observed.

Modeling using resonant tunneling theory suggested that the observed effects on images of changes in bias voltage, tunneling medium, and tip-sample distance could best be explained if only the Fe atoms (not the oxygen atoms) on hematite were being imaged in the range of conditions used. In this interpretation, the different spacings did not represent Fe versus O atoms but different probing of Fe atoms. Work by Eggleston and Hochella (1992) and Eggleston (1999) showed that STM imaging needs to be conducted under a wide range of conditions and combined with modeling, and that STM is a powerful technique for

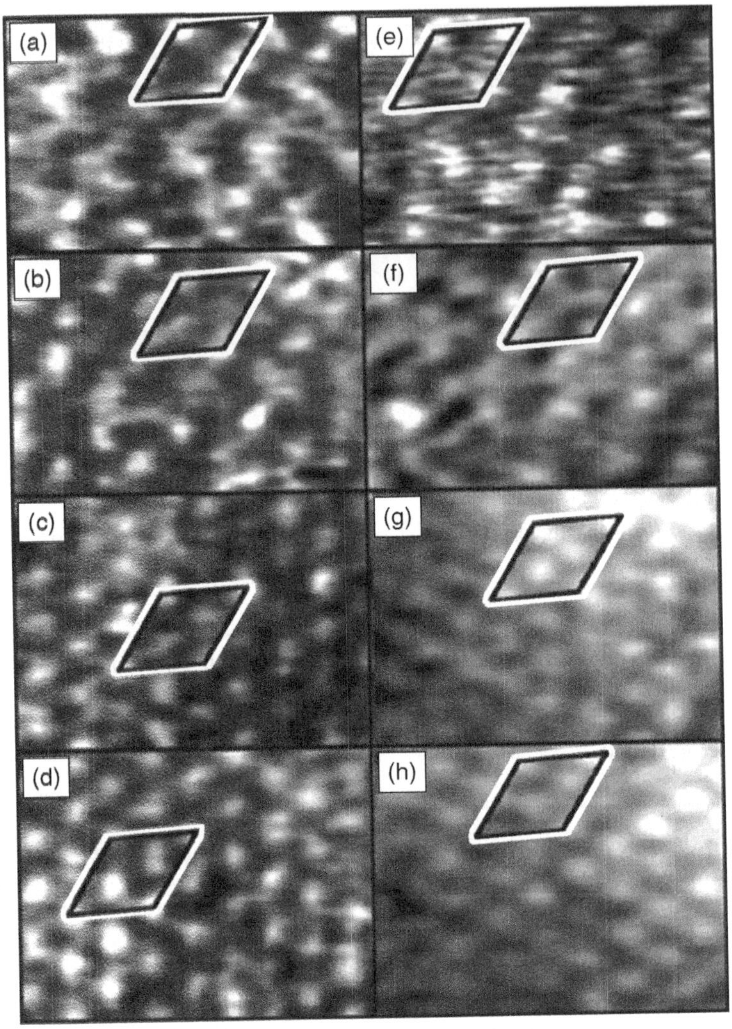

FIGURE 4.7 STM images of the hematite (001) surface at positive bias voltage (+1.15 V) and different tunneling currents (a = 5 nA, b = 10.7 nA, c = 19.7 nA, d = 28.0 nA, e = 4.8 nA, f = 10.1 nA, g = 19.0 nA, h = 32.2 nA); vertical scale variable. Example unit cells are marked. From Eggleston, 1999 (with permission).

increasing our understanding of surface electronic structure. Although STM has been used for more than two decades, our ability to interpret STM images of environmental surfaces is still evolving, as is our understanding of the hematite surface.

AFM and Interfacial Forces

Atomic force microscopy (AFM), which is also known as scanning force microscopy (SFM), was developed as an offshoot of STM in the mid-1980s. AFM works by rastering a sample underneath a sharp tip that is attached to or part of a cantilever (Figure 4.8). A variety of interfacial forces causes the tip to deflect as different surface features pass beneath it. By

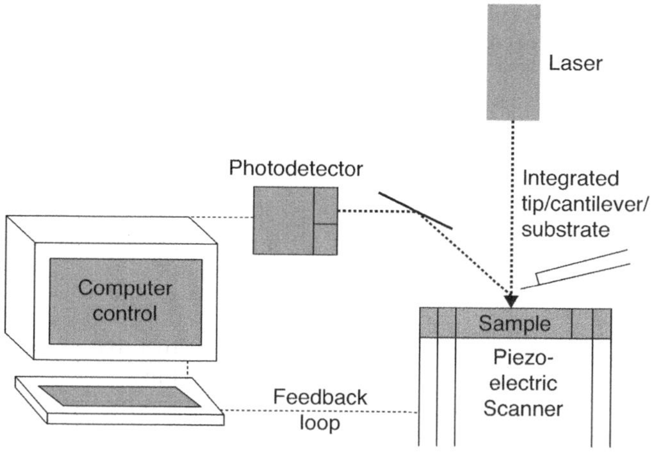

FIGURE 4.8 Schematic of an atomic force microscope.

monitoring this deflection, a 3-D map of the sample surface can be constructed. Although the tip may be used to map out a variety of surface/interface properties, the most commonly used modes of AFM result in an image of surface microtopography.

AFM has become an important technique for studies of surface structure and reactivity of environmental samples. AFM is for the most part nondestructive. Under most operating conditions, micron- to nanometer-scale resolution is easily attainable; under ideal operating conditions, molecular- to atomic-scale resolution may be achieved. Surfaces may be imaged in air or immersed in liquids, which include aqueous solutions. Vacuum AFMs are also available. Little or no sample preparation is needed. A nanometer- to micron-scale portion of the surface may be imaged repeatedly/continuously in solution so that the reaction progress can be monitored.

In most early research, AFM did not seem to produce true atomic-scale images of complex mineral surfaces because atomic-scale defects were not apparent. However, Ohnesorge and Binnig (1993) demonstrated that AFM could provide true atomic-scale resolution on the mineral calcite, including atomic-scale defects, using small attractive forces of $\sim 10^{-11}$ N. In addition to atomic-scale imaging, AFM has been used for a wide variety of environmental surface applications, such as probing forces at the mineral–water interface, measuring directly the kinetics of mineral growth and dissolution, imaging sorbed macromolecular organic substances, and determining particle size, micromorphology, and microtopography of clays.

Figure 4.8 shows a schematic of an AFM. A sample is mounted on a piezoelectric tube, which allows the sample to be scanned precisely under a sharp tip that is attached to or part of a cantilever. Deflection of the cantilever as the sample is scanned under it can be monitored by a variety of mechanisms. The most common technique is an "optical lever" that consists of laser light reflected off the end of the cantilever toward a photodiode detector. In constant-force (or "height" mode) mode imaging, a feedback loop is used to adjust the height of the sample to keep the cantilever deflection constant during sample scanning. In the constant-height (or "force" mode) mode, the response of the z-piezo is more sluggish so that the height of the sample remains more nearly constant, and the cantilever changes its deflection as the sample is scanned under it. As the surface is scanned, the computer keeps

TABLE 4.1 Examples of some fundamental intramolecular and intermolecular forces that occur between molecules or colloidal particles.

Type of force		Dependence of energy on distance (d)	Energy (kJ/mol)	Range (nm)
Molecular	Intramolecular (ionic or covalent)	1/d	100 s	< 1
	London dispersion	$1/d^6$	1 to 3	0.5 to 5
	H-bonding	$1/d^3$	15 to 20	0.5 to 3
	Dipoles	$1/d^3$	5 to 10	0.5 to 3
Colloidal	Electrostatic	e^{-d}	10 to 100	10 s to 100 s
	van der Waals	1/d	1 to 5	5 to 10
	Solvation	$\sim e^{-d}$	1 to 10	< 5
	Hydrophobic	$\sim e^{-d}$	1 to 5	10 s to 100 s

After Maurice and Lower, 2008.

track of the position of the piezoelectric scanner, and individual scan lines are compiled into an image.

A variety of forces that come into play between the tip and the sample is summarized in Table 4.1. The discussion that follows will focus on contact-mode AFM, which is the most commonly used imaging mode.

The total forces (Ft) between the tip and the sample result from the sum of various attractive (Fa) and repulsive (Fr) forces. One context for understanding the AFM tip/sample interactions is to consider the *Lennard-Jones* function, which describes the change in intermolecular potential energy (ϕ) that occurs as two uncharged particles, such as atoms or molecules (on tip and sample), are brought closer together (Figure 4.9):

$$\phi = 4\varepsilon[(\sigma/r)^{12} - (\sigma/r)^{6}], \tag{4.4}$$

where

σ = approximately the atomic or molecular diameter (distance of closest approach),
ε = minimum value of the potential energy or the depth of the potential energy "well"
r = separation distance.

Referring to Figure 4.9, as the particles are brought closer together from relatively distant separations, the $(1/r)^6$ van der Waals term describes the slow change in attractive forces. As the particles are brought very close together, the $(1/r)^{12}$ term describes the strong Born repulsion that occurs when the electron clouds strongly repel one another because of the Pauli exclusion principle.

The term "van der Waals forces" has been used with different meanings in the literature. The van der Waals (vdW) interaction forces are long-range, relatively weak attractive forces. The origin of the vdW forces is quantum mechanical in nature. They result from a variety of interactions, primarily induced dipole and quadrupole interactions. The vdW forces are nonlocalized; they are spread out over many atoms.

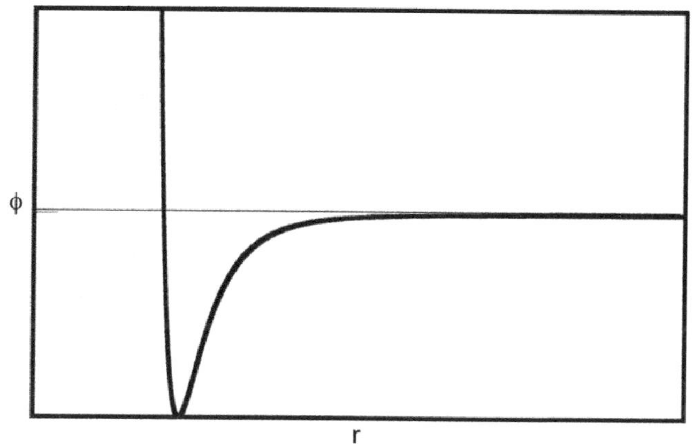

FIGURE 4.9 Changes in interaction potential energy with distance r, graphically representing the Lennard-Jones potential energy function.

The "atomic force" or Born repulsion is a result of the Pauli exclusion principle, and it is the primary repulsive force at close approach. The magnitude of this force is difficult to predict without a detailed understanding of surface structure.

Capillary adhesion is another attractive force that occurs during imaging in air. This capillary force results from formation of a meniscus made up of water and organic contaminants sorbed on the surface of the tip and the sample (Figure 4.10), and it has been estimated to be on the order of 100 nN or greater. When the tip and the sample are completely immersed in liquid, a meniscus does not form, and the capillary forces are absent. Some tips and samples may have hydrophobic properties; in which case, hydrophobic interactions must also be taken into consideration.

Water near surfaces is structured. When the tip and the sample are brought into close contact during force microscopy in solution or humid air, repulsion occurs as the structured water molecules on the surfaces of the tip and the sample are pushed away.

In aqueous solutions, electrical double-layer forces, which have been discussed in detail above and may be either attractive or repulsive, are present near the surfaces of the tip and the sample. These electrostatic forces also need to be considered.

Lateral frictional forces occur as the sample is rastered beneath the tip. At forces to ~30-nN repulsive force (Warmack et al., 1994), a linear relation should hold between the

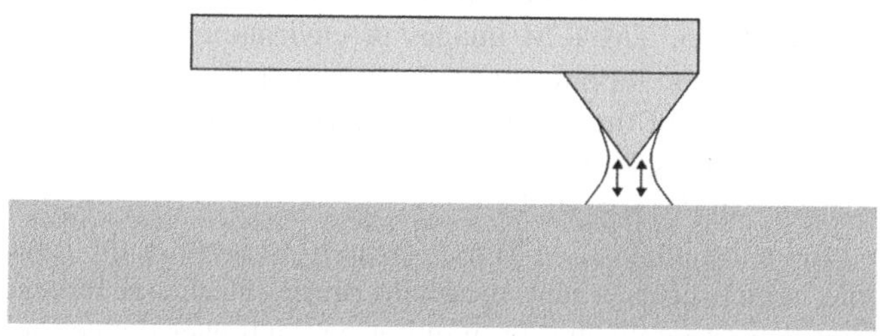

FIGURE 4.10 Schematic illustration of the strong capillary force that tends to drive the tip and the sample together during AFM imaging in air. This can result in damage to the sample surface and can make atomic-scale resolution in air difficult or impossible to achieve.

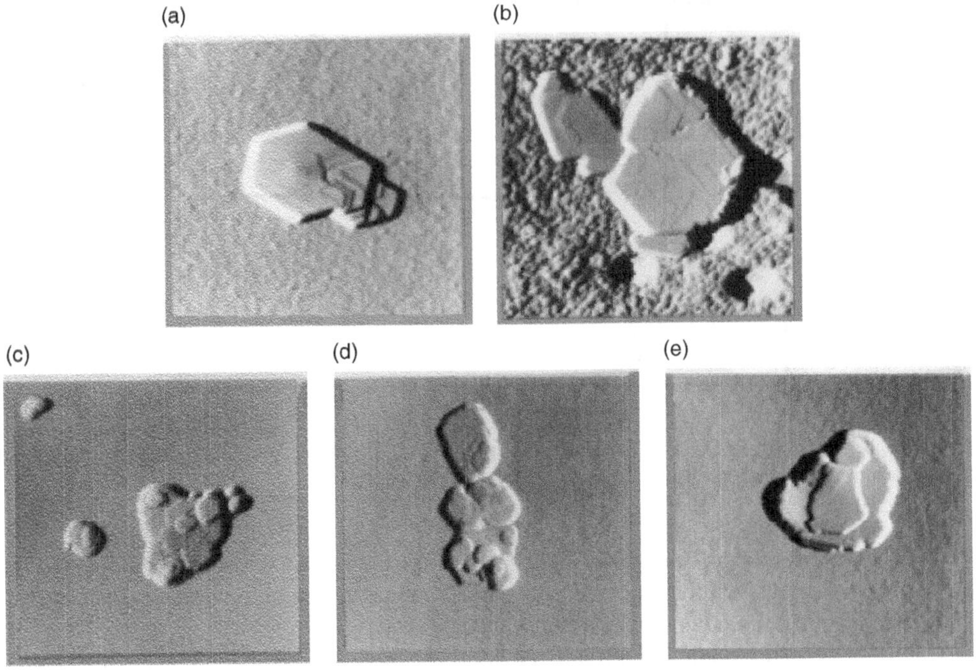

FIGURE 4.11 TMAFM images of kaolinite particles. (a) and (b) are well-ordered kaolinite KGa-1b, and (c)–(e) are poorly ordered kaolinite KGa-2. Dimensions of each image on a side: (a) 900 nm, (b) 1.25 μm, (c) 2.5 μm, (d) 1.3 μm, and (e) 1.0 μm. From Sutheimer et al., 1999.

lateral force F_L and the force normal (vertical) to the surface F_V, with a proportionality constant equal to the coefficient of friction μ. Frictional forces vary on an atomic scale, and with temperature, scan velocity, and relative humidity, as well as tip and sample materials.

The size and shape of the tip may affect images, and a wide variety of potential artifacts in AFM imaging have been discussed in detail (e.g., Maurice and Lower, 2008). The limitations imposed by the finite dimensions of the tip can sometimes be compensated for by modeling; at other times, they may severely limit the usefulness of the technique. For example, the aggregates of nanoparticles (Chapter 13) or sharp features such as the terminations of goethite particles can sometimes be inaccessible to AFM because of overwhelming artifacts.

In many instances, soft and easily deformable surfaces can be imaged more simply using tapping-mode AFM (TMAFM) in which the tip oscillates or 'taps' at a high frequency during imaging. Examples of TMAFM images of environmental surfaces are shown in Figures 4.11 and 4.12. As the tapping tip contacts the surface, the frequency may shift depending on the properties of the underlying material. This shift may be recorded and used in phase imaging (Figures 4.13), which highlights portions of a surface that are softer or stickier versus harder and/or less sticky.

In addition to producing images, AFM may be used to monitor the forces between the tip and the sample, which gives a new means to probe interfacial forces (Figure 4.14). Force measurement data are collected by monitoring cantilever deflection as the sample approaches, makes contact with, and then retracts from the cantilever. The raw data of cantilever deflection are not in units of force; rather, the cantilever deflection is monitored by a photodiode detector and therefore has units of volts. Furthermore, the distance

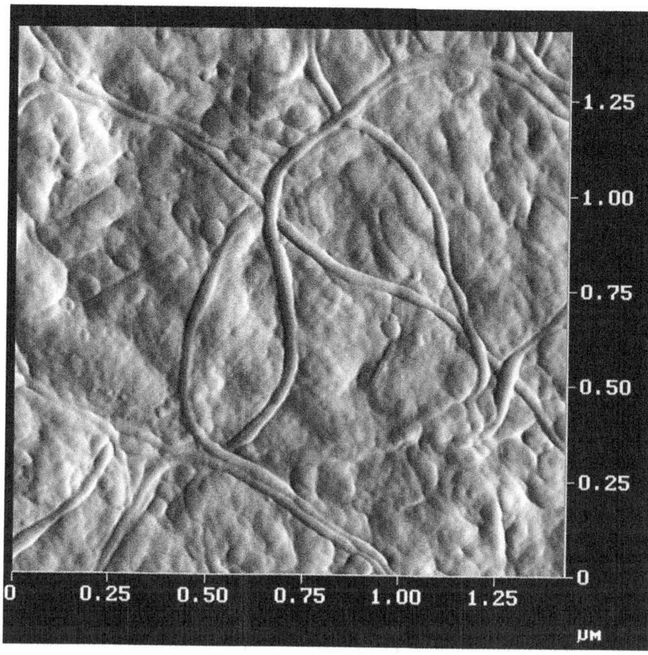

FIGURE 4.12 TMAFM image of fibrous attachment features in a biofilm formed by *P. mendocina* bacteria. Image is 1.5 μm on a side. From Forsythe et al., 1998 (with permission).

(or separation) between the tip and the sample is not measured directly. Rather, the AFM measures the displacement of the piezoelectric scanner that supports the sample. A so-called "force curve" plot of cantilever deflection (in diode volts) and corresponding piezo displacement (in meters) must be manipulated to produce a force-distance curve, which is a plot of force of interaction in Newtons versus separation between the sample and the cantilever in meters (Figure 4.14).

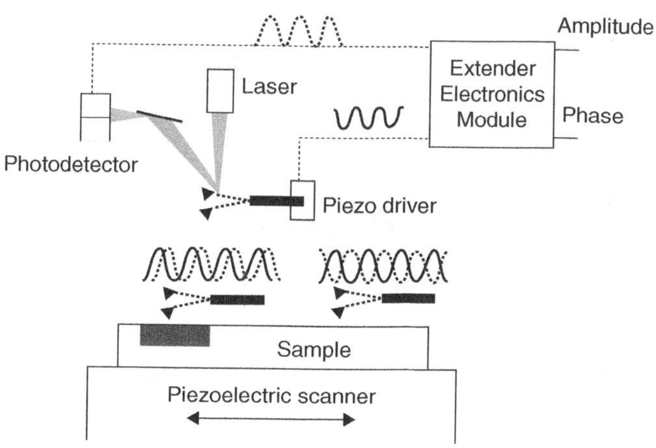

FIGURE 4.13 Schematic illustration of AFM phase imaging. From Forsythe et al., 1998 (with permission).

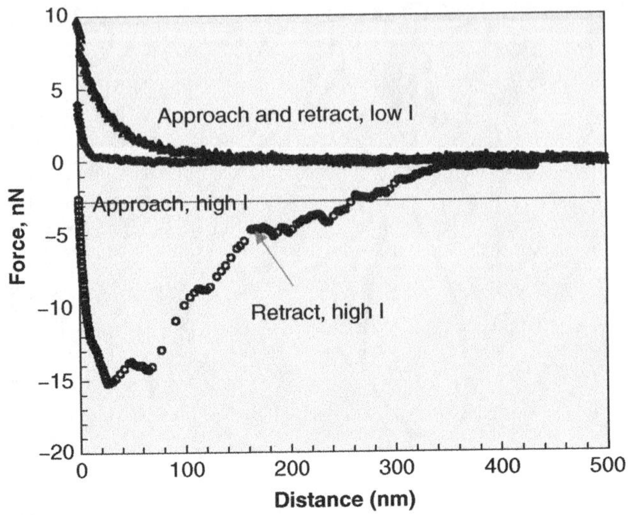

FIGURE 4.14 AFM force-distance curves between *Escherichia coli* and the (001) surface of muscovite in aqueous solution in high and low ionic strength. Modified from Lower et al., 2000 (with permission).

In some cases, researchers have made custom "tips" to measure specialized types of interfacial forces. Figure 4.15 shows a cantilever with a glass bead that has been coated with bacteria so that forces can be measured between the bacterial surfaces and other substances, such as minerals.

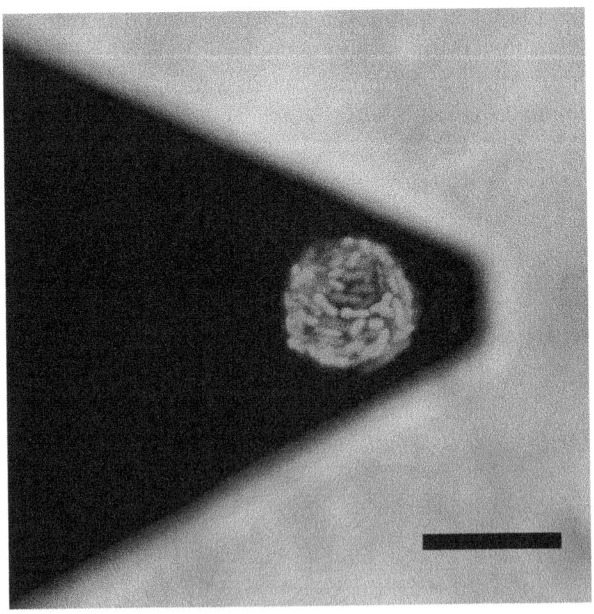

FIGURE 4.15 A specialized AFM tip consisting of a glass bead to which living bacterial cells have been attached. The scale bar is 10 μm long. From Lower et al., 2000 (with permission).

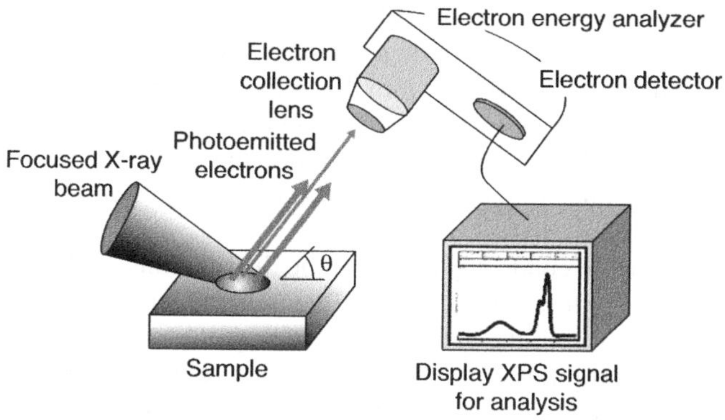

FIGURE 4.16 Schematic of an XPS instrument. Revised from: http://en.wikipedia.org/wiki/Image:System2.gif.

X-Ray Photoelectron Spectroscopy (XPS)

XPS is an UHV spectroscopic technique that measures the elemental composition, chemical state, and electronic state of elements at the surface of a material. XPS spectra are obtained (Figure 4.16) by irradiating the sample with an X-ray beam while simultaneously measuring the kinetic energy (KE) and the number of electrons that escape from within 1–10 nm of the surface. XPS is a widely used surface analytical technique because it is usually nondestructive (although surface charging can lead to damage in the scan area), and because it can provide a wealth of information about the composition of surfaces and aspects such as the oxidation state of elements.

Soft X-rays are used, which means that they are relatively low energy. When the soft X-rays impinge on the sample surface, photoelectrons are ejected from valence and core levels of atoms at and near the surface. An electron-energy analyzer is used to measure the kinetic energy of the ejected photoelectrons.

Because the energy of a particular X-ray wavelength is known, we can determine the electron binding energy (BE) of each of the emitted electrons by

$$E_{binding} = E_{photon} - E_{kinetic} - \Phi, \quad (4.5)$$

where

$E_{binding}$ = energy of the electron emitted from one electron configuration within the atom
E_{photon} = energy of the X-ray photons being used
$E_{kinetic}$ = kinetic energy of the emitted electron as measured by the instrument
Φ = work function of the spectrometer (not the material).

Because it is only the electrons in the top 10–12 nm or less of the surface that make it into the vacuum and the detector, XPS is a *surface* analytical technique.

XPS can be used to analyze the surface chemistry of a material with little or no sample preparation. In many instances, new fracture surfaces are created and analyzed. Sometimes, surfaces are precleaned. Among chemists, XPS is often called ESCA, which is an acronym for electron spectroscopy for chemical analysis.

XPS cannot detect hydrogen or helium. Elements heavier than H and He can be detected at concentrations above 0.05 to 1.0 atomic percentage, depending on the element and the sample. Typical sample areas are from 1 mm down to 30 μm in diameter, although some beam sizes are smaller and some instruments can detect composition at as high as 1-μm resolution on the surface. XPS can be applied to conducting, semiconducting, and insulating surfaces, although techniques to prevent surface charging are often needed for insulating surfaces.

Surface mapping may be conducted to determine distributions of elements at a surface; this process can be more or less complex for different instruments. Limited depth profiling can be accomplished by tilting the sample, or profiling at greater depths can be accomplished by etching with an ion beam, although this may result in artifacts of composition versus depth. Some XPS instruments are capable of mapping to produce images that show distribution of a particular element.

Some important components of an XPS system include (Figure 4.16) the following elements: (1) a source of Mg or Al Kα X-rays, (2) an UHV chamber and pumps, (3) an electron collection lens, (4) an electron-energy analyzer, and (5) a moderate vacuum chamber for sample introduction, along with a sample stage and various manipulators for moving the sample stage through the instrument and then moving samples for analysis.

XPS has been used in many geologic and environmental applications. For example, Tingle et al. (1990) applied XPS to determine the composition of organic compounds on the surfaces of olivine particles. Liu et al. (2009) applied XPS survey spectra and narrow spectra, along with high-resolution TEM (HRTEM), as part of a study of how dissolution rate varies with the particle size, shape, and aggregation state of PbS nanoparticles. Figure 14.17 shows an HRTEM image of a PbS nanoparticle, along with an XPS survey scan and a narrow scan around the Pb4f peak.

BET Surface Area Measurements

The surface area is a fundamental parameter in mineral–water interface geochemistry, and good surface-area measurements are key to many studies that involve processes such as adsorption and dissolution. A commonly used approach for measuring surface area is the BET method, which is named after Stephen Brunauer, Paul Hugh Emmett, and Edward Teller (Brunauer, Emmett, and Teller, 1938). The method works by measuring gas adsorption and desorption to/from a surface, and by applying a formula based on theoretical consideration to calculate surface area from these measurements.

The Langmuir theory discussed in Chapter 6 deals with monolayer molecular adsorption. In the case of gases such as N_2, multiple layers may adsorb. Thus, the BET method considers multilayer gas molecule adsorption, with the following assumptions: (1) gas molecules physically adsorb on a solid in essentially infinite layers, (2) no interaction occurs between the different adsorption layers, (3) the *Langmuir theory* can be applied to each layer, and (4) except for the first layer, the tendency for molecular adsorption or desorption is independent of how many layers are underlying.

To conduct BET surface-area measurements, ambient gas molecules must be removed from the sample surface. The solid (usually a powder) samples are placed in a vessel and pretreated with heat, vacuum, and/or flowing gas to remove adsorbed gases. Next, the sample is cooled under a vacuum to cryogenic temperature. An adsorptive gas (usually nitrogen, but sometimes argon or krypton) is then introduced into the system incrementally. After each addition, the pressure is allowed to equilibrate and the quantity of adsorbed gas

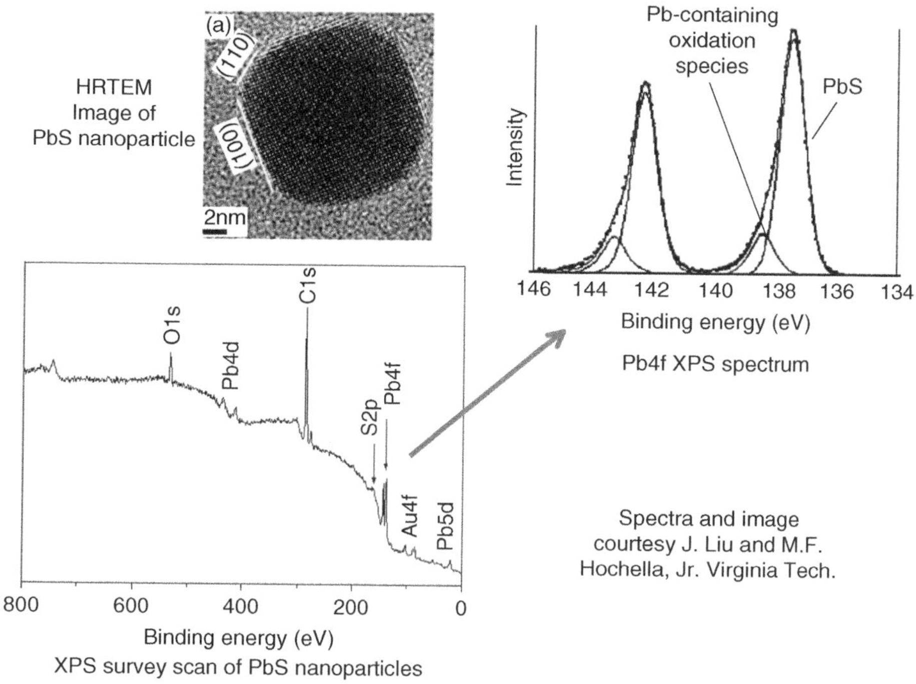

FIGURE 4.17 HRTEM and XPS analysis of PbS nanoparticles. Focusing on $Pb4f_{5/2}$ and $Pb4f_{7/2}$. In $Pb4f_{7/2}$ spectra, the strong emission at binding energy 137.3 – 137.5 eV is assigned to PbS. The other component at a binding energy near 138.5 eV can be attributed to Pb-containing oxidation species, which have been observed on both fresh and oxidized galena surfaces. This work shows how XPS can provide information not only about which elements are present at a surface and their (relative) concentrations but also about the binding state or environment of individual elements such as Pb. Adapted with permission from L. Liu and M. F. Hochella, Jr (Virginia Tech.).

is calculated. An adsorption isotherm is constructed by plotting the gas adsorbed at each pressure at a constant temperature. The data from the adsorption isotherm can be linearized by plotting $p / [v(p_0 - p)]$ on the y-axis and $\phi = p / p_0$ on the x-axis, which yields a BET plot (Figure 4.18). Here, v is the adsorbed gas quantity (generally, in volume units cm^3), p is the equilibrium pressure (generally, in Pa = pascal), and 0_0 is the saturation pressure. The linear relationship is maintained only at $0.05 < p/p_0 < 0.35$. The monolayer adsorbed gas quantity v_m and the BET constant c_{BET} are calculated from the slope A and the intercept I (Figure 4.19) using the following equations:

$$v_m = 1/(A + I) \tag{4.6}$$

and

$$c_{BET} = 1 + A/I. \tag{4.7}$$

The total surface area $S_{BET,\,Tot}$ is calculated as:

$$S_{BET,\,Tot} = (v_m Ns)/V, \tag{4.8}$$

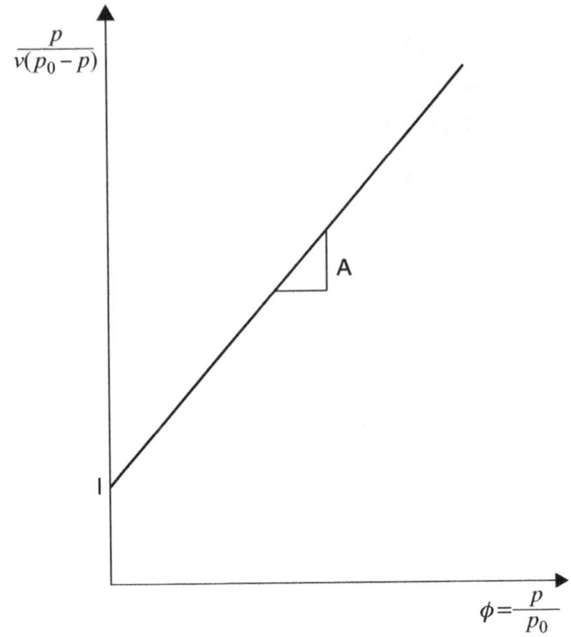

FIGURE 4.18 An example of a BET plot.

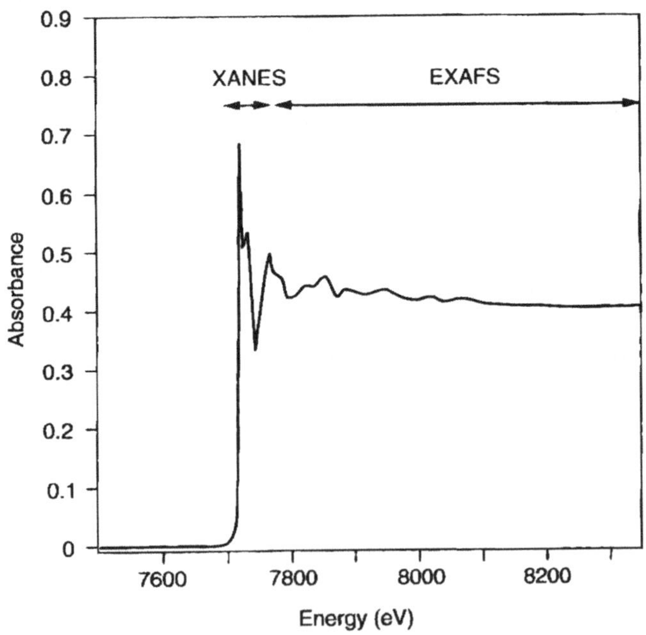

FIGURE 4.19 An example showing components of the XAS spectrum. From Brown et al., 1995 (with permission).

and the specific surface area S_{BET} is calculated as

$$S_{BET} = S_{total}/a, \tag{4.9}$$

where

N = Avogadro's number (6.02×10^{23} atoms $mole^{-1}$),
s = adsorption cross section (for $N_2 = 16.2 \times 10^{-20}$ m^2 $atom^{-1}$),
V = molar volume of adsorbent gas (22,414 cm^3 $mole^{-1}$)
a = weight of sample solid (g).

BET equations are used to calculate the quantity of gas required to form a monolayer at the mineral surface, and the surface area can be calculated from that information and the area covered by each gas molecule. In some instances, the surface area is calculated based on gas desorption.

BET-measured surface areas can be compared with specific surface areas (A_s) that are calculated based on the following geometric considerations:

$$A_s = \text{crystal surface area}/(\text{density} \times \text{volume}). \tag{4.10}$$

According to Cornell and Schwertmann (2003), an agreement between the two methods is considered reasonable if the results are within 10–20%; greater discrepancies can occur if the crystals have porosity.

For porous materials, gas molecules may absorb in the pores, which needs to be taken into account (Davis and Kent, 1990). For nanoparticles that aggregate into larger porous clusters, BET measurements may underestimate the specific surface area. Some minerals, such as goethite, can undergo phase transformations as the temperature increases, so that the degassing conditions may need to be modified for such temperature-sensitive samples (e.g., lower temperature or longer degas time).

Some Synchrotron-Based Techniques

XAS: EXAFS, XANES, and Some Related Techniques X-ray absorption spectroscopy (XAS) is a term used to refer to a family of synchrotron-based techniques. Although it is typically a bulk technique, XAS can be used to probe surface complexes when the element of interest is present only at the surface. Two types of XAS include extended X-ray absorption fine structure (EXAFS) and X-ray absorption near edge structure (XANES), with the designations referring to their positions on the XAS spectra (Figure 4.19). By using grazing-incidence techniques, XAS can be specifically surface sensitive or interface sensitive.

Accelerating electrons from a synchrotron (Figure 4.20) produce X-rays (Figure 4.21). When monochromatic X-rays hit the sample, many processes occur, the most important of which for XAS is the ejection of a core photoelectron. The energy of the ejected photoelectron is equal to that of the absorbed photon minus the binding energy of the initial core state. The ejected photoelectron interacts with electrons associated with the surrounding nonexcited atoms. The processes of scattering of the ejected photoelectron by surrounding atoms is shown in Figure 4.21. The resulting spectrum has oscillations caused by the wave-like interactions of the photoelectron, which occur because an electron has not

FIGURE 4.20 The synchrotron at Argonne National Lab, Advanced Photon Source (APS). From www.anl.gov.

only a particle but also a wave nature. The XAS spectra can provide information about the local coordination environment of the element of interest, which includes atomic distances and the number (and sometimes identity) of the nearest and next neighbors.

Spectra are displayed as graphs of the absorption coefficient of a given material as a function of energy. XANES spectra occur when the kinetic energy of the incident beam is up to 50 eV above the binding energy E_b of the core electron of the element of interest. The EXAFS spectrum occurs at higher energies (Figure 4.19) (Brown et al. 1995). Often, EXAFS data are presented as Fourier transforms of the original spectra, which can make comparison of spectra more straightforward.

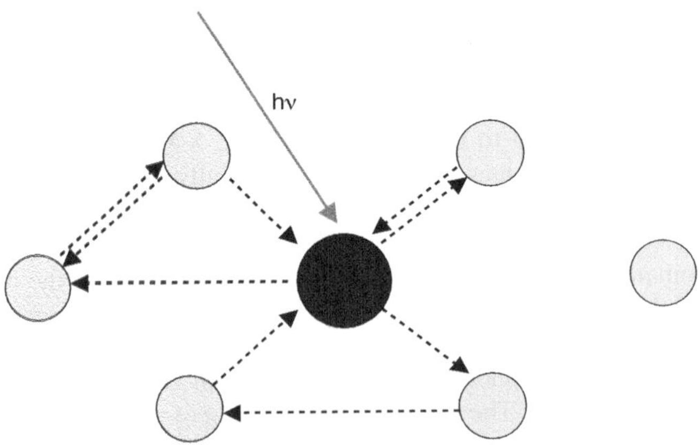

FIGURE 4.21 A schematic illustration of the photoelectron scattering processes involved in XAS.

FIGURE 4.22 A flow-through cell set up in a beamline at the APS for aqueous geochemical experiments. Courtesy of B. Mishra, University of Notre Dame.

Computer programs are needed to calculate atomic scattering factors and to calculate both single and multiple scattering contributions. XAS does not require UHV, and samples may be run in solution or on hydrated thin films of particles. For example, a fluid-cell attachment to the end of a beamline for conducting XAS of species in solution is shown in Figure 4.22. This advantage is important for applications to environmental samples.

XAS techniques can be extremely powerful for investigating environmental surface interactions, particularly metal adsorption to mineral or microbial surfaces. Disadvantages include (1) the need to travel to a synchrotron and to plan experiments according to available synchrotron time; (2) the fact that if multiple sorption modes exist (such as adsorption at edge versus basal-plane sites), it can be difficult to sort out the geometries and relative abundances of the different sorption structures; and (3) these techniques require considerable expertise and have a variety of potential pitfalls. Over time, investigators have studied increasingly complex samples, getting closer and closer to natural environments. Fenter et al. (2002) provided a review of XAS, but synchrotron-based spectroscopies and microscopies are evolving so quickly that the reader is encouraged to do a library or web-based search if the latest information is needed.

Although the standard spot size of XAS analysis has typically been in the mm range, advanced techniques are permitting analysis spot size in the μm range. Figure 4.23 shows the coupling of micro-X-ray-fluorescence (μ-SXRF) spectroscopy with microextended X-ray absorption fine structure (μ-EXAFS) spectroscopy to probe the association of Zn with different components of a soil.

In recent years, several related techniques have been applied to environmental surfaces (see Problem 1, below). Resonant anomalous X-ray reflectivity (RAXR) is proving to be particularly useful for investigating sorption phenomena (Chapter 7). RAXR is an X-ray technique that combines the interface specificity of high-resolution X-ray reflectivity with the element specificity of X-ray absorption spectroscopy (Fenter et al., 2007). As described by Fenter et al. (2007), the reflected intensity is measured as a function of

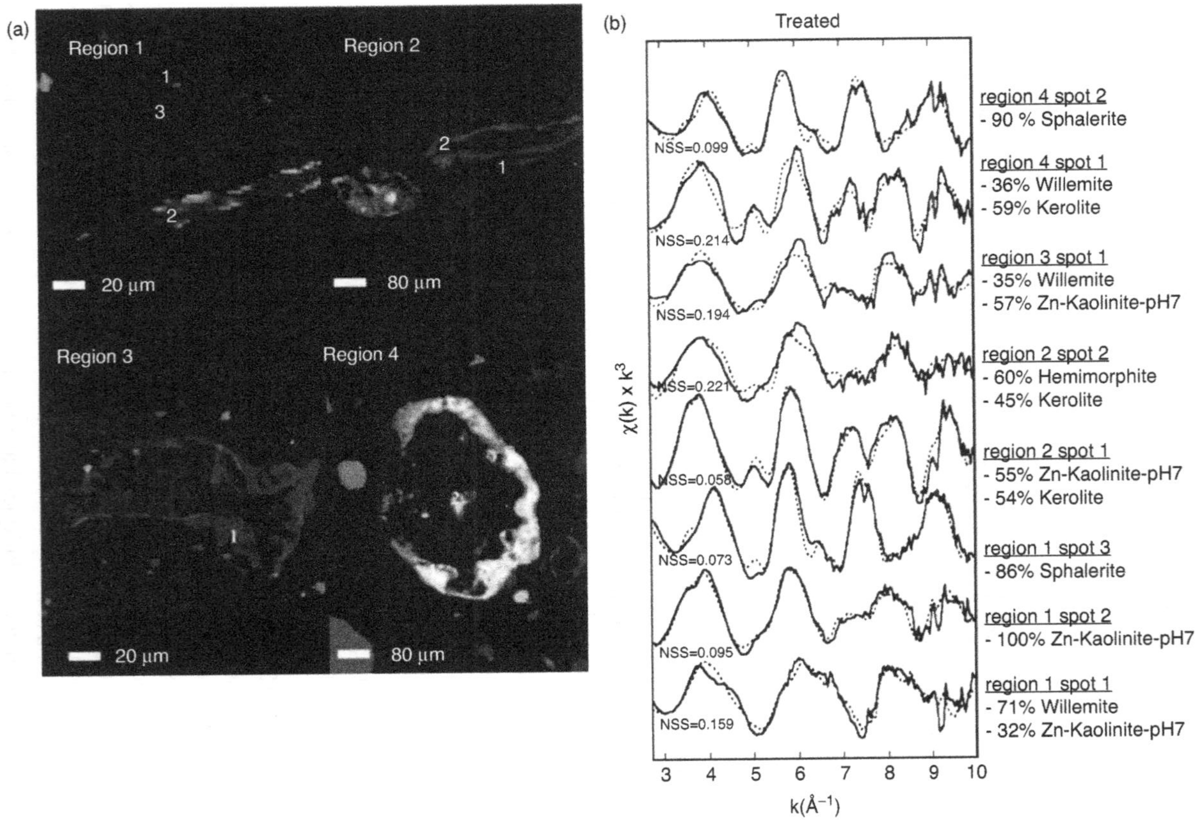

FIGURE 4.23 Coupling of μ-SXRF mapping (left) and μ-EXAFS analysis (right) to probe geochemical associations in a soil sample. A web based copy of the original article shows color mapping indicative of metal distributions (red = iron, green = copper, blue = zinc) can be found at ag.udel.edu/soilchem/2005publications.html. From Nachtegaal et al., 2005 (with permission).

energy near the absorption edge associated with the cation at fixed momentum transfer. The RAXR spectra are modulated at the characteristic absorption edge of the resonant species of interest, thus providing element-specific information while maintaining the interfacial specificity associated with surface scattering. One advantage of RAXR over EXAFS is that RAXR is better at distinguishing the distributions of different sorption complexes that occur simultaneously at a surface.

Using RAXR, Fenter et al. (2007) showed that Rb^+ adsorbs to the basal-plane surface of muscovite mica in a partially hydrated state, whereas Sr^{2+} adsorbs in both fully and partially hydrated states. Catalano et al. (2008) showed that arsenate adsorbs by both inner sphere (relatively strong, specific, and direct binding to the surface) and outer sphere (weaker, electrostatically driven, with water between the adsorbed species and the surface) surface complexation (Chapter 6) to hematite and corundum at pH 5 in 0.01 M NaCl.

Soft and Hard X-Ray imaging Techniques In addition to spectroscopy, synchrotron-based imaging and mapping techniques are becoming increasingly important for mineral–water interactions, including microbial and organic. These techniques are evolving in spot size and sensitivity.

Hard X-ray spectroscopy and *spectromicroscopy* uses high-energy X-rays with energies greater than 2 keV, offers small spot-size spectroscopy (60–90 nm spot size improving to 30 nm), and 1 attogram (1×10^4 atoms) sensitivity. The advantages of high-energy X-ray microprobes include better elemental sensitivity than electron microprobes as well as the ability to image hydrated samples, to perform spectroscopy experiments in small spots to probe spatial distribution of valence states of an element, to probe composition within versus at the surface of structures, and to perform microdiffraction experiments (K. Kemner, ANL, personal communication). This technique is excellent for heavy metals. An example of mapping of many elements on a bacterial surface is shown in Figure 4.24.

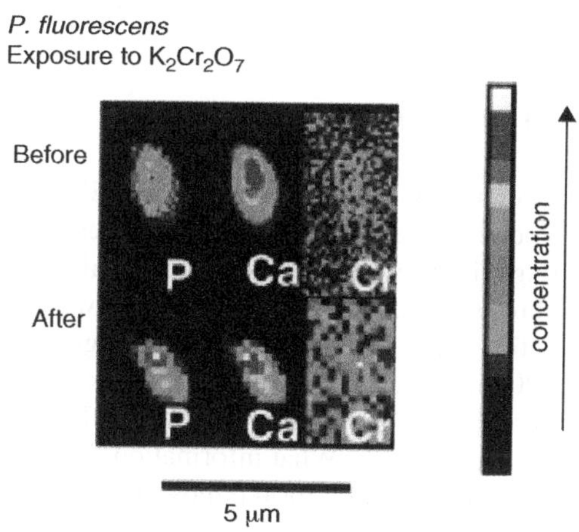

FIGURE 4.24 Spatial distribution of P and Ca on *P. fluorescens* before and after exposure to potassium dichromate solution. Modified for clarity from Kemner et al., 2004 (with permission).

Soft X-ray spectroscopy and *spectromicroscopy* uses lower energy (<2 ke V) X-rays. This technique combines high spatial resolution with elemental discrimination and an ability to image under different experimental conditions. A 20-nm spatial resolution has been demonstrated. This technique is excellent for analysis of organic functional groups.

MICROSCOPIES FOR BIOFILM IMAGING

Biofilms are accumulations of bacteria, exopolymers, and minerals (Chapter 12). Two primary means for imaging biofilms are epifluorescence microscopy (EPI, see images, above) and confocal scanning laser microscopy (CSLM). EPI is an optical microscopy technique in which objectives of 40, 60, or 100× are used, providing mm-scale resolution that varies with wavelength, contrast, and other imaging parameters. In-solution imaging is common, even using flow cells. An example of in-solution epifluorescence imaging of a biofilm is shown in Chapter 12. A wide range of fluorescent stains is available, such as acridine orange and LIVE/DEAD, the latter allowing one to distinguish cells with intact versus disrupted cell walls. With some microscopes, the potential for 3-D reconstruction exists from "optical sections" made as the stage moves in and images are computer analyzed. EPI can be coupled with transmission techniques (brightfield, polarizing, etc.) and other fluorescence methods, such as fluorescent in-situ hybridization (FISH).

CSLM employs a pair of pinhole apertures to limit specimen focal plane to a confined volume ~1 mm in size. Optical sections are made along the z-axis, and deconvolution techniques are applied. The advantages over conventional optical microscopy include controllable depth of field, elimination of out-of-focus information, and ability to perform serial optical sections. Examples of CSLM images are shown in Chapter 12.

Many other techniques of biofilm imaging and biogeochemical analysis are also available, and the "toolbox" is growing continuously. For example, Majors et al. (2005) described coupling of optical microscopy and nuclear magnetic resonance spectroscopy (MRS) to monitor metabolite concentrations in *Shewanella oneidensis* strain MR-1 biofilms growing in fumarate and lactate medium. Such coupling of microscopic and analytical tools is important to many aspects of surface and interface processes in the environment.

QUESTIONS FOR FURTHER THOUGHT

1. Many synchrotron techniques are available beyond EXAFS, XANES, and X-ray imaging. Examples include X-ray reflectivity (e.g., Fenter, 2002), resonant anomalous X-ray reflectivity (e.g., Park et al., 2005; Fenter et al., 2007), grazing incidence EXAFS (e.g., Waychunas et al., 2005), X-ray standing waves (e.g., Templeton et al., 2001), crystal truncation (e.g., Eng et al., 2000), and grazing-incidence techniques (e.g., Trainor et al., 2006; Waychunas et al. 2005). Pick one of these techniques and read at least one paper that used it. In a short essay, briefly describe the technique and how it was used by the researchers. What information did the technique provide? As these techniques are evolving, search for the most recent papers that you can find by these authors.
2. Kai Seigbahn was awarded the Nobel Prize in Physics for his work developing XPS, in 1981. Gerd Binnig and Heinrich Rohrer were awarded the Nobel Prize in Physics for development of the STM in 1986. Write a brief essay addressing the following: Do

you think that these awards were deserved? Why or why not? Support your arguments with discussion of what these instruments provide and how they are used, drawing from the literature and/or the Internet, and citing references appropriately.

3. The BET surface area of a 2.1-g sample of ferrihydrite is determined by N_2 gas adsorption at 77 K, and the following data are obtained:

p (kPa)	3.2	7.8	14.6	21.2	27.3	36.2
v (cm^3)	76	98	114	120	134	158

The saturation vapor pressure p_o is 91.0 kPa.

(a) Plot the adsorption isotherm and calculate the BET surface area of ferrihydrite.

(b) What is the BET constant, c_{BET}?

(c) Does the value of BET specific surface area seem to be appropriate for ferrihydrite?

(d) The ferrihydrite sample used has individual particles that are approximately spherical with a radius of 1.0 nanometers (1 nm $= 10^{-9}$ m) and a density of 4.8 g cm^{-3}. Compare the geometric surface area with the BET measured surface area.

(e) Comment on the comparison of surface area determinations and what it might tell one about ferrihydrite.

4. For each goethite sample described in Table 3.2:

(a) use length and aspect ratio data to compute geometric surface area, assuming a smooth cylinder shape as an approximation.

(b) Compare with BET surface areas.

(c) What does this comparison suggest about the samples?

5. Read the following two papers from the literature. Compare and contrast the understanding of hematite surface (electronic) structure presented by these two papers. Then, search in the literature for at least one more recent paper that contributes to our understanding of hematite surface structure and/or electronic surface structure. What do these readings tell you about how the science of surface mineralogy is evolving and how the scientific community works to address complex questions and problems? Write a short essay, including references.

Eggleston, C.M. 1999. The surface structure of alpha-Fe_2O_3 (001) by scanning tunneling microscopy: Implications for interfacial electron transfer reactions. *American Mineralogist* 84, 1061-1070.

Becker U., Hochella M.F., Jr., and Apra, E. 1996. The electronic structure of hematite {001} surfaces: Applications to the interpretation of STM images and heterogeneous surface reactions. *American Mineralogist* 81, 1301-1314.

6. The AFM has been used to measure forces between bacterial surfaces and mineral surfaces. Two approaches are exemplified by the work of Drs. Steven Lower and Terri Camesano. Read one or more papers for each author (e.g., Lower et al., 2000, 2001a and b; Camesano et al., 2000 and Camesano and Logan, 2000) and compare their approaches. Write a short essay describing the strengths and potential pitfalls of the different approaches.

7. A muscovite mica sample is freshly cleaved in air and then analyzed by XPS. The elemental concentrations (% atomic) measured are shown below. Assuming that the C is primarily "adventitious carbon":
   (a) How did the surface composition compare with what would be expected from the standard formula for bulk muscovite, $KAl_2(Si_3Al)O_{10}(OH)_2$?
   (b) What, if anything, does the XPS analysis suggest about potential isomorphic substitution in this sample? (Hint: You may wish to ratio each element to Si; do not attempt to correct for adventitious carbon in this exercise). The other cleavage surface was placed in deionized water open to the atmosphere for about 1 minute, blown dry with ultrapure air, and analyzed by XPS. The K % atomic decreased to $\sim 3.5 \pm 0.1$.
   (c) What could have happened to the sample to decrease the K surface concentration so quickly?

Element	Atomic %
Al	12.0 ± 0.2
Si	13.2 ± 0.2
O	55 ± 1.2
K	4.2 ± 0.1
Fe	1.8 ± 0.1
C	12.1 ± 1.2
Ca	0.02 ± 0.01
F	1.1 ± 0.1
H	not detectable by XPS
Mg	none detected
Na	0.50 ± 0.2

8. An EXAFS analysis is run on a divalent metal adsorbed to kaolinite at pH 7 and 9. Adsorption isotherms indicate adsorption at both pH values. The results suggest primarily inner-sphere adsorption at one pH value and primarily outer-sphere adsorption at the other pH value. The identity of each "neighbor" atom, the number of these atoms (N), and the distance (R, in Å) are given below. (see Chapter 6.)
   (a) Based on the data below, at which pH (7 or 9) is outersphere adsorption most likely?
   (b) At which pH is inner-sphere adsorption most likely?
   (c) Explain your reasoning. (Note: Si and Al have similar atomic weights and could not be distinguished from one another by EXAFS).

	Neighbor	N	R (Å)
pH = 7	O	5.9 ± 0.2	2.27 ± 0.01
	Si or Al	none detected	
pH = 9	O	4.9 ± 0.3	2.26 ± 0.01
	Si or Al	1.0	3.34 ± 0.04

FURTHER READING

Fenter, P., Rivers, M., Sturchio, N., and Sutton, S. (eds). 2002. Applications of Synchrotron Radiation in Low Temperature Geochemistry and Environmental Science. Reviews in Mineralogy and Geochemistry, vol. 49, J. Russo and P.H. Ribbe, series eds. The Geochemical Society and the Mineralogical Society of America, Washington, D.C., 579 p.

Perry, D.L., ed. 1990. Instrumental Surface Analysis of Geologic Materials. VCH Publishers, Inc, New York, 373 p.

Vaughan, D.J., and Patrick, R.A.D. (eds). 1995. Mineral Surfaces. Chapman and Hall, London, 370 p.

5

SURFACES AND INTERFACES

This chapter introduces many basic concepts of surfaces and interfaces that will be developed throughout the text. One key concept is how to define a surface or an interface both generally and specifically for different applications. The main focus of this chapter is mineral surfaces and the mineral–water interface, although many concepts apply to other types of surfaces and interfaces, such as those that involve bacteria or amorphous solids.

WHAT IS A SURFACE? WHAT IS AN INTERFACE?

According to Parks (1990),

> A surface is the exterior boundary of a condensed phase in a vacuum or gaseous environment.
>
> An interface is the boundary between two condensed phases.

Many different types of binary interfaces are possible, which include gas–liquid, gas–solid, liquid–liquid, liquid–solid, and solid–solid. The liquid–solid interface is the primary realm of mineral–water interface geochemistry. This text also discusses gas–solid interfaces and ternary gas–solid–liquid interfaces, which are representative of atmospheric and soil environments. Various other ternary interfaces, such as between bacteria, minerals, and aqueous solutions, are also important to the environment.

According to the International Union of Pure and Applied Chemistry (IUPAC) Gold Book,

> A physical surface is "That atomic layer of a sample which, if the sample were placed in a vacuum, is the layer 'in contact' with the vacuum; the outermost atomic layer of a sample."

Environmental Surfaces and Interfaces from the Nanoscale to the Global Scale, by Patricia A. Maurice

The study of environmental surfaces begins with thermodynamic considerations. For a reaction to occur spontaneously, the change in *Gibbs energy* (ΔG_r) must be negative, which means that the system must go to a lower overall energy state. The formation of an interface involves a change in the energy of a system. For an interface between two phases to be stable, the free energy of formation (ΔG_f) of the interface has to be positive. If it were negative, then the interface would tend to expand continuously until eventually the materials became completely dispersed into one another.

We can apply some thermodynamic concepts to surfaces and interfaces; for example, in this text, we shall discuss *surface free energy*, which includes *surface tension*. However, the study of surfaces and interfaces generally requires concepts from kinetics, and interfaces are often metastable systems.

THE CHALLENGES OF DEFINING SURFACES AND INTERFACES

Although formal definitions exist for *surface* and *interface*, how does one define or describe a given surface or an interface in practice? These are interesting and nontrivial questions.

One way of defining a surface is by examining how one property changes with space along a transect across a surface or interface region. If different properties (for example, concentrations of different chemical species) were measured, then the exact location in (three-dimensional 3-D) space of the surface or interface might be defined differently for the different properties.

Figure 5.1 shows how the concentration of scandium (Sc) changes across a grain boundary in polycrystalline alumina, as determined by scanning transmission electron spectroscopy. The impurity Sc occurs in preferential association with the grain boundary, and it illustrates the point that it can be difficult to define a surface or an interface exactly. In reality, often "interfacial regions" exist rather than exact finite planes.

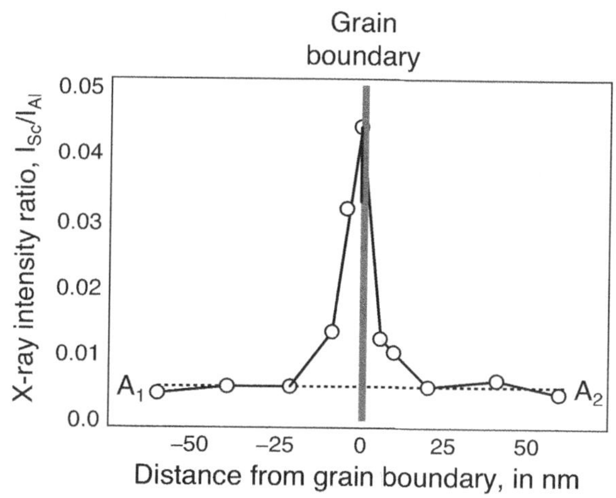

FIGURE 5.1 Relative concentration of Sc near a grain boundary. After Parks, 1990 (with permission), who cites data from Li and Kingery, 1984.

SURFACES ARE COMPLEX

Environmental surfaces are complex structurally, compositionally, and (micro) topographically, and they require an interdisciplinary approach, as discussed in the following sections.

Relaxation and Reconstruction

Examining the mineral surface from the mineral point of view, the presence of a surface, which has different bonding than the bulk of the mineral, can alter the near-surface structure and composition.

A lack of full coordinative bonding at the mineral surface may lead to *relaxation* and/or *reconstruction*. The atoms at the surface of a mineral experience a different bonding environment than the atoms in the bulk of the mineral. In a vacuum, the atoms at the surface of a clean mineral are left with dangling bonds, so an imbalance of forces may exist on atoms at the surface. The lack of a proper coordination environment for surface atoms can result in higher energy. The surface responds to minimize the surface-associated free energy. As a result, most surfaces in a vacuum have been shown to relax inward toward the bulk (movement mostly perpendicular to surface), so that the spacings between the first and second atomic layers are decreased by up to ~15% (Hochella, 1990). In *surface relaxation*, no change occurs in the periodicity parallel to the surface or to surface symmetry (Figure 5.2).

Sometimes, the positions and bonding of the surface atoms change in more profound ways for the atoms to achieve more favorable bonding environments. The surface coordination of atoms at the surface, along with bond angles and bond distances, may be different from those of the bulk mineral. The actual repeat units at the surface—the surface unit cells—may change. This more profound structural change at the surface is termed *surface reconstruction* (Figure 5.2). At least some changes to bonding environments may extend down several atomic layers into the crystal lattice.

The surface structure in air or aqueous solution also can be distinct from the bulk structure. Eng et al. (2000) studied the surface structure of α-Al_2O_3 under humid conditions allowing for the surface to be hydroxylated. A model of the surface structure as compared with unreconstructed bulk structure is shown in Figure 5.3. A layer of oxygens from water is included in the model of the hydrated surface (lower part of figure) appearing about 2 Å

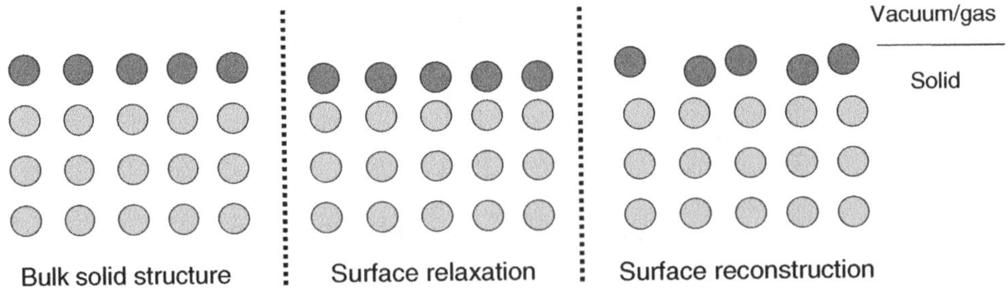

FIGURE 5.2 Highly schematic representations of surface relaxation and reconstruction in the air or a vacuum. Relaxation is primarily a vertical movement of atoms at the surface. Reconstruction is more radical, and sometimes it involves breaking and formation of bonds that may result in a different unit cell at the surface than in the bulk.

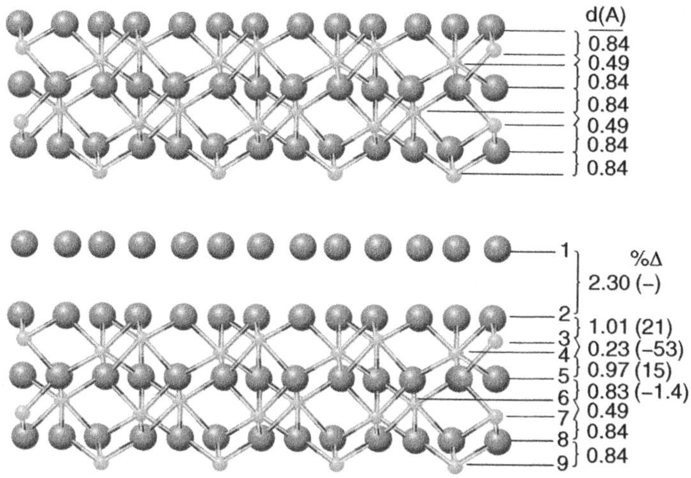

FIGURE 5.3 Model showing the surface structure of α-Al_2O_3 unreconstructed (top) and reconstructed, in air (bottom). On the right-hand side of each, the atomic spacings between different atoms (large spheres are Al, whereas small are O) are shown, and to the right of that in the bottom figure is the percent difference between the two structures (+ means expanded and − contracted). A layer of water of hydration is also shown at the top of the structure in air. Modified from Eng et al., 2000 (with permission).

above the mineral surface. Expansion and contraction occurs that extends several Å down into the structure. Overall, the processes of relaxation and reconstruction may lead to the *surface* being more of a *surficial region* than an infinitely thin plane.

Surface Sites

Let us consider the surface of a mineral in water—what might appear at first glance to be a simple case. First, imagine taking a piece of an oxide mineral (such as Fe or Al oxide), splitting it apart to form a fresh surface, and then dropping the particle into water. If we could view the surface at the atomic scale, what would we "see?"

At the atomic to molecular scale, the mineral surface would have a variety of surface sites: $>$M—OH, $>$M—O^-, $>$M—OH_2^+, and, so on, where M stands for a metal such as Fe or Al and $>$ represents at the surface. The original bonds existing in the mineral structure would be broken. To provide full coordination of the metal centers, various types of protonated and unprotonated hydroxy groups would be present at the surface, with different proportions depending on pH.

For an oxide mineral, the deprotonation reactions at the surface can be expressed as

$$>\text{M—OH}^0 \rightleftarrows >\text{M—O}^- + \text{H}^+ \qquad (5.1)$$

$$>\text{M—OH}_2^+ \rightleftarrows >\text{MOH}^0 + \text{H}^+. \qquad (5.2)$$

A single surface site that can undergo both deprotonation and protonation, therefore having both acid and base behaviors, is termed *amphoteric*.

The equilibrium constant expression for Equation 5.1 is written

$$\frac{\{>\mathrm{M{-}O^-}\}a_{\mathrm{H+}}}{\{>\mathrm{M{-}OH^0}\}} = \mathrm{K_1 e^{(-\Delta ZF\psi/RT)}}, \tag{5.3}$$

and that for Equation 5.2

$$\frac{\{>\mathrm{M{-}OH^0}\}a_{\mathrm{H+}}}{\{>\mathrm{M{-}OH_2^+}\}} = \mathrm{K_2 e^{(-\Delta ZF\psi/RT)}}, \tag{5.4}$$

where

K_1, K_2 = equilibrium constants
$\{\}$ = activity of surface species
F = Faraday constant
ψ = surface potential
R = ideal gas constant
T = absolute temperature
ΔZ = change in charge for the species at the surface.

The K values for different sites can be determined from potentiometric titrations of suspensions containing the mineral of interest, with data analysis by computer programs such as that of Turner and Fein (2006).

Surface Microtopography

Given entropy considerations alone, an extended surface would be highly unlikely to be atomically flat. Even a mineral like muscovite, which has excellent cleavage, always has some imperfections on cleavage surfaces. Surfaces have a variety of microtopographic features, such as *flat terraces*, *steps*, *kinks*, *terrace vacancies*, and *adatoms* (Figure 5.4). Atoms associated with the different sites have different coordination environments; hence, they have different energetics and potentially different reactivities with respect to processes such as adsorption, mineral growth and dissolution, and catalysis. In many instances, microtopographic features such as steps and kinks control surface reactivity rather than terraces because of their different surface electronic structure and chemical behavior. From a stereochemical standpoint, the geometries of the various sites could affect reactivity if, for example, certain molecules would not react with a given configuration for reasons of shape or fit. Together with composition and atomic-scale structure, the microtopographic

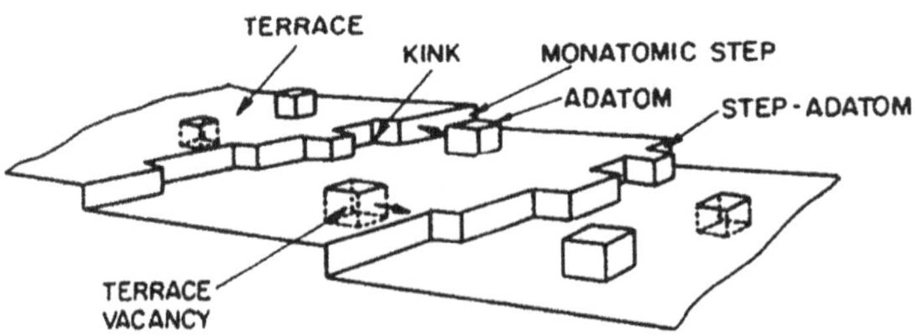

FIGURE 5.4 Some atomic- to nanometer-scale structures that would be present on a heterogeneous surface. From Hochella, 1990 (with permission).

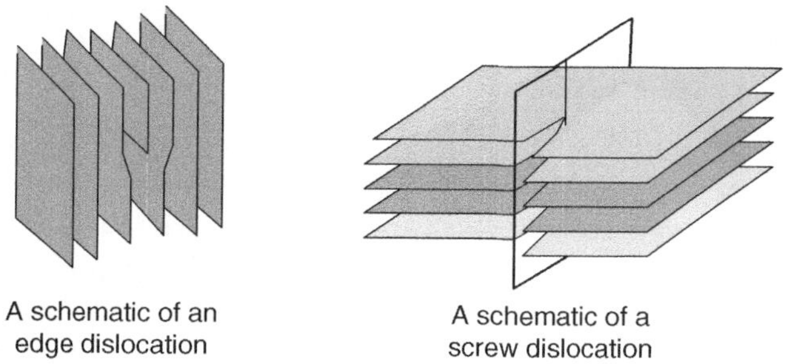

FIGURE 5.5 Schematic illustrations of an edge dislocation and a screw dislocation in a crystal.

features on surfaces often control reactivity with respect to processes such as mineral growth and dissolution.

Dislocations are types of crystallographic defects that occur within a crystal structure. When they intersect a surface, the features may greatly affect surface reactivity. In an *edge dislocation*, an extra partial plane of atoms is introduced midway through the crystal, distorting nearby planes of atoms (Figure 5.5). The distortion results in stresses that are described by *Burger's vectors*, which give the magnitude and direction of the distortion. *Screw dislocations* form when one portion of a crystal is out of alignment with another in a way that can be visualized by imaging slicing a stack of paper perpendicular to the stack halfway through and then lifting one side relative to the other (Figure 5.5). The growth of a new surface from a screw dislocation will result in a surface that looks like a spiral ramp.

How does one define such a complex surface, both conceptually and quantitatively? The surface is clearly not a flat plane, but it is a complex 3-D topographic configuration (Figure 5.4). Steps on the basal-plane surface of hematite particles are shown in Figure 5.6.

Various procedures to prepare surfaces for experiments can induce additional micro-topography. For example, the surface of a feldspar fragment that was polished to make it appear flat on a macroscopic scale or by optical microscopy is shown in Figure 5.7. Features include striations, low patches, and fragments likely left over by the polishing. It is also possible that the atomic-scale structure of the surface layers was altered, but such structural changes could not be determined at this resolution.

Surface Free Energy

Surfaces have free energy associated with them, which is known as *surface free energy,* or for most applications (such as water-rock interactions) as *interfacial free energy*. This surface free energy has a profound impact on the structure and reactivity of a surface. The surface free energy of minerals is linked to surface microtopography; one such link has to do with how crystals grow.

When minerals grow in open void spaces, the system works to minimize the surface free energy of the individual grains, Which generally leads to crystals with faceted surfaces. Pierre Curie (1885) showed that when the surface free energy of a crystal was minimized, or at an equilibrium situation, the perpendicular distance from a given crystal face to the center of the crystal divided by the surface free energy of that face would be constant for all faces of the crystal. This concept is known as the *Wulff rule* or the *Gibbs-Curie-Wulff theorem* (Markov, 1995). Gibbs (1906a,b) showed that the equililbrium form of a mineral is such that the total surface area times the surface free energy is minimized (Sunagawa, 2005).

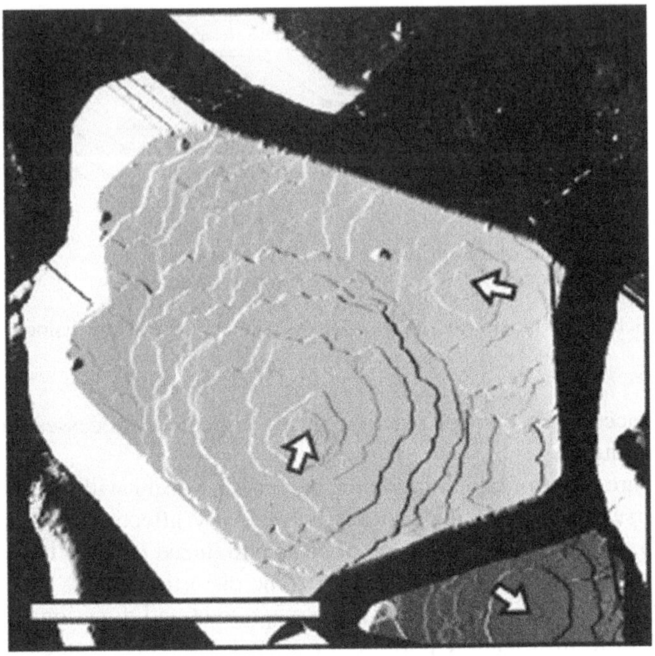

FIGURE 5.6 AFM image in air showing steps along growth features on hematite particle surfaces. The scale bar = 10 μm. From Rosso et al., 2003 (with permission).

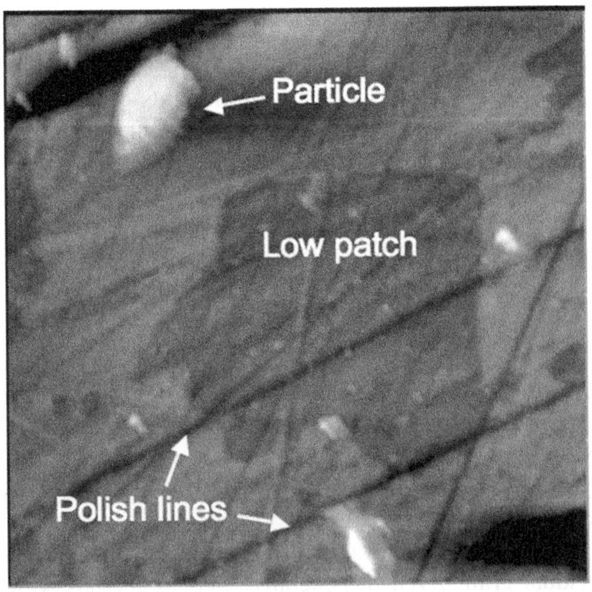

FIGURE 5.7 AFM image of polished feldspar showing a low patch caused by to uneven polishing, polish striations, and some surface litter (fine particles). Image is 4 μm on a side. In this image, lightness/darkness represents topographic high or low, respectively. Revised from Nugent et al., 1998a.

For symmetrically nonequivalent crystal faces, differences in microscale structures and microtopographies are reflected in the different surface free energies. From a microtopographic standpoint, surfaces are classified as F for "flat," K for "kinked," or S for "stepped." *F-surfaces* grow by mechanisms of layer spreading, whereas *S-* and *K-surfaces* form by piling up of steps generated on two (leading to a stepped surface) or three (leading to a kinked surface) F-faces (Jamtviet and Meakin, 1999). Mineral nucleation and growth as related to surface free energy is discussed in greater detail in Chapter 9.

Water Near Surfaces

The presence of a mineral surface may perturb the properties of the solution at and near the interface. The presence of a solid surface can alter the *dielectric constant* of water near the surface (Brown et al., 1995). Many studies have shown that the structure of water in the vicinity of a solid surface, including in the interlayers of clays, can be different from that of water in the bulk liquid environment (e.g., Fripiat et al., 1982; Sposito and Prost, 1982; Michot et al. 2002). Michot et al. (2002) pointed out that the structure of bulk water is "due to a delicate balance between the ordering effect of hydrogen bonds and the disordering effect of thermal motion." The presence of the surface perturbs the hydrogen-bond network of bulk water.

The structure of water at and near the (100) surface of MgO has been studied with a variety of approaches. Marmier et al. (1998) used molecular dynamics (MD; see below) modeling tied to spectroscopic measurements to investigate the structure of a single

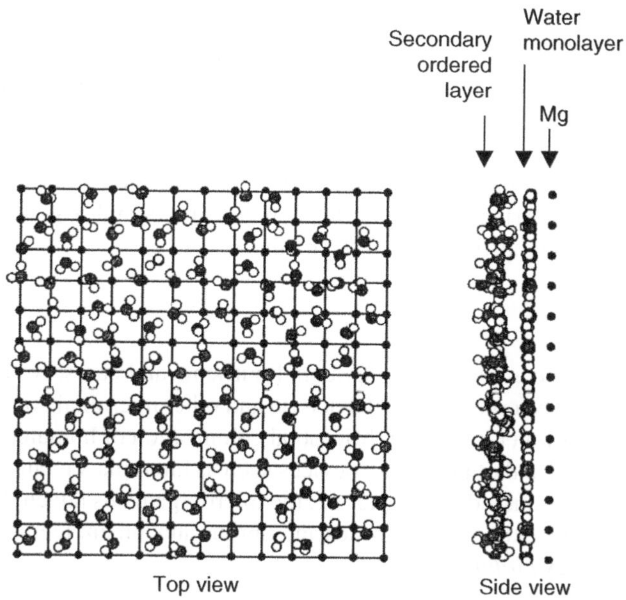

FIGURE 5.8 Molecular dynamics simulation of water adsorption on and near the (100) face of MgO at 150 K. Top view (left) and side view (right). As can be seen from the side view, a planar-oriented monolayer of water is associated with the Mg^{2+} sites (dark solid circles) and a secondary structured region of several water layers is also present. Modified from Marmier et al., 1998 (with permission).

monolayer of water on MgO (Figure 5.8). They found that the water *monolayer* was strongly structured by the Mg^{2+} cations at the MgO surface, with one H_2O molecule lying flat above each surface Mg site and with weak hydrogen bonds formed between molecules. As additional water was added to the model, the flat monolayer remained, but a secondary ordered region developed that extended 10–15 Å from a solid surface, with an apparently ice-like structure at 150 K (Marmier et al., 1998). As temperature was increased to 300 K, the thickness of this secondary structured region decreased and became more of a liquid-like structure.

For the surfaces of environmental interest: kaolinite, montmorillonite, quartz, gibbsite, and calcite, the structure of at most two to four layers of water are influenced by the presence of the surface, up to 10–15 Å (Michot et al., 2002). This short-range effect on the water structure has implications for (1) the shear plane that occurs during electrophoretic mobility measurements and (2) the measurement of adsorption enthalpies, as such measurements are dominated by displacement of water by an adsorbing ion or molecule (an *adsorbate*) (see Chapter 6). Chapter 6 focuses on the fact that mineral surfaces often are charged and that this charge brings about changes to the distributions of ions in solution near surfaces.

Dynamic Surfaces

Surfaces are dynamic. A surface exposed to air immediately becomes contaminated with hydrocarbons and other carbon-containing compounds to form a layer of *adventitious carbon*. Surfaces in air often become covered by at least a monolayer and often many layers of water, depending on the relative humidity (e.g., Brown et al., 1999). Minerals such as galena or magnetite could oxidize more or less quickly when exposed to air. Photoreactive surfaces could undergo redox reactions when exposed to light. In solution, a surface could undergo an array of reactions, such as dissolution and/or (re)precipitation. For example, atoms or molecules might dissolve from one region, diffuse across the surface, and reattach or precipitate in another location. After immersion in an aqueous solution, the mineral surface may become depleted in certain constituents or develop coatings of amorphous or crystalline materials (Figure 5.9). Such processes can make it difficult to define surface and the interface. Thus, a 3-D geometrically complex surface would be not only spatially variable but also temporally dynamic.

Bacterial Substrates

Because (mineral) surfaces can contain essential nutrients and can serve as physical substrates, surfaces tend to be colonized with bacteria and other microbial components (Figure 5.10), especially when water is present. Microbial colonization of human teeth, which promotes cavity formation, and of prosthetic devices such as hip replacements are examples of this phenomenon. As discussed in Chapter 12, the microorganisms attach through a variety of mechanisms, exuding exopolysaccharides and fibrous attachment features (Figure 5.11), and often form complex *biofilms*.

Fractal Properties of Surfaces and Environmental Particles

Because surfaces and interfaces are often heterogeneous and highly irregular, they often shown *fractal* properties. A natural *fractal* (Mandelbrot, 1982) typically (1) shows a fine structure that occurs at arbitrarily small scales, (2) is too irregular to be described easily by

FIGURE 5.9 Kaolinite grains on the surface of a weathered feldspar sample from the Navajo Sandstone aquifer (Arizona). This scanning electron microscope (SEM) image is about 25 μm across. The origin of the kaolinite could be from feldspar weathering or other processes in the study system. Modified from Zhu et al., 2006 (with permission).

traditional Euclidean geometry, and (3) has a *statistically* self-similar structure or behavior at different scales. Natural fractal systems can be said to display "quasi" or "statistically" self-similar structures or scale-invariant properties in a statistical sense (e.g., Senesi and Boddy, 2002). An object such as a mineral surface, an aggregate of colloids, or a biofilm can be defined as having a fractal nature when it displays self-similar behavior in a given property at a length scale that covers at least one order of magnitude of the property that is being considered or measured. Fractals obey a power law for a given property (p):

$$p \propto \nu^{\alpha}, \tag{5.5}$$

where ν is the variable measured and the exponent α can be related to the fractal dimension D (e.g., Senesi and Boddy, 2002).

Solid fractal objects in the environment may be classified as *mass fractals*, *surface fractals*, or *pore fractals*. For a *mass fractal*, both the actual object and its surface or interface are fractal; for a *surface fractal*, only the surface or interface of the object has a fractal nature; and for a *pore fractal*, the interface and the pore structure are together fractal (Pfeifer, 1987). These three classes of fractal objects are discussed in detail by Okuda and Senesi (1998).

As an example of an application to natural surfaces and interfaces, Sposito (2004) showed that if an aggregate of colloidal particles is fractal, then the fraction p(r) of colloids making up the aggregate that occur within a distance r from the colloid placed at the center of mass can be described by

$$\mathrm{p(r)} \sim \mathrm{r^D}, \tag{5.6}$$

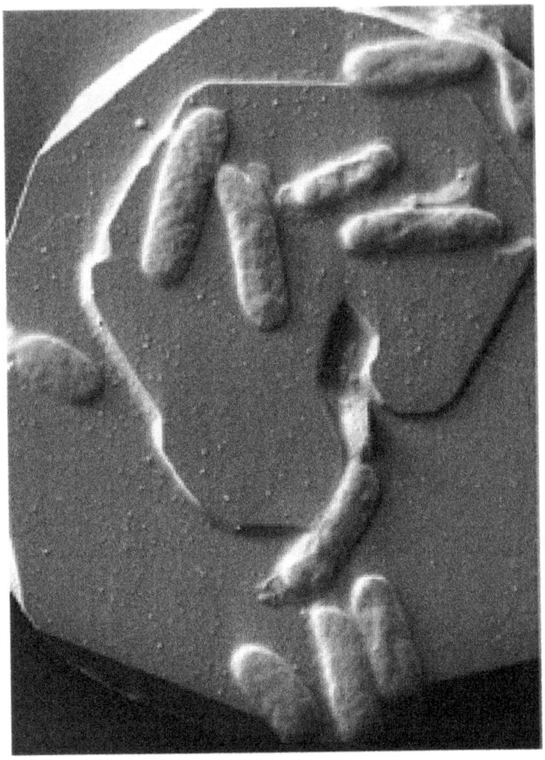

FIGURE 5.10 An SEM image of *Shewanella putrefaciens* CN32 cells on the surface of a tabular hematite particle from Alice Dohnalkova, Pacific Northwest National Laboratory (downloaded from http://picturethis.pnl.gov/picturet.nsf/by+id/AMER-5TX3DA). The *S. putrefaciens* bacterial cells gain energy from the oxidation of lactate coupled to reduction of Fe(III) from the hematite.

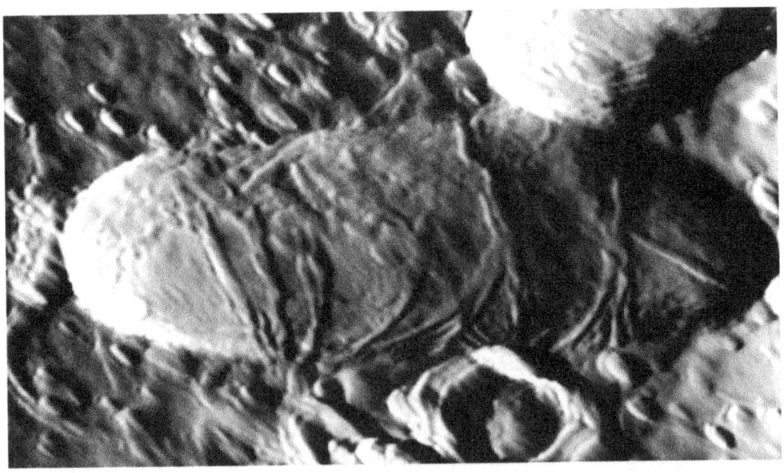

FIGURE 5.11 An atomic force microscope (AFM) image of *Pseudomonas mendocina* bacterium attached to a mineral surface with fibrous attachment features. The oblong bacterium in the center of this image is about 1.5 μm long.

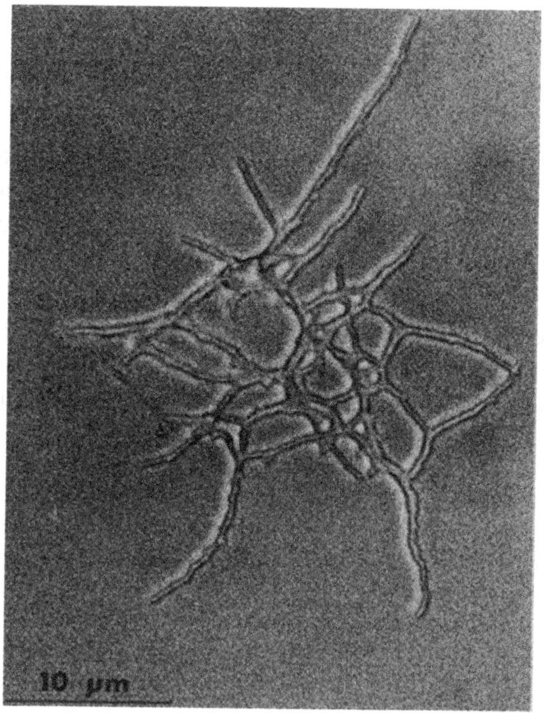

FIGURE 5.12 A photograph of a colony of *S. griseus* 16 hours after inoculation, forming a mass fractal. From Obert et al., 1990 (with permission).

where $r >>> r_0$ and $1 < D < 3$. Here, D is the *mass* or *cluster fractal dimension* of the aggregate and is a measure of how completely the colloids in the aggregate fill the 3-D space. A close-packed arrangement of colloids would have D close to 3, whereas a very open arrangement with significant pore space would have D closer to 1 (Sposito, 2004).

Fractal geometry has been applied to investigations of the macromolecular morphology, surface characteristics, and aggregation processes of natural organic molecules known as humic substances (Senesi and Boddy, 2002). Fractal analysis also has been applied to the growth patterns of microbial colonies. For example, Figure 5.12 shows a mass fractal formed by *Streptomyces griseus* (Obert et al., 1990). In Chapter 10, we shall learn that fractal geometry is being applied to feldspar surface area and reactivity (Brantley et al., 1999).

Interdisciplinary Topic of Study

The study of environmental surfaces and interfaces involves approaches from many fields, such as mineralogy, surface chemistry, solution chemistry, microbiology, kinetics, and thermodynamics. Some important properties of surfaces and interfaces include the following:

1. Mineral surface structure and composition are often distinct from the bulk of the mineral.
2. Mineral surfaces are heterogeneous in both structure and composition, and they often (but not always) show fractal properties.

3. Inhomogeneities in structure and composition (impurities) often determine reactivity.
4. Mineral surface structure generally is not a simple termination of the bulk (Hochella, 1990).
5. Surfaces and interfaces have free energy associated with them, and this free energy plays a major and often dominant role in controlling structure and reactivity.
6. Proximity of a mineral–water interface can affect both the underlying mineral and the nearby solution.
7. Mineral–water interfaces are dynamic environments.
8. It can be difficult or impossible to predict mineral–water interface phenomena based simply on properties of the bulk mineral and the bulk solution.
9. Mineral surfaces are often covered with organic matter and bacteria.

Finally, as we shall see throughout this text,

10. Investigating and understanding environmental surface and interface interactions requires combining approaches from fields such as solution chemistry, (micro)biology, mineralogy, thermodynamics, kinetics, and surface science to form a unique science of environmental surfaces and interfaces.

SURFACE FREE ENERGY AND SURFACE EXCESS

The discussion above introduced the concepts that mineral surfaces are complex and that it can be challenging to define a surface or an interfacial region at the atomic scale. One useful means of dealing with the finite thicknesses of surfaces was developed by Gibbs and described by Parks, 1990. For thermodynamic purposes,

> Gibbs defined a surface as a mathematical plane or dividing surface of zero thickness in the region of the actual physical surface, and a given surface property as the net positive or negative excess in the vicinity of the surface, relative to the same property in the bulk.

Surface excess describes the positive or negative accumulation of certain chemical constituents such as cations, anions, or organic molecules:

$$\Gamma \equiv \int_{A_1}^{A_2} (c_x - c_b)dx, \tag{5.7}$$

where c_x is the measured concentration of some species at any point x, in the concentration profile, and c_b is the bulk concentration of that species, far from the interface, where it is not influenced by the presence of the surface. A_1 and A_2 are points at any given (arbitrary) positions on either side of the interface (Parks, 1990). For the example of Sc associated with grain boundaries shown in Figure 5.1, once the y-axis were converted to Sc concentration, integrating the area under the plot would give a total surface excess concentration of Sc.

Sposito (2004) defines relative surface excess $n_i^{(j)}$ for components in solution at the mineral–water interface as

$$n_i^{(j)} = n_i - (n_j x_i / x_j), \tag{5.8}$$

where $n_i^{(j)}$ is the excess moles of substance i in the reacted mixture, relative to the content of a reference substance j in the mixture and to the composition of the aqueous solution far from the interface. The values x_i and x_j are the mole fractions of substance i or j in the solution (the supernatant once the particles and solution have been separated). The relative surface excess can be positive (indicating adsorption), zero (no surface excess), or negative (desorption), depending on the relative behaviors of substances i and j. For most mineral surfaces in aqueous solution, most applications take j to be liquid water.

The concept of a surface as an imaginary plane of zero thickness is of particular usefulness for many modeling exercises on pollutant fate and transport in porous media. However, it also can introduce limitations or error into both conceptual and quantitative models.

If we were to take a glass of water and spill some on a *hydrophobic* or "water fearing" table, we would find that the water does not spread over the surface but tends to ball up. The balls of water would coalesce into larger flattened spheres. The water is taking on the shape that minimizes the surface area of the water in direct contact with the hydrophobic surface. This observation, and many more, suggests that surfaces have some energy associated with them, and that this *surface free energy* may be a driving force for certain physical or chemical transformations that act to minimize the surface free energy.

When a change to a system involves formation of new surface or interface, this component of the free energy must be taken into consideration. As discussed above, an interface will only be stable if the free energy of formation of the interface is positive. The *surface or interfacial free energy* γ is defined as (Adamson, 1982):

$$\gamma \equiv (\delta G/\delta A)_{P,T}(\text{pressure, temp, remain constant}), \tag{5.9}$$

where

G = Gibbs free energy
A = surface area.

For a system with a surface, the total free energy of the system comprises the bulk free energy (G_b) or the sum of the molar free energies of formation of its constituents, i, $\sum_i \mu_i n_i$, (recall $\mu_i = \delta G/\delta n_i)_{T,P}$ (chemical potential) plus the total excess free energy of formation of all surfaces and interfaces j (G_s). The surface, or more properly the interfacial free energy, is $\sum_j \gamma_j A_j$.

Thus, total free energy is equal to the bulk free energy plus the interfacial free energy

$$G_{tot} = \sum_i \mu_i n_i + \sum_j \gamma_j A_j \tag{5.10}$$

$$= \gamma A + \sum_i \mu_i n_i. \tag{5.11}$$

For surfaces in water, the interfacial free energy or interfacial tension has two primary components, because of van der Waals forces (including London interactions; $\gamma_{H_2O,L}$) and hydrogen bonding ($\gamma_{H_2O,H}$)

$$\gamma = \gamma_{H_2O,L} + \gamma_{H_2O,H}, \tag{5.12}$$

with about one third of the interfacial tension caused by van der Waals attraction and the rest caused by hydrogen bonding (Stumm and Morgan, 1996). The concepts of surface and interfacial free energy are key to the discussion of surface phenomena, such as mineral growth and dissolution and adsorption, which are discussed in subsequent chapters.

SURFACE TENSION AND RELATED PHENOMENA

For liquids and sometimes other substances, it is customary to apply the term *surface tension* for interfacial free energy. The surfaces of liquids have an elastic force associated with them, which is known as *surface tension*, that is caused by the attraction of molecules below the surface of the liquid for those at the surface. This molecular attraction tends to restrict the tendency of a liquid to form a large interface with another substance. The surface tension determines the tendency for surfaces to interact with one another. Although the term *interfacial tension* is more appropriate for most environmental applications (e.g., liquid–air, solid–air, liquid–liquid, and nonvacuum), surface tension is commonly used as a general term that encompasses interfaces.

Many surface and interface properties can be related to surface tension. Separation of oil and water is an excellent example of interfacial tension. Oil and water are immiscible, with water being polar in nature and oil nonpolar. If we pour a small amount of oil onto water, the oil will immediately begin to ball up into flattened spherical shapes, and the shapes will eventually coalesce. Together, the change in shape and the coalescence act to minimize the area of contact between the oil and the water and, hence, to minimize the interfacial free energy.

Surfactants and Micelles

A *surfactant* (from "surface active" molecule), which is a type of molecule that has a polar (*hydrophilic* or "water loving", shown as round structures in Figure 5.13) head and a nonpolar (*hydrophobic* or "water fearing") tail, can decrease the surface tension, which allows fine droplets of the oil to become disseminated in the water. The surfactants will line up at the surface of water so that the polar ends face the water and the nonpolar tails face toward the air. *Micelles* form when surfactant molecules in water group together into a generally spherical aggregate with the polar ends facing outward toward the water and the nonpolar tails facing each other (Figure 5.13). The shape of the micelle is determined by geometric factors, including the area of the polar head group (α_0), the volume of the nonpolar (alkyl) chain (V), and the maximum length to which the alkyl chain can extend (l_c). The following rules are obeyed for micelle formation:

$$\text{spherical micelles } V/(a_0 l_c) < 1/3, \quad (5.13)$$

$$\text{nonspherical micelles } 1/3 < V/(a_0 l_c) < 1/2, \quad (5.14)$$

$$\text{vesicles or bilayers } \tfrac{1}{2} < V/(a_0 l_c) < 1, \quad (5.15)$$

inverted or reverse micelles (which form in a nonpolar organic solvent)

$$1 < V/(a_0 l_c) \quad (5.16)$$

(Burda et al., 2005).

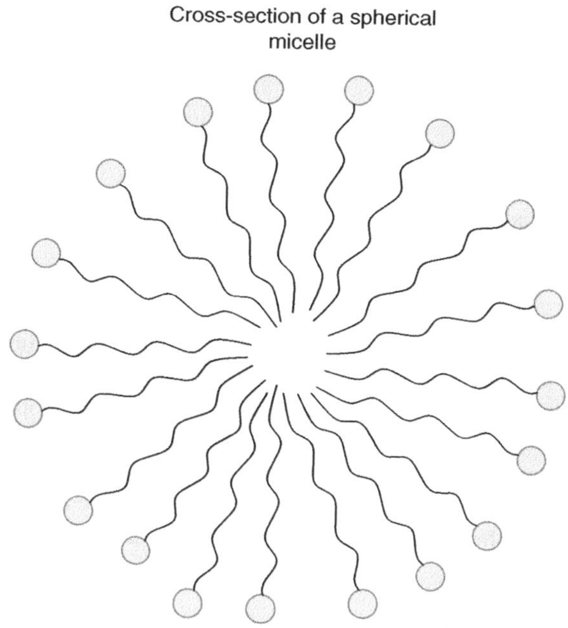

FIGURE 5.13 A cross section of a spherical micelle with hydrophobic tails pointing inward and hydrophilic heads pointing outward toward an aqueous solution.

In all these cases, the micelle is acting to mimimize surface free energy. Micelles, and particularly reverse micelles, are often used to grow nanoparticles because of the small controlled environment within the micelle.

Contact Angle

When a solid surface comes together with an interface between a gas and a liquid, the surface tension defines the angle of the gas/liquid interface relative to the solid surface (Figure 5.14). This *contact angle* is defined by the properties of the solid, the gas, and the liquid. The angle is generally measured using a goniometer.

The contact angle θ is related to interfacial tension according to the *Young equation* (Adamson, 1982):

$$\cos\theta = \frac{\gamma_{SV} - \gamma_{SL}}{\gamma_{LV}} \tag{5.17}$$

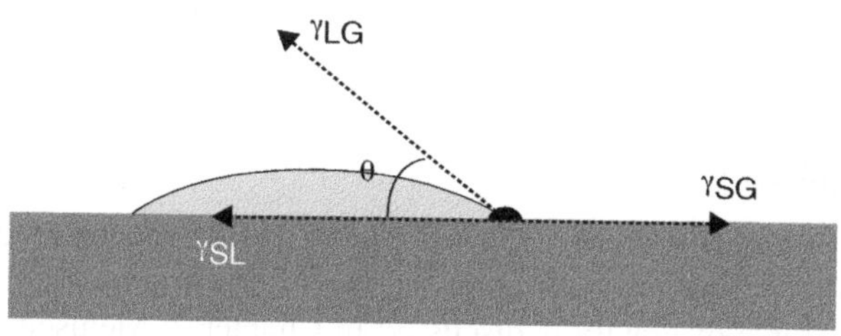

FIGURE 5.14 Schematic illustration of contact angle related to surface (or interfacial) tension.

where

γ = interfacial tension
SV = solid–vapor
SL = solid–liquid
LV = liquid–vapor.

In the case of a finite contact angle and a spreading coefficient of $S_{L/S}$,

$$S_{L/S} = \gamma_{SV} - \gamma_{LV} - \gamma_{SL}, \tag{5.18}$$

which means that the interfacial tension between the solid and the liquid as well as the liquid and the vapor needs to be small for spreading of the liquid over the solid surface to occur (Adamson, 1982). One way of achieving this result for an oil droplet on a hydrophilic surface is by adding a surfactant. The contact angle between water and a clean and fully hydroxylated alumina surface approaches zero because the water spreads out readily on the hydrophilic surface. In contrast, the contact angle of water on the basal surface of graphite is >90°, as this surface is hydrophobic (Healy and Fuerstenau, 2007).

The Young-Laplace Equation

The *Young-Laplace equation* provides more insight into commonly observed phenomena such as droplet shape, soap bubbles, and meniscus formation. If the pressure on one side of a surface is different from the pressure on the other side, then the surface tension forces must balance the forces caused by pressure. To accomplish this, the surface takes on a curved shape. The Young-Laplace equation helps to show the relationship among pressure difference (ΔP), surface tension (γ), and radii of curvature (R_x, R_y) of the surface

$$\Delta P = \gamma(1/R_x + 1/R_y), \tag{5.19}$$

where

ΔP = pressure difference on either side of the surface
γ = surface (or interfacial) tension
R_x and R_y = radii of curvature of the axes that are parallel to the surface.

For a sphere, $R_x = R_y$, and the Young-Laplace Equation reduces to

$$\Delta P = 2\gamma/R. \tag{5.20}$$

For a large planar surface, $R_x = R_y = \infty$, and $\Delta P = 0$, which means that the pressure does not change across the planar surface.

Meniscus and Capillarity

The concepts of a *meniscus* (Figure 5.15) and *capillarity* are encountered several times in this text. The capillary fringe that forms at the top of the water table as water is pulled upward against the force of gravity is discussed in Chapter 2. Meniscus forces that form between water adsorbed to the surface of an AFM tip and a sample surface are discussed

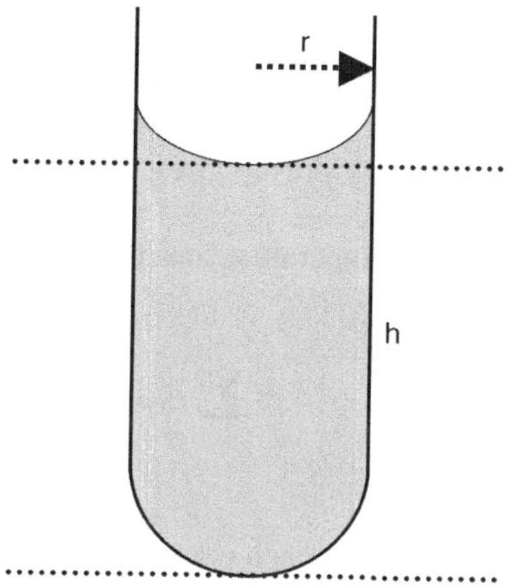

FIGURE 5.15 A concave meniscus formed in a capillary tube.

in Chapter 4. A meniscus occurs because of interactions between a surface and a fluid, when air is present. If the liquid is compatible with the surface such that the liquid adheres to the surface, then the liquid will act to maximize the contact area with the surface, thus decreasing the overall surface free energy. In a narrow tube, the surface tension can act to draw a liquid up along the edges of the tube and against the pull of gravity. This process is known as *capillary action*.

The Young-Laplace equation applies to formation of a meniscus. In a cylindrical tube, the meniscus will take on a hemispherical shape, with $R_x = R_y = r$ (the radius of the cylinder). The height to which the liquid will rise in the column is given by

$$h = \frac{2\gamma_{la}\cos\theta}{\Delta\rho gr}, \qquad (5.21)$$

where

h = height to which the liquid is lifted (the bottom of a concave meniscus) (in m)
γ_{la} = liquid–air surface or interface tension (in mN m^{-1} or mJ m^{-2})
$\Delta\rho$ = difference in density between the liquid and the gaseous phase
r = radius of the capillary tube (m)
g = acceleration because of due to gravity (9.807 m s^{-2})
θ = contact angle between the liquid and the sides of the tube.

In this equation, $\Delta\rho gr = \Delta P$. The top of the water column within the tube will form a curved, concave meniscus, where the center is lower than the edges in contact with the tube surface (Figure 5.15).

From geometrical considerations, the weight of the liquid column in a cylindrical tube is proportional to the square of the radius of a tube. But the contact length between the liquid and the tube is proportional to the radius of the tube. Thus, a narrow tube will draw up a column of liquid to a greater height than will a wide tube made of the same material.

The Gibbs Equation

As defined by Gibbs, surfaces and interfaces have free energy, heat content, entropy, and composition, as do bulk phases. The surface excess properties are denoted with the subscript s in this context. The discussion below presupposes the Gibbs definition of a surface as explained above.

For a surface alone, at constant temperature and pressure, the change in surface free energy during adsorption is given by

$$dG_s = \gamma dA + \sum_i \mu_{s,i} dn_{s,i} \tag{5.22}$$

(Parks, 1990).

At equilibrium, the chemical potential of each component i is the same in every phase by definition of equilibrium, so that a single μ_i represents i throughout the entire system.

The Gibbs adsorption equation for a system at constant temperature is as follows:

$$d\gamma = -\sum_i (n_{s,i}/A) d\mu_i = -\sum_i \Gamma_i d\mu_i \tag{5.23}$$

(Gibbs, 1931; Chattoraj and Birdi, 1984; Parks, 1990), where the change in surface molar concentration ($n_{s,i}$) or, more properly, the surface excess (Γ_i) of some component i during adsorption is related to the change in surface or interfacial tension. For a solid surface in aqueous solution, a decrease in surface free energy occurs at the surface whenever anything adsorbs at the surface. Another way of understanding this concept is that a species will only adsorb to a surface if this adsorption causes a decrease in surface free energy. In Chapter 9, we shall learn see that sometimes components can attach to a surface, for example, during mineral growth, and increase the surface microtopography, hence, surface free energy. But, this is offset by the decrease in overall system free energy afforded by a decrease in supersaturation.

Another form of the Gibbs adsorption equation at constant temperature is

$$\Gamma_i = -1/\mathrm{RT}\ (d\gamma/d\ln a_i)_{\mathrm{T,P}}. \tag{5.24}$$

A plot of surface tension versus log activity of adsorbate can be used to determine surface excess (Figure 5.16) (Stumm and Morgan, 1996)

SOME APPROACHES TO SURFACE AND INTERFACE MODELING

Numerous approaches are available for modeling surfaces and interfaces in vacuum, air, and/or aqueous solution. *Surface complexation modeling*, which is primarily used to study adsorption of ions from aqueous solutions, is discussed at length in Chapter 6. The *Monte Carlo simulation* approach, which can be applied to processes such as adsorption, absorption, and mineral growth and dissolution, is discussed in Chapter 9.

One of the most sophisticated approaches for molecular-scale modeling, which includes surface and interface applications, is molecular dynamics modeling (MD). In MD modeling (e.g., Cygan, 2001; Kirkpatrick et al., 2005), the forces on individual atoms are evaluated using an energy forcefield. Initial positions and velocities of atoms and/or molecules are

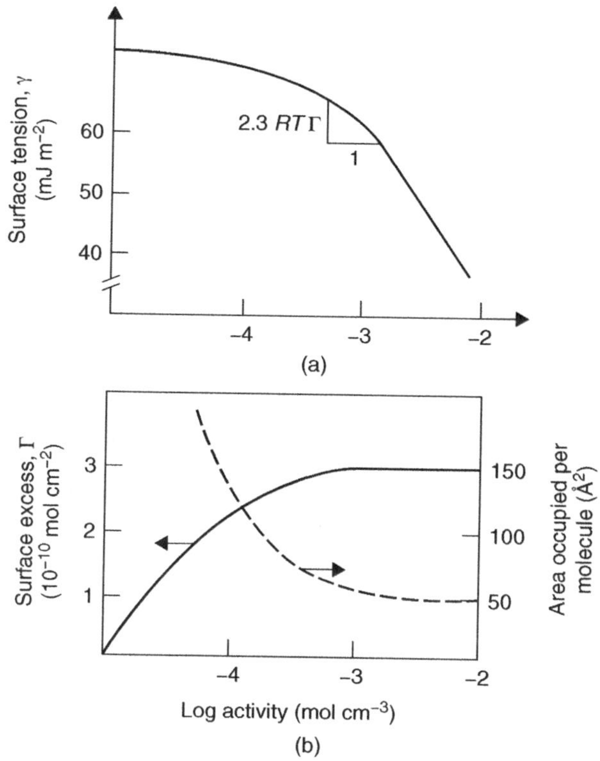

FIGURE 5.16 Using the Gibbs adsorption equation, a plot of surface tension γ versus log activity (or concentration) of adsorbate can provide the surface excess Γ. The area that is occupied by an adsorbed molecule or ion can also be calculated. Modified from Stumm and Morgan, 1996 (with permission).

assigned by the modeler. Following a short time step, generally on the order of a femtosecond ($= 10^{-15}$ s), new atomic positions are calculated based on Newton's classic equation of motion. By modeling over a large number of time steps, a time-dependent trajectory can be calculated of all atomic motions (Cygan, 2001), and a computerized movie can often be constructed.

One of the principal applications of MD modeling in environmental surface and interface chemistry has been the structure and dynamical properties of water at and near mineral surfaces (Figure 5.17). The MD approach is particularly helpful for H-bonding networks (Kirkpatrick et al., 2005). A "typical" MD modeling cell for layered mineral–fluid interfaces, in which the model mineral structures are cleaved, and the intervening layers filled wholly or in part by water with or without dissolved ions, is shown in Figure 5.17.

MD modeling for surfaces and interfaces requires significant computer run times. Given the complexity of environmental surface and interface interactions, particularly in water, and the sophistication of the MD modeling techniques, only a relatively small number of environmental geochemists and mineralogists currently use the technique to its full (and growing) capacity. MD modeling has been applied to studies of, for example, metal adsorption to kaolinite, focusing on interactions with different types of sites on the kaolinite surface (e.g., Vasconcelos et al, 2007; see Chapter 7).

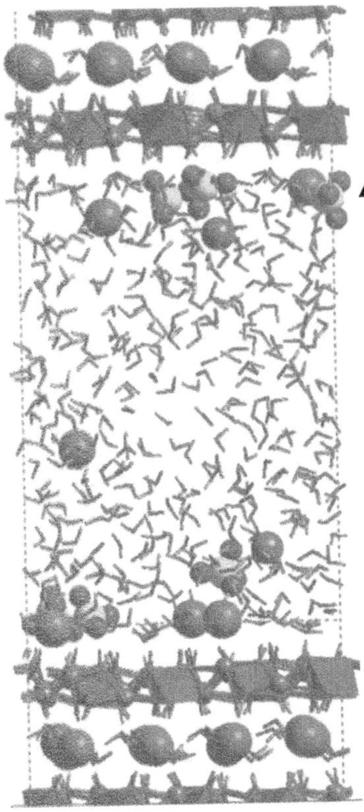

FIGURE 5.17 A "typical" MD modeling cell for environmental surface and interface interactions. The crystal structures are hydrocalumite, which is a layered double hydroxide mineral. The region in between has water molecules along with Cl^- SO_4^{2-} and Na^+ ions. The water layer is ~30 Å thick in this model. Modified from Kirkpatrick et al., 2005 (with permission).

Case Study: Bacteria–Mineral–Gas Interactions in the Vadose Zone

The concepts and equations related to surface energy are used on a daily basis by environmental scientists and engineers. An example is work by Wan et al. (1994) on the effects of air bubbles in porous media on the transport of microorganisms in the *vadose zone*.

First, the relative surface *hydrophobicity* of different bacterial strains was estimated by contact angle measurements (Table 5.1) at the junction of a drop of aqueous solution, a bacterial surface, and air, using a goniometer. Second, the surface charge state of each

TABLE 5.1 Bacteria and surface measurements from Wan et al.'s (1994) experiments on bacteria.

Bacterial Strain	Size (μm)	Contact angle (°)	Zetal potential (mV)
P. cepacia 3N3A	1.3 × 0.3	24.7 ± 3.1	−12.1
Arthrobacter sp. strain S-139	1.0 × 0.8	77.1 ± 2.5	−56.3
Arthrobacter sp. strain ZAL001	1.1 × 0.8	48.2 ± 2.4	−46.2

Modified from Wan et al., 1994.

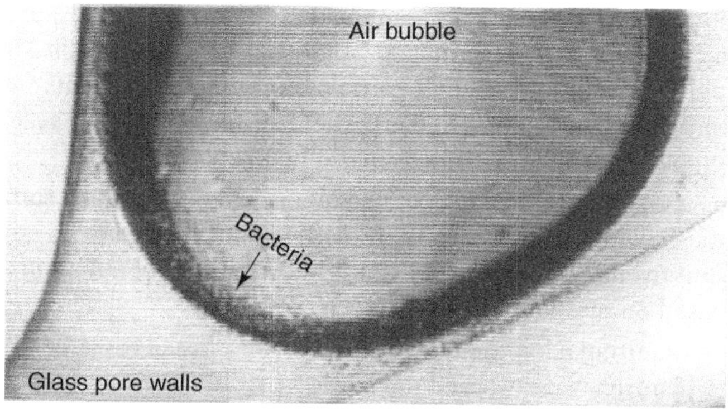

FIGURE 5.18 This figure shows how bacteria preferentially attach along the perimeter of a bubble because of surface and interfacial forces. From Wan et al., 1994 (with permission).

bacterial strain in a given solution was investigated through electrophoretic mobility measurements, which monitor the movement of a charged particle in a solution with an applied potential gradient to yield zeta potential (see Chapter 4). The results in Table 5.1 show that the more hydrophobic bacteria were also less mobile (smaller absolute value of zeta potential), as would be expected.

Next, distributions of bacteria were monitored by optical microscopy in fluid cells consisting of glass plates with etched porous media with different ratios of air versus water in the spaces. Finally, columns of quartz sand at different saturations were used to measure bacterial mobility as a model for a subsurface environment, such as a soil or a sediment.

Results showed that the more hydrophilic bacteria tended to attach preferentially along the edges of air bubbles, rather than at the interface between the glass and the aqueous solution (Figure 5.18). Both hydrophobic and hydrophilic bacterial cells were observed along the edges of the air bubbles, as well as at the glass/solution interface. The more hydrophobic cells had greater affinity for both types of interface than did the more hydrophilic cells, because hydrophobic substances tend to separate from water. Sorption of bacteria onto the gas–water interface (edges of air bubbles) was so strong as to be "essentially irreversible." The bacteria also tended to move with the gas bubbles in column experiments. Wan et al.'s (1994) work demonstrated that the presence of air bubbles can be important to bacterial mobility in porous media because of *surface tension* effects.

QUESTIONS FOR FURTHER THOUGHT

Answer questions 1–7 in paragraph form. Use equations and/or figures to help in formulating your answers wherever possible.

1. Imagine that you are blowing bubbles with a bubble wand. Two bubbles meet. What would you expect their contact to look like if the bubbles were (a) the same size or (b) different sizes? Explain your reasoning based on surface/interface theory.
2. Imagine that you are an environmental engineer and that you have been asked to provide some insight from your knowledge of surface and interface science on cleaning up an oil spill in San Francisco Bay. What surface science concepts would be helpful to know for the clean-up?

3. You are preparing to run an experiment on dissolution of the feldspar mineral, albite. To run the experiment, you grind up a feldspar sample into silt-sized fragments. Why would you want to use small particles for your experiments? What might the grinding procedure have done to the particle surfaces and their reactivity? Write a few paragraphs on what the surfaces might be like.
4. Most of us are probably used to seeing a meniscus formed at the surface of water in a tube or flask, with the meniscus shape being concave upward. Is it ever possible to have a meniscus that is convex (or concave downward) for a fluid in a tube or flask? Explain based on reasoning from surface science.
5. A question requiring additional library and/or web-based research: The 2007 Nobel Prize for Chemistry was awarded to Gerhard Ertl for his work in surface chemistry. Write an ~ 4-page (double-spaced, typed) summary of his work, focusing on what new things he learned about surface chemistry and the techniques he used. Cite a minimum of three references (Internet or library based). Write this summary in a style that would be suitable for a publication that explains science to the educated public, such as a monthly popular science periodical. You may include figures or photos as long as they are properly referenced.
6. The formation of gas (or vapor) bubbles involves surface/interface chemistry. A study of how best to serve champagne suggested that servers should always wipe the champagne glass with a clean, dry cloth immediately before pouring the champagne. Provide at least two potential explanations for this phenomenon, based on concepts of surface and interface science. Explain your reasoning.
7. For the example of Sc concentrations across grain boundaries, how would one apply the Gibbs definitions of surface, interface, and surface excess? Make a sketch or photocopy the Sc example and explain how these terms apply in this case. Then, take the more complex case of bacteria attaching to the surface of an air bubble. How might one define the surface and interface in this case? There is no right or wrong answer; give your reasoning.
8. Calculate the capillary rise of water in a glass capillary tube at STP if the capillary tube width radius is (a) 0.1 versus (b) 1 mm. The water has a contact angle of $\theta_w = 0°$, $\rho_w = 1000$ kg m^{-3}, and $\gamma_w = 0.078$ J m^{-2} (c) If a plant leaf has fine capillaries of radius $= 10^{-8}$ m, and the cell walls are hydrophilic, would you expect the capillary rise to be important in the area just below the leaf surface (below where evaporation is occurring)? Explain. Note: ρ_{air} at STP ~ 1.2 kg m^{-3}.
9. Calculate the "capillary rise" of Hg in a glass capillary tube at STP if the capillary tube width radius is (a) 0.1 versus (b) 1 mm. The Hg has a contact angle of $\theta_{Hg} = 140°$, $\rho_{Hg} = 13.5$ g cm^{-3}, and $\gamma_{Hg} = 485$ mNm^{-1}. The capillary rise of Hg is distinctly different from that of water, as calculated in problem 8. (c) Explain why the capillary rise is different. (Note: In this problem, you will need to do some unit conversions, which is realistic given that data often are provided in different units in the literature).
10. Explain how observations of capillary rise might be used to measure the surface tension of a liquid. What might need to be considered in such measurements?
11. (a) At STP, what is the internal pressure inside a spherical water droplet of the following sizes (diameter): 2 nm, 10 nm, 100 nm, 1 μm, 0.1 mm, 1.0 mm, and 10 mm? (b) What does this tell you about the pressure inside a water droplet as it approaches the size of a molecule?

FURTHER READING

Adamson, A.W., and Gast, A.P. 1997. Physical Chemistry of Surfaces 6th ed. Wiley-Interscience, New York, 808 p.

Parks, G. 1990. Surface energy and adsorption at mineral/water interfaces: An introduction. In Hochella, M.F., Jr., and White, A.F. (eds), Mineral-water Interface Geochemistry, vol. 23. Mineralogical Society of America, Chantilly, VA.

Hochella, M.F., Jr. 1990. Atomic structure, microtopography, composition, and reactivity of mineral surfaces. In Hochella, M.F., Jr., and White, A.F. (eds) Mineral-Water Interface Geochemistry, vol. 23. Mineralogical Society of America, Chantilly, VA.

6

THE CHARGED INTERFACE AND SURFACE COMPLEXATION

Mineral surfaces in water have charge properties. Depending on the solution conditions, they may be positively charged, negatively charged, or neutral. Additionally, the surfaces may have acid/base properties caused by hydrolysis reactions at surface sites. This chapter deals with controls and consequences of surface charge properties. It discusses surface complexation reactions and modeling approaches that are used primarily to characterize adsorption on mineral surfaces but sometimes are applied to microbial surfaces.

SOME EVIDENCE FOR SURFACE CHARGE

How do we know that mineral surfaces often are charged?

First, if an oxide mineral such as quartz or hematite is put in aqueous solution, the pH of the solution changes quickly, unless the mineral happens to have equal protonation/deprotonation at the solution pH. This led researchers to suspect surface hydrolysis reactions. Protons are consumed or released by protonation/deprotonation of surface sites (see Equations 6.3 and 6.4, below). This protonation/deprotonation at the surface would lead to a component of surface charge.

Second, mineral surfaces exhibit electrokinetic phenomena. These include (1) *electrophoresis*, which is the motion of charged colloidal particles relative to stationary liquid in an applied constant electric field; (2) *electroosmosis*, which is the movement of liquid relative to a stationary charged surface in an applied electric field; (3) *sedimentation potential*, which refers to the electric field that is created when the charged particles move relative to the stationary liquid; and (4) *streaming potential*, which refers to the electric field that is created when liquid is caused to flow along a stationary charged surface.

An example of *electrophoretic mobility* data is shown in Figure 6.1, for attachment or specific adsorption of anionic (negatively charged) phosphate to what is otherwise a

Environmental Surfaces and Interfaces from the Nanoscale to the Global Scale, by Patricia A. Maurice

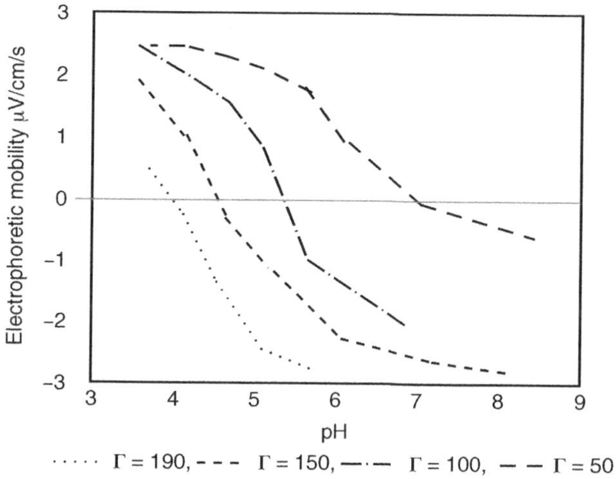

FIGURE 6.1 Electrophoretic mobility of goethite with different amounts of adsorbed phosphate. From Tejedor-Tejedor and Anderson, 1990 (with permission).

positively charged goethite mineral surface under the study conditions. With increasing surface coverages (Γ) by phosphate, the curves shift to the left and the pH at which the mobility goes to zero then changes direction shifts to the left.

Electrophoretic mobility, μ (in $m^2s^{-1}V^{-1}$), is defined by the equation

$$\mu = v/E, \tag{6.1}$$

where v is the steady-state velocity (in m s^{-1}) of a charged particle moving in response to an electric field of magnitude E (in Volts m^{-1}).

Third, ions of either one sign or another (cations or anions) often accumulate preferentially near surfaces in solution, which suggests that the surface has the opposite charge.

Fourth, contact angle measurements on surfaces often suggest charged or at least polar surfaces, although not all mineral surfaces are charged and/or polar.

Fifth, measurements of forces between two mineral surfaces immersed in aqueous solution often can only be explained by considering electrostatic forces.

SOURCES OF MINERAL SURFACE CHARGE

Detailed reviews of the sources of surface charge are provided by Sposito (1989, 2004). The overall charge density on a particle is σ_p. The unit of σ is moles of charge per kilogram solid ($mmol_c$ kg^{-1}) or Coulombs per meter squared of surface area (C m^{-2}). One of the most important components of σ_p is the *intrinsic surface charge density* σ_{in}, which is inherent to the particle itself. Two sources of σ_{in} are presented as follows:

1. *Structural surface charge density* σ_o. The permanent structural charge on a mineral comes about primarily by isomorphic substitution, for example, of Al(III) for Si (IV), and is characteristic of 2:1 clays such as montmorillonite. For clays, σ_o (in $mmol_c$ kg^{-1}) can be calculated using the equation

$$\sigma_o = -(x/M_r) \times 10^3, \tag{6.2}$$

where x is the layer charge and M_r is the relative molecular mass, which can be calculated from the mineral formula and the relative molecular mass of each element in the mineral formula (Sposito, 1989). The magnitude of σ_o can be estimated from chemical composition and X-ray diffraction (XRD) data or measured using the Cs surface excess method of Chorover and Sposito (1995).

2. *Proton surface charge density* σ_H. This is caused by proton association/dissociation reactions at the surface

$$>S{-}OH + H^+ <\text{----}>\ >S{-}OH_2^+ \tag{6.3}$$

$$>S{-}OH + OH^- <\text{----}>\ >S{-}O^- + H_2O. \tag{6.4}$$

The value of σ_H can be positive, negative, or zero depending on the solution conditions and the properties of the solid.

Note that σ_H refers to proton charge only at the surface, not in the diffuse layer. The diffuse layer is the zone of solution near the surface where ion distribution is affected by surface charge. The value of σ_H is measured by titration, taking into consideration any hydrolysis reactions related to potential mineral dissolution and release of components such as Fe or Al to solution.

The intrinsic surface charge density is the sum of permanent structural charge density plus proton surface charge density

$$\sigma_{in} = \sigma_o + \sigma_H. \tag{6.5}$$

Besides the intrinsic surface charge density, two other components of the density of surface charge on a particle are related to the ions associated with the surface from solution. Ions that bind directly to the surface form *inner-sphere complexes*. The *surface charge density* of *inner-sphere complexes* σ_{is} is equal to the net total surface charge of the ions, other than H^+ and OH^-, which has formed inner-sphere complexes with the surface functional groups. In addition to these inner-sphere complexes, ions exist that are more weakly or loosely associated with the surface, known as *outer-sphere complexes*. The *surface charge density* of *outer-sphere complexes* σ_{os} is equal to the net total surface charge of ions that have formed outer-sphere complexes with surface functional groups. A principal difference between inner-sphere and outer-sphere complexes is that inner-sphere complexes directly attach to surface sites, whereas outer-sphere complexes have at least one water molecule between them and the surface.

The net total particle charge σ_p can be expressed mathematically as

$$\sigma_p = \sigma_{in} + \sigma_{is} + \sigma_{os} \tag{6.6}$$

$$= \sigma_o + \sigma_H + \sigma_{is} + \sigma_{os}. \tag{6.7}$$

The charge on the particle surface σ_p may be positive, negative, or zero. To maintain overall charge neutrality, the ions in the solution (not in inner-sphere or outer-sphere complexes) must balance the surface charge. These ions are the "diffuse ion-swarm" ions. To maintain surface charge balance,

$$\sigma_p = -\sigma_d \tag{6.8}$$

(Sposito, 2004).

POINTS OF ZERO CHARGE

Since the late 1950s, geochemists have recognized that different minerals have different pH values at which the surface goes from positively charged to uncharged to negatively charged. The pH at which the charge reverses (and is zero) has become known as the pH of the *point of zero charge*. Because the pH of the point of zero charge changes with the solution conditions, different terms have developed over time.

1. *The point of zero charge* (p.z.c.) is the pH value of a solution at which the total net particle charge vanishes, $\sigma_p = 0$ (Sposito, 1984). It is important to understand that any given mineral, such as hematite or goethite, does not have just one p.z.c. associated with it. This is because of differences in purity, aspect ratio—which controls how much of each different face is presented to solution—and other factors. The p.z.c. is often measured by electrophoresis; however, such measurements can be problematic if some components of the diffuse ion swarm move with the particle during the measurements (Sposito, 2004).

   Kosmulski (2006) defines the "pristine" p.z.c. (p.p.z.c.) of an oxide or aluminosilicate mineral as the p.z.c. obtained in the presence of alkali halides, nitrates, or chlorates in the absence of other solutes. Table 6.1 provides some ranges of p.z.c. values reported based on electrophoretic mobility measurements.
2. The *point of zero net proton charge* (p.z.n.p.c.) (sometimes termed the point of zero net proton condition) is defined as $\sigma_H = 0$. This value is measured by potentiometric titration, accounting for the protons at the surface being titrated versus other changes to H^+ and OH^- over the course of the titration because of changes to speciation of ions in solution, dissolution, and so on. Some values for various minerals of interest in environmental surface chemistry are provided in Table 6.2.
3. The *point of zero net charge* (p.z.n.c.) is defined by $\sigma_{in} = 0$; equivalently, the point where

$$\Delta q = q_+ - q_- = \sigma_{is} + \sigma_{os} + \sigma_D = 0, \qquad (6.9)$$

   where

   q_+ = the adsorbed cation charge
   q_- = the adsorbed anion charge (Chorover and Sposito, 1995).

TABLE 6.1 Some ranges of pH of p.z.c. values reported in the literature through 2005, based on data in Kosmulski (2006).

Solid	Range
Al_2O_3	6.4–9.5[1]
$Al(OH)_3$	9.2–11.5
Hematite	5.8[2]–10
Goethite	7.8–9.3
Manganite	6.3
SiO_2[3]	1.5–2

[1]Kosmulski (2006) also reports values of 4–6 measured on different individual faces.
[2]Lowest value is a natural hematite sample with Al, Si, Mn impurities; most other values are around 8.
[3]Primarily small particles and amorphous; some values reported as "<2".
[4]The range of values is specific as measured by electrophoretic mobility.

TABLE 6.2 Point of zero net proton charge for selected solid phases.

Solid phase	p.z.n.p.c.
Corundum α-Al_2O_3	9.1
Gibbsite, α-$Al(OH)_3$	8.5 ± 0.5
Hematite, α-Fe_2O_3	8.4 ± 0.1
Goethite, α-FeOOH	7.7 ± 0.2
Amorphous	8.0 ± 0.1
Fe hydroxide	
Quartz	1.8[2]
Birnessite δ-MnO_2	2.2 ± 0.7
Feldspars	2.2 ± 0.2
Kaolinite	5.0–5.4[1]
Montmorillonite	2.5

[1]Schroth and Sposito (1997).
[2]From Brantley and Chen (1995), referencing Parks, 1967.
[3]From compilation of Zelazny et al. (1996) unless otherwise noted.

In manuscripts, values are often given as the pH at which the point occurs, with subscript symbols, either with or without the periods; $pH_{p.z.c.}$ or pH_{pzc}, and $pH_{p.z.n.p.c.}$ or pH_{pznpc}, and $pH_{p.z.n.c.}$ or pH_{pznc}.

The *point of zero salt effect* (p.z.s.e.) is the point of zero charge that remains consistent with ionic strength (I). It is the point where $d\sigma_H/dI = 0$. This point is determined by measuring σ_H by titration in the presence of an indifferent electrolyte at several ionic strengths. When the curves of σ_H versus pH intersect, the intersection point is the p.z.s.e. In practice, the curves do not always intersect at a single point. The *isoelectric point* (IEP) is often measured by electrophoretic mobility, and sometimes it is taken to be equivalent to p.z.c. However, this is not always the case as portions of the diffuse swarm may potentially migrate with particles in electrophoretic mobility experiments (Sposito, 2004).

Whenever a reader encounters a point of zero charge in a publication, it is important to try to verify how the measurement was made and how the term is being used. For example, is it a p.p.z.c. or a p.z.n.p.c.? Any particular mineral sample may have different points of zero charge than other samples of the same mineral reported in the literature. Different mineral samples vary in their purity. For example, natural kaolinite samples may contain small amounts of carbonate, Fe(hydr)oxides, anatase, illite, smectite, organic matter, and other components. Mineral particles, whether collected in the field or grown in the laboratory, may have variable surface composition because impurities often are concentrated at surfaces. Particles with different aspect ratios have different proportions of one crystallographic face versus another exposed to solution, which can affect surface charge. Pretreatment of samples, such as to remove adsorbed organics, can affect charge properties. For particles that undergo dissolution over the course of titrations to detemine point of zero net proton charge, the effects of this dissolution are not always handled consistently. Measured points of zero charge can also vary with temperature. For example, Oleson and Sahai (2008) report that the pH values of p.z.c.s of quartz, rutile, and corundum are about half a log unit lower at 55°C than at 25°C.

Case Study: The Surface Charge Properties of Kaolinitic Soils

Anderson and Sposito (1991) and Chorover and Sposito (1993) developed a method for measuring structural charge on clays by Cs^+ adsorption. They chose Cs^+ because it tends to sorb to phyllosilicate clays at least in part by inner-sphere adsorption mechanisms (because of low hydration enthalpy) and because it generally has the highest affinity for clays of the Group I alkali monovalent cations (Chapter 7).

The overall methodology is based on the equations of "surface charge balance" (Sposito, 1989)

$$\sigma_p = \sigma_o + \sigma_H + \sigma_{IS} + \sigma_{OS} \tag{6.7}$$

$$\sigma_P = -\sigma_d \tag{6.8}$$

and

$$\sigma_{in} = \sigma_o + \sigma_H = -(q_+ - q) = -\Delta q, \tag{6.10}$$

where

q_- = adsorbed cation charge
q_- = adsorbed anion charge, irrespective of surface speciation (Sposito, 1989).

At the pH corresponding to the p.z.n.p.c., $\sigma_H = 0$, and $\sigma_o = -\Delta q$. This is the theoretical basis for the experimental approach.

The σ_o is determined by measuring the surface excess of Cs—a cation that is highly selective to the permanent charge sites—that remains following a Li^+ exchange reaction. Because of its high hydration energy, Li^+ tends to have the lowest affinity of the Group I alkali monovalent cations for clay surfaces.

The value of σ_H is determined by simultaneous proton titration and LiCl background electrolyte adsorption measurements. The proton titration must take into account the species released by mineral dissolution over the course of the titration (e.g., in the case of kaolinite, Al), background ion (LiCl) adsorption, and charge balance. By combining proton titration and LiCl adsorption data, proton surface charge densities and points of zero charge of soil particles could be measured independently.

The methods were applied by Chorover and Sposito (1995) to kaolinitic tropical soils and by Schroth and Sposito (1997) to Georgia kaolinites. Using this approach, and accounting for Al release resulting from dissolution, Schroth and Sposito (1997) measured the p.z.n.p.c. of Georgia kaolinites and the permanent structural charge density (σ_o) values as reported in Chapter 3. For kaolinitic tropical soils, the surface charge behavior indicated contributions by kaolinite, organic matter, and a small amount of 2:1 clay minerals to the surface charge (Chorover and Sposito, 1995).

According to Equation 6.10, a test of surface charge balance and, therefore, of the integrity of such measurements can be obtained by plotting Δq (mmol$_c$ kg^{-1}) versus σ_H (mmol$_c$ kg^{-1}). As shown in Figure 6.2, such a plot should have a slope equal to -1 and intercepts equal to $-\sigma_o$, independent of ionic strength.

SORPTION TERMINOLOGY

When a mineral surface is present in solution, an accumulation of ions or molecules occurs in association with the surface in a process known as *sorption*. In some cases, precipitation of a new phase occurs on the mineral surface. Knowledge of the mechanism of sorption is essential for building predictive models of reactive transport of contaminants such as heavy metals and radionuclides and for understanding nutrient cycling.

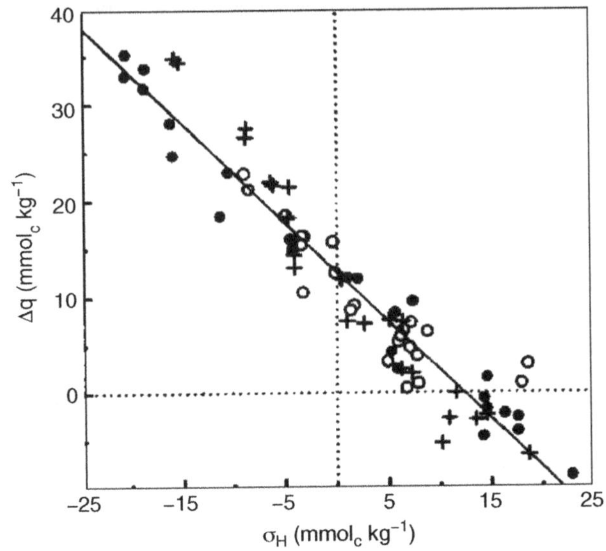

FIGURE 6.2 Balance of surface charge plot for variable-charge kaolinitic tropical soil suspended in 1-, 5-, and 10-mM LiCl solutions at variable pH. After Chorover and Sposito, 1995 (with permission).

The term *sorption* is used when the mechanism by which a given species accumulates at an interface is unknown or to refer to several processes. Figure 6.3 is a schematic illustration of the different modes of sorption. *Adsorption* refers to a process in which individual species or dimers, trimers, and so on accumulate in a two-dimensional fashion at an external surface. *Absorption* refers to accumulation that goes into a particle, such as diffusion into the crystal lattice, or absorption into micropores or interlayers within a structure. *Precipitation* refers to a 3-D accumulation process on a surface. In many instances, it can be difficult to discern whether precipitation in the presence of a surface has resulted from nucleation on the surface (*heterogeneous nucleation*) or in the surrounding solution (*homogeneous nucleation*). *Coprecipitation* occurs when a precipitate includes chemical species from the aqueous solution along with species released by dissolution of the mineral sorbent. Cations and/or anions of similar size and charge may coprecipitate or codeposit to form an actual *solid solution*. *Diffusion* of ions into a surface may also result in solid solution formation.

Species that adsorb directly onto a surface, forming chemical bonds with the surface, are known as *inner-sphere complexes* (Figures 6.3 and 6.4). The generic term *specific adsorption* is often used to describe inner-sphere surface complexation. As an example

$$>\text{Fe—OH} + \text{M}^{2+} \rightleftarrows\ >\text{Fe—OM}^{+} + \text{H}^{+}, \tag{6.11}$$

where M stands for a metal and > indicates a surface site. From Equation 6.11, it is clear that binding of this metal cation will alter the surface charge density. Anions can also adsorb specifically (Figure 6.1).

The charge on mineral surfaces affects the distribution of ions near the surface. Negatively charged ions will be attracted to a positively charged surface, and postively charged ions will be attracted to a negatively charged surface. Ions that adsorb relatively weakly by electrostatic attraction, with water molecules between the adsorbing species and the surface, form *outer-sphere* complexes (Figure 6.4).

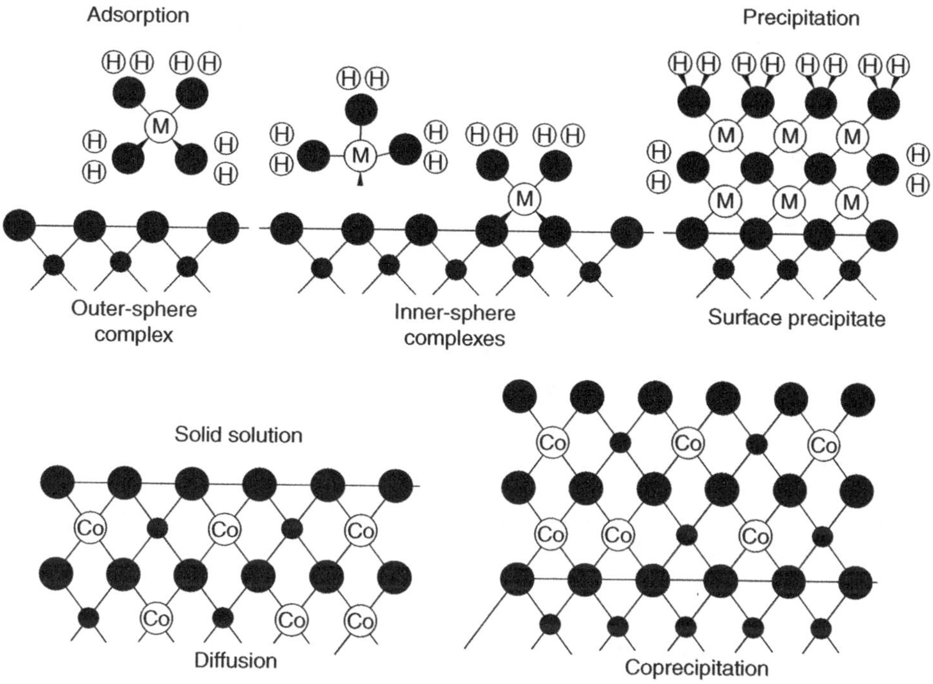

FIGURE 6.3 Schematic illustration of sorption at the mineral–water interface, illustrating Co sorption to calcite. From Brown et al., 1995 (with permission).

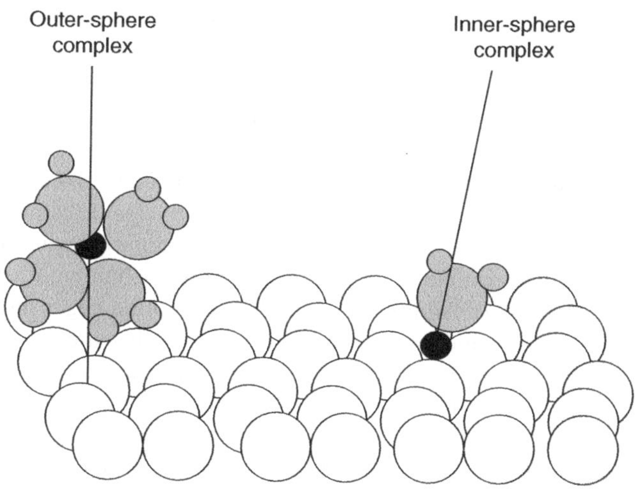

FIGURE 6.4 Schematic illustration of inner-sphere versus outer-sphere complexes. Inner-sphere complexes are directly bound to the surface, whereas outersphere complexes do not bind directly to the surface and have water between the ion and the surface.

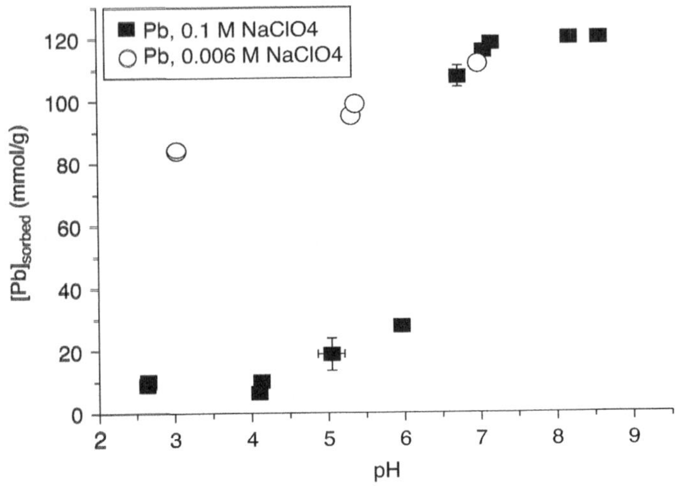

FIGURE 6.5 Sorption edges of Pb on montmorillonite at two different ionic strengths. From Haack and Maurice, unpublished data, University of Notre Dame.

A *sorbate* is a substance that is sorbed; a *sorbent* is the surface that the sorbate is sorbed to. In this text, the term *sorptive* is used to describe the substance whose sorption behavior is being investigated, when it is in solution. This is not a standard use of the term but helps to distinguish something that can sorb when in solution, versus at the surface.

A *complex*, as used in this text, is a dissolved species that exists because of the association of a cation and an anion or a neutral molecule. Surface complexes may also occur. A *ligand* is an anion or a neutral molecule that can combine with a cation to form a complex.

The plots of adsorption customarily have been given specific names. When the equilibrium concentration of adsorptive in solution is plotted on the x-axis and the concentration of the adsorbate normalized to grams (or more commonly meters squared surface area) of adsorbent is plotted on the y-axis, the plot is called an *adsorption isotherm* (data at constant temperature, pH, ionic strength). A plot with a concentration of adsorbate per unit adsorbent on the y-axis and pH on the x-axis is called an *(ad)sorption edge* if the plot is an s-shaped curve (as commonly encountered for many cations; see Figure 6.5) and an *(ad)sorption envelope* if the shape is the reverse of an s (as commonly encountered for many anions; see Figure 6.6) (Sposito, 1984). Additional examples of sorption edges and envelopes are given in Chapters 7 and 8. Sometimes, sorption follows neither a clear envelope nor an edge; for example, sorption may increase with pH in one pH range and decrease in another pH range. An example is shown in Figure 6.7 and another in Chapter 7, Figure 7.16.

Let us return to the discussion at the start of Chapter 5, on how to define a surface. If adsorption occurs, then how are the surface and interface defined? Is the adsorbed material included in the surface? If there is precipitation of a new phase on an underlying surface (known as *heterogeneous nucleation*), then how are the surfaces and interfaces defined? Much depends on the particular application. We often need to think in terms of surficial or interfacial regions, which means that surfaces have finite, but poorly defined, thicknesses. For some applications, a precise mathematical means of describing a surface is needed.

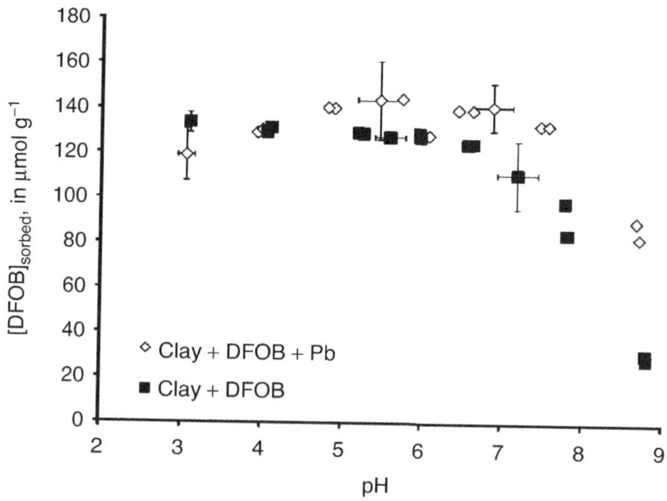

FIGURE 6.6 Sorption envelope of the microbial organic ligand desferrioxamine B (DFOB) on montomorillonite in 0.1-M $NaClO_4$. Note also the sorption of Pb plus DFOB, which follows neither a clear envelope nor an edge. From Haack and Maurice, unpublished data, University of Notre Dame.

CATION EXCHANGE CAPACITY

The cation exchange capacity (CEC) is used to describe the ability of a mineral to sorb and exchange cations. Swelling clays and micas that have substantial isomorphic substitution that leads to negative permanent structural charge density often have high exchange capacities. Other minerals, such as many Mn oxides, may also have high CECs, and natural organic matter adsorbed to a surface may contribute to the CEC. The CEC has been defined

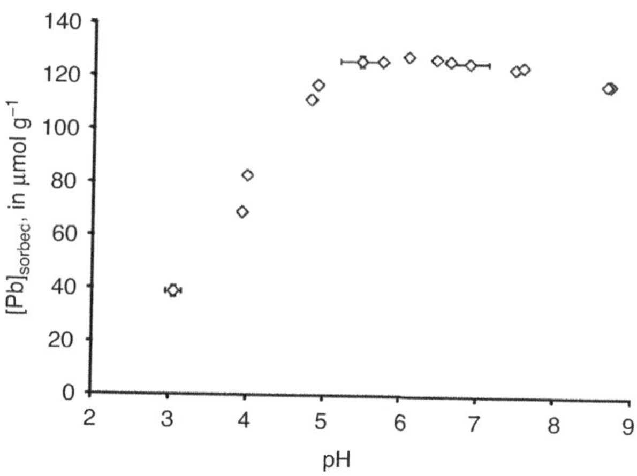

FIGURE 6.7 Sorption versus pH for Pb to montmorillonite in the presence of DFOB. Although the curve looks primarily like a sorption edge, the small decrease in sorption above pH 6 is more indicative of a sorption envelope. The sorption of Pb seems to be affected by the presence of DFOB. From Haack and Maurice, unpublished data, University of Notre Dame.

in various ways. It is important to consider CEC not only as an excess of cations but also as a deficit (negative adsorption) of anions, such that

$$CEC = M^{X+}_{excess} + A^{x-}_{deficit}, \tag{6.12}$$

where M^{x+} and and A^{x-} are the cations and anions, respectively (Sumner and Miller, 1996). The anion component is often small enough to be neglected, which makes CEC easier to calculate.

An example of a cation exchange reaction between two monovalent ions, K^+ and Na^+, is

$$\text{K-X }(s) + Na^+(aq) \rightleftarrows \text{Na-X }(s) + K^+(aq), \tag{6.13}$$

where X represents the exchanging solid, such as a clay mineral.

Cation exchange can sometimes be a component of weathering reactions, such as in the incongruent dissolution of muscovite, which has K^+ in the interlayer, to form vermiculite, which has Ca^{2+} in the interlayer:

$$\begin{gathered}\text{muscovite} \hspace{20em} \text{vermiculite}\\ K_2[Si_6Al_2]Al_4O_{20}(OH)_4(s) + 0.8Ca^{2+}(aq) + 1.3Si(OH)_4^0(aq) \rightleftarrows 1.1Ca_{0.7}[Si_{6.6}Al_{1.4}]\\ Al_4O_{20}(OH)_4(s) + 2K^+(aq) + 0.4OH^-(aq) + 1.6H_2O(l).\end{gathered} \tag{6.14}$$

This weathering reaction can occur in calcareous, poorly drained soils (Sposito, 1989).

An example of cation exchange of ions with different charge is that of Ca^{2+} for Na^+:

$$\text{2Na-X}(s) + Ca^{2+}(aq) \rightleftarrows \text{Ca-X}_2(s) + 2Na^+(aq), \tag{6.15a}$$

which can be written equivalently as

$$Na_2X(s) + Ca^{2+}(aq) \rightleftarrows CaX(s) + 2Na^+(aq), \tag{6.15b}$$

where X represents the exchanger. For a reversible reaction, the equilibrium constant (K_{ex}) in terms of species activities is

$$K_{ex} = \frac{(\text{Ca-X}_2)(Na^+)^2}{(\text{Na-X})^2(Ca^{2+})} \quad \text{for 6.15a.} \tag{6.16a}$$

and

$$K_{ex} = \frac{(CaX)(Na^+)^2}{(Na_2X)(Ca^{2+})} \quad \text{for 6.15b.} \tag{6.16b}$$

The value of K_{ex} would be the same in Equations 6.16a and 6.16b, because the reactions are the same, only they are written with a different notation for exchange sites.

The activity coefficients of species in solution can be calculated using methods of Chapter 1, but the activities of sorbed species are much more difficult to determine. One method is to use a *rational activity coefficient*, which is defined as

$$[MX] = \lambda_{MX} N_{MX}, \tag{6.17}$$

where M is the cation, MX is the adsorbed cation, λ_{MX} is the adsorbed species activity coefficient, and N_{MX} is the mole fraction of X sites occupied by the given cation. For the example in Equation 6.16, $N_{NaX} + N_{CaX} = 1$.

In many instances, the activity corrections are neglected, units of concentration are used for solution species, the mole fraction is used for the cations associated with the surface, and a *conditional exchange constant* is $^{c}K_{ex}$ used, which is sometimes also known as the selectivity coefficient D. This concept works well for an equation written in the format of Equation 6.15b

$$^{c}K_{ex} = \frac{(N_{CaX})(mNa^{+})^2}{(N_{Na_2X})(mCa^{2+})} \quad \text{for 6.15b.} \tag{6.18}$$

However, $^{c}K_{ex}$ is not truly a constant, because it can vary as a function of the ratio of one cation to the other on the surface. This variation occurs because of interactions between the sorbed cations. Moreover, $^{c}K_{ex}$ is only useful when no other exchanging cations are present. $^{c}K_{ex}$ tends to work best for major cations sorbing to sorbents with numerous sites, such as smectitic clays and zeolites.

In many instances, exchanging surfaces have higher preference for divalent cations such as Mg^{2+} and Ca^{2+} than for monovalent cations, particularly Na^{+}. In determining cation exchange, both the preference and the cation concentration in solution are important.

Sumner and Miller (1996) discuss the use of different cation exchange expressions and methods of measuring CECs, which include problems inherent in these measurements. McBride (1994) provides a detailed discussion of cation exchange, which includes rates of exchange. CECs tend to be different for monovalent versus divalent cations. The CEC tends to increase with pH. The CEC values of smectites $< 0.1\ \mu m$ in diameter tends to be less than for larger smectite particles (Schoonheydt, 1995).

The CECs of swelling clays tend to be high because both external surfaces and interlayers can undergo cation exchange. Smectite minerals have a CEC of 70–120 cmoles kg^{-1}, whereas nonswelling kaolinite has a much lower CEC of 1–15 cmoles kg^{-1} (McBride, 1994). Although micas can have considerable permanent structural charge, they tend to have much lower CECs because the K^{+} fixed in the interlayer prevents interlayer exchange from occurring. Weathering can expose frayed edges (Chapter 7) for further exchange. In soils, the natural organic matter (Chapter 8) can also contribute to the CEC, as it has the ability to sorb and exchange cations. The CEC is sometimes given in terms of millimoles charge per gram (or centimoles per kg) but other times in terms of mEq $100g^{-1}$.

One application of cation exchange is in water softening processes to treat Ca^{2+} and Mg^{2+} -rich "hard" water. Originally, zeolites were used as cation exchangers, but synthetic resins have become far more common. The cation exchange resin is saturated with Na^{+} ions. As the hard water is passed through the resin (R), the Ca^{2+} and Mg^{2+} ions exchange with the Na^{+} ions and remain sorbed to the resin because of their stronger attachment. An example reaction is

$$Ca(HCO_3)_2 + Na\text{-}R \rightleftarrows Ca\text{-}R + 2NaHCO_3. \tag{6.19}$$

This process helps to prevent Ca- and Mg-carbonate deposits from forming in plumbing, on dishware, and so on, and it helps to make laundry detergents more effective. But, it also adds Na to water, with potential biomedical consequences.

In the discussion on cation exchange above, a simplified approach was given in which activity corrections were neglected. More sophisticated approaches have also been developed, using specific exchange constants and increasing the ratio of mole fractions sorbed to different *n*th powers (e.g., Langmuir, 1981, 1997). In general, these are known as *power exchange functions*. For cases in which the cations exchanging are both of the same valence (i.e., "homovalent"),

$$\text{A (aq)} + \text{BX (s)} \rightleftarrows \text{B(aq)} + \text{AX(s)}, \tag{6.20}$$

$$K_{ex} = (\text{B})/(\text{A})(\text{AX}/\text{BX})^n, \tag{6.21}$$

where the dissolved species are in activities and AX and BX are in units of equivalent fractions. Equation 6.21 can be transformed to

$$\log (\text{B})/(\text{A}) = n \log (\text{BX}/\text{AX}) + \log K_{ex}, \tag{6.22}$$

which is in the form of the equation for a line. By plotting log (BX/AX) on the x-axis and log (B)/(A) on the y-axis, the slope provides the value of n and the intercept the value of log K_{ex}. The resulting values are specific to a particular sorbent and set of exchanging cations, range of cation concentrations, and pH. This approach is empirical and does not provide a broader understanding, but it does provide a means for modeling binary cation exchange, especially for well characterized solids. Data are lacking for trace concentrations of sorbate cations, and many systems contain multiple exchangeable cations, so that applicability to field situations is limited. A similar approach can be used for cations that are not homovalent, but it does not include charge balance and is also not mechanistic. Langmuir (1997) provides a tabulation of values of K_{ex} and n for various cation exchange reactions on surfaces such as montmorillonite, ferric oxide, and several soils at specific pH ranges.

SORPTION ISOTHERMS

Sorption data are often displayed by plotting the relationship between the bulk equilibrium solution activity (or concentration) of the sorptive on the x-axis and the amount sorbed per unit surface area (or g solid) on the y-axis, at constant temperature. As described above, the resulting plots are known as *sorption isotherms*.

Sposito (1984) provided a now classic description and accompanying summary figure of the shapes of isotherms that commonly occur. These shapes are the S-curve, L-curve, H-curve, and C-curve, which are shown in Figure 6.8.

The *S-curve isotherm* implies that the affinity of the adsorbent for that of the adsorbate is less than that of the aqueous solution. The *L-curve isotherm* occurs when there is high affinity of the adsorbate for the surface at low surface coverage, but the surface sites ultimately become filled. The *H-curve isotherm* suggests that the adsorbate has a very high affinity for the surface. The *C-curve isotherm* can be produced by either (1) constant partitioning of the adsorptive between the surface and the solution or (2) proportionate increase in the amount of surface that functions to adsorb as adsorption increases (Sposito, 1989).

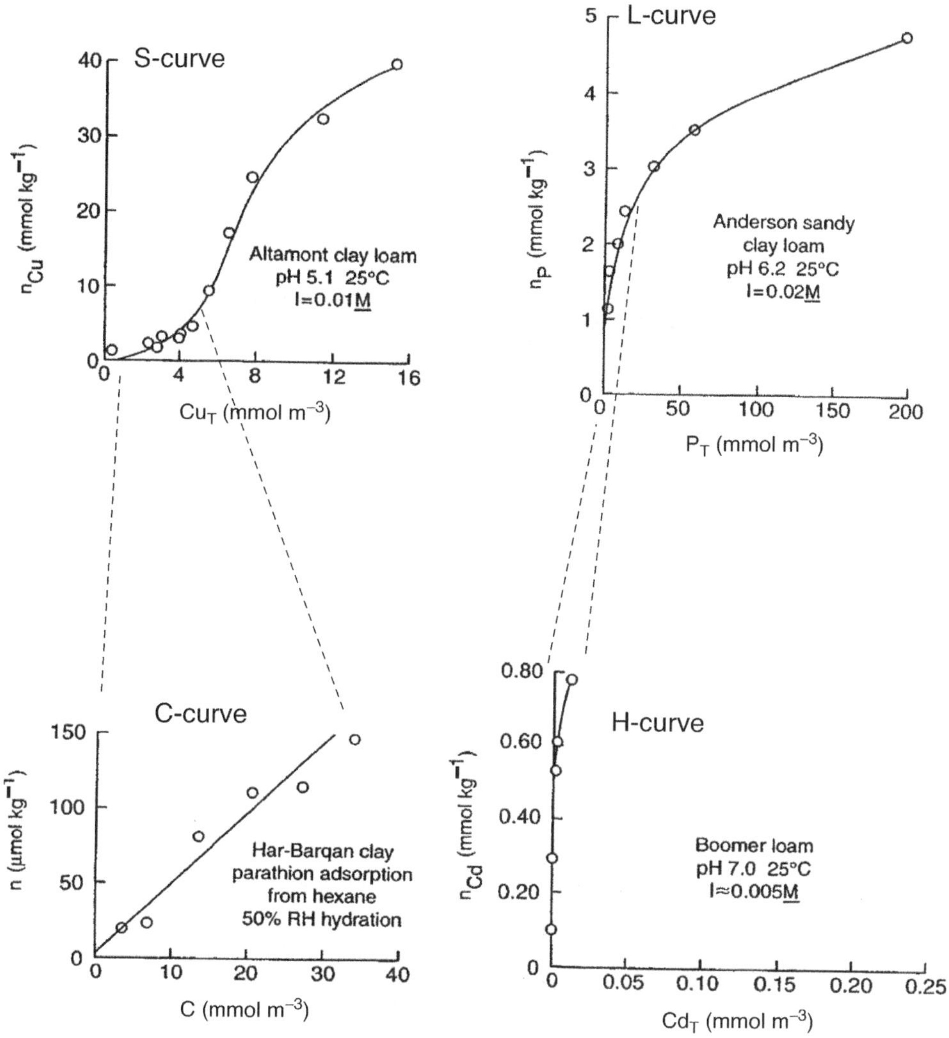

FIGURE 6.8 The different broad types of adsorption isotherms. Note that the C-curve resembles the first, low-affinity part of the S-curve, and the H-curve resembles the first, high-affinity part of the L-curve. Individual adsorption curves are from Sposito, 1984 (with permission).

Other curve shapes are also possible for sorption isotherms. For example, sometimes an overall L-shaped isotherm will obtain not just one plateau, but two or several. This can happen, for example, when hydrophobic organic molecules adsorb roughly parallel to the surface at low surface coverages but perpendicular to the surface (allowing more dense packing) at higher surface coverages. Multiplateau curves can also occur when a mixture of compounds adsorbs, such as in a sample of natural organic matter (Chapter 8) (Figure 6.9). There is some debate as to whether isotherms can occur that are concave downward. It has been proposed that concave downward isotherms can occur when a component reaches a high enough concentration to precipitate or form a micelle in solution

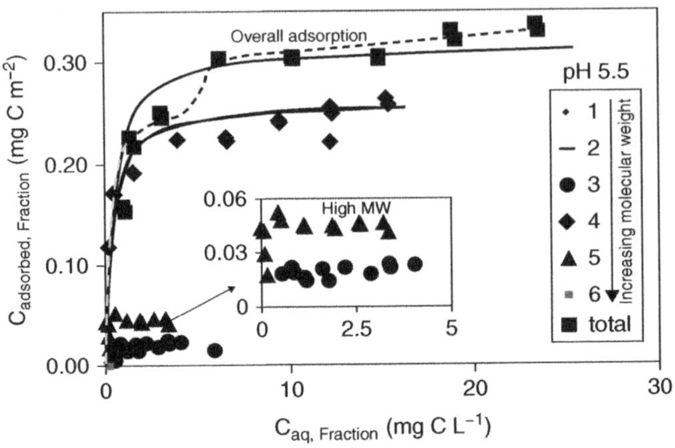

FIGURE 6.9 Adsorption of natural organic matter to goethite shows complex isotherm behavior. Although the curve of total organic matter adsorbed can be fit to an L-shaped isotherm, often a hint of two plateaus is visible. This results from the summation of isotherms for different-size fractions having different shapes. For example, the high-molecular-weight component isotherm is concave downward.

that no longer associates with the surface; however, such processes have been under debate particularly with respect to whether they can represent equilibrium scenarios. Adsorption from solutions with multiple adsorbates can result in concave downward isotherms because of competition or other interactive processes (Figure 6.9).

Sometimes, an isotherm will be described as an H- or C-curve without realizing that only the first few data points of an overall L-curve isotherm have been collected. The curve could be a different shape when higher surface coverages are included. Data points can have large enough error bars that it is difficult to determine whether a curve is, for example, S- or L-shaped based on the first few points.

Many (but not all) adsorption experiments are accomplished by measuring the concentration of the adsorptive in solution before a solid is added (initial solution concentration c_{in}), and then separating the solid by filtration and/or centrifugation from the solution and measuring the concentration of the adsorptive remaining in solution after adsorption (equilibrium solution concentration c_{eq}). The "concentration" adsorbed (c_s) is then calculated from the difference

$$c_s = c_{in} - c_{eq}, \tag{6.23}$$

from which moles per unit weight (or surface area) can be calculated. At low adsorptive ion or molecule concentrations, if there is fairly high affinity, then most of the adsorptive ion or molecule is removed from solution by adsorption. In such a case, the concentration of the adsorptive ion or molecule in solution can often be at or near the detection limits of aqueous analytical methods. Further along the isotherm, especially if an adsorption plateau is reached, the amount adsorbed can represent only a small fraction of the total adsorptive component in solution. In this case, the difference between adsorptive ion or molecule concentration in the original and reacted solutions can be so small as to be hard or impossible to distinguish. Special care must be taken for data at both ends of the isotherm.

The shape of an isotherm does not provide unambiguous information on the sorption mechanism (Kitchener, 1965). Spectroscopic and/or other more direct evidence is needed.

ADSORPTION ISOTHERM EQUATIONS

Different isotherm treatments can be applied to compare adsorption for different adsorbents and adsorbates. They do not provide much insight into adsorption mechanism but can be useful for comparing different adsorbates, adsorbents, effects of different electrolytes, and other factors.

The Langmuir Isotherm Equation

One of the most commonly used isotherms was developed in 1916 by Irving Langmuir and is known as the *Langmuir equation*, applied for adsorption

$$q = \Gamma = \frac{K_L q_m C}{1 + K_L C,} \tag{6.24}$$

where

q = adsorbed density (mg m^{-2} or mg g^{-1}) on the surface
q_m = maximum adsorbed density (mg m^{-2} or mg^{-1})
K_L = adsorption equilibrium constant (L mg^{-1})
C = equilibrium concentration (more properly, activity, although concentration is often used instead) in solution (of the same component that is adsorbing) (mg L^{-1}).

This equation can be linearized by simple rearrangement to

$$q = -q/K_L C + q_m. \tag{6.25}$$

If q/C is plotted on the x-axis and q is plotted on the y-axis, then the slope $m = -1/K_L$ and the intercept $b = q_m$.

The equation is often stated in an equivalent form

$$q/C = (q_m\text{-}q)K_L. \tag{6.26}$$

The value of q_m is determined by the isotherm plateau, whereas K_L is typically determined from q_m and the first few data points at low amounts of adsorption.

Irving Langmuir's research in surface chemistry, which includes isotherm treatments, was recognized by the 1932 Nobel Prize in Chemistry.

Figure 6.10, based on data from Tipping (1981), shows a Langmuir isotherm for natural organic matter (humics) adsorption to goethite, along with a linearization of the Langmuir equation.

Application of the Langmuir equation is based on several assumptions, which may not hold true in many instances of adsorption.

1. Adsorption occurs at adsorption sites, each of which has a specific surface area associated with it.
2. Adsorption occurs up to formation of a single, uniform layer or "monolayer" on the surface.

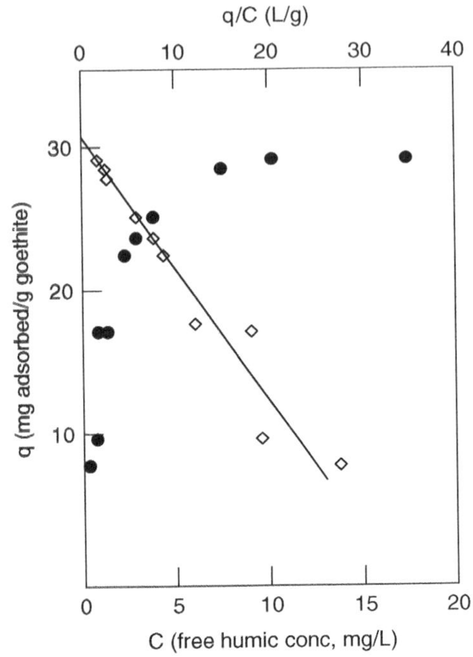

FIGURE 6.10 Adsorption of humics (natural organic molecules) to goethite at pH 7 with Langmuir linearization. Note that this isotherm is in terms of g sorbent rather than m^2 surface area. Goethite BET surface areas range from around 10–90 $m^2\ g^{-1}$. After and using data from Tipping, 1981 (with permission).

3. The energy of adsorption is independent of how much is adsorbed, meaning that adsorption at one site does not affect adsorption at another nearby site.
4. Differences in adsorption behavior with loading are caused by differences in adsorbate–solid interactions, and not by lateral interactions between neighboring adsorbates.

Just because adsorption data fit well to the Langmuir equation does *not* imply that the assumptions listed above hold true.

The Freundlich Isotherm Equation

Other isotherm treatments, such as the *Freundlich isotherm*, can also be applied especially when there is a poor fit to the Langmuir equation. The *Freundlich equation* is

$$q = \Gamma = K_f C^n, \tag{6.27}$$

where

q = sorbate sorption density (in mg m^{-2} or sometimes mg g^{-1})
C = equilibrium concentration of sorptive ion or molecule remaining in solution after sorption (in mg L^{-1})
n = constant expressing isotherm curvature
K_f = *Freundlich isotherm coefficient.*

In this case, the log of the surface concentration of adsorbate q or Γ is plotted against the log of concentration of the adsorptive ion or molecule in solution, C.

The Frumkin Isotherm Equation

When organic molecule sorption is dominated by hydrophobic interactions, the sorption isotherm data are generally fit to a *Frumkin isotherm equation* (or sometimes a Freundlich isotherm equation). The *Frumkin equation* is (Schindler, 1990)

$$\Theta/(1\text{-}\Theta) = \mathrm{BC}\ e^{2\alpha\Theta}, \qquad (6.28)$$

where

Θ = fraction of adsorbent surface covered by adsorbate
C = concentration of the adsorptive molecule in aqueous solution
B = adsorption constant
α = interaction coefficient.

The exponential factor in the equation above accounts for lateral interactions that occur between the adsorbed molecules (Schindler, 1990).

In plotting an adsorption isotherm and applying one of the isotherm equations, it is important to determine whether adsorption is at equilibrium, or at least at an apparent steady state. This is generally done using (1) kinetic experiments to determine whether the amount of adsorbate on the surface of the adsorbent has stopped changing over time and (2) reversibility experiments, because equilibrium implies reversibility. Hysteresis is often observed in reversibility experiments, and it is up to the researcher to determine the potential cause(s) of hysteresis and the implications for the sorption phenomena. The various isotherm equations can fit reasonably well to a wide array of data. The fit does not tell us much about actual sorption behavior or mechanisms, but it can help in comparing different systems.

THE DOUBLE LAYER, GOUY-CHAPMAN THEORY

A few basic terms need to be defined before discussing the details of models for the charged interface. The term *electrical potential* (ψ) refers to the potential energy per unit of charge, and the difference in electrical potential between two points is the voltage (V). The *capacitance* (C) is a measure of the amount of electric charge stored for a given electric potential. A common form of capacitor for charge storage is a set of parallel plates. For parallel plates with charge +Q and −Q, and a voltage difference V between the plates,

$$\mathrm{C} = \mathrm{Q/V}. \qquad (6.29)$$

Gouy and Chapman devised a model for a charged interface in solution that included charge associated with the particle itself as well as a dissociated charge in the diffuse layer; together, these make up the *electronic double layer* (EDL; Figure 6.11). Several assumptions of the model include (1) the surface is flat; (2) the charge is distributed uniformly over the surface; (3) the charged ions in solution are treated as point charges distributed near the

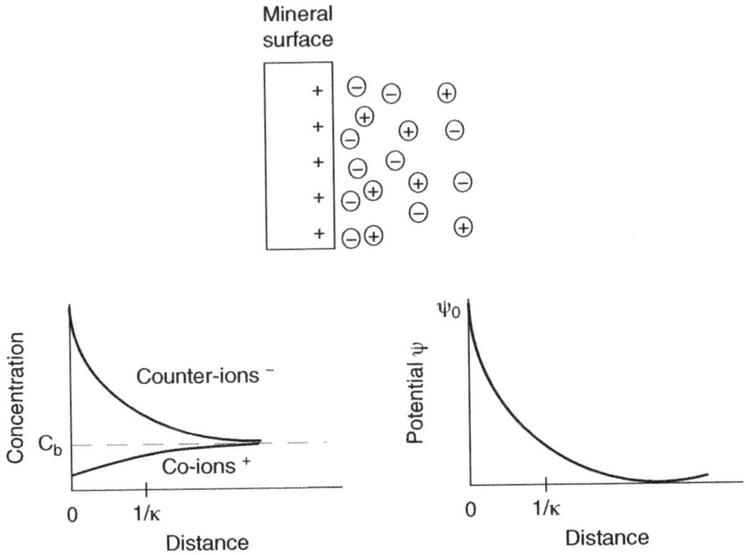

FIGURE 6.11 The basic elements of Gouy-Chapman theory.

surface according to the Boltzmann distribution; (4) the solvent influences the double layer only by its dielectric constant, which is taken to be constant across the double layer; and (5) the ions only associate with a surface through electrostatic interactions.

According to Gouy-Chapman theory, the charge density at the surface σ_p in C m^{-2} (Coulombs per meter squared) is related to the electrical potential at the surface ψ_o in volts through the following equation, which is valid for a symmetrical electrolyte (1:1 or 2:2) (Dzombak and Morel, 1990):

$$\sigma_p = (8RT\varepsilon\varepsilon_o \quad c \times 10^3)^{1/2} \quad \sinh\,(Z\psi_o F/2RT), \tag{6.30}$$

where

R = molar gas constant (8.314 J mol^{-1} K^{-1})
T = absolute temperature (K)
ε = dielectric constant of water (= 78.5 at 25°C, dimensionless)
ε_o = permittivity of free space (8.854 × 10^{-12} C^{2}J^{-1}m^{-1} or C V^{-1} m^{-1})
c = molar electrolyte concentration in the bulk solution (M)
Z = ionic charge or valence on the symmetrical electrolyte (e.g., 1 for Na, Cl)
F = Faraday constant (96.490 C mol^{-1}).

At 25°C,

$$\sigma_p = 0.1174\, c^{1/2} \sinh\,(19.46\, Z\psi_o) \tag{6.31}$$

(Stumm, 1992).

At low potential, this equation can be linearized as

$$\sigma_p = \varepsilon\varepsilon_o\kappa\psi_o, \tag{6.32}$$

where the double-layer thickness κ^{-1} (in meters) is defined by

$$\kappa = [(2F^2I \times 10^3)/\varepsilon\varepsilon_o RT)]^{1/2}, \tag{6.33}$$

where I is the ionic strength (Stumm, 1992). For an aqueous solution of a symmetrical electrolyte at 25°C, Equation 6.32 becomes (Dzombak and Morel, 1990)

$$\sigma_p = 2.5I^{1/2}\ \psi, \tag{6.34}$$

and Equation 6.33 becomes

$$\kappa = 0.329 \times 10^{10}\ (cZ^2/\text{mol L}^{-1})^{1/2} \tag{6.35}$$

in units of m^{-1}, which can be used to calculate the double-layer thickness for solutions of different molar concentrations (c) (Shaw, 1992).

Recall that

$$\sigma_p = -\sigma_d, \tag{6.8}$$

so that the charge on the surface is balanced by the diffuse ion charge.

The double-layer thickness, which is also known as the *Debye length*, is κ^{-1}. Hence, as solution I increases, κ increases and κ^{-1} decreases, so that the double-layer thickness decreases as I increases. The fact that double-layer thickness tends to decrease with increasing I has important ramifications for the stabilities of colloidal suspensions and the attachment of colloidal particles to porous media, as discussed in the section on Derjaguin, Landau, Verwey, and Overbeek (DLVO) theory, below. The charge on an ion is important; for example, trivalent cations such as Fe(III) and Al(III) are often used in water treatment to promote flocculation. A schematic illustration of the factors involved in Guoy-Chapman theory is shown in Figure 6.11.

Note that the surface potential ψ_o is related to both I (because κ is a function of I) and the surface charge density (σ_p). Hence, if the double layer is compressed (κ increases) and the surface-charge density remains constant, then the surface potential must decrease. However, all three parameters may change during double-layer compression or expansion.

Electrostatic forces and *van der Waals forces* are two of the most important forces that act between surfaces in solution, but many other forces are involved as well. Interfacial forces between surfaces in air and solution are discussed in greater detail in the section on atomic force microscopy, in Chapter 4. Interfacial forces that affect colloid stability are discussed in the section on DLVO theory at the end of this chapter.

BEYOND GOUY-CHAPMAN: SURFACE COMPLEXATION MODELS

Over time, the basic EDL model has been expanded to

1. Recognize the fact that counterions in the diffuse layer are not point charges but rather have finite size
2. Distinguish among adsorbed ions in terms of hydration and bond type, such as whether inner sphere or outer sphere or nuances within these adsorption types

3. Account for the changes in surface charge that occur during adsorption of charged ions
4. Account for the variations of the dielectric constant of water at and near the interface
5. Account for induced dipole/multipole effects and van der Waals forces
6. Account for the work associated with replacing water at the mineral surface after adsorption.

The evolution of models for the charged interface and specifically for adsorption is an ongoing process. A series of electrostatic models and surface complexation models have been developed.

The *Gouy-Chapman model* generally does not fit well to adsorption data when the double layer is compressed. The local concentrations of ions near the surface become far too large because of the assumption of point charges, which neglects to take into account ionic diameter. Stern (1924) expanded the model by dividing the interface into two parts, the first consisting of a compact layer of ions adsorbed at the surface (now called the *Stern layer* or the *β plane*) and the second consisting of the diffuse layer (the *Gouy layer*). In the Stern treatment, surface H^+ and OH^- are incorporated in the charge associated with the solid and other adsorbed ions are placed at a plane separated from the solid by a short distance. The *Stern plane* is located about one hydrated ion radius distant from the surface. The Stern model helped to compensate at least in part for the problem that ions adsorbed directly to the surface have real dimensions, which was a shortcoming in the original Gouy-Chapman model. In the Stern EDL model, the particle surface charge is balanced by the charge in the Stern layer plus that of the diffuse layer. The surface charge density σ_p and the charge in the Stern plane are balanced by the charge in the diffuse swarm, so that

$$\sigma_p\ (\sigma_o + \sigma_H) + \sigma_\beta \text{ (the } \beta \text{ or Stern plane, essentially } = \sigma_{is}) + \sigma_d \text{ (dissociated charge, essentially outersphere } \sigma_{os} \text{ but also any in the diffuse swarm)} = 0. \quad (6.36)$$

The major difference is that the original Gouy-Chapman model treated all the ions in solution with a single function. In the Stern model, the ions closely and tightly associated with the surface in the Stern layer are treated differently from the ions that are diffusely associated.

Grahame (1947) refined the Stern model even more to distinguish between specifically adsorbed ions, which were taken to adsorb at an *inner Helmholtz plane* and outer sphere adsorbed ions, which adsorb at *an outer Holmholtz plane* (the Stern plane). Going beyond the Gouy-Chapman assumption of uniform charge distribution, an ion adsorbed in the inner Helmholtz plane will affect the surface-charge distribution in its immediate vicinity, which imposes a potential on the ion itself. This effect is known as "discreteness of charge" (Shaw, 1992).

What has become known as the *Gouy–Chapman–Stern–Grahame* model of the EDL assumes that electrical potential decays linearly with distance between the charged plates from the actual surface to the β or Stern plane and then from the β or Stern plane to the diffuse layer according to the following equations (Davis and Kent, 1990):

for the distance between charged plates representing the surface (o) and the β plane

$$\Psi_o - \Psi_\beta = \sigma_p / C_1, \quad (6.37)$$

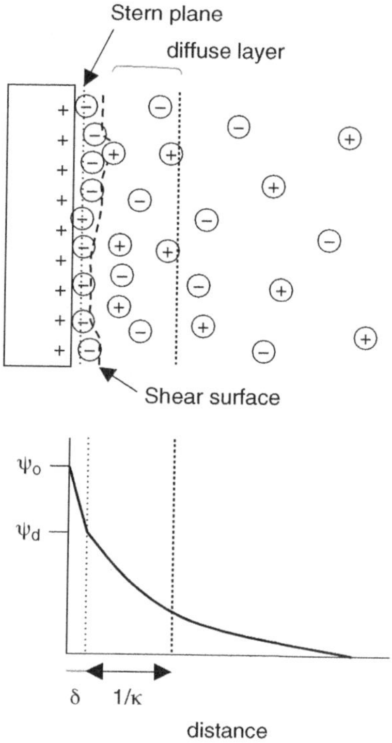

FIGURE 6.12 (top) Schematic illustration showing the structure of the electric double layer according to Stern. (bottom) How potential changes with distance at and away from the surface.

and between the β plane and the diffuse swarm (d)

$$\Psi_\beta - \Psi_d = (\sigma_p + \sigma_\beta)/C_2 = -\sigma_d/C_2, \tag{6.38}$$

where C_1 and C_2 represent the integral capacitances of the interfacial layers. The exponential decay of charge in the diffuse layer is treated as in Gouy-Chapman theory (Figure 6.12).

The value of Ψ_d is generally estimated based on electrokinetic measurements. When particles move in suspensions, such as during electrophoretic mobility measurements, adsorbates in the Stern layer travel with the particle. The exact location of the shear plane that defines what moves with the particle and what stays behind with the solution is difficult to assign precisely. This is because the properties of water near surfaces are such that some components of the diffuse ion swarm may travel with the particle during electrophoretic mobility experiments. This property can complicate interpretation of experimental results and Ψ_d calculations. The shear plane is generally taken to lie just beyond the zone of closest approach of the dissociated ion charge (Davis and Kent, 1990) or a little more than one hydrated ion radius from the surface. Thus, the shear zone is approximately at the edge between the Stern plane and the diffuse layer.

Returning to the discussion in Chapter 5, on how to define a surface and an interface—if we were to define a surface now, we might make that surface the outer boundary of the mineral, or we might include the Stern layer, or perhaps the innermost part of the diffuse swarm, depending on the particular application at hand.

Over time, advanced models were developed, along with quantitative computer programs that describe sorption. A key addition was the observation that ions can directly bind to the surface forming bonds with surface sites. Swiss professors Paul Schindler (e.g., Schindler and Kamber, 1968; Schindler and Gamsjager, 1972), Werner Stumm (e.g., Stumm et al., 1970), and their colleagues developed and advocated the *surface complex model* (SCM), in which specifically adsorbed (inner-sphere) ions are treated like surface-complexed H^+ and OH^-, as part of the solid phase. In this model, surface complexation reactions occur in which adsorption is inner sphere, at specific functional groups on the mineral surface. The equilibria for surface complexation reactions are treated similarly to equilibria for species in solution. However, electrostatic effects at the interface must be taken into consideration. The need to take into consideration the effects of inner-sphere adsorption on surface charge was recognized. Although it is possible to measure empirical, conditional stability constants for surface complexation reactions, these measures are different from intrinsic stability constants because of electrostatic effects at the surface. Surface charge and surface potential result from interactions involving surface functional groups and including adsorption.

The SCM approach differs from the Stern model in that specific chemical interactions occur at the surface rather than purely electrostatic effects. In SCM, the amount of surface charge that can be developed at an oxide surface is restricted by the number of surface sites. This limitation is not observed within original the Gouy-Chapman theory. Many different surface complexation models have evolved, but they all rely on the same fundamental concepts (Dzombak and Morel, 1990)

1. Sorption on oxide minerals occurs at specific coordination sites.
2. Sorption reactions can be described quantitatively using mass law equations.
3. Sorption reactions affect surface charge.
4. Effects of surface charge on sorption can be accounted for by applying a correction factor derived from EDL theory to the mass law constants for the various surface reactions.

Like the models used in aquatic chemistry to describe equilibria in solution, these surface equilibria models are solved using a set of simultaneous equations (Dzombak and Morel, 1990; Davis and Kent, 1990) that include

1. Mass action laws for all of the surface reactions, which require equations and their equilibrium constants for surface reactions
2. Molar mass balance equation for surface sites
3. Equation for computation of surface charge
4. Set of equations representing the constraints imposed by the model on interfacial structure, which is is an added nuance.

Dzombak and Morel (1990) describe the amphoteric surface reactions of oxide surfaces in detail, as related to acid-base oxide titrations. Consider a surface with sites >XO, and amphoteric surface reactions

$$>XOH_2^+ \rightleftarrows >XOH^o + H^+ \qquad K_{a1}^{app} \tag{6.39}$$

$$>XOH^o \rightleftarrows >XO^- + H^+ \qquad K_{a2}^{app}, \tag{6.40}$$

where $K^{app}{}_{a1}$ and $K^{app}{}_{a2}$ are the apparent equilibrium acidity constants. If the proton is the only specifically adsorbing ion present in the solution, then the proton surface charge density σ_H is given by

$$\sigma_H = (F/AS)[(>XOH_2^+) - (>XO^-)] = F[\Gamma_H\text{-}\Gamma_{OH}], \tag{6.41}$$

where

F = Faraday constant
A = specific surface area in $m^2\ g^{-1}$
S = solid concentration in $g\ L^{-1}$
Γ_H and Γ_{OH} = experimental sorption densities in mol m^{-2} of H^+ and OH^-.

The values of Γ_H and Γ_{OH} are often measured by a potentiometric acid-base titration in a solution of inert electrolyte and closed to the atmosphere to exclude CO_2 so that

$$\Gamma_H - \Gamma_{OH} = (1/AS)[C_A - C_B - (H^+) + (OH^-)], \tag{6.42}$$

where C_A and C_B are the molar concentrations (moles L^{-1}) of acid or base added, respectively (Dzombak and Morel, 1990). If the mineral is soluble, then H^+ and OH^- uptake or release related to dissolution and/or metal species in solution, must also be taken into account (e.g., Chorover and Sposito, 1993, 1995). In many instances, this approach is refined by taking into consideration titration of a similar solution in the same reaction vessel but in the absence of the oxide, and substracting to account for effects such as caused by impurities and the reaction vessel walls.

The species (variables) present in this titration system are H^+(aq), OH^-(aq), $>XOH^\circ$, $>XOH_2{}^+$, and $>XO^-$. The mass law equations that describe the system include (neglecting activity coefficients, for simplicity)

$$(OH^-)(H^+) = K_w \tag{6.43}$$

$$\frac{(>XOH^o)(H^+)}{(>XOH_2^+)} = K_{a1}^{app} \tag{6.44}$$

$$\frac{(>XO^-)(H^+)}{(>XOH^o)} = K_{a2}^{app}, \tag{6.45}$$

where the mass law constants are "apparent" (app) because they are functions of surface charge, which varies with pH and ionic strength. The apparent constants can be related to intrinsic constants (K^{int}) through a coulombic correction factor

$$K^{app} = K^{int} e^{(-\Delta ZF\psi/RT)}, \tag{6.46}$$

where

ΔZ = net change in the charge of the surface species
F = Faraday constant
R = ideal gas constant
T = absolute temperature in K
ψ = surface potential.

The proton mole balance equation (TOTH) is

$$\text{TOTH} = C_A - C_B = (H^+) - (OH^-) + (>XOH_2^+) - (>XO^-) \tag{6.47}$$

and

$$q = \text{excess acid} = C_A - C_B - (H^+) + (OH^-), \tag{6.48}$$

$$\Gamma_H - \Gamma_{OH} = q/AS. \tag{6.49}$$

The proton surface charge density,

$$\sigma_H = F[\Gamma_H - \Gamma_{OH}] = q(F/AS) = (F/AS)[(>XOH_2^+) - (>XO^-)] \tag{6.50}$$

(Dzombak and Morel, 1990).

In reality, most oxides are soluble over some component of the titration pH range, and the release of metal ions must be taken into account.

If we consider a generalized reaction for metal adsorption to one surface site (Schindler, 1990)

$$>S\text{—}OH + M^{z+} \rightleftarrows >S\text{—}OM^{(z-1)+} + H^+, \tag{6.51}$$

the intrinsic macroscopic stability constant for this surface reaction is given by

$$K_s^{int} = \frac{a_{>SOM}\, a_H}{a_{>SOH}\, a_M}, \tag{6.52}$$

where a_i refers to activity of the subscripted species i and > indicates a surface species.

Surface concentrations can be measured, but then one needs surface activity coefficients (y) to obtain activities of surface species. On additional equation manipulation, and to account for the charged properties of surfaces (the exponential component in the equation below)

$$K_s^{app} = K_s^{int} \frac{y_{>SOH}}{y_{>SOM}} e^{-((Z-1)F\psi/RT)}, \tag{6.53}$$

where Z refers to charge, K_s^{app} is the experimentally obtainable stability constant (apparent or conditional stability constant), and K_s^{int} is the intrinsic stability constant (subscript s refers to surface). The terms to the right of K_s^{int} are terms that contribute to deviation from ideal behavior (Schindler, 1990).

The difficulty posed by this approach is that one needs to determine the intrinsic stability constant from the measured stability constant. Also, one needs to determine surface species' activity coefficients. In practice, the surface species' activity coefficients are often taken to be 1.

The properties of water near a surface are different from what they are in the bulk. The water molecules in the interfacial region have a low dielectric constant. Because of this, the hydrolysis of adsorbed metal ions and the protolysis (or breaking off a proton) of adsorbed anions may occur more readily and at different pH values than occurs in the bulk solution.

There are different ways of treating the differences in the dielectric constant of water at surfaces, and this is the subject of ongoing research.

Specific models that have been developed, including: the *constant capacitance model* (CCM), the *diffuse double layer* (DDL) *model, the triple layer model* (TLM), and the *MUSIC/charge distribution* (CD) *model*, which is a variation of the TLM. These models each have their own nuances because they assume different geometric features of the oxide/water interface (Dzombak and Morel, 1990). A major difference is in the location and hydrated state of the sorbed ions at the surface and the spatial distribution of the counterions in solution. In applying surface complexation models, adsorption data, such as an adsorption edge or envelope, are often plotted and compared with a curve predicted by the model.

All surface complexation models described here contain correction factor(s) to account for the effect of surface charge on surface complexation. These factors are basically in the form given by Schindler (1990) $e^{-(F\psi/RT)}$. Surface-site densities are needed, which can be determined or estimated using a variety of different approaches, all of which have some associated uncertainty that can limit the reliability of the models. Values of capacitance (C) are needed to relate the surface charges (σ_o) and surface potentials (ψ_o) with those of the various planes. The values of C are determined differently for the different models. Surface protonation constants are determined by titration, with various methods for optimizing fits to titration data.

Constant Capacitance Model (CCM)

The CCM is applicable to systems at high, constant ionic strength. At high ionic strength, the EDL that consists of the coordinative surface charge and the dissociated counterion charge, can be approximated by a parallel plate capacitor (Figure 6.13). The potential decreases linearly away from the surface. Only one surface plane is present, which contains surface OH^-, H^+, and specifically or inner-spherically bound ions. This model can be used to model adsorption, but it is only suitable for systems with high constant ionic strength and is thus not applicable to many problems that involve adsorption in environmental systems, for example in low-ionic-strength lakes and streams. For some experimental systems, particularly involving clays, high ionic strength might cause particles to aggregate and settle out of solution, thus confounding the experimental process.

The Diffuse Double Layer (DDL) Model

This model differs from the constant capacitance model in several key ways. This model has two surface planes: the surface plane for adsorption of H^+, OH^-, and all specifically adsorbed solutes, and a diffuse layer plane that represents the closest distance of approach for counterions (Figure 6.13). The electrical potential at the beginning of the diffuse layer is equal to the surface potential. Gouy-Chapman theory (exponential decay in the diffuse layer) is applied for the surface-charge potential in the diffuse layer. This model is applicable to a wider range of ionic strengths and has been improved in various ways over time.

Triple Layer Model (TLM)

In the TLM, which was developed primarily by James Davis (U.S. Geological Survey) and James Leckie (Stanford University), specifically adsorbed ions are placed as partially

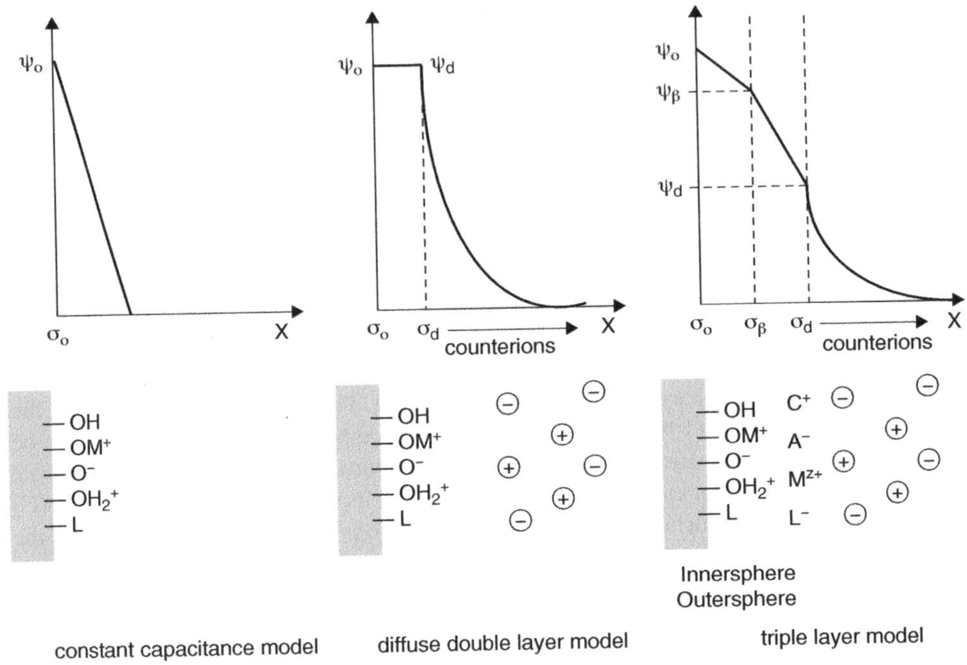

FIGURE 6.13 Idealized schematics of electrical potentials for the diffuse-layer, constant capacitance, and triple-layer surface complexation models. Based on discussion in Davis and Kent (1990) and Goldberg (1998).

solvated ions at a plane of closest approach, along with protons and hydroxides (Figure 6.13). More weakly adsorbed species are located in a second plane. Then, the diffuse charge is distributed in a third zone. This model thus distinguishes between weakly bound and strongly bound ions. Beyond the two layers of strongly and weakly bound ions, the Gouy-Chapman diffuse layer extends. For the two inner layers, the Stern-Grahame model is used for the charge-potential relationships. The Gouy-Chapman model is used for the diffuse layer. The TLM requires the most parameters to be input, and it is the most versatile for modeling complex systems under a variety of conditions.

Charge Distribution CD/MUSIC Model

The CD/MUSIC model was developed by professors Willem H. van Riemsdijk and Tjisse Hiemstra of Wageningen Agricultural University in the Netherlands (e.g., Hiemstra and Van Riemsdijk, 1996) and their coworkers. It is a type of triple-layer model. The CD model relies on the Pauling bond valence concept of charge distribution of ions over coordinating ligands for innersphere adsorption. According to the Pauling concept, ions tend to distribute their charge equally over the coordinating ligand; however, for ions at a surface, there can be an asymmetric distribution of charge, which is taken into consideration in the CD model (Hiemstra and Van Riemsdijk, 1996). According to the CD model, only part of the attached ligand would be incorporated into the structure at the surface; the other part would be in the Stern Layer. Hence, the CD model uses two different electrostatic planes for inner-sphere adsorbed ions. For outer sphere adsorption, the ions ("ion pairs" in

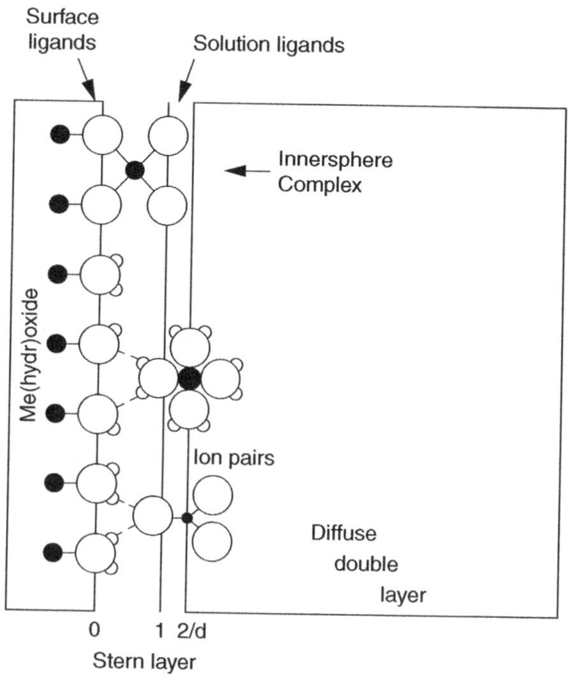

FIGURE 6.14 A schematic illustration of the locations of outersphere and innersphere complexes as represented in the CD model. From Hiemstra and Van Riemsdijk, 1996 (with permission).

Figure 6.14) are treated as point charges, as in previous models. Outer sphere ions are placed in a third plane, removed even more from the surface. In using the CD model, spectroscopic data are important for assigning where an ion resides. Figure 6.14 illustrates the different positions of inner sphere versus outer sphere complexes.

The basic CD model treats the surface functional group as a singly coordinated oxygen atom with either one or two protons, which results in the protonating/deprotonating surface sites that are described with a single pKa value. The more complex MUSIC model uses different equilibrium constants for different types of sites on a surface.

MODEL VERIFICATION AND VALIDATION

Surface complexation modeling is just one form of modeling that is important to environmental surfaces. In a broader sense, it is worth considering: Why are models important? Why does one create a model? How are the models used?

Broadly, models can be either conceptual or quantitative. Quantitative models are equation based and generally computer driven. Construction of quantitative computer models is a goal of much of geochemistry and environmental engineering, because quantitative models are needed for good predictive ability. Equation-based models can be used to compare data quantitatively with model predictions, to test sensitivity of a system to changes in different parameters, to simulate new scenarios, and ultimately to test whether our conceptual models provide good descriptions of real phenomena. Sometimes legal reasons for modeling are indicated. For example, a contaminated site cleanup might be required to ensure that no pollutants in an aquifer reach a nearby river in 10,000 years. Reactive transport models would be needed to help support strategies in such a case.

For any model, model verification and validation (V&V) are crucial. Verification is about ensuring the "simulation is right" whereas validation is about ensuring the "right simulation" (Xiang et al., 2006). Although neither process guarantees absolute confidence in the model predictions, they can often help to point out potential problems or pitfalls in modeling and can help us to converge on a sound modeling approach. V&V typically includes (1) face validity, which involves asking scientific experts to review the conceptual model for scientific plausibility; (2) cross-simulation comparison using different models; (3) empirical validation, which involves comparison of simulation output with actual empirical observations; and (4) visualization of the behavior of the simulation, which can be another useful technique for validation of a simulation (Xiang et al., 2006).

Case Study: Incorporating the Work Associated with Removal of Water During Adsorption into the TLM

Modeling of adsorption at the charged interface is a dynamic and evolving field. For example, Dmitri Sverjensky and his colleagues (e.g., Sverjensky and Fukushi, 2006) recently showed that it is essential to incorporate the electrostatic work associated with desorption of water dipoles (Figure 6.15) in models treating anion adsorption by ligand

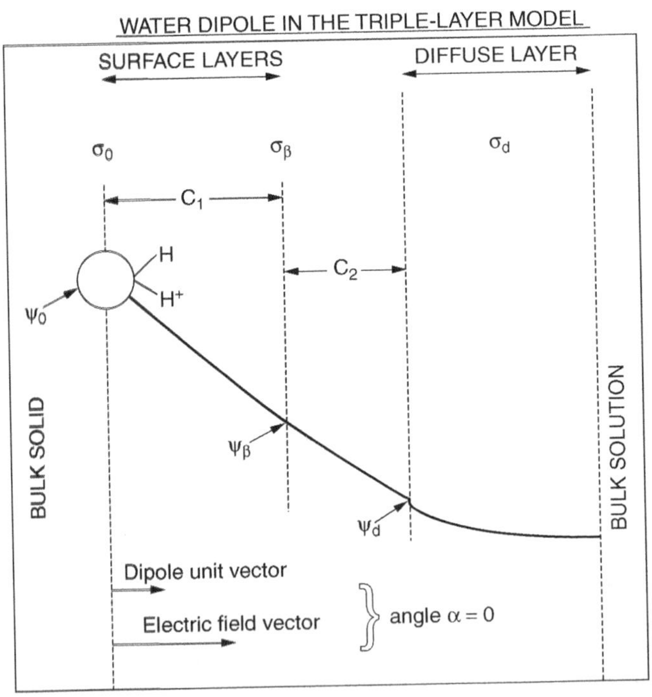

FIGURE 6.15 Representation of a water dipole adsorbed at a surface site in the TLM, neglecting to show other ions, for simplicity. From Sverjensky and Fukushi, 2006 (with permission).

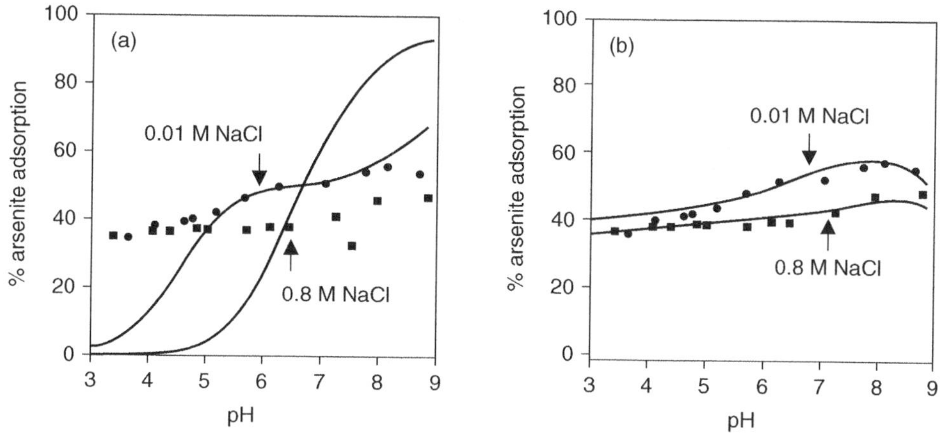

FIGURE 6.16 Modeling of arsenite sorption to aluminum hydroxide. In (a), the classic triple-layer model is used, and there is poor agreement between the model curves and the actual data for arsenite adsorption at two different ionic strengths. In (b), the revised triple-layer model that accounts for the water dipole work fits experimental data extremely well. The models used for (a) and (b) differ in the use of a different electrostatic factor for the binuclear surface complex and also use different log *K* values for sorption, which include the consideration of the water dipole. From Sverjensky and Fukushi, 2006 (with permission).

exchange. They showed that TLM results for anion adsorption match data, which includes spectroscopic data, more closely when the work associated with desorption of water dipoles is incorporated quantitatively.

An example of their work, on arsenite sorption to the aluminum hydroxide mineral bayerite, is shown in Figure 6.16. Surface spectroscopic methods had shown that arsenite forms both a binuclear, bidentate inner-sphere complex and an outer-sphere complex to bayerite. This information was used to construct a surface complexation model. The model was compared with adsorption data as a function of pH and ionic strength, and it was found to fit the data much better when the work associated with the water dipole was included (Figure 6.16).

Sverjensky and Fukushi's work illustrates the V&V methods of *cross-simulation comparison* (the classic TLM and the revised model) and *comparison with empirical analytical results*. By publishing the new model and results for several applications, they have distributed to the scientific community for *face validation* (which also occurred during the review process before publication). The plots provided are a means of *visualizing* the model results.

DLVO THEORY AND COLLOID ATTACHMENT IN POROUS MEDIA

Colloidal particles have diameter that ranges from 1 to 10,000 nm, and they can form stable suspensions under many conditions. These small particles are subject to translational diffusion and Brownian motion and most are too small to be strongly affected by gravity. The interactions that control colloid stability were investigated quantitatively and separately by Derjaguin and Landau (1941) as well as Verwey and Overbeek (1948). The theory that has developed for colloid stability has become known as the DLVO theory. DLVO theory is used to understand and model the aggregation/coagulation behavior of colloidal particles and the interactions of colloids with sediments and soils in the subsurface.

According to classic DLVO theory, interactions between colloids or colloids and larger particles are primarily caused by electrostatic forces and attractive van der Waals forces. The stability of a particle in suspension is dependent on its total potential energy V_T

$$V_T = V_A + V_R + V_S, \tag{6.54}$$

which is the sum of

V_A = potential energy from attractive van der Waals forces

V_R = potential energy from repulsive, electrical double layer forces (e.g., Hogg et al., 1966);

V_s = potential energy from the solvent, which is generally small and often neglected in these calculations.

In DLVO theory, the stability of a colloidal system is determined by the sum of the attractive van der Waals and electrical double-layer forces (repulsive between particles of like charge) that exist between the particles as they approach one another through Brownian motion. For charged particles of like sign that are approaching one another, the energy barrier resulting from the repulsive electrostatic force prevents the particles from approaching one another closely and adhering. If the particles can collide with sufficient energy to overcome the repulsive barrier, then the attractive van der Waals force will allow them to adhere to one another strongly, which causes the suspension to become unstable and the colloidal particles to flocculate out.

For suspensions in which particles repulse one another strongly, the dispersion will resist flocculation so that the colloidal suspension will be stable. In the absence of strong repulsion, the colloidal suspension will be unstable, and the particles will flocculate out of suspension.

For like particles, the value of V_T as the sum of different V_A and V_R and as a function of different particle separation distances (r) is shown in Figure 6.17. In Figure 6.17, $V_R(1)$ and $V_T(1)$ would be for a lower ionic strength solution where the double-layer thickness would be greater and $V_R(2)$ and $V_T(2)$ would be for a higher ionic strength solution where the double layer would be more compressed. The attractive van der Waals potential energy V_A decreases as an inverse power of r. The repulsive electrostatic potential energy V_R acts over the thickness of the double layer (κ^{-1}) and increases as an approximately exponential function of r. A more positive value of V_T promotes suspension stability. The potential energy maximum tends to occur at intermediate r where electrostatic repulsion is relatively more important. At short and long distances of approach, van der Waals interaction forces tend to predominate.

Overall, the colloidal suspension should be stable when the potential energy maximum is large relative to the thermal kinetic energy kT of the particles. In general, a higher ionic strength solution tends to promote colloid instability and aggregation, given that the potential energy maximum that results from summation of the two curves is depressed. Particle flocculation often can be observed when low-ionic-strength, colloid-carrying rivers come together with high-ionic-strength waters in rivers and in estuaries. An example is the

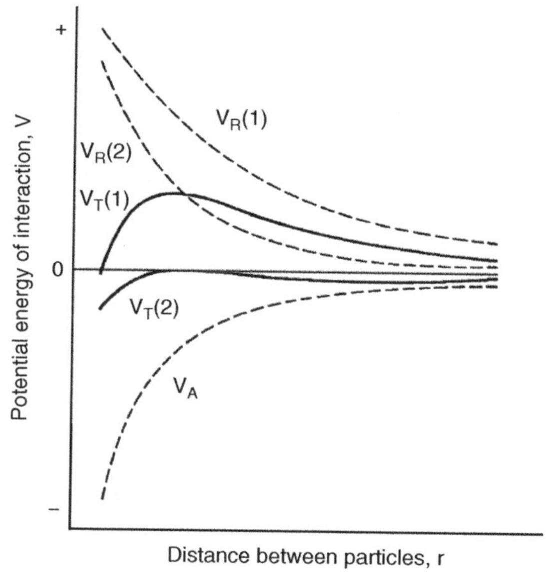

FIGURE 6.17 Potential energy of interaction (V_T) curves as a function of distance r, as the sum of attractive (V_A) and repulsive (V_R) energies, for two systems with lower (1) and higher (2) ionic strength. Note that these curves neglect the strong Born repulsion that occurs at very close approach and that is not considered in classic DLVO theory. The Born repulsion is shown in the schematic in Figure 4.9. After Shaw, 1992 (with permission).

confluence of the low-ionic-strength Rio Negro and the high-ionic-strength Rio Solimões to form the Amazon River, where colloids from the Rio Negro quickly settle out of suspension.

DLVO theory has been a mainstay of colloid research. Yet, many subsequent studies have shown that the forces accounted for in DLVO theory often are not sufficient to describe observed interactions between surfaces (e.g, Israelachvili and Adams, 1978) or colloid mobility and aggregation/flocculation behavior. One example is the very high repulsive forces that occur during very close approach as the electron orbitals associated with atoms on one particle begin to overlap with those of the other particle. These forces are known as *Born repulsion forces* and are a result of the *Pauli exclusion principle*.

In a study of the detachment of polystyrene latex colloids from a packed column of glass beads, Bergendahl and Grasso (1999) considered the total interaction energy (V_T or ΔG_T) between the colloid and glass bead surfaces as the sum of electrostatic (V_R) and van der Waals (V_A) interactions as in DLVO theory, but added Born repulsion (V_B) and Lewis acid-base (V_{AB}) interactions energies

$$V_T = V_A + V_R + V_B + V_{AB}. \tag{6.55}$$

A negative value of V_T is indicative of attraction, which will lead to attachment of colloids to beads, whereas a positive value indicates repulsion, which will lead to lack of attachment or detachment of the colloids from the beads. Detailed equations for each of the components of the interaction energy are provided by Bergendahl and Grasso (1999); refinement of equations and approaches is ongoing in the field of colloid chemistry and transport.

Extensions to the DLVO theory are needed to account for additional forces and interactions that involve environmental *colloids*, such as hydrogen bonding and the related hydrophobic effect, hydration pressure, non–charge-transfer Lewis acid–base interactions, and steric interactions (Grasso et al., 2002). Adsorption can alter surface properties and needs to be considered as well. The surface roughness of particles can cause deviations of experimental results from predictions of both DLVO and "extended" or EDLVO theory. The effects of surface roughness on colloid mobility through porous media are discussed in more detail in Chaper 13, with respect to colloids and nanoparticles. As pointed out in Chapter 5, assumptions of perfectly planar surfaces can make modeling easier but can lead to incorrect model predictions. How one defines a surface or an interface is essential for accurate predictive modeling of attachment/detachment from grains in porous media. The fact that water is flowing in a column experiment or an aquifer also needs to be considered.

Understanding interactions at charged particle surfaces is a continuously evolving field. On the one hand, scientists and engineers strive for simple models. But on the other hand, the natural world is complex and simple models seldom provide the flexibility needed to account for complex interactions. Thus, the mineral–water interface remains a frontier that will continue to provide opportunities for exploration for many years into the future.

QUESTIONS FOR FURTHER THOUGHT

1. Using the library and the Internet, compare different point of zero charge values and descriptions given by different authors; for example, you may use many texts and manuscripts referenced in this chapter. You may also look up information on websites. Be careful to provide detailed references. Are the terms being used consistently and correctly? Can the values be compared directly? What sort of ranges are apparent for different minerals, both within a given author's tabulation and between authors? Write a paper several pages long discussing the values of points of zero charge given by different authors, taking into consideration the questions raised above. What do you think may be contributing to the differences in values provided by different authors? What are some of the points of disagreement or confusion? [Note: Sposito's *The Surface Chemistry of Natural Particles* goes into points of zero charge in great detail.]
2. If a smectite sample has the following formula, what is the range of σ_0 if the layer charge varies from 0.4 to 1.2 (Sposito, 2004)?

$$Na_{0.65}(Si_{7.85}Al_{0.15})(Al_{3.04}Fe_{0.46}Mg_{0.50})O_{20}(OH)_4$$

3. Answer the following questions in a few sentences, using figures as needed to help explain your reasoning.
   (a) The wall of a glass beaker contains >SiOH groups, and glass p.z.n.p.c. is similar to that of quartz. Why should we not use glass reaction vessels for studies involving dissolved trace metals?
   (b) If an anion such as sulfate were to adsorb inner-spherically, as does phosphate, then should adsorption of that anion shift the point of zero charge of hematite to a higher or lower pH? Explain your reasoning, including what factors might need to be considered.
   (c) Particles of hematite are placed in an aqueous solution whose initial pH is 2.0. Would you expect the solution pH to rise or fall? Explain your reasoning.

4. Would you expect ionic strength to affect electrophoretic mobility of oxide particles? Why or why not?
5. Considering Figure 6.12, how might you go about defining the surface and the interface? Give you reasoning.
6. (a) Plot how the thickness of the double layer in nanometers (1 nm $= 10^{-9}$ m) would change for a particle in NaCl solution from 0.0001M to 0.1 M at 25°C.
   (b) Use your plot to comment on the effects of electrolyte concentration on double-layer thickness.
7. Using Guoy-Chapman theory, explain why a trivalent cation such as Al^{3+} or Fe^{3+} is more likely to promote particle aggregation and colloid instability than a divalent or monovalent ion.
8. Calculate the electrophoretic mobility of a colloidal particle that travels 20 mm in 1 hour in a tube 100 mm long with a total potential difference across the length of the tube of 250 V.
9. One sample of smectite clay is added to a sample of an "average" river water and a separate sample is reacted with ocean water. Which suspension would you expect to be least stable with respect to aggregation and settling of the clay? Explain your reasoning, including equations and/or figures as needed.
10. A mineral in suspension has an intrinsic surface charge density of -13.6 mmol$_c$ kg^{-1} and inner-sphere and outer-sphere adsorbed ion charges of $+1.2$ mmol$_c$ kg^{-1} and $+2.1$ mmol$_c$ kg^{-1}, respectively. The mineral has specific surface area $= 100$ m^2 g^{-1}.
    (a) What are σ_{in}, σ_{is}, and σ_{os} in units of Coulombs m^{-2}?
    (b) What is the diffuse ion swarm charge, σ_d, in the vicinity of the particle?
    (c) If $\sigma_o = -10.8$ mmol$_c$ kg^{-1}, then what is the value of σ_H?
11. Calculate surface potential ψ (in V) at 25°C for a particle suspension in 0.01 M ionic strength solution of an inert electrolyte when Z $= 1$ (1:1) and $\sigma_p = 6.5 \times 10^{-2}$ Coulombs m^{-2}.
12. The cation exchange reaction on a montmorillonite clay proceeds by:

$$Ca^{2+} + Na_2X \rightleftarrows 2Na^+ + CaX$$

    (a) If the clay is in equilibrium with a solution with Na and Ca concentrations the same as in mean ocean water, what is the ratio of mole fraction N_{CaX} to N_{Na_2X}? For this problem, assume that a conditional equilibrium exchange constant or selectivity coefficient can be used, where D $=$ $^cK_{ex} = 0.25$ for the reaction as written above, and for simplicity also assume that any other cations can be neglected.
    (b) The same clay is in equilibrium with a groundwater with [Ca] $= 80$ mg L^{-1} and [Na] $= 8.1$ mg L^{-1}. What is the ratio of N_{CaX} to N_{Na_2X} for this groundwater?
    (c) The same clay is in equilibrium with a surface water with [Ca] $= 1.2$ mg L^{-1} and [Na] $= 1.0$ mg L^{-1}. What is the ratio for this surface water?
    (d) What does this exercise tell you about the cation sorption to montmorillonite in different environments? What does this problem tell you about what controls the relative sorption of Ca versus Na to montmorillonite in different environments?

13. A surface water with 1 mmol each Na and Ca is in equilibrium with a clay with $D = K_c = 1$, using the same reaction and mass action as in question 12.
    (a) What is the ratio of mole fraction N_{CaX} to N_{Na_2X}?
    (b) For this clay, if the equilibrium concentration of Na is 0.01 M, at what solution concentration of Ca will Ca and Na be sorbed in equal number of moles?

$$Ca^{2+} + Na_2X \rightleftarrows 2Na^+ + CaX$$

14. The cation exchange of a montmorillonite clay proceeds according to:

$$K^+ + NaX \rightleftarrows KX + Na^+,$$

and $D = {}^cK_{ex} = 1$ for the reaction as written.
    (a) If the equilibrium solution concentrations of Na and K are both 0.001 M, what is the ratio of N_{KX} to N_{NaX}?
    (b) If the concentration of Na is 0.002 M and the concentration of K is 0.001 M, what is the ratio of N_{KX} to N_{NaX}?
    (c) If the concentration of Na is 0.0005 M and the concentration of K is 0.001 M, what is the ratio of N_{KX} to N_{NaX}?
    (d) What does this suggest to you about the sorption preference for cations of the same valence (homovalent cations), when $D = K_c = 1$? For this problem assume no other cations in solution.
15. A clay has a CEC of 0.1 $mol_c\ kg^{-1}$ and is fully Na-exchanged. Ten grams of the clay are placed in a solution with 0.01 M Na and 0.02 M K (and no other cations) and allowed to equilibrate. For the exchange reaction

$$K^+ + NaX \rightleftarrows KX + Na^+$$
$${}^cK_{ex} = 0.80 = \frac{(m_{Na})(KX)}{(m_K)(NaX)},$$

where m_{Na}, m_K, (KX) and (NaX) are in units of moles L^{-1}.
    (a) What will be the final solution concentrations in moles L^{-1} at equilibrium of Na and K?
    (b) The final surface concentrations of Na and K in moles cation kg^{-1} solid? Assume neglible anion exchange capacity for the clay.
16. Langmuir (1997) gives values of $K_{ex} = 0.25 \pm 0.05$ and $n = 1.25 \pm 0.06$ at pH 4.1 to 8.2 for Ca^{2+} (7.9 to 9×10^{-4} M range) and Pb^{2+} (4 to 890×10^{-7}M range) exchange on Ca-montmorillonite:

$$Ca^{2+} + PbX \rightleftarrows Pb^{2+} + CaX.$$

If the dissolved concentrations of Ca and Pb in equilibrium with the montmorillonite are 8.0×10^{-4} M and 1.0×10^{-6} M, respectively, then what is the ratio of mole fractions PbX/CaX?
17. Dzombak and Morel (1990) discuss diffuse-layer model applications to the hydrous ferric oxide (HFO) surface, with extensive data compilations. They give values of the

coulombic correction factor $e^{(-F\psi/RT)}$ for different pH and ionic strength values for a 1:1 electrolyte:

	I (M)			
	1.0	0.1	0.01	0.001
pH = 4.00	$10^{-1.58}$	$10^{-2.45}$	$10^{-3.11}$	$10^{-3.56}$
pH = 5.00	$10^{-1.47}$	$10^{-2.12}$	$10^{-2.56}$	$10^{-2.85}$
pH = 6.00	$10^{-1.15}$	$10^{-1.57}$	$10^{-1.85}$	$10^{-2.01}$
pH = 7.00	$10^{-0.64}$	$10^{-0.88}$	$10^{-1.02}$	$10^{-1.07}$
pH = 8.00	$10^{-0.06}$	$10^{-0.09}$	$10^{-0.10}$	$10^{-0.10}$

Based on this data set, plot

(a) how the value of ψ changes with ionic strength at each of the pH values and

(b) how the value of ψ changes with pH at I = 0.1 M at 25°C. Note: IF you use F = 96,480 C mol^{-1}, R = 8.314 J mol^{-1} K^{-1} and T = 298.15 K, the ψ will be in volts (1 VC = 1 J). (Hint: As an example to check your calculations, for pH 6 and I = 0.1, ψ = 0.093 V.)

(c) How does the thickness of the double layer (in nm) change as a function of I from 0.001M to 0.1 M?

18. Based on the data set and values for ψ calculated in problem 17, calculate the surface charge density σ values in C m^{-2} for pH 6, and I = 1.0, 0.1, 0.01, and 0.001 M. Assume that protons are the only surface potential determining ions. (Note: Typically, to determine sinh of a number X on a calculator, press HYP followed by SIN and then key in X).

19. According to Dzombak and Morel (1990), the values of intrinsic equlibrium acidity constants for HFO (FeOOH) at 25°C are $K_{a1}^{int} = 10^{-7.29}$ and $K_{a2}^{int} = 10^{-8.23}$. You have a sample of HFO with 600 m^2 g^{-1} specific surface area (value used by Dzombak and Morel, 1990) and a suspension that contains 0.001 M HFO (where the molarity is in terms of surface sites). The binding site density for adsorption of protons, cations, and anions on HFO = 0.2 mol/mol Fe (Dzombak and Morel, 1990). Your suspension is at pH 5, and in an inert 1:1 electrolyte with I = 0.01 M. Assuming that the only ion that adsorbs to the surface in this suspension is H^+ (and that HFO is insoluble), so that the mole balance for the surface can be written in terms of surface sites (Γ = $>FeOH_2^+ + >FeOH^\circ + > FeO^-$),

(a) calculate the surface molar concentrations of each of these three sites.

(b) What percent of the total surface sites does each of these three sites correspond to at pH 5 and I = 0.01M (1:1 electrolyte)?

(c) Repeat the calculations (a) and (b) for a similar suspension at pH 8 and I = 0.01M.

(d) How does the distribution of positively charged, negatively charged, and neutral sites change with pH from 5 to 8?

(e) Do your results make sense? [Hint: Use the values of ψ calculated in previous problem. Solve simultaneous equations using the expressions for the two intrinsic surface acidic constants, the expression for pH, and the equation for surface mole balance. Use the Gouy-Chapman relationship for σ and assume activity coefficients = 1]. If you have access to a computer program that can model surface speciation using the diffuse layer approach, compare the results of your hand calculations with those of the computer program.

20. The same system as in Problem 19 has 5×10^{-7} M total of a metal M(II) at pH = 6.0 and I = 0.01 M (1:1 inert electrolyte). If M(II) adsorption occurs at only one site on the surface and the intrinsic surface complex constant for M(II) adsorption is $10^{1.02}$, what percent of the total M is adsorbed to HFO? (Note: This problem is hypothetical and is not representative of any specific metal studied by Dzombak and Morel, 1990; it is simplified to only encompass one surface site, whereas for HFO there are two sites for most metals). The metal exists in solution under these conditions only as the M^{2+} species. Use the diffuse-layer model approach and ignore activity coefficients. (Hint: The equation for surface mole balance now contains the $>FeOM^+$ species, and you can write a mole balance for total M(II) = M^{2+} (aq) + $FeOM^+$).

$$K^{int}_{M,s} = 10^{1.02} = \frac{(>FeOM^+)(H^+)}{(>FeOH^o)(M^{2+})} e^{(-F\psi/RT)}.$$

21. A river laden with colloidal particles such as Fe(hydr)oxides and clays enters an estuary to the ocean.
    (a) What would you expect to happen to the particles?
    (b) Explain your reasoning using equations where possible.

FURTHER READING

Chorover, J., and Sposito, G. 1995. Surface charge characteristics of kaolinitic tropical soils. *Geochimica et Cosmochimica Acta* 59, 875–884.

Davis, J.A., and Kent, D.B. 1990. Surface complexation modeling in aqueous geochemistry. In Hochella, M.F., Jr., and White, A.F. (eds), Mineral Water Interface Geochemistry, Reviews in Mineralogy, vol. 23. The Mineralogical Society of America, Chantilly, VA, pp. 177–279.

Dzombak, D.A., and Morel, F.M.M. 1990. Surface Complexation Modeling: Hydrous Ferric Oxide. John Wiley and Sons, New York, 393 p.

Sposito, G. 1989. The Chemistry of Soils. Oxford University Press, Oxford, UK. Chapters 7 and 8.

Sposito, G. 2004. The Surface Chemistry of Natural Particles. Oxford University Press, Oxford, UK. Chapter 1.

Stumm, W. 1992. Chemistry of the Solid-Water Interface: Processes at the Mineral-Water and Particle-Water Interface in Natural Systems. Wiley Interscience, New York. Chapter 3.

7

SORPTION: INORGANIC CATIONS AND ANIONS

In 1965, J. A. Kitchener wrote:

> Despite its wide occurrence and practical importance, adsorption at the interface between a solid and an aqueous solution still presents one of the most complicated problems in surface chemistry.

Although we have come very far in understanding sorption processes, these words still are true today, especially as more techniques evolve and more complexities are discerned. This chapter discusses metal and inorganic anion sorption by drawing on the material in Chapters 5 and 6 and demonstrates how a variety of approaches and methods can be applied.

To understand controls on and mechanisms of inorganic ion sorption, a basic understanding of intermolecular and surface forces is needed. Israelachvili (1992) identifies the following three classes of intermolecular forces: (1) purely electrostatic, resulting from Coulombic forces between charges; (2) polarization forces that develop from induced dipole moments; and (3) quantum mechanical forces that cause to covalent or chemical bonding.

The sorption of inorganic ions involves mostly electrostatic interactions, of primary importance for outer-sphere sorption, and formation of covalent bonds with the surface, of primary importance for inner-sphere sorption. In some instances, hydrogen bonds, which are a special type of dipole–dipole interaction with strength intermediate between purely covalent bonds and electrostatic interactions, may form between the sorbate and the sorbent. Solvation or (in water) hydration forces also need to be considered in ionic sorption (Sverjensky and Fukushi, 2006).

Although sorption is generally thought of as immobilizing an ion or molecule, in some cases, mobility can be enhanced when the sorption is to (or forms) mobile colloids, including nanoparticles (McCarthy and Zachara, 1989) (Figure 7.1). This chapter focuses on sorption

Environmental Surfaces and Interfaces from the Nanoscale to the Global Scale, by Patricia A. Maurice

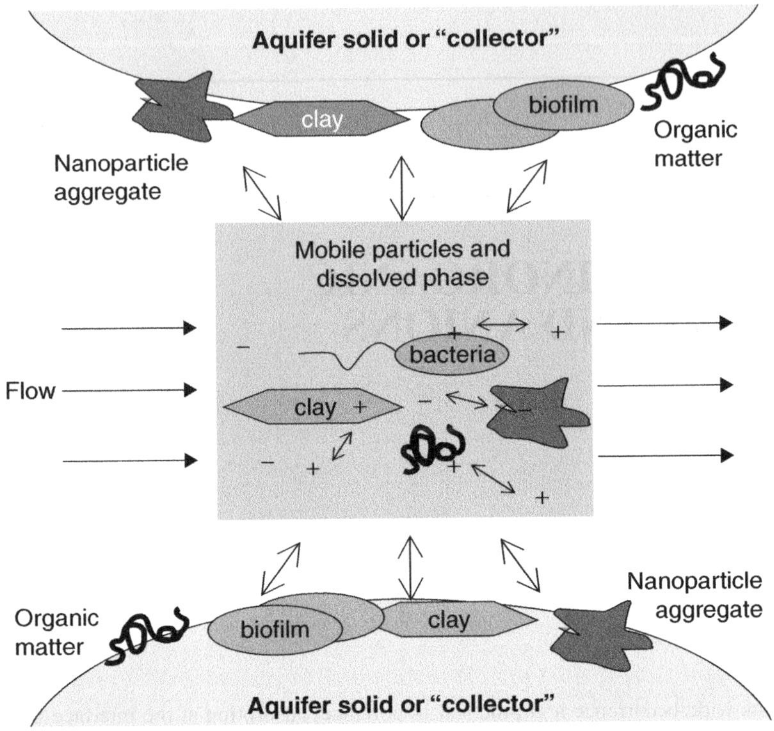

FIGURE 7.1 A schematic of how sorption can affect fate and transport of organic and inorganic species through porous media.

to mineral surfaces, with discussion of sorption to bacteria and biofilms reserved primarily for Chapter 12.

A TYPICAL SORPTION EXPERIMENT DESIGN

Many controls on sorption to mineral surfaces are indicated, such as (1) the properties of the surface, including density and nature of surface sites, surface charge properties, micromorphology, presence of impurities, or surface coatings; (2) the properties of the solution, including pH, ionic strength, identity of the electrolyte, temperature, concentration of sorptive ion or molecule and of other species that may interact, for example, by forming ternary surface complexes (Chapter 8) or competing for sorption sites; (3) the speciation/complexation properties of the sorptive ion both in solution and at the surface, such as hydrolysis and valence state; (4) the saturation state of the sorbent (relative to potential dissolution) and the sorbate (relative to potential precipitation) and of the two together (relative to potential coprecipitation); and (5) the redox conditions, along with the potential for redox reations to occur between the sorbate and the sorbent or between different species in solution. Redox reactions may be enhanced in the presence of light for systems that have the potential for photocatalysis. In the field, or if laboratory experiments are conducted over days to weeks without sterile conditions, microbial processes may occur that can complicate results and interpretation.

To build models of how sorption affects solute fate and transport, information about sorption mechanism is essential. A key question is whether sorption is by adsorption, absorption, precipitation, or coprecipitation. For adsorption (and absorption), it is important to characterize whether the complex is inner sphere or outer sphere. Generally, adsorption is thought to be outer sphere if it only occurs to surfaces of opposite charge. Thus, outer sphere adsorption of a cation, for example, would only occur above the point of zero charge of the mineral adsorbent. Outer-sphere adsorption is generally thought to be far more sensitive to ionic strength effects than is inner-sphere adsorption. If adsorption is found to shift the pH of the point of zero charge appreciably, then inner-sphere adsorption is suspected. These generalizations do not always hold true. For example, many ions may adsorb both inner spherically or outer spherically. Independent evidence, commonly spectroscopic, is always needed to determine adsorption mechanism(s) under a given set of conditions.

A laboratory sorption experiment typically is conducted by reacting a sorptive ion or molecule with a sorbent and by measuring the amount of sorption either directly or indirectly, based on changes to solution concentration. A "typical" experimental design for an (ad)sorption experiment is shown in Figure 7.2. An important component of a sorption

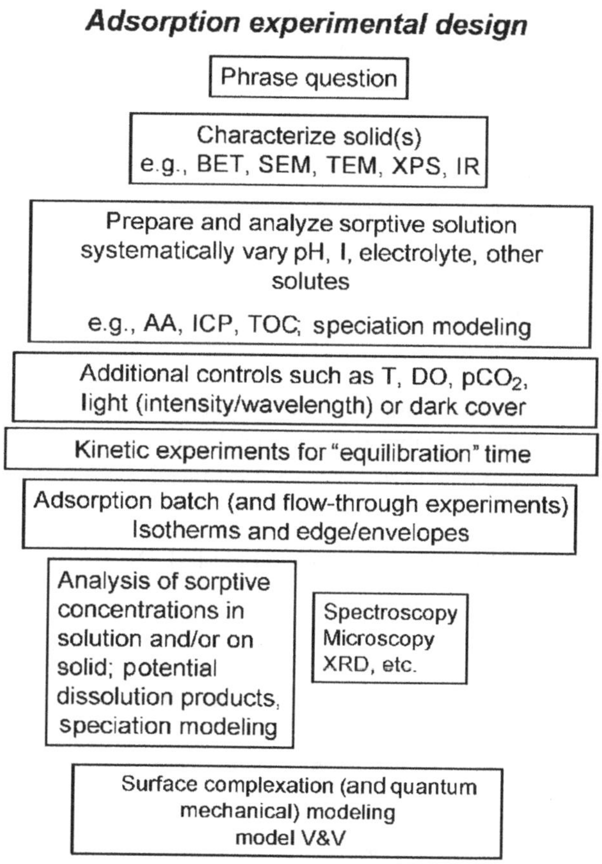

FIGURE 7.2 A schematic of a typical adsorption experiment. Generally, the process flows from top to bottom, but sometimes the sequence is altered for specific purposes. BET = surface area measurement, SEM = Scanning Electron Microscopy, TEM = Transmission Electron Microscopy, XPS = X-ray Photoelectron Spectroscopy, IR = Infrared spectroscopy, AA = Atomic Absorption, ICP = Inductively-Coupled Plasma emission spectrophotometry, TOC = Total Organic Carbon analysis, T = Temperature, DO = Dissolved Oxygen, pCO_2 = atmospheric carbon dioxide concentration, V&V = Verification and Validation.

experiment is the use of a variety of methods, including spectroscopic and microscopic, to help account for different controls and to determine sorption mechanism(s). Few projects follow all the components given in this generalized design. But over time, many converge on this approach. Although this design is for experiments that involve mineral (or amorphous solid) sorbents, experiments are increasingly focusing on organic and/or biotic components such as biofilms.

METAL CATION SORPTION

Metal cation adsorption to mineral surfaces generally follows an adsorption edge, in which adsorption increases with increasing pH (Figure 7.3). Adsorption can be based on electrostatic attraction or inner-sphere surface complexation. In some instances, a given metal cation may adsorb to a given surface primarily by electrostatic attraction under some conditions (of, for example, pH, ionic strength, absorptive concentration) and by surface complexation under other conditions, or by both simultaneously.

When metal–cation adsorption occurs through electrostatic attraction alone, the following two different situations are possible:

1. The adsorbed metal ion is part of the diffuse layer and is not associated with any specific surface site
2. Adsorption is based on ion pairing with deprotonated surface hydroxyls, as illustrated in the equation below.

$$>S{-}O^- + M(H_2O)_6^{z+} \rightleftarrows >S{-}O^-(H_2O)M(H_2O)_5^{z+} \quad (7.1)$$

(Schindler, 1990).

In both (1) and (2), the adsorption forms an outer-sphere complex with at least one water molecule between the surface and the adsorbed metal ion.

Surface inner-sphere complexation involves reaction of deprotonated hydroxyls with metal cations according to the following generalized reaction:

$$>S{-}O^- + M(H_2O)_6^{z+} \rightleftarrows >S{-}O - M(H_2O)_5^{(z-1)} + H_2O \quad (7.2)$$

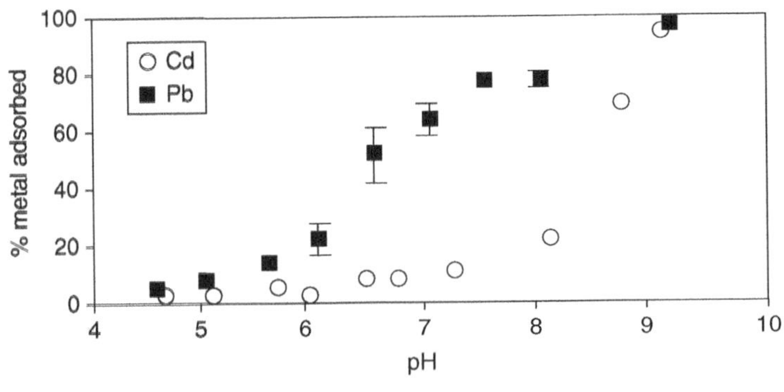

FIGURE 7.3 Adsorption edges for Cd and Pb (11.1 μM) adsorption to kaolinite in 0.1-M $NaClO_4$ at 22°C. Note that this edge is in terms of percent metal adsorbed rather than concentration per unit surface area. Based on Hepinstall et al., 2005.

(Schindler, 1990), in which the metal cation becomes directly bound to the mineral surface. Note that Equations 7.1 and 7.2 are for metals with a coordination number of 6, although analogous reaction schemes can occur for metals with different coordination. These equations are also generalized; different nuances have been observed over time.

As for many aspects of environmental surface and interface chemistry, it is difficult to generalize about (ad)sorption phenomena. Although some trends for cation adsorption are given here, it should be recognized that they are broad generalizations and that many exceptions exist. Adsorption of alkali and alkaline earth elements (Groups I and II of the periodic table) occurs in many instances by formation of outer-sphere complexes. However, inner-sphere adsorption may occur for some ions and to some surfaces. For clays with permanent structural charge, the likelihood of inner-sphere complexation decreases for more hydrating cations (Chapter 2).

The *alkali metals* (Group I), which include Li, Na, K, Rb, Cs, and Fr, tend to lose one electron to form singly charged cations. The ability to exchange solvation water molecules defines the inner-sphere sorption affinity for the siloxane interlayer surface of smectite/montmorillonite. The sequence for selectivity of inner-sphere sorption to smectites, therefore, tends to be

$$Cs^+ > Rb^+ > K^+ > Na^+ > Li^+.$$

K^+ and Cs^+ are only weakly hydrating and thus are more likely to lose their waters of hydration and form inner sphere bonds with the siloxane surface of smectites than are Na^+ and Li^+. K^+ in mica interlayers tends to hold the layers tightly together and to prevent swelling. The selectivity may be the reverse for minerals such as iron oxides that do not have the permanent structural charge of smectite.

Cs^+ is the highest the affinity and Li^+ the least affinity for charged clays such as smectite. The differences in hydration and sorption behavior form the basis for using Cs^+ to probe permanent structural charge and Li^+ (along with Cl^-) for other surface charge components of clays (Chorover and Sposito, 1995; see Chapter 6).

The *alkaline earth metals* are a series of elements in Group II of the periodic table, which include Be, Mg, Ca, Sr, Ba, and Ra. These elements tend to lose two electrons to form doubly charged cations. The affinity sequence for sorption to clays with permanent charge tends to be

$$Ba^{2+} > Sr^{2+} > Ca^{2+} > Mg^{2+}$$

(Sposito, 1989). Ca^{2+} and Mg^{2+} are strongly hydrating and tend to sorb outer spherically to the permanently charged clays.

Trace-metal adsorption frequently occurs by formation of inner-sphere surface complexes (e.g., Brown and Parks, 2001). Divalent trace metal adsorption affinities have a tendency to follow approximately the Irving-Williams order (Sposito, 1989)

$$Cu^{2+} > Ni^{2+} > Co^{2+} > Fe^{2+} > Mn^{2+}$$

but with many exceptions. This series also tends to be followed for ligand exchange of metals in solution. The reasons for this series/trend include (1) across the periodic table (in the opposite direction shown from the series above), effective nuclear charge increases because three–dimensional (3–D) orbitals are poorly shielding; (2) the metal ionic radius decreases, resulting in increased Lewis acidity and stronger metal–ligand bond; and (3) there is increased stabilization from the ligand field stabilization energy.

Over time, different experimental studies for trace metals have resulted in different measured affinity sequences. Some metals may sorb by different mechanisms to different surfaces or under different conditions. At some pH and concentration values, (surface) precipitation may occur, which if not recognized could greatly affect results. Other factors such as temperature, ionic strength, identity of the background electrolyte, initial concentration of the (ad)sorptive ion, and purity of mineral (ad)sorbents can also result in different observed sorption affinities for various trace metals. "Equilibration time" is also important, as fast initial (ad)sorption can sometimes be followed by slower changes to (ad)sorption over the course of days to weeks to months (e.g., Fischer et al., 2007). In some cases, the differences in affinities between the metals are small and can overlap.

Cornell and Schwertmann (2003) compared some published observations of metal adsorption to *goethite*

$$Cu > Pb > Zn > Cd > Co > Ni > Mn,$$

to *hematite*

$$Pb > Cu > Zn > Cd > Co > Ni > Mn,$$

and to *ferrihydrite*

$$Pb > Cu > Cd > Zn > Ni > Ca.$$

The differences in observed affinities can be small in some instances, as shown in Figure 7.4.

Brown et al. (1999) compared sorption edges of Co(II) with α-SiO_2, TiO_2, and γ-Al_2O_3 (Figure 7.5; data for γ-Al_2O_3 not shown). In the case of α-SiO_2, sorption occurs well above the point of zero charge of the surface; whereas for the other sorbents, there is far more sorption below and/or around the point of zero charge. The authors point out that the different behaviors can be explained at least in part by different mechanisms or modes of sorption, as probed by X-ray absorption spectroscopy (XAS) methods. Extended X-ray

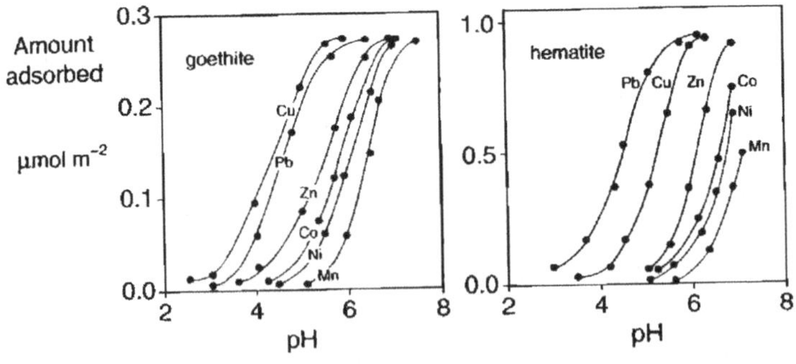

FIGURE 7.4 Adsorption edges of some heavy metal cations on hematite and goethite. From McKenzie, 1980 (with permission).

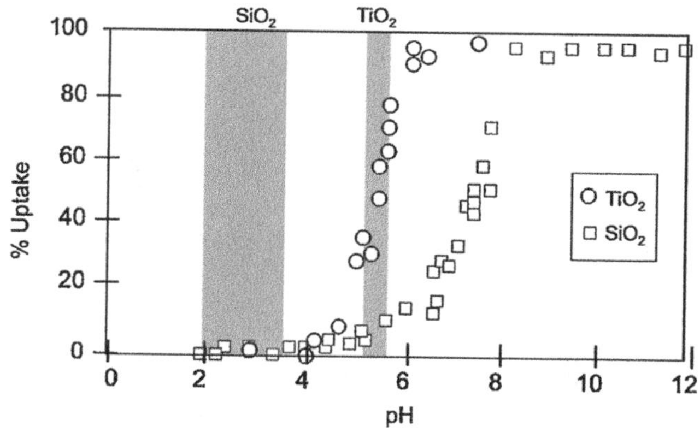

FIGURE 7.5 Co sorption edges to SiO_2 and TiO_2. Bands show approximate point of zero charge range for these two surfaces. After Brown et al., 1999 (with permission).

absorption fine structure (EXAFS) results showed that Co formed multinuclear complexes or surface precipitates on α-SiO_2 and TiO_2, but that Co formed both multinuclear complexes and inner-sphere (bidentate) complexes on γ-Al_2O_3. This example shows the importance of distinguishing uptake by adsorption versus by other processes such as precipitation.

The study of metal affinities to oxide surfaces is an evolving field, particularly as spectroscopic techniques are applied that allow determination of adsorption complex structures of various metal–mineral sorbent combinations (Brown et al., 1999), along with different modeling approaches (e.g., Ponthieu et al, 2006). As we shall learn in Chapter 13, nanoparticle oxides can have such variable properties, particularly with regard to surface chemistry and reactivity, that much needs to be learned about sorption to these smallest of particles.

Stipp et al. (1992) investigated Cd uptake by calcite from aqueous solution and analyzed the reacted mineral surface using XPS, Auger electron spectroscopy (AES), and low-energy electron diffraction (LEED). Their results indicated that initial Cd adsorption to calcite is followed by incorporation into the solid, which eventually results in solid solution of Ca- and Cd-carbonate.

The Complexity of Cation Adsorption

Numerous examples illustrate the complexity of metal cation sorption processes. A few are provided here.

Cd Adsorption to Kaolinite Cd adsorption to kaolinite is an example of the observation that a given metal cation may adsorb to a given mineral by different mechanisms under different conditions. Cd is a trace metal that is used in batteries and certain pigments, particularly for plastics. It tends to exist as the Cd^{2+} cation in aqueous solution over a broad pH range, but it can precipitate as a carbonate or hydroxide mineral at high pH. Vasconcelos et al. (2008) investigated the adsorption of Cd(II) to kaolinite as a function of pH and Cd loading at 0.1 M ionic strength, combining batch adsorption experiments with EXAFS analysis. The adsorption of Cd increased with increasing pH, forming an adsorption edge, as shown in Figure 7.6.

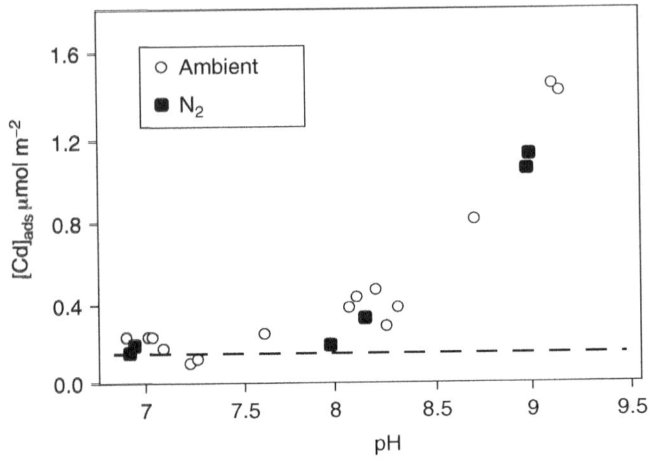

FIGURE 7.6 Adsorption edges for Cd on kaolinite in the presence of CO_2 (ambient) and in samples purged with N_2 to prevent carbonate formation. Revised from Vasconcelos et al., 2008 (with permission).

EXAFS was used to probe the coordination environment of Cd(II) at different pH conditions. Kaolinite displays two general classes of sites that have the potential to interact with adsorptive cations: sites on the basal (001) faces and sites at the particle edges (Sposito, 1989; Brady et al., 1996) (see Figure 3.5). In terms of cation adsorption, the Si–O–Si sites on the basal plane surfaces are generally considered to be unreactive (Sposito, 1984), although they may adsorb nonpolar organics. A small amount of structural charge from isomorphic substitution can create diffuse negatively charged sites that allow at most a minor amount of cation exchange on kaolinite. Molecular modeling by Vasconcelos et al. (2007) showed that Cd(II) may associate with the ditrigonal cavities associated with the siloxane basal-plane surfaces (Figure 7.7). Sites on the basal plane where a hydroxide group is bridged between two Al atoms (Al–OH–Al) also can permit cation adsorption.

In addition to basal-plane sites, three types of edge sites include sites where the OH group is bonded to a single Al atom (Al–OH), sites where the surface oxygen bridges Al

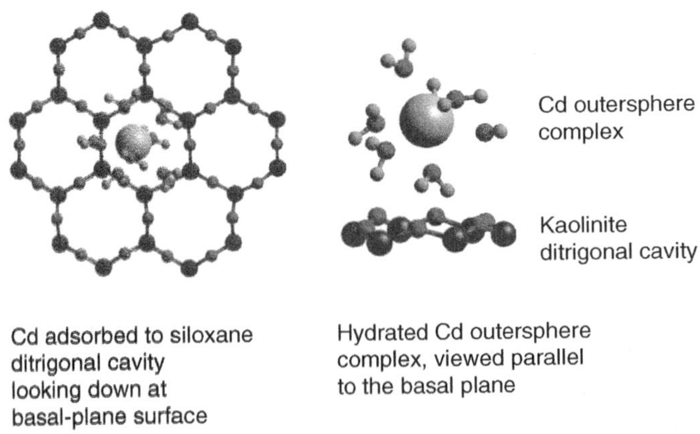

FIGURE 7.7 Cd sorption to the basal-plane surface of kaolinite as proposed by Vasconcelos et al. (2007, 2008). Models courtesy of I. Vasconcelos, Univ of Notre Dame.

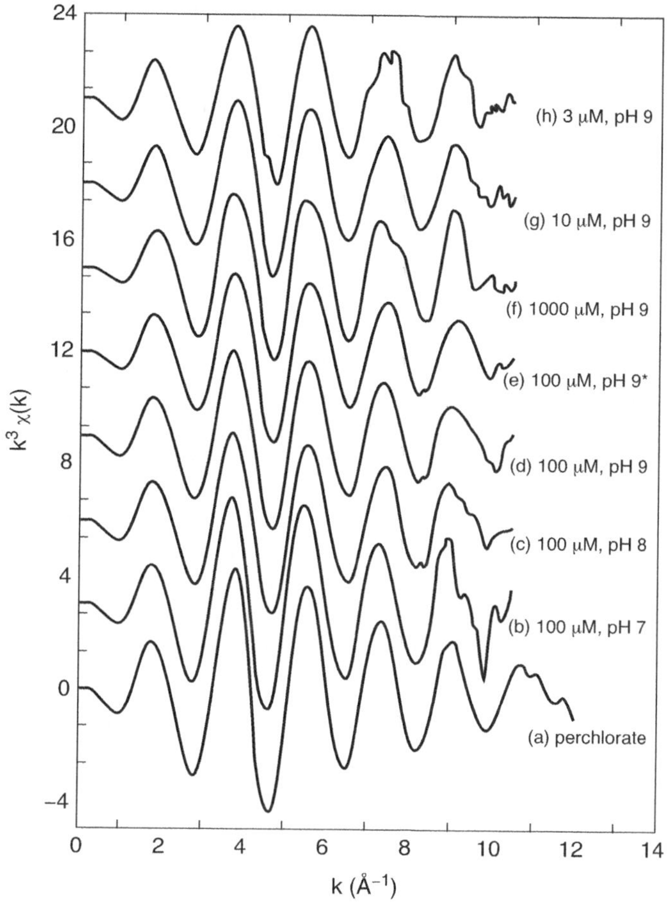

FIGURE 7.8 EXAFS data for adsorption of Cd(II) compared with a Cd perchlorate standard. Changes to the curves can be related to inner-sphere versus outer-sphere adsorption, as described in the text. From Vasconcelos et al., 2008 (with permission).

and Si (Al–O–Si), and sites that consist of surface oxygens bonded to one Si atom (Si–O, which may be protonated at the surface). Of these, the Al–OH and Si–O sites seem to be most reactive for cation adsorption (Sposito, 1984).

EXAFS data for adsorption of Cd(II) at different initial Cd concentrations and/or pH values, compared with a Cd perchlorate standard, are shown in Figure 7.8. At pH 7 ($I = 0.1$ M), the extent of Cd(II) adsorption was small (Figure 7.6). The EXAFS results indicated that the local environment surrounding the adsorbed Cd(II) was similar to that of hydrated Cd(II) as probed by a Cd perchlorate standard. The EXAFS results at pH 7 were consistent with outer-sphere adsorption of Cd(II), perhaps on the kaolinite basal-plane surface. Outer-sphere adsorption to the basal-plane surface was also suggested by molecular dynamics similations. The point of zero net proton charge (p.z.n.p.c.) of kaolinite is ~5.0, so kaolinite would have an overall negative surface charge at pH 7, whereas Cd(II) has a divalent positive charge.

At pH 8, the extent of Cd adsorption to kaolinite was greater, and the EXAFS results showed that Cd(II) was bound by a combination of outer-sphere and inner-sphere adsorption. At pH 9, much more Cd adsorption occurs, and the EXAFS results indicated predominance of inner-sphere adsorption. Whether this adsorption was to Al or Si edge sites could not be determined from the EXAFS spectra. No Cd-mineral precipitation was observed under the study conditions in the EXAFS results. Overall, the results showed that the adsorption mechanism can change with pH, even for a single metal-mineral-ionic-strength combination. EXAFS was essential for determining the adsorption mechanism as either inner- or outer-sphere adsorption with no precipitation.

Co Sorption to Kaolinite: Long-Term Experiments Most cation sorption experiments are conducted by allowing the adsorptive ion to react with the sorbent for only a few hours or at most a few days. However, in nature, sorbates interact with sorbents for long periods of time, and the minerals may dissolve extensively. An example is illustrated by the work of Thompson et al. (1999) on Co sorption to kaolinite. Co is a trace metal that is used in many applications, particularly in metal alloys, and it can be toxic to aquatic organisms. Thompson et al. (1999) allowed Co to react with kaolinite for several months, during which time kaolinite dissolution occurred. Based on solution composition analysis combined with EXAFS and TEM, the authors concluded that Co initially adsorbed to kaolinite in the form of hydroxy-bridged polymers and multimers. However, over time, Co coprecipitated as a hydrotalcite-like soid containing Al derived both from kaolinite dissolution and aluminum hydroxide dissolution. Al hydroxide formed during the initial sample preparation and pH adjustment portions of the experiment. The resulting Co-containing precipitates were nanoscale (see Chapter 13). The reaction of Co with aluminum released to solution by kaolinite dissolution is as follows:

$$6Co^{2+} + Al_2Si_2O_5(OH)_4 + 15H_2O \rightleftarrows Co_6Al_2(OH)_{16}(H_3SiO_4)_2 + 12H^+. \qquad (7.3)$$

The process continues over time, as kaolinite dissolution slowly progresses.

Cs Sorption to Micas Another example of the complexities of cation sorption is that of Cs^+ (cesium) sorption to weathered edges of micas (McKinley et al., 2004). Cs is an alkali metal. ^{137}Cs is a pollutant of considerable concern in the subsurface sediments at the Department of Energy's Hanford Site. Micas are phyllosilicate minerals whose structures consist of an octahedral layer sandwiched between two tetrahedral layers. Al and Si may be replaced by isomorphic substitution of cations of less positive charge, and the resulting overall negative charge is balanced by interlaminar weakly hydrated K^+. The K^+ fits into ditrigonal cavities on surfaces of adjacent tetrahedral layers, and holds the layers together. McKinley et al. (2004) investigated the sorption of Cs^+ to biotite and muscovite micas using SEM, TEM, and X-ray microprobe analysis. Cs^+ is weakly hydrated and similar in size to K^+, and the ability of micas to sorb Cs^+ has been documented in the past.

Binding sites for Cs^+ sorption to mica (Figure 7.9) include not only exchange sites on the phyllosilicate surfaces including sites along accessible edges and the basal plane and interlayer sites in which Cs^+ can replace K^+ , but also selective frayed edge sites that form after removal of K^+ from interlayers during weathering.

These frayed edge sites were found to be particularly important to Cs^+ sorption.

The region where the frayed edge sites meet the unweathered interlayer provides a favorable environment for binding weakly hydrated cations such as K^+, Cs^+, and NH_4^+ that can displace more strongly hydrated and weakly held cations such as Na^+. Cs^+ also penetrated the mica interior by diffusion of Cs^+ along crystal defects, cracks, or perhaps altered zones where fluids had previously penetrated the crystals.

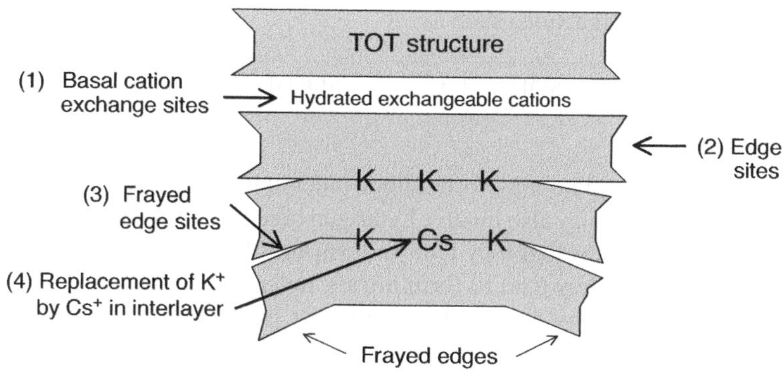

FIGURE 7.9 Schematic illustration of Cs adsorption sites as (1) cation exchange sites in the interlayers, (2) sites on the edges of the mica sheets, (3) frayed edge sites, and (4) replacement of K^+ in interlayer sites within the portion of the mica not affected by mineral dissolution (i.e., within the interior of the grains). TOT = tetrahedron–octahedion–Tetrahedron. Based on McKinley et al., 2004.

INORGANIC ANION ADSORPTION

Adsorption of anions generally follows an adsorption envelope, in which adsorption decreases with increasing pH (see Figure 7.11, below). As a surface becomes more negatively charged, the outer-sphere adsorption of anions becomes electrostatically unfavorable. In addition, surface deprotonation from increasing pH can inhibit inner-sphere anion adsorption. Inorganic oxyanion adsorption often involves a two-step *ligand-exchange* reaction (Sposito, 1984). This can be illustrated as

$$>\text{M—OH} + \text{H}^+(\text{aq}) \rightleftarrows >\text{M—OH}_2{}^+, \tag{7.4}$$

$$>\text{M—OH}_2{}^+ + \text{L}^{l-} \rightleftarrows >\text{M—L}^{1-l} + \text{H}_2\text{O}, \tag{7.5}$$

with the first protonation step not always necessary for all oxyanions. Some oxyanions, particularly nitrate, may adsorb by forming outer-sphere surface complexes with the protonated

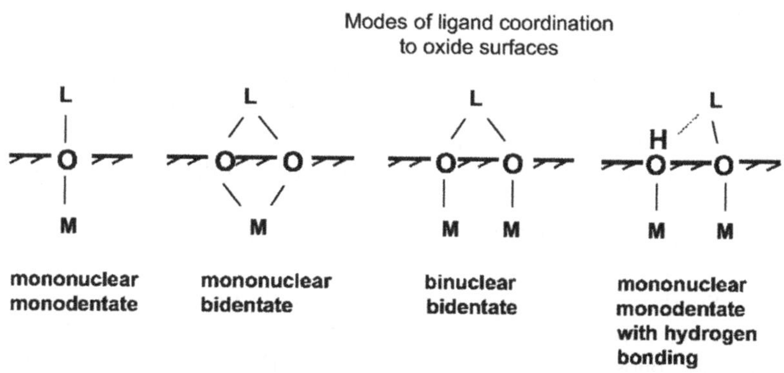

FIGURE 7.10 Models of ligand coordination to oxide surfaces, where M represents the metal site at an oxide surface and L is the adsorbing ligand. Note that hydrogen bonding may also be involved, as shown at the far right side of the figure.

surface hydroxyl group in reactions such as

$$>SOH_2^{+} + L^{l-} \rightleftarrows >SOH_2L^{l-} \quad (7.6)$$

(Sposito, 1984).

Some anions such as sulfate may form either inner-sphere or outer-sphere complexes. Adsorption of some anions may also involve hydrogen bonding. Anions have been classified as high affinity, which means that they tend to form mostly inner-sphere complexes; low affinity, which means that they tend to form mostly outer-sphere complexes; or moderate affinity (Sposito, 1984). Many factors that can affect metal cation adsorption may also influence inorganic anion adsorption.

For inner sphere adsorption of inorganic ions such as sulfate, phosphate, and arsenate, an important question is whether the adsorption is mononuclear or binuclear and monodentate or bidentate (Figure 7.10). The adsorption geometry is not always easy to determine, especially because the adsorption mechanism may change with pH, ionic strength, concentration of the adsorbate, and the nature of the adsorbent. Sometimes hydrogen bonding can also be involved.

Phosphate Adsorption

Phosphate (PO_4^{3-}) is an important nutrient, and the adsorption of phosphate to mineral surfaces such as goethite tends to be inner sphere and strong. Thus, phosphate is considered to be a "high affinity" anion. Such strong adsorption can limit phosphate bioavailability and affect cycling in ecosystems. An overabundant supply of phosphate as fertilizers can lead to algal blooms and *eutrophication.* Adsorption may decrease phosphate mobility into waters, thus helping to prevent eutrophication.

Adsorption of phosphate to goethite decreases with increasing pH (e.g., Parfitt and Atkinson, 1976; Tejedor-Tejedor and Anderson, 1990; Arai and Sparks, 2001). Because of its high affinity, phosphate can often compete with other anions such as sulfate, selenite, and arsenate/arsenite for adsorption. Identification of the adsorption complex(es) has been the subject of much debate.

One question that has developed involves the ionic strength dependency of phosphate adsorption. An observed ionic strength dependency generally is believed to indicate outer-sphere complexation. This is because outer-sphere complexation would be affected by the strength of electrostatic forces and the thickness of the double layer, whereas inner-sphere complexation should be less sensitive. Spectroscopic evidence indicates that phosphate adsorbs inner spherically to Fe oxide. Yet, Arai and Sparks (2001) noted an ionic strength dependency at a pH above ~6.5 (Figure 7.11) in the NaCl electrolyte. They suggested that it might not be ionic strength per se that is affecting adsorption, but rather the ability of coadsorbing Na ions to neutralize the surface charge effect of the negative phosphate ions

$$\equiv Fe—OH + Na^{+} + HPO_4^{2} \rightarrow \equiv Fe—O—PO_3H \cdots\cdots Na^{+} + H_2O. \quad (7.7)$$

Elzinga and Sparks (2007) applied attenuated total reflection fourier transform infrared (ATR-FTIR) spectroscopy, including use of D_2O (heavy water, containing deuterium) and H_2O, to investigate phosphate adsorption to hematite. D_2O allowed investigation of the P–OH stretching vibrations, which would shift for D_2O if protons are involved in the binding of phosphate. These authors found that various inner-sphere complexes formed at

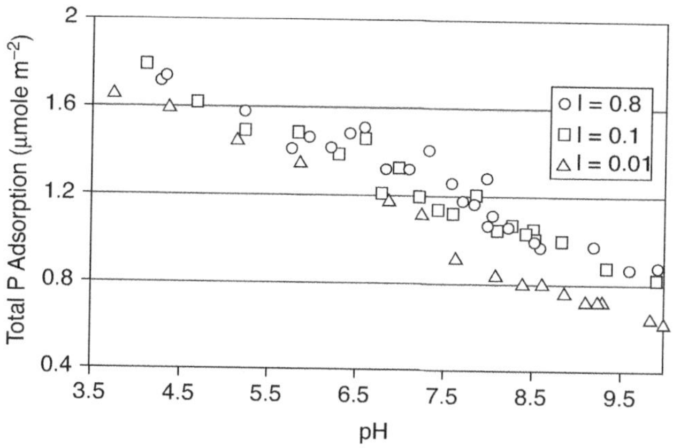

FIGURE 7.11 The effects of ionic strength on adsorption of P to ferrihydrite. From Arai and Sparks, 2001 (with permission).

the hematite surface, with relative importance varying with pH and surface coverage. The exact nature of the complexes could not be determined unambiguously at all pH values. Structures that were consistent with the experimental results are shown in Figure 7.12. They include monodentate mononuclear, monodentate binuclear, and monodentate mononuclear with hydrogen bonding and either one or two protons associated.

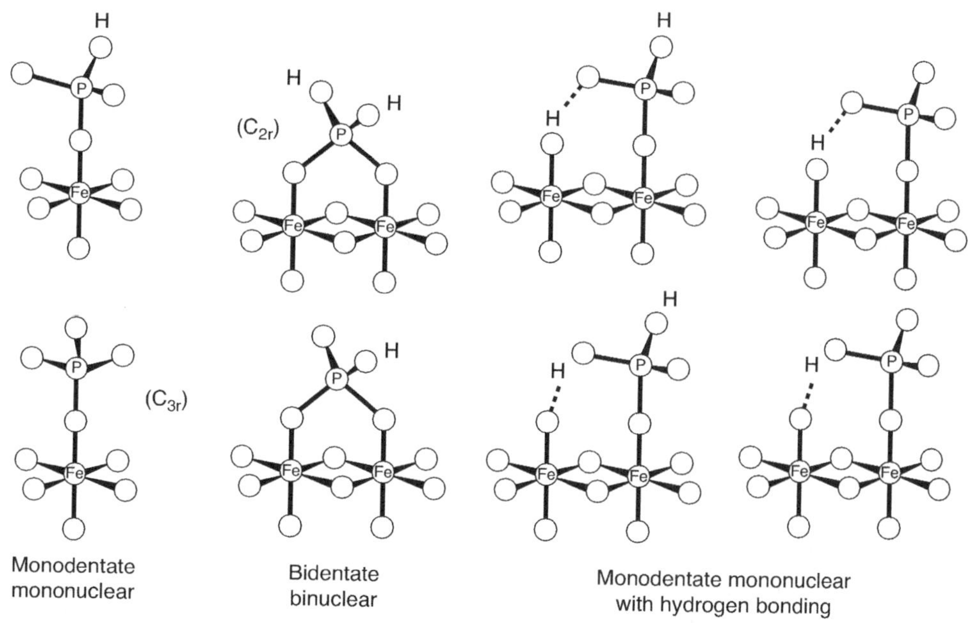

FIGURE 7.12 Some potential phosphate surface complexes on hematite. Based on Elzinga and Sparks, 2007 (with permission).

Nitrate Adsorption

Like phosphorus, nitrogen is an important nutrient, and it is also released to the environment as nonpoint source pollution in fertilizer. Nitrogen can occur in many forms (see, for example, Chapter 12); here, we consider adsorption of nitrate, with the understanding that hydrobiogeochemical redox reactions may lead to the formation of other species such as nitrite (NO_2^-) and nitrogen (N_2) gas that can behave very differently with respect to sorption and transport phenomena. In general, nitrate adsorption to oxides and clays is considered to be very weak, forming weak outer-sphere complexes or "ion pairs." This is one reason why metal adsorption experiments often add the metal as a nitrate salt, use KNO_3 or $NaNO_3$ to set ionic strength, and add nitric acid to adjust pH. Nevertheless, the presence of "swamping electrolytes" such as NO_3 needs to be taken into consideration in evaluation of conditional (not intrinsic) surface complexation constants for metal adsorption, for use in surface complex models (SCMs) such as the triple layer model (e.g., Smith and Jenne, 1991).

Nitrate contamination in drinking water is of concern because bacteria that can convert nitrate to nitrite are present in the digestive tracts of infants (but not of older children and adults), which leads to a dangerous medical condition known as "blue baby syndrome." Because of the low adsorption affinity for mineral surfaces, natural attenuation by adsorption is not viable in most locations. Thus, nitrate removal by engineered techniques such as manmade adsorbents and reactive barriers, including various bioremediation strategies, is an important area of research.

Sulfate Adsorption

Adsorption of sulfate also has been studied extensively because of environmental implications that involve, for example, soil responses to acid precipitation and mobility of sulfate in acid mine drainage waters (see Chapter 14). Sulfate is a "moderate affinity" anion, and its adsorption may occur by a variety of inner-sphere mechanisms as well as outer spherically (e.g., Sposito, 1984; Wijnja and Schulthess, 2000b; Peak et al., 1999). Two lines of evidence have suggested that sulfate adsorbs outer spherically to oxides: (1) sulfate adsorption is sensitive to ionic strength, with adsorption generally increasing as ionic strength decreases; and (2) sulfate generally does not adsorb above the p.z.c. of an oxide, as the surface would be negatively charged (Charlet et al., 1993).

Peak et al. (1999) used in situ ATR-FTIR to show that sulfate forms both inner sphere and outer sphere complexes on goethite at pH < 6, but only outer sphere complexes at pH > 6. In the pH range where inner-sphere and outer-sphere adsorption occurs, outer-sphere adsorption becomes relatively more important with decreasing ionic strength (Peak et al., 1999). As stated by Paul et al. (2006), "The dependence of sulfate adsorption on pH and surface charge alone creates *a complicated network* of potential reaction pathways. . ." This is another example of the fact that indirect evidence such as ionic strength dependency or sorption behavior relative to p.z.c. is not sufficient to determine adsorption mechanism, particularly when adsorption may be both inner sphere or outer sphere for a given ion.

Carbonate Sorption

Carbonate adsorption to oxides has been the subject of much recent research, particularly because carbonate introduced from $CO_2(g)$ can be overlooked in adsorption experiments. Wijnja and Schulthess (2002) found that carbonate may enhance, decrease, or have no apparent effect on the adsorption of moderate affinity anions such as selenate and sulfate on goethite, depending on the carbonate concentrations. Wijnja and Schulthess (2000a) observed that carbonate enhanced adsorption of selenate and sulfate on aluminum oxide at pH 6–8, but it had no effect on adsorption of the high-affinity anion phosphate. Carbonate can from complexes with many metal and radionuclides, thus influencing adsorption, and some metals or radionuclides may precipitate as carbonate solids.

IMPORTANCE OF REDOX STATE AND VALENCE TO INORGANIC ION ADSORPTION

For elements that undergo redox transformations, the valence state can determine adsorption behavior, along with solubility. This topic is complex; some examples from Cr, Np, U, Se, and As are described briefly here.

Chromium

Cr occurs primarily as the stable and mobile hexavalent Cr(VI)-containing chromate anion (CrO_4^{2-}) under highly oxidizing conditions; but under less highly oxidizing and acidic conditions; it occurs as less mobile Cr(III) in $Cr(OH)^{2+}$ or Cr^{3+}. Cr(III) tends to sorb strongly to most metal(hydr)oxide phases above pH 4 (Fitts et al., 2000).

Fitts et al. (2000) used XAFS to investigate the sorption of Cr(III) to γ-Al_2O_3. They found that during the first 2 hours of the sorption experiment, Cr(III) was taken up rapidly, forming a monomeric inner-sphere complex with at least two surface functional groups (Figure 7.13). Subsequently, Cr(III) continued to be taken up from solution, but much more slowly. The XAFS analysis at reaction times from 2 hours to 1 week revealed an increase in scattering contributions from neighboring Cr(III) atoms, which indicates a progression from hydroxo-bridged Cr(III) dimers at the γ-Al_2O_3 surface to higher order polymeric precipitates. This progression is shown schematically in Figure 7.13.

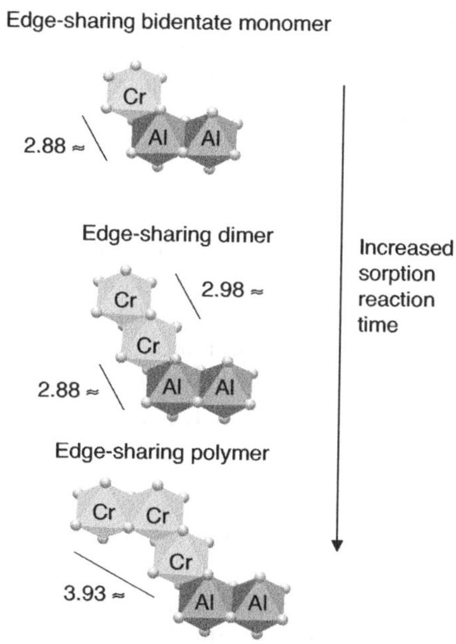

FIGURE 7.13 The progression of Cr(III) hydroxo species sorption to γ-Al_2O_3 from monomeric adsorption to dimer formation to polymer formation. Distances in Å. Revised from Fitts et al., 2000 (with permission).

Redox reactions that involve Fe or Mn oxides, organic matter, and/or microorganisms can play an important role in Cr speciation and mobility. An example of Mn(II) reduction of mobile Cr(VI) to less mobile Cr(III) in an acid mine drainage stream is discussed in Chapter 14.

Neptunium

Neptunium, which is a man-made byproduct of nuclear fission, can occur as dissolved Np^{3+}, Np^{4+}, NpO_2^{+}, or NpO_2^{2+} (Gorman-Lewis et al., 2005), with Np(V) the dominant solution valence state. Because Np(V) tends to be mobile and ^{237}Np has a long half-life, this actinide has the potential to be a mobile and long-lived radioactive contaminant. Gorman-Lewis et al. (2005) showed that Np(V) species adsorbed reversibly to the soil bacterium, *Bacillus subtilis*, at pH above ~4. Their results suggested reduction of Np(V) to Np(IV) at low pH and 0.1 M ionic strength, with the reduced species sorbing irreversibly. The irreversibility of sorption suggests potential precipitation. Bacteria therefore may play a role in Np mobility in porous media.

Uranium

Uranium occurs primarily in mineral form as relatively immobile $U(IV)O_2$ and in dissolved form as mobile $U(VI)O_2^{2+}$. As discussed in Chapter 12, some bacteria can reduce more mobile U(VI) to less mobile U(IV). Many studies have investigated U(VI) sorption to a variety of mineral surfaces. Wazne et al. (2003) applied a combination of batch adsorption experiments, zeta potential measurements, and FTIR spectroscopy to investigate U(VI) adsorption to ferrihydrite. They showed that U(VI) adsorbed to ferrihydrite in the absence of carbonate at pH 6 as cationic UO_2^{2+} to form the surface complex, $>SO\text{-}UO_2^{+}$. However, U(VI) adsorbed as an anionic U(VI) carbonate complex in the presence of carbonate at this pH. The presence of carbonate decreased adsorption, as shown in Figure 7.14. At pH 6, this decrease in adsorption was likely caused by competition between adsorbed carbonate and the U(VI) complex, as well as the fact that U(VI) carbonate complexes have less affinity for the surface than do the various U(VI) hydroxide species present at this pH.

Um et al. (2007) showed that U(VI) sorbed primarily to Fe(III)(hydr)oxide coatings on sands from the Department of Energy's Hanford Site, but that adsorption decreased at high pH because of the formation of anionic aqueous uranium–carbonate complexes. At high pH, U(VI) was also incorporated into precipitating calcium carbonate. This example shows how for many species (e.g., U, Pb, Co), formation of carbonate complexes and of carbonate precipitates can occur at intermediate to high pH and needs to be considered in sorption studies. Curtis et al. (2006) investigated the incorporation of SCM for U(VI) in reactive transport models, as described in Chapter 14. Chapter 12 gives an example of how bacteria may affect uranium mobility by enhancing calcium carbonate precipitation and uranium sequestration.

Selenium

Selenium can be toxic, and its mobility depends on redox conditions. Hayes et al. (1987) suggested that selenate (SeO_4^{2-}) adsorbed as a relatively weak outer-sphere complex to goethite, whereas selenite (SeO_3^{2-}) appeared to adsorb as an inner-sphere complex. The adsorption of selenate was more ionic-strength dependent than that of selenite, and the authors argued that outer-sphere adsorption should be more dependent on ionic strength.

Manceau and Charlet (1994) suggested based on EXAFS data that selenate forms inner-sphere complexes to goethite and hydrous ferric oxide, but subsequent authors found varied

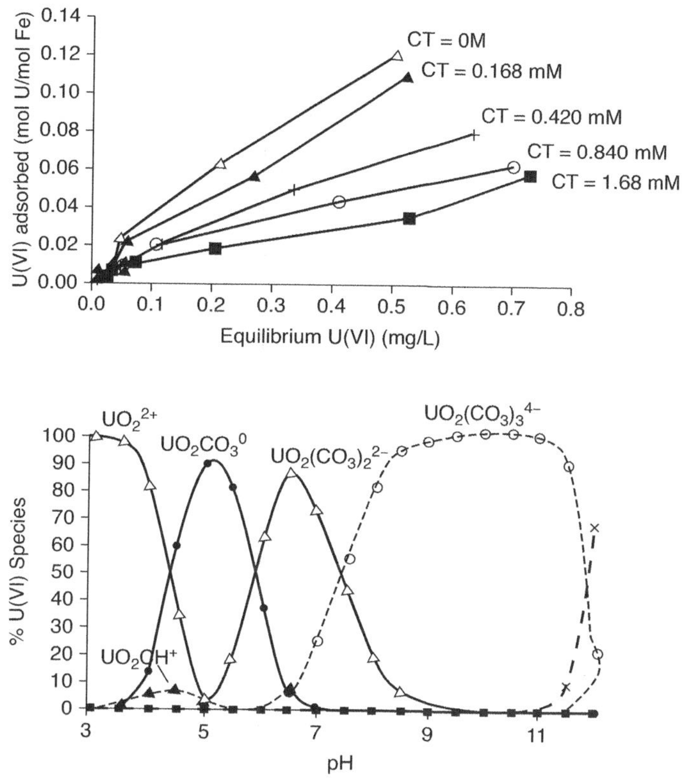

FIGURE 7.14 Effect of carbonate on U(VI) adsorption on ferrihydrite at pH 6 in 0.01-M NaCl. CT is the total carbonate concentration. (top) Sorption isotherms at different CT values. (bottom) Speciation of U(VI) at a total concentration of 1 mg $L^{-1} = 4.20 \times 10^{-6}$ M in the presence of CO_2. The $UO_2(OH)_2^0$ species plots along the bottom of the graph. In the absence of CO_2 (not shown here), UO_2^{2+} and a variety of $(UO_2)_xOH_y$ species are present. Revised from Wazne et al., 2003 (with permission).

results. Peak and Sparks (2002) applied EXAFS and ATR-FTIR to show that selenate can form both inner-sphere and outer-sphere complexes on goethite. The "continuum" of adsorption mechanisms on goethite was affected by pH and ionic strength, with more inner-sphere adsorption at lower pH and at higher ionic strength. Wijnja and Schulthess (2000b) applied Raman spectroscopy to selenate adsorption on goethite (Figure 7.15). Their results agree in general with the EXAFS results, which suggests that selenate adsorbs as both an outer-sphere and an inner-sphere surface complex at pH ~7, and that the inner-sphere complex becomes predominant at lower pH. The inner-sphere complex seems to be monodentate (Wijnja and Schulthess, 2000b). This study again demonstrates that adsorption mechanisms may differ depending on solution conditions and that spectroscopic methods are needed to determine mechanism(s) under different conditions.

Case Study: Arsenic Speciation and Mobility

Arsenic (As) is a pollutant of concern in many locations throughout the world. Its occurrence can be natural or caused by human activities such as mining, smelting, and use of arsenical biocides. In the Bengal delta, an estimated 30 million people obtain their drinking

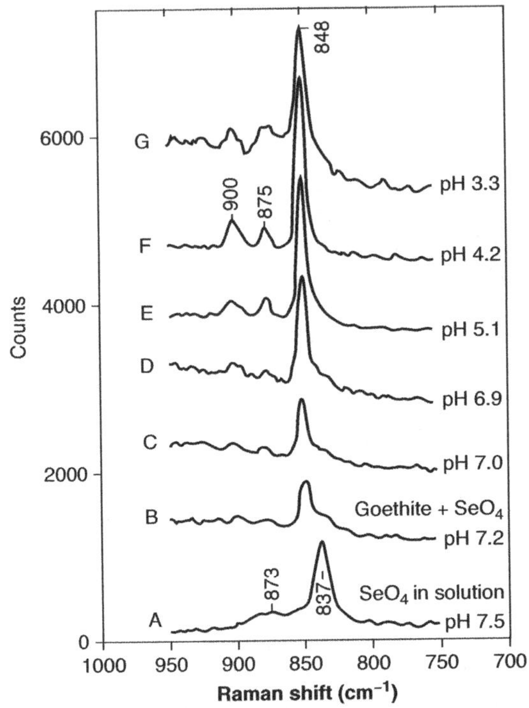

FIGURE 7.15 Comparison of the Raman spectra of adsorbed SeO_4 (B–G) with the spectrum of free SeO_4 in solution (A) highlights differences in number and position of various bands. The similarity between the shoulder band at 837 cm^{-1} at pH above 6.9 and the band of the free SeO_4 anion suggests outer-sphere adsorption. However, the band in the spectrum of free (aqueous) SeO_4 at 837 cm^{-1} is shifted to 848 cm^{-1}, and the split of bands at 875 and 900 cm^{-1}, which progresses as pH decreases, is consistent with a monodentate inner-sphere selenate surface complex. From Wijnja and Schulthess, 2000b (with permission).

water from wells with naturally elevated As concentrations. Because As is a carcinogen, concern develops when As concentrations in water exceed the U.S. Enviornmental Protection Agency's (EPA's) drinking water standard of 10 $\mu g\ L^{-1}$ (Smith et al., 2002).

Both the mobility and the toxicity of As are controlled to a large extent by its speciation, particularly with respect to oxidation state (see review by Smedley and Kinniburgh, 2002). The complex biogeochemical cycling of arsenic, which includes processes such as sorption, precipitation, and biotic and abiotic redox transformations, is illustrated schematically in Figure 7.16 (Reisinger et al. 2005).

Arsenic can exhibit different sorption behavior in the +III versus +V oxidation states. Particularly with Al-based sorbents, As(III) is poorly sorbed relative to As(V) (Dixit and Hering, 2003), although direct competition was not observed between As(III) and As(V) (Goldberg, 2002). With Fe (hydr)oxides as the sorbent, the relative affinities of As species change with pH (Figure 7.17). Dixit and Hering (2003) observed that sorption of As(V) to Fe(hydr)oxides was favored at pH $< \sim$5–6 whereas As(III) sorption was favored at pH $> \sim$7–8. Note that the adsorption-versus-pH data shown in Figure 7.17 include both sorption envelopes and curves that are more complex than straightforward envelopes nor edges (see Chapter 6). The sorption behaviors of As(III) and As(V) are also affected differently by the presence of phosphate as a competing sorbate (Dixit and Hering, 2003).

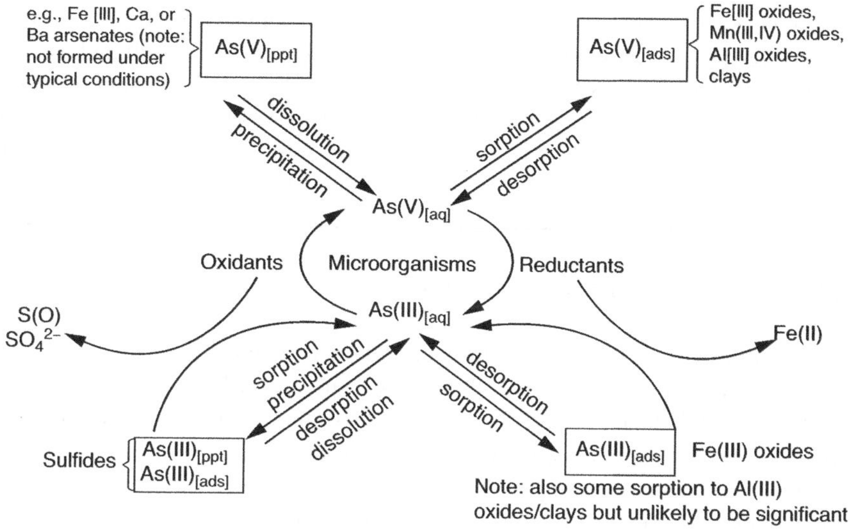

FIGURE 7.16 The complex adsorption and cycling behavior of As. The immobilized As species are in boxes. Revised from Reisinger et al., 2005 (with permission).

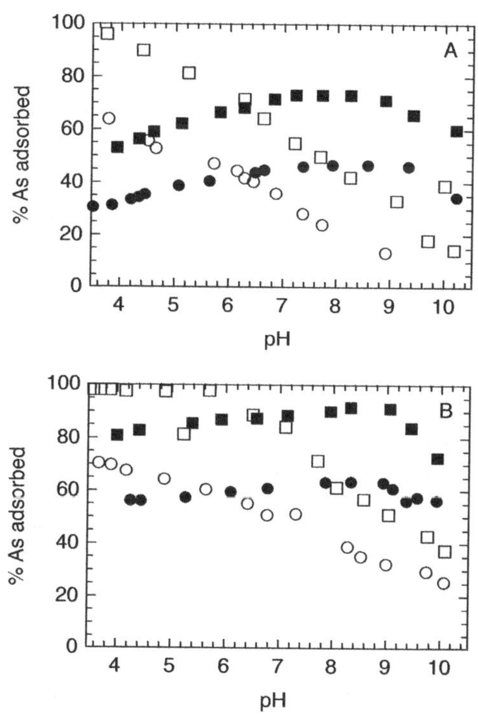

FIGURE 7.17 Comparison of sorption of As(V) (open symbols) and As(III) (closed symbols) on (a) HFO and (b) goethite as a function of pH. Total As concentration = 100 μM (circles) and = 50 μM (squares). From Dixit and Hering, 2003 (with permission).

For many years, the conventional wisdom was that arsenate ($As(V)O_4^{3-}$) adsorption to Fe and Al oxides was inner sphere. However, recent modeling by Fukushi and Sverjensky (2007) and resonant anomalous X-ray reflectivity (RAXR) analysis by Catalano et al. (2008) is demonstrating that arsenate adsorption to Fe and Al oxides may be both inner sphere and outer sphere. This example shows how our understanding of sorption is continuing to evolve as the scientific community works to develop and apply new approaches.

QUESTIONS FOR FURTHER THOUGHT

1. Which would you expect to be a stronger sorbent of metals in an acid rock drainage stream: ferrihydrite or birnessite? Such a stream would have very low pH, likely around 2 (see Chapter 14 for more on acid rock drainage). Explain your answer.
2. How would one go about distinguishing between surface adsorption and precipitation for a metal such as Cu reacting with an Fe oxide surface? For the case of precipitation, how would one distinguish between precipitation occurring on the Fe oxide surface (heterogeneous) versus directly from solution (homogeneous)?
3. How might one go about distinguishing between inner-sphere and outer-sphere adsorption of Cs^+ on kaolinite? Write a paper several pages long outlining a set of experiments, including broad discussion of methods and what they would contribute to understanding.
4. Can ionic strength effects on adsorption be used to determine definitively whether adsorption of a cation is inner sphere or outer sphere? Explain your reasoning.
5. Using the data below on Cd adsorption to kaolinite,
   (a) plot the percent adsorbed and the mmol m^{-2} adsorbed, both as a function of pH.
   (b) Is the resulting plot an adsorption edge or an adsorption envelope?
   0.0516 g of kaolinite with 12.1 $m^2\ g^{-1}$ surface area were used in each batch reaction vessel except for the control. The experiments were run with total solution volume = 25 mL.
   Cd adsorption experiment data [in the absence of the organic ligand desferrioxamine-B (DFO-B); see Chapter 8 questions for comparison] in mg Cd L^{-1}. Assume that the mean [Cd] values at pH 5.8 are representative of the initial [Cd] for all reps (not all reps given, to save space in table). The values given as rep1, rep2, and rep3 are the [Cd] remaining in solution following adsorption steady state.

Sample	pH	[Cd] rep 1	[Cd] rep 2	[Cd] rep 3
Control (no kaolinite)	4-9	1.03	1.02	1.02
1	4.60	1.03	1.02	1.01
2	4.96	0.99	0.98	0.99
3	5.59	0.97	0.97	0.97
4	5.90	1.00	0.99	1.00
5	6.40	0.96	0.96	0.97
6	6.70	0.95	0.95	0.97
7	7.21	0.93	0.93	0.93
8	8.06	0.80	0.81	0.80
9	8.7	0.33	0.28	0.30
10	9.1	0.30	0.04	0.04

6. A question for library research. Cr sorption to Fe(hydr)oxides is complex, involving different sorption mechanisms and redox processes. Write a short paper describing what we currently know about Cr sorption to Fe(hydr)oxides, and how different methods and approaches have contributed to this understanding. How has our understanding of Cr sorption to Fe(hydr)oxides changed over time?
7. A student runs a batch adsorption experiment of Cd (using $Cd(NO_3)_2$) on a kaolinite sample with BET surface area 12.0 $m^2\ g^{-1}$ at pH 7.5 in 0.1 M ionic strength solution (using $NaClO_4$ as the background electrolyte). In all, 25 mL of total solution or suspension are present in each reaction vessel, and the kaolinite suspension concentration is 2.064 g L^{-1} (except in controls without kaolinite). Dissolved Cd concentrations in the initial or "control" (no kaolinite added) and in three replicates at each initial Cd concentration (i.e., solution concentrations following adsorption and removal of kaolinite particles by filtration) are given below.
   (a) Calculate the concentration of adsorbed Cd in mg Cd m^{-2} for each equilibrium concentration (control-rep), and
   (b) plot mean adsorbed values in mg m^{-2} with standard error on the y axis and mean equilibrium Cd concentration (in mg L^{-1}) and standard error on the x-axis.
   (c) Based on the Langmuir equation, calculate the values of K_L and q_m.
   (d) Using the Freundlich equation, calculate K_f.
   (e) For the Freundlich treatment, what value of n best fits the isotherm? See Chapter 6 for the isotherm equations.

Cd concentrations, in μmoles per liter

Sample Data points	Initial or control conc. (uM)	Eqm. conc. rep 1 (uM)	Eqm. conc. rep 2 (uM)	Eqm. conc. rep 3 (uM)
Cd 1	0.738	0.002	0.004	0.009
Cd 2	3.100	0.075	0.083	0.117
Cd 3	8.790	2.199	2.195	2.255
Cd 4	17.120	8.100	8.121	7.890
Cd 5	28.705	17.060	16.838	17.160
Cd 6	35.210	23.801	23.210	23.043

FURTHER READING

Brown, G.E., and Parks, G.A. 2001. Sorption of trace elements on mineral surfaces: Modern perspectives from spectroscopic studies, and comments on sorption in the marine environment. *International Geology Review* 43, 963–973.

Elzinga, E.J., and Sparks, D.L. 2007 Phosphate adsorption onto hematite: An in situ ATR-FTIR investigation of the effects of pH and loading level on the mode of phosphate surface complexation. *Journal of Colloid and Interface Science* 308, 53–70.

Reisinger, H.J., Burris, D.R., and Hering, J.G. 2005. Remediating subsurface arsenic contamination with monitored natural attenuation. *Environmental Science and Technology* 39, 458A–464A.

8

SORPTION: ORGANIC COMPOUNDS

Organic matter is ubiquitous in aquatic and terrestrial environments at and near the Earth's surface. Many surfaces in nature are coated with organic matter, from "adventitious carbon," to microbial biofilms, to natural organic ligands such as humic and fulvic acids. Organic compounds are used commercially and agriculturally, for example, as antibiotics, solvents, surfactants, pesticides, and herbicides, and these synthetic compounds can become widely distributed in the environment. Organic compounds are so diverse, with such a wide range of composition and properties such as volatility, solubility, metal binding behavior, and bioavailability, that we can only discuss a few key properties and behaviors in this chapter.

Adsorption of organic matter to mineral surfaces can greatly affect the properties and reactivity of the mineral–water interface. For example, adsorption of natural organic matter to Fe oxides has been shown to affect surface charge and particle aggregation/flocculation behavior. Any gardener knows that organic matter plays an essential role in creating a good soil environment for plant growth. The organic "humus" helps to hold individual soil mineral particles together to form clusters or aggregates, which is essential for soil structure, nutrient cycling and bioavailability, and proper balance of water retention versus flow.

Many mechanisms are available by which organic compounds may adsorb to mineral surfaces, with different mechanisms coming into play for different organic molecules or even for a single molecule under different conditions (e.g., pH or surface loading). Some mechanisms of organic sorption include (Sposito, 1984) *hydrophobic interactions, electrostatic attraction, surface complexation, ligand exchange, water bridging, cation bridging, hydrogen bonding,* and *cation exchange*. In addition to adsorption, some organic compounds can *absorb* in the interlayers of swelling clays such as montmorillonite and other porous minerals. Hydrophobic organic compounds also can either *absorb* into interlayers or *partition* into natural organic matter (NOM) in soils and sediments.

This chapter begins with a basic introduction to organic chemistry. Next, some organic compounds of interest to environmental interface chemistry are introduced. Finally, we

---

Environmental Surfaces and Interfaces from the Nanoscale to the Global Scale, by Patricia A. Maurice

discuss sorption first of simple organic ligands, surfactants, and natural organic matter; and then of a variety of pollutant organic compounds, such as polycyclic aromatic compounds, substituted nitrobenezenes, atrazine, and volatile organic compounds.

A BRIEF INTRODUCTION TO ORGANIC CHEMISTRY

To understand the sorption of organic compounds to mineral surfaces, a brief introduction to the structures and properties of organic compounds of environmental interest is needed. Although a strict nomenclature for organic compounds exists, most environmental applications use more loose or "trivial" names and terms. The structures shown in this text are depicted as two-dimensional representations, but the actual structures are three-dimensional. For most simple organic structures shown here, C-atoms are designated; otherwise, the locations of C-atoms are taken to be at the bends.

The element carbon has the symbol C and the atomic number 6. It exists as stable isotopes ^{12}C and ^{13}C as well as radioactive ^{14}C, which has a half-life ($t_{1/2}$) of ~5700 years. *Allotropes* or natural elemental forms of C include diamond and graphite. The first shell (1s) of C has two electrons. The second shell (2s, 2p) has only four electrons, whereas it needs eight to be full. Because of this, a C atom can form four *covalent bonds*. A single covalent bond consists of two shared electrons, one from each of two bonded atoms. The ability of C to form highly stable C-to-C bonds is responsible for the millions of known natural and synthetic C compounds and ultimately for the complexity of processes that make up the living or *organic* world.

Along with the natural allotropes of C, a few types of compounds such as the carbonates, simple oxides of C, and cyanides are considered to be inorganic. Organic C compounds are a class of chemicals whose molecules contain C, with the terminology derived historically from C-containing molecules associated with life. Compounds with only C and H are known as *hydrocarbons*. *Heteroatoms* may also be present on organic compounds, including but not limited to O, N, S, P, and the halogens Cl, F, Br, and I.

Simple straight- or branched-chain organic compounds are known as *aliphatic*. Aliphatic compounds with no double bonds are known as *alkanes*; *alkenes* have at least one double bond, and *alkynes* have at least one triple bond. Examples of aliphatic compounds are ethane, which is a colorless, odorless gas at standard temperature and pressure, and isobutane (methylpropane), which is used as a common refrigerant and a propellant in aerosol sprays (Figure 8.1).

Aromatic compounds have structures with conjugated double bonds so that electrons can move through a region that contains more than two atoms and are *delocalized*. This delocalization of electrons can give aromatic compounds interesting light absorption and redox properties. The best-known aromatic compounds are based on the *benzene* structure, which consists of a ring of six C-atoms that contains three conjugated double bonds. The structure can be represented in many different ways: with or without hydrogen atoms, and with every other bond doubled or with a ring that represents the delocalization of electrons in the conjugated double (π) bonds (Figure 8.2).

Organic molecules may have a variety of different functional groups, which can strongly affect their properties and reactivities. Most functional groups tend to make organic compounds more hydrophilic by imparting a more polar nature. On the benzene ring structure, the positions of different functional groups or components relative to a specified group (Figure 8.3) are referred to as ortho (*o*-), meta (*m*-), and para (*p*-).

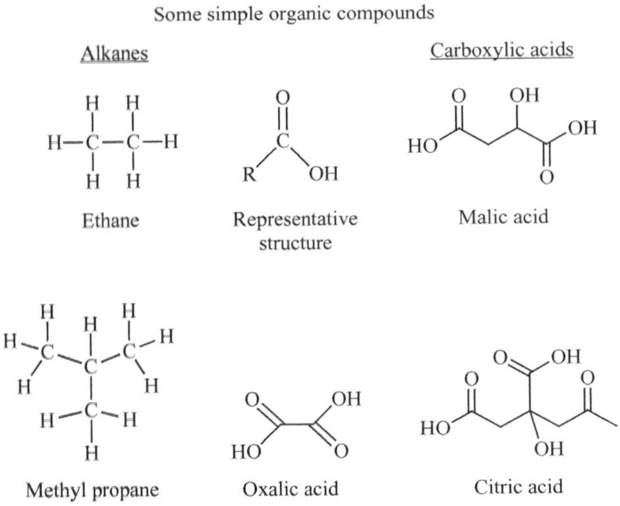

FIGURE 8.1 Some simple organic carbon molecules.

The carboxylic functional group —COOH results in weak acids with pK_as generally ~4.0–5.0, but dependent on the nature of the organic backbone and the presence and relative locations of other functional groups and heteroatoms of different electronegativities. A representative structure of a carboxylic acid functional group, where R may be H or an organic group is:

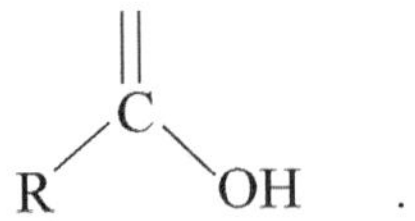

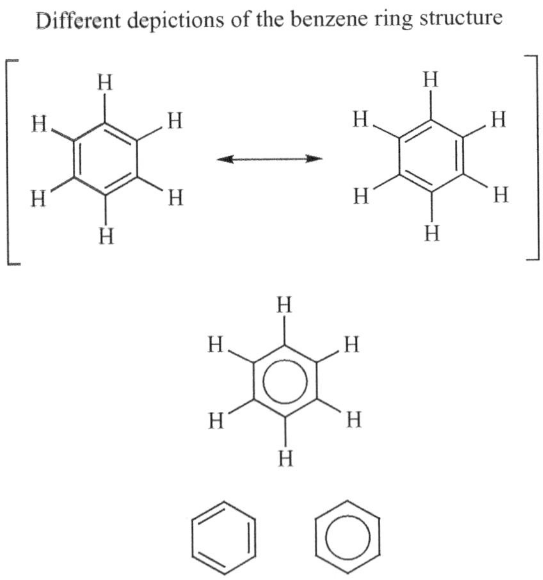

FIGURE 8.2 Some different ways of depicting the structure of the benzene ring.

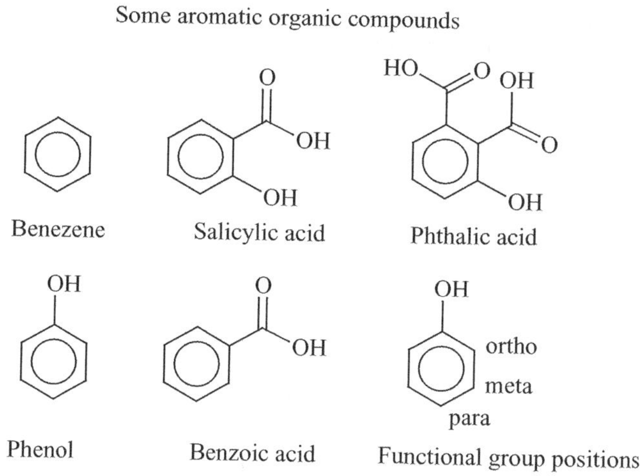

FIGURE 8.3 Some aromatic organic carbon compounds.

Some examples of carboxylic acids shown in Figure 8.1 include the dicarboxylic oxalic acid, the dicarboxylic malic acid, and the tricarboxylic citric acid.

Additional aromatic carboxylic acids include the monocarboxylic benzoic acid, dicarboxylic phthalic acid, and salicylic acid, which has a carboxyl group ortho to a phenolic group (Figure 8.3).

Carboxylic functional groups can be important for metal binding and for formation of surface complexes at metal sites on mineral surfaces (e.g., Al and Fe surface sites). For example, oxalate forms strong inner-sphere complexes with Fe and Al oxides.

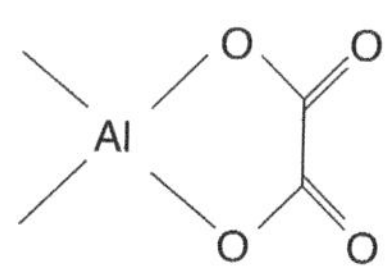

The hydroxyl group —OH can be attached to aliphatic molecules to form alcohols. The acidity constants of aliphatic alcohols tend to be very high ($pK_a > 14$), so that these functional groups tend not to be deprotonated. Carbohydrates contain multiple —OH functional groups. The common sugar glucose ($C_6H_{12}O_6$) can occur in either chain (acyclic) or ring (cyclic form), the latter through hydrogen bonding. A glucose ring structure is:

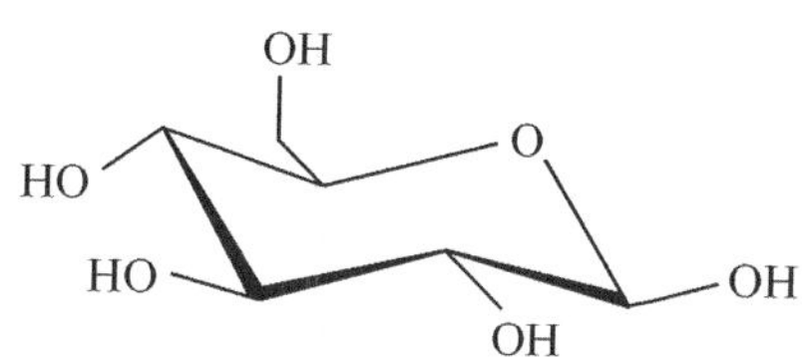

A phenol occurs when an hydroxyl (—OH) group is attached to an aromatic benzene ring. Phenols can have a wide range of pK_as, from ~2 to 12, with many at ~10

TABLE 8.1 Some dissociation constants (pK_as) at 25°C of some common organic acids.

	pK_a
acetic	4.8
chloroacetic	2.85[1]
trichloroacetic	0.70[1]
oxalic	1.3, 4.3
citric	3.2, 4.8, 6.4 (COOH groups)
malic	3.4, 5.13 (COOH groups)
phenol	10.0
phthalic acid	3.0, 5.4
benzoic	4.2
salicylic	3.0 (COOH), 13.4 (OH)

Source: Martell and Smith, 1976.
[1]*Source*: Handbook of Chemistry and Physics, online version.

(Perdue, 1985). Salicylic acid contains a carboxyl group next to (ortho to) a phenolic group (Figure 8.3), functions as a plant hormone, and has important metal chelation abilities. Salicylate can form particularly strong surface complexes on Fe and Al (hydr)oxides and can greatly enhance dissolution rates.

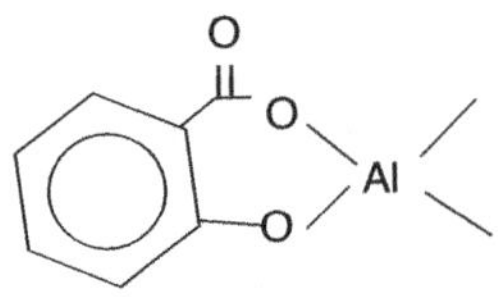

The acidity constants (given in negative log form as pK_a) of a variety of organic acids are tabulated in Table 8.1. An example of how the presence of other heteroatoms on a molecule can alter the pK_a of a given functional group is provided by comparing the pK_as of acetic acid (= 4.8)

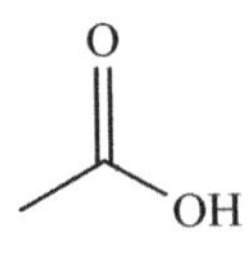

with those of chloroacetic acid (= 2.8)

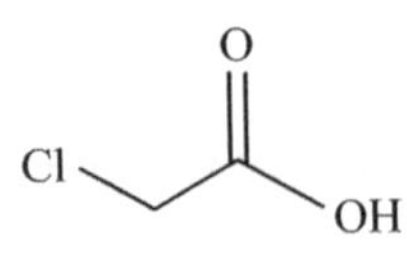

and trichloroacetic acid (= 0.7)

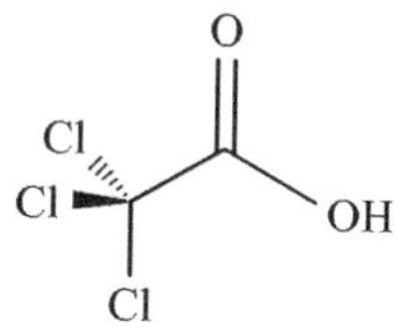

Many other important functional groups exist, which lead to different classes of organic compounds. Esters are similar to alcohols, but the H on the —OH group is replaced by an alkyl group designated (R) to form, for example, a carboxylic acid ester

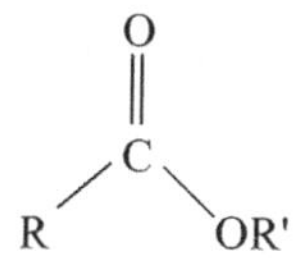

Esters tend to be more hydrophilic (see below) than corresponding hydrocarbons but less hydrophilic than corresponding alcohols.

Amines contain a basic N atom with a lone pair of electrons, and include, for example, amino acids. The general amine structure is

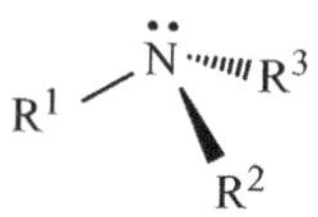

A thiol group on an organic compound consists of a sulfur atom and a hydrogen atom, —SH, with generalized structure

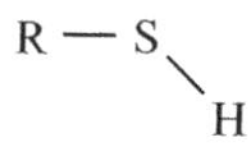

Sulfonates contain the SO_3^- group, and are common as synthetic surfactants or detergents

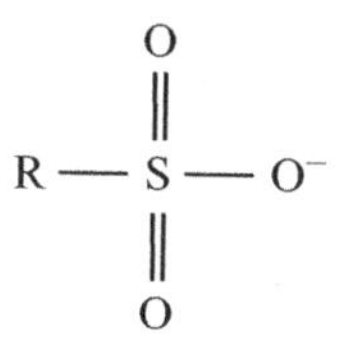

The hydroxamate group, which forms the basis for many organic metal chelating compounds known as siderophores, contains both N and O

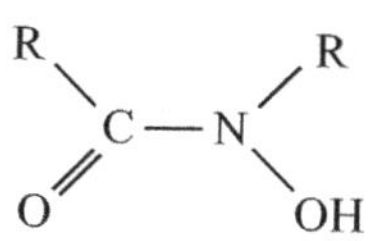

When it binds Fe, a proton is released

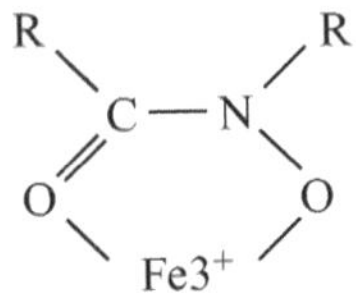

$+ \ H^+$ (8.1)

In trihydroxamate siderophores, the Fe is bound strongly to three hydroxamate groups.

SOME ORGANIC COMPOUNDS OF INTEREST IN ENVIRONMENTAL SURFACE CHEMISTRY

In discussing environmental surface and interface interactions, a variety of different kinds or classes of organic compounds is important. Here, we introduce just a few. We begin by discussing simple organic ligands along with polymers and surfactants, which include both natural and synthetic compounds, and then humic and fulvic acids, which are by definition naturally occuring. Next, we consider three types of organic compounds that are generally synthetic (human-made) and often toxic: polycyclic aromatic hydrocarbons (PAHs), substituted nitrobenzenes (SNBs), and volatile organic compounds (VOCs).

Polymers

An organic polymer is a macromolecule composed of repeating units bound by covalent bonds. The term *polymer* often brings to mind plastics, but there are many natural and synthetic polymers ranging from polysaccharide, polypeptide and polynucleotide biopolymers to the common hydrophilic polymer polyacrylamide, to the common hydrophobic polymer polyethylene. Polystyrene sulfonates are polymers that are often used as standards for calibrating size exclusion chromatography applications for determining the molecular weight distributions of humic and fulvic acids (see below; Zhou et al., 2000).

Polymers often readily adsorb to mineral surfaces (e.g., Israelachvili, 1992). Some important considerations for adsorption include how open or coiled the polymer structure is at the surface, interactions between nearest neighbor adsorbed polymer molecules, and potential bridging of neighboring particle surfaces by polymers (see Figure 8.4). Polymer theory is often applied to humic and fulvic acids (see below), but it should be recognized that humics and fulvics can have distinct properties that require unique approaches.

Organic Surfactants, Including Fatty Acids

A surfactant is a "surface active" agent molecule that behaves as a wetting agent, lowering the surface tension between two liquids and enhancing spreading. In organic chemistry, surfactants most often are amphiphilic, having a hydrophobic "head" and a hydrophilic "tail." Organic surfactants are often classified into four primary groups: anionic, cationic, nonionic, and *zwitterionic* (dual charge).

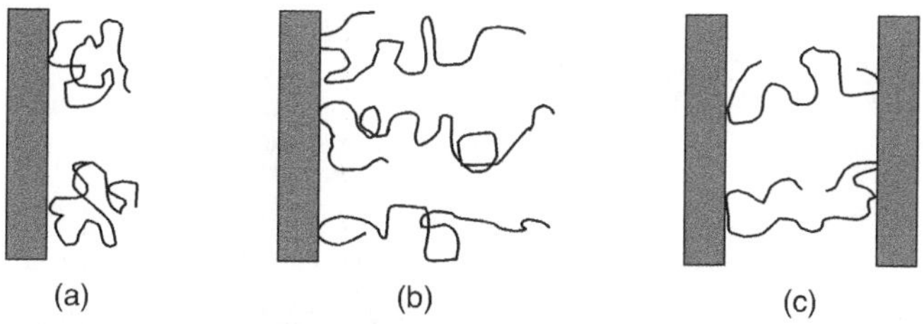

FIGURE 8.4 Some aspects of polymer sorption. Left is at low surface coverage where the molecules do not overlap, middle is a high surface coverage where neighboring molecules overlap and affect conformation, and right shows how molecules may bridge two surfaces. After Israelachvili, 1992.

A *fatty acid* is a surfactant that consists of a (hydrophilic) carboxylic acid, often with a long, unbranched (hydrophobic) aliphatic tail, that is either saturated or unsaturated. The term "saturated" indicates that all of the C-atoms (except for the one associated with the carboxylate group) contain as many H-atoms as possible, meaning that no double bonds or other functional groups are involved. Carboxylic acids with as few as four C-atoms (butyric acid) are considered to be fatty acids. Fatty acids have long been used in the ore industry as part of the process of concentrating oxide and carbonate ores by flotation. Lauric acid is a frequently studied fatty acid with structure:

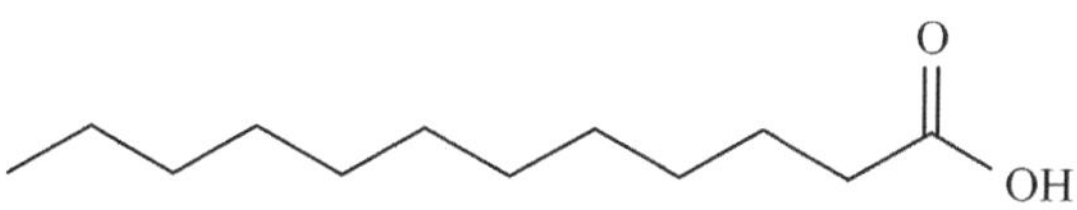

Humic Substances

Aiken (1985) defines humic substances as "naturally occurring, biogenic, heterogeneous organic substances that can generally be characterized as being yellow to black in color, of high molecular weight, and refractory." The two soluble forms of humic substances, humic acid and fulvic acid, are operationally defined in terms of their ranges of solubilities. Humic acids are not soluble in water below pH 2 but are soluble under less acidic conditions, whereas fulvic acids are soluble even below pH 2. Humic and fulvic acids are complex mixtures of molecules that exhibit a variety of structures and reactivities. Often, the term NOM is used to encompass primarily the humic and fulvic acids. Humic and fulvic acids can impart a tea-brown color to a stream when present at concentrations above a few mgC L^{-1} and can have natural surfactant and foaming properties (Figure 8.5).

NOM is ubiquitous in aquatic and terrestrial environments, and it plays an important role in many biogeochemical processes, affecting the fate and transport of metals, radionuclides, and hydrophobic organic compounds (Cabaniss et al. 2000 and references therein); helping to control the penetration of potentially harmful ultraviolet (UV) radiation in surface waters (Scully et al., 1995); and serving as a primary carbon source to microorganisms (Findlay et al., 1986). NOM can present problems in drinking water production by clogging filtration systems, serving as a C source to bacteria within distribution systems, and leading to disinfection byproducts when Br and Cl react with humic substances (Allpike et al., 2005).

FIGURE 8.5 A humic/fulvic-rich stream. Note the dark color imparted by the organic acids and the natural surfactant that creates a foam on the surface of the stream.

Humic and fulvic acids have amphiphilic characteristics, which means that they contain hydrophobic moieties in the form of a carbon skeleton, along with hydrophilic functional groups, the most important of which are often carboxyl and phenolic or hydroxyl. Titrations show a broad spectrum of acidic pK_as that span the range of common carboxylic and phenolic/hydroxyl organic acids (e.g., Perdue, 1985), and center around pH 5.0 and pH 10.0, respectively. Common oxygen-containing functional groups include carboxyl, CO_2H; phenolic —OH; alcoholic —OH; quinoid C=O; ketonic C=O; and methoxy, —OCH_3. Humic and fulvic acids also may contain varying amounts of N- and S-functional groups, including heterocycle N, —NH_2, amino-acid N and S (cysteine-like) (McKnight, 1989), and sulfonyl —SO_3H (McKnight and Thurman, 1989). Because humic and fulvic acids are complex mixtures of molecules, their properties are difficult to quantify, and the effects of individual properties such as molecular weight and aliphatic/aromatic ratios are difficult to distinguish. Natural humic substances generally have at least some Fe, Al, Mn, and/or other metals associated with them that cannot be fully removed by isolation and other processes such as passing the organic sample through a chelating resin.

Some proposed structures of fulvic acid are shown in Figure 8.6. A "typical" molecular weight distribution of aquatic NOM as related to properties and behavior is shown in Figure 8.7.

Polycyclic Aromatic Hydrocarbons (PAHs)

PAHs are organic compounds that consist of fused aromatic rings and that do not contain heteroatoms or carry substituents. They can occur in oil and coal and as byproducts of the burning of fossil fuels. Many are carcinogenic, mutagenic, and/or teratogenic, which means that they may cause birth defects. They tend to be hydrophobic, and the larger compounds are less volatile and less water soluble.

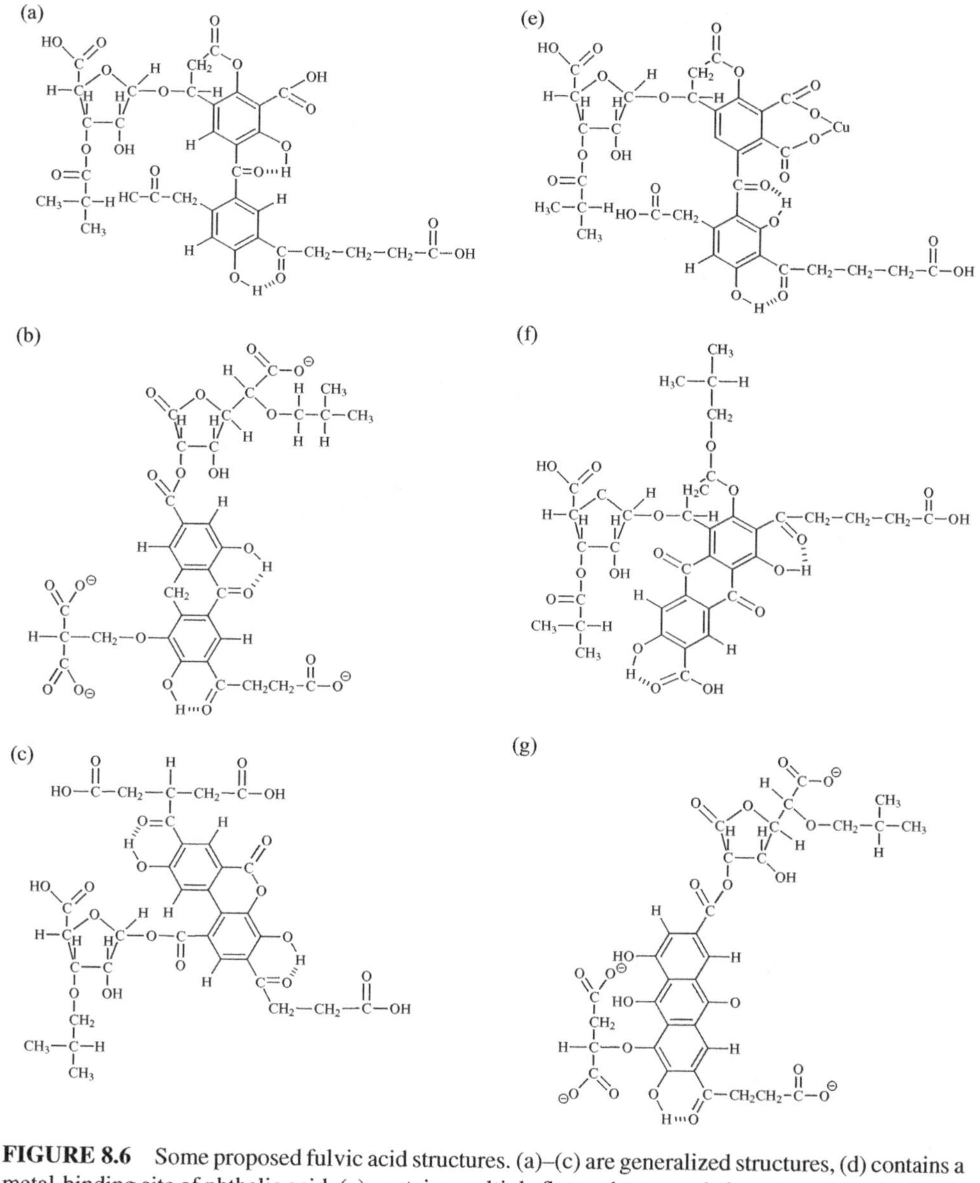

FIGURE 8.6 Some proposed fulvic acid structures. (a)–(c) are generalized structures, (d) contains a metal-binding site of phthalic acid, (e) contains multiple fluorophores, and (f) contains a semiquinone free radical. From Leenheer et al., 1989.

The extensive π bonding of PAHs leads to characteristic UV absorbance and fluorescence spectra. An example of a PAH is anthracene:

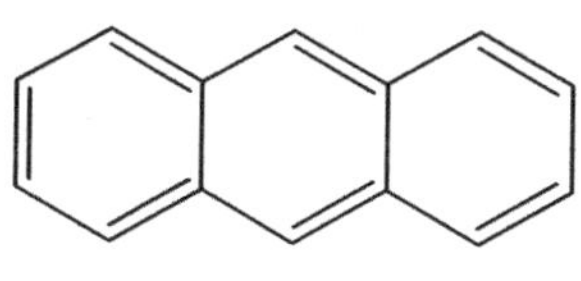

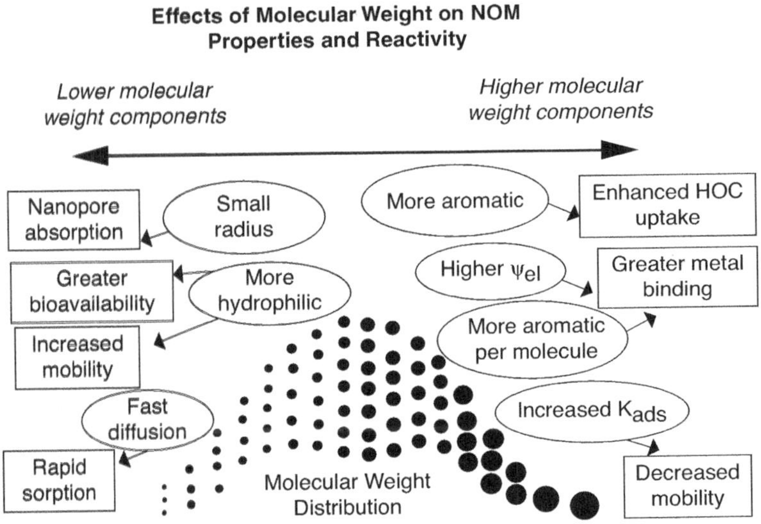

FIGURE 8.7 How properties and reactivity of NOM vary with molecular weight. From Cabaniss et al., 2000.

which is derived from coal tar and is used in pesticides and wood preservatives. Because of its hydrophobicity, anthracene tends to adsorb only very weakly or not at all to Fe oxides and kaolinite, except in the presence of adsorbed organic matter (e.g., Angove et al., 2002). PAHs tend to remain behind on beaches long after oil spills, and their removal through biodegradation processes, sometimes in combination with surfactants and emulsifiers, is an important topic of study. Some other PAHs include benzopyrene, fluoranthene, and coal tars.

Substituted Nitrobenzenes (SNBs)

SNBs are hydrophobic organic compounds (HOCs) that tend to be insoluble in aqueous solutions. Nitrobenzene itself is highly toxic and readily absorbs through the skin. The (substituted) nitrobenzene structure is shown in Figure 8.8.

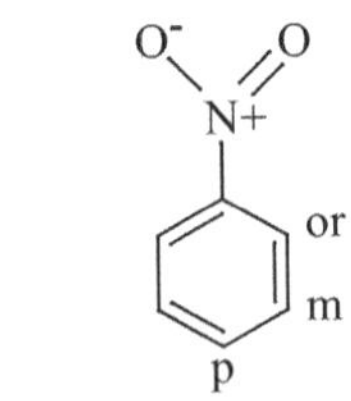

SNBs,
or = ortho,
m = metal,
p = para

FIGURE 8.8 The substituted nitrobenzene (SNB) structure.

Various substituted nitrobenzenes are used as pesticides, solvents, and explosives, and they are widely distributed pollutants. They tend not to adsorb to iron oxides and non-swelling clays, but they can absorb into smectite interlayers or partition into natural organic matter. The presence of electron-withdrawing groups tends to enhance absorption, whereas the presence of electron-releasing groups tends to decrease absorption (Boyd et al., 2001).

Volatile Organic Compounds (VOCs)

VOCs have high enough vapor pressures under normal conditions to vaporize significantly and enter the atmosphere. A wide range of organic molecules, such as aldehydes, ketones, and other light hydrocarbons, tend to occur as VOCs. The exact use of this terminology by agencies such as the U.S. Environmental Protection Agency (EPA) varies, with important environmental and regulatory implications. VOCs tend to vaporize from surface waters, but they can accumulate to relatively high concentrations in groundwaters. Because many VOCs are toxic, often carcinogenic, or mutagenic, their presence in domestic waters can be of special concern. For example, groundwater contamination by the industrial solvent and suspected carcinogen trichloroethylene (TCE) formed the basis of a well-known legal proceeding that occurred in Woburn, Massachusetts, in the 1980s, as depicted in the movie *A Civil Action* (1998). Some important VOCs include

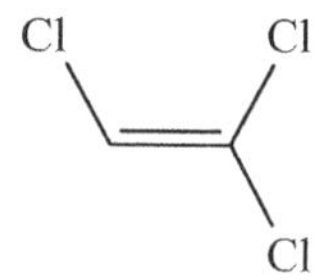

carbon tetrachloride

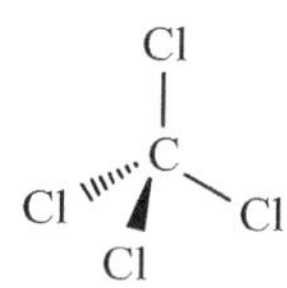

and vinyl chloride

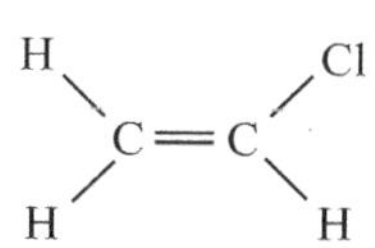

SORPTION OF SIMPLE ORGANIC LIGANDS, SURFACTANTS, AND NATURAL ORGANIC MATTER

Adsorption of Simple Organic Ligands

The adsorption of simple organic ligands with carboxyl, hydroxyl, and/or phenolic functional groups has been studied extensively, particularly with respect to how adsorption contributes to mineral dissolution. Adsorption of simple organic ligands such as oxalate, citrate, and salicylate can be driven electrostatically to form weak outer-sphere complexes, or through ligand exchange to form stronger inner-sphere complexes. The outer-sphere adsorption can be described by the equation

$$>S{-}OH_2^+ + L^- \rightleftarrows >S{-}OH_2^+\,L^-, \tag{8.2}$$

and the inner sphere adsorption can be described by the equation

$$>S{-}OH + HL \rightleftarrows >S\text{-}L + H_2O, \tag{8.3}$$

where L refers to the organic ligand and >S refers to the mineral surface site. These two modes of adsorption cannot be distinguished based on reaction stoichiometry alone (Schindler, 1990). Rather, spectroscopic methods are needed. Adsorption of simple organic ligands may also involve hydrogen bonding (Schindler, 1990). Overall, the sorption mode and strength of sorption depend on a variety of factors, such as the shape of the organic molecule, the position of a functional group with respect to any other functional groups that may exist on the molecule, and the polarity and ease by which each functional group ionizes. If an organic ligand binds to a dissolved metal, then this complexation may either enhance sorption as a ternary surface complex (see below) or inhibit sorption.

For bidentate ligands such as oxalate, an inner-sphere surface complex is often bidentate mononuclear, as

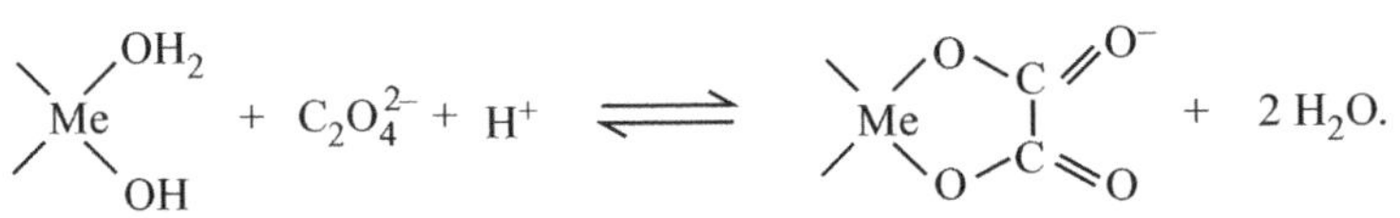

(8.4)

Adsorption of simple organic ligands generally decreases with increasing pH. When adsorption to a mineral of relatively high pH p.z.c., such as hematite or goethite, is driven primarily by electrostatic interactions, adsorption will occur primarily in the pH range between the pK_a of the acid and the point-of-zero charge (p.z.c.) of the solid (Figure 8.9). In this pH range, the organic ligand is negatively charged and the mineral surface overall is positively charged (Figure 8.9). Significant adsorption beyond this pH range implies

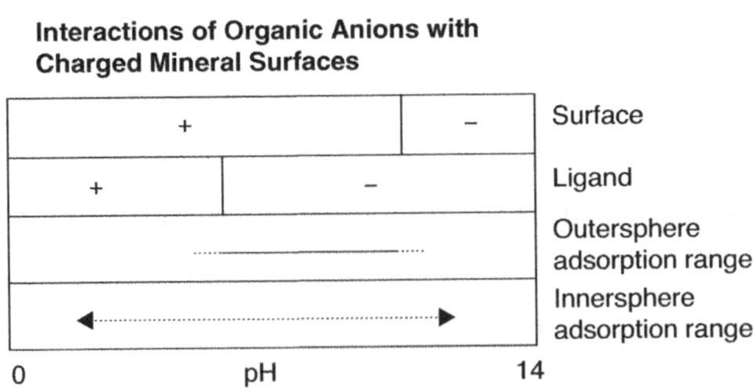

FIGURE 8.9 Schematic Schindler diagram showing charge considerations in the interactions between organic anions and charged mineral surfaces.

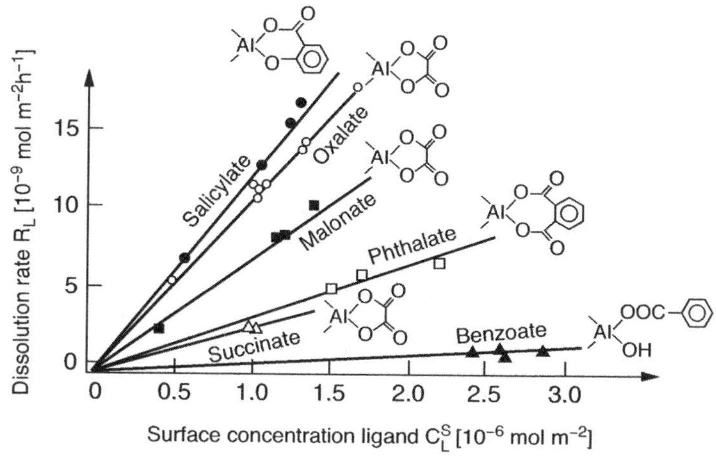

FIGURE 8.10 Adsorption of simple organic ligands enhances surface-controlled mineral dissolution. From Furrer and Stumm, 1986 (with permission).

inner-sphere adsorption. Inner-sphere adsorption of simple organic ligands is also promoted by low pH, because of protonation of surface sites.

Many simple organic ligands, such as citrate, oxalate, and salicylate, can enhance mineral dissolution upon adsorption, and this dissolution process needs to be considered in adsorption experiments. In addition to altering the surface, dissolution may release metals to solution that may form complexes with the organic ligand, shifting equilibria. Ligands that adsorb more strongly, with higher surface concentrations, tend to enhance surface-controlled mineral dissolution rates (Chapter 1) to a greater extent (Figure 8.10).

Adsorption of Anionic Surfactants, Fatty Acids

Anionic surfactants have been of interest in surface chemistry for many decades. Recently, interest has renewed because soil microorganisms have been found to produce biosurfactants (e.g., Rosenberg and Ron, 1999) and because of templating and control of growth of nanoparticles (e.g., Hamley, 2005). At low surface coverages, the electrostatic interaction between the polar function group of the surfactant and the mineral surface is dominant in controlling adsorption (Rea and Parks, 1990). Adsorption by electrostatic interactions will occur at pH values at which the surface is positively charged, which for some minerals such as goethite is over a broad pH range. Multivalent cations can enhance anionic surfactant adsorption either through surface charge effects, when the surface would otherwise be negatively charged, or through coadsorption of a cation-surfactant complex (Rea and Parks, 1990).

At higher anionic surfactant concentrations, hydrophobic interactions between the nonpolar portions of neighboring adsorbed molecules will act to stabilize sorption (Rea and Parks, 1990). When nonpolar entities interact at a surface, these interactions—sometimes termed "hydrophobic bonding"—occur from changes in the structure of water near the organic molecules, along with other entropic considerations. Stabilization may also occur through van der Waals interactions.

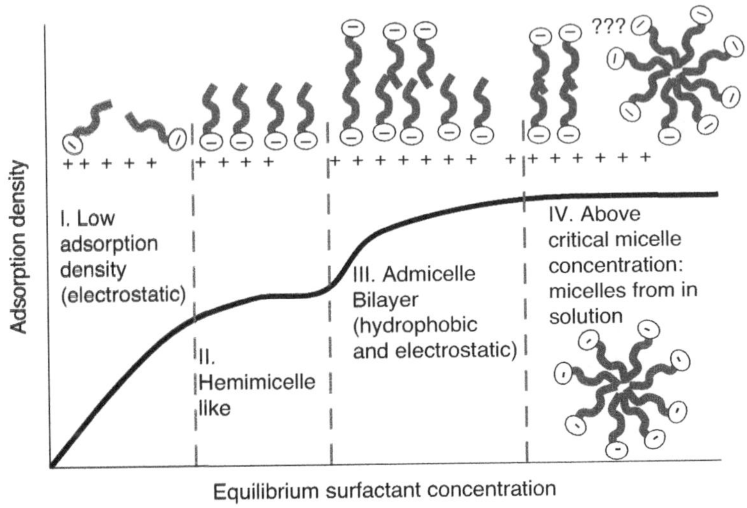

FIGURE 8.11 How surfactant structure and sorption vary with surfactant concentration. Above the critical micelle concentration (CMC), the micelles in solution may compete with the surface and/or the micelles themselves may sorb to the surface.

Aggregation of sorbed hydrophobic molecules can result in structures similar to hemimicelles at a surface (Schindler, 1990). At still higher surface coverages, a second layer of surfactant molecules can form, which creates a *bilayer* with anionic groups facing the mineral surface and the solution and hydrophobic components sandwiched between. These structures are called *admicelles* (Figure 8.11). Formation of micelles in solution has the potential to inhibit adsorption. A schematic of how anionic surfactant concentration may affect adsorption behavior is shown in Figure 8.11. Anionic surfactant adsorption to oxide surfaces can lead to an increase in surface protonation (e.g., Lee and Koopal, 1996), which has the potential to affect not only adsorption but also mineral dissolution (e.g., Carrasco et al., 2008).

For fatty acids, the longer the C-chain that forms the hydrophobic tail, generally the stronger the lateral interactions, at least up to a point. Ulrich et al. (1988) reported that the ΔG_{ads} of fatty acids adsorption to Al_2O_3 dropped from -17.3 kJ mol^{-1} for C_2H_5COOH to -27 kJ mol^{-1} for $C_{11}H_{23}COOH$, in agreement with increased sorption stabilization with increased C-chain length.

Liang and Morgan (1990) showed that the stability of a hematite colloidal suspension first decreased and then increased with loading of fatty acids as the organic compounds decreased the charge on the hematite particles (+ hematite charge, − fatty acid charge) but then formed hemimicelle-like structures on the hematite surface at higher fatty acid coverages. Carrasco et al. (2008) showed that the electrophoretic mobility of boehmite (γ-AlOOH) was reversed after sorption of the sodium dodecyl sulfate (SDS; Figure 8.12).

Sorption of Cationic Surfactants

Many principles described above for anionic surfactants have analogs for cationic surfactants. *Cationic surfactants* are being considered as a means of modifying zeolites, which are

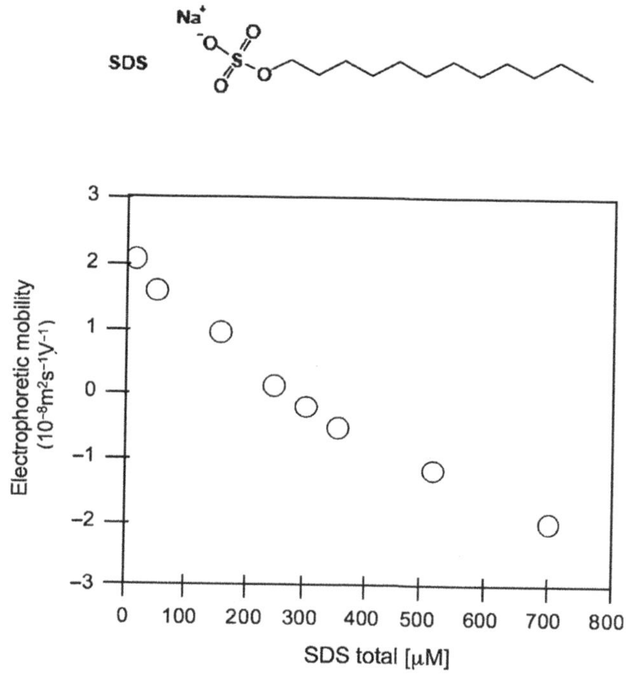

FIGURE 8.12 (top) The structure of the surfactant SDS. (bottom) Adsorption of the anionic surfactant SDS changes the electrophoretic mobility (EM) of boehmite. The error bars are not shown here but are the size of the data points or slightly larger. Based on and using data from Carrasco et al., 2008 (with permission).

used to remove contaminants in water treatment. The cationic surfactants can promote sorption of nonpolar organic contaminants, such as benzene, toluene, and chlorinated solvents, for remediation of these contaminant compounds (Figure 8.13). The removal efficiency is greatest when the cationic surfactant concentration is low enough that admicelles do not form (Scherer et al., 2000). Above this concentration, the positive charge exposed to solution inhibits nonpolar organic compound sorption. At higher surface coverages, there is potential for immobilization of anionic pollutants such as chromate (CrO_4^{2-}) and selenate (SeO_4^{2-}). When cationic surfactants are sorbed to zeolites, the ability of the zeolite to take up metal cations is not impaired, perhaps because the surfactant does not enter the internal pore space of the zeolite, and thus it does not block cation exchange sites (Scherer et al., 2000).

Sorption of Phospholipid Surfactants: Biomedical Implications

Phospholipids are a major component of all biological membranes, together with other components such as glycolipids, cholesterol, and proteins. Phospholipids are natural surfactant structures that have a polar, hydrophilic head, and a nonpolar, hydrophobic tail. How these molecules aggregate to form layered structures is an important area of research in geomicrobiology and biomedicine. Phospholipid aggregation behavior is referred to as *polymorphism*.

Dipalmitoylphosphatidylcholine (DPPC; $C_{40}H_{80}NO_8P$) is a phospholipid that is the major constituent of pulmonary surfactant and that has been studied as a model biological

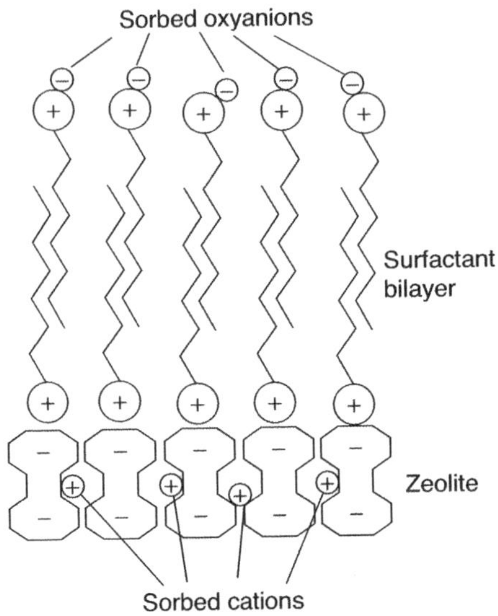

FIGURE 8.13 Schematic illustration of how modification of a zeolite by sorption of a cationic surfactant can allow for removal of oxyanions such as chromate and selenate from water while permitting the zeolite to remove cations. From Scherer et al., 2000 (with permission).

membrane constituent. Oleson and Sahai (2008) investigated the sorption of DPPC on quartz, rutile, and corundum, as a function of pH, ionic strength, and presence or absence of Ca^{2+}. Sorption was affected by interactions between the hydrophobic groups as well as electrostatic interactions between mineral surface sites and the negatively charged phosphate ester component (—R(PO_4^-)R′—) of the DPPC molecules. On negatively charged surfaces, the maximum sorption of DPPC occurred as two bilayers, whereas on positively charged surfaces, up to four bilayers formed. These processes are shown schematically in Figure 8.14. The presence of Ca^{2+} tended to promote additional bilayer sorption beyond the second bilayer.

Adsorption of Humic And Fulvic Acids (NOM)

NOM adsorption to oxides and aluminosilicates tends to follow an adsorption envelope, with adsorption decreasing as pH increases (Tipping, 1981). An example is shown in Figure 8.15.

Because NOM sorption involves chemical (e.g., ligand exchange) as well as physical (e.g., hydrophobic interactions) processes, the free energy of sorption (ΔG_{ads}) contains several terms (Schlautman, 1992):

$$\Delta G_{\text{ads}} = \Delta G_{\text{chem}} + \Delta G_{\text{elec}} + \Delta G_{\text{hyd}}, \tag{8.5}$$

where

ΔG_{chem} = chemical component
ΔG_{elec} = electrostatic component
ΔG_{hyd} = hydrophobic component of sorption.

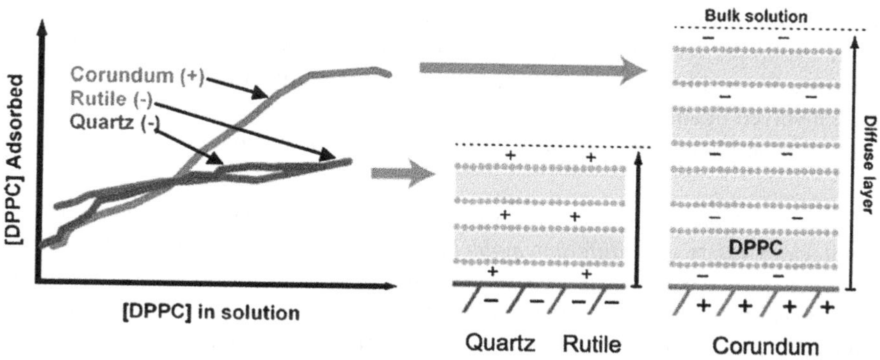

FIGURE 8.14 Sorption isotherms and schematic for DPPC sorption to corundum, rutile, and quartz. Under the conditions shown here, the corundum surface would be positively charged, whereas the rutile and quartz surfaces would be negatively charged. The DPPC formed two bilayers on the negatively charged surfaces but up to four bilayers on the positively charged corundum surface. Bilayer formation and sorption were driven primarily by interactions between the hydrophobic components but with modification by electrostatic forces. From Oleson and Sahai, 2008 (with permission).

The relative importance of different sorption mechanisms is likely to vary with the specific compositions of the humic and fulvic acids, the solution conditions such as pH, ionic strength, metal content, and the surface properties of minerals, such as metal coordination at the mineral surface and point-of-zero net proton charge (p.z.n.p.c.).

Direct, unambiguous spectroscopic evidence for binding mechanism(s) is rare but growing. Liang and Morgan (1990) showed that the electrophoretic mobility of hematite particles decreased to zero and then changed direction with increased sorption of humic and fulvic acids, which indicates that sorption occurred even when the charges of the surface and the sorbing molecules were of the same signs. Thus, sorption must involve specific chemical interactions as well as electrostatic interactions.

Although *ligand exchange* seems to be the most important mechanism for sorption of humic and fulvic acids on iron (hydr)oxides, the relative contributions of other sorption mechanisms such as hydrophobic interactions may vary with solution conditions and with

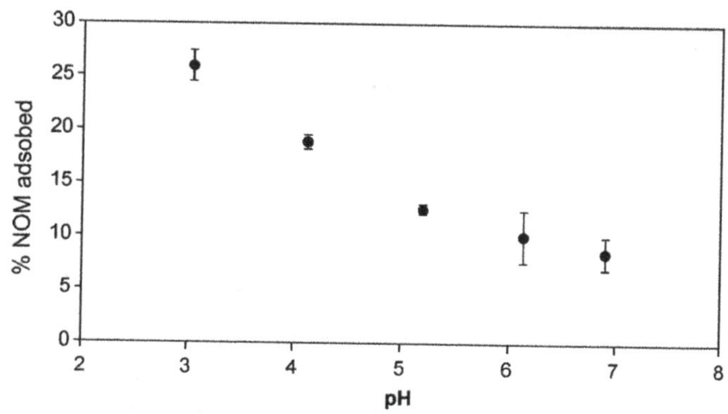

FIGURE 8.15 Adsorption envelope of NOM to goethite. Note that percent adsorption rather than mg m^{-2} is plotted. From Maurice et al., 2004.

the composition of the NOM. Humic and fulvic acids are likely to have more than one functional group per molecule so that a given molecule may form more than one surface complex(es). Steric hindrance may limit the number of functional groups on any one molecule involved in adsorption.

Humic substances tend to exhibit the following sorption behaviors.

1. *Given that humic and fulvic acids are polydisperse, which means that a variety of molecules exists with different molecular weights (and other properties), fractionation may occur over the course of adsorption, with the more readily adsorbed components preferentially removed from solution.* Studies have suggested preferential adsorption of, for example, intermediate or high-molecular-weight fractions (e.g., Tomaic and Zutic, 1988; Meier et al., 1999; Zhou et al., 2001), aromatic moieties (e.g., McKnight et al., 1992), and hydrophobic components (e.g., McKnight et al., 1992; Gu et al., 1995). Zhou et al. (2001) studied humic/fulvic adsorption to goethite. At low surface coverages, no competition for surface sites and no fractionation was observed. However, as surface coverage increased, higher molecular weight components preferentially adsorbed, so that the organic matter that remained in solution had lower molecular weight (Figure 8.16).
2. *Because large, polyfunctional humic and fulvic molecules may be expected to adsorb with a variety of configurations, attainment of adsorption steady state may be slow, as molecules rearrange on the surface.*

   Zhou et al. (2001) observed that NOM adsorption did not come to steady state in terms of amount adsorbed (Figure 8.17) or the average molecular weight (M_n) of NOM remaining in solution until nearly 24 hours. This example shows how adsorption kinetics may be slow as molecules of different sizes arrange at the surface and packing density alters.
3. *Because of the large size and complex functionality of humic and fulvic acids, steric factors related to shape or three-dimensional structure may affect adsorption.* The

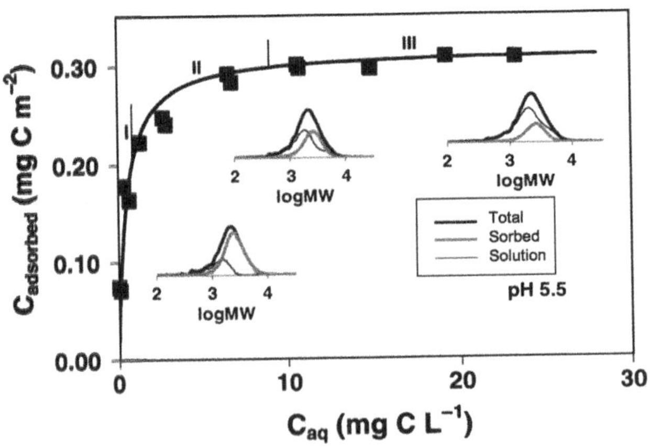

FIGURE 8.16 Fractionation of NOM during adsorption to goethite at pH 5.5. The molecular weight distributions at low, intermediate, and high surface coverages (I, II, III) are shown. Preferential adsorption of high to intermediate molecular weight components leave lower molecular weight components in solution. Based on data in Zhou et al., 2001.

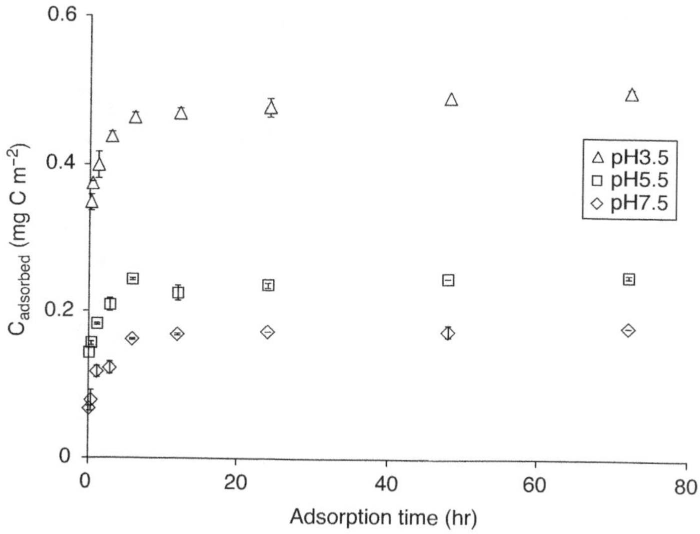

FIGURE 8.17 Change in amount of NOM adsorbed over time during adsorption experiment on goethite, which indicates slow adsorption kinetics as molecules rearrange on the surface. Based on data in Zhou et al., 2001.

degree of steric hindrance of adsorption may be influenced by the overall size of the molecules, as well as by molecular flexibility and by relative positions of functional groups. Although simple organic acids (Stumm and Furrer, 1987) can provide valuable clues to coordination chemistry of humic and fulvic acids, the simple organic acids cannot account for complex steric effects.

4. *Humic and fulvic acids may not lie flat on an oxide surface, and different confirmations and aggregate structures may occur.* Namjesnik-Dejanovic and Maurice (2001) showed that humic and fulvic acids can sorb to mineral surfaces as spherical or linear structures, or as ring-shaped or toroidal aggregate structures (Figure 8.18). In general, humic and fulvic acids seem to be more coiled at lower pH but are more open at higher pH. However, other factors such as ionic strength and the presence of trivalent metals can affect conformation and aggregate structure. If long segments of humic and fulvic acids extend into solution, then these segments may form bridges to the surfaces of nearby particles (e.g. Israelachvili, 1985; Amal et al., 1992), and thus play a role in the formation of aggregates (Clapp et al., 1991).
5. *As amphiphilic (=amphipatic) molecules, the relative importance of different sorption mechanisms may change with solution conditions and the nature of the surface.* For example, at relatively high pH, adsorption may be driven by electrostatics, whereas at lower pH, when the negative charge on the NOM molecule decreases during protonation, hydrophobic interactions may become more important.
6. *Adsorption may be irreversible, with hysteresis in adsorption/desorption curves.* Extraction of humic acids from soils often includes an HCl/HF treatment to remove tightly bound clay minerals (e.g. Schnitzer, 1978). The need for this procedure is strong evidence for irreversible sorption of some soil NOM.

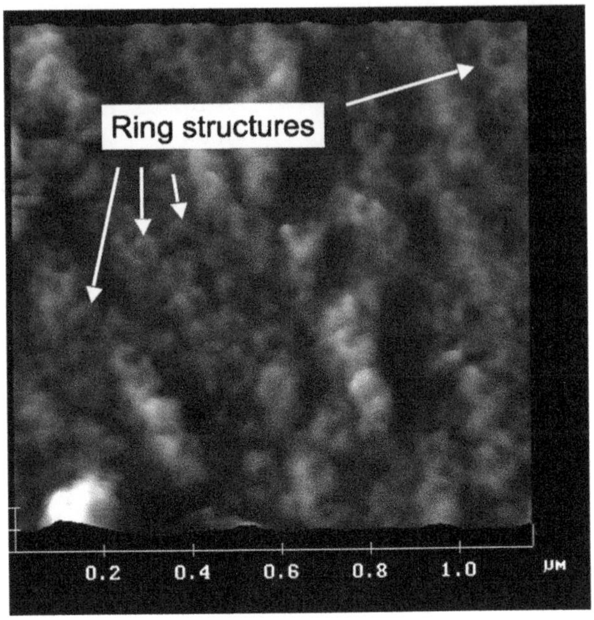

FIGURE 8.18 In-solution atomic force microscopy (AFM) image of natural organic matter (humics/fulvics) adsorbed to mica, showing aggregate toroidal or ring structures. Sample preparation and image parameters as described by Maurice and Namjesnik-Dejanovic, 1999.

METAL–LIGAND COADSORPTION: TERNARY SURFACE COMPLEXES

Metals and organic ligands often are associated as complexes in aqueous solution, and they may coadsorb as ternary complexes on mineral surfaces. Schindler (1990) described two types of ternary surface complexes (TSCs), depending on whether the metal or the ligand is directly coordinated with the surface

Type A: >S—O—M—L

Type B: >S—L—M.

Subsequent research has shown the TSCs may involve inner-sphere and/or outer-sphere adsorption; for example, in a Type A TSC, the metal may be innerspherically bound to the surface but the ligand either inner-spherically or outer-spherically bound to the metal.

Determining the geometry of a ternary surface complex requires the use of spectroscopic methods such as extended X-ray absorption fine structure (EXAFS), Fourier transform infrared spectroscopy (FTIR), or electronic spin resonance (ESR). For example, Mishra et al. (2007) used EXAFS and X-ray absorption near-edge structure (XANES) to show that Type A ternary surface complex forms between the kaolinite surface, Pb, and the trihydroxamate siderophore desferrioxamine B (DFO-B) at $pH > 6.0$. Cheah et al. (1999) used a combination of ESR, XAFS, and FTIR to show that Cu forms with bipyridine an inner-sphere, mononuclear, monodentate Type A ternary surface complex on the amorphous Si surface (Figure 8.19).

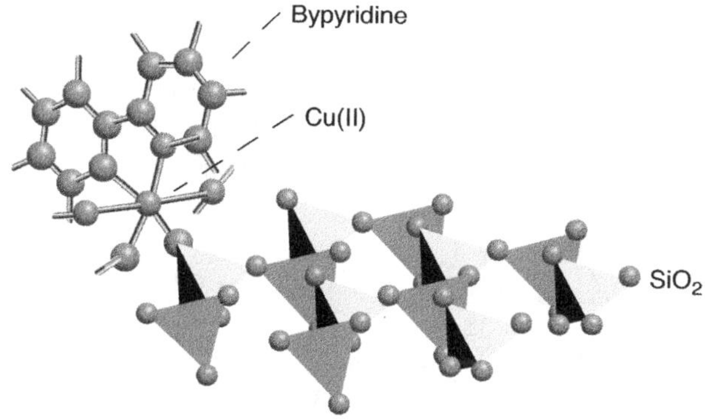

FIGURE 8.19 A proposed Type A TSC of Cu(II) and bipyridine on the amorphous SiO_2 surface. Adapted from Cheah et al., 1999 (with permission).

Boily et al. (2005) used a combination of FTIR, EXAFS, and surface complexation modeling (SCM) to investigate coadsorption of Cd and phthalate on goethite. They found that Cd forms corner-sharing, inner-sphere complexes at the goethite surface over a wide pH range. Depending on the pH, phthalate may be bound inner-spherically or outer-spherically to the adsorbed Cd to form a Type A TSC, depending on the pH. In both cases, a TSC forms, but the nature of the binding is different (Figure 8.20).

For large and complex organic ligands such as humic and fulvic acids, it is possible that both Type A and Type B TSC could occur on different parts of a natural organic molecule, or for different components in an NOM mixture. Because all NOM samples have some metals naturally associated with them that cannot be removed fully without altering the molecular structure, TSCs always need to be considered in NOM adsorption studies, even in the absence of added metals. TSCs are not limited to metals and organics; they may also occur with metals and inorganic ligands.

SORPTION OF SOME ORGANIC POLLUTANTS

Our discussion of the sorption behaviors of some pollutant organic compounds begins with review of some fundamental properties, such as vapor pressure, solubility, and density. Next, empirical means of characterizing organic compounds which includes the octanol-water partition coefficient and Hammett constants are discussed. Some examples of sorption behavior are provided. This section is followed by discussions of the K_d approach to hydrophobic organic compound transport in porous media and sorption of VOCs to activated carbon.

Vapor Pressure, Solubility, and Density

Several properties of pollutant organics such as vapor pressure, solubility in water, and density, may not at first glance seem to be important to adsorption phenomena. But, they can help to control organic matter concentrations and other behaviors important to fate and transport processes of many pollutant organic compounds.

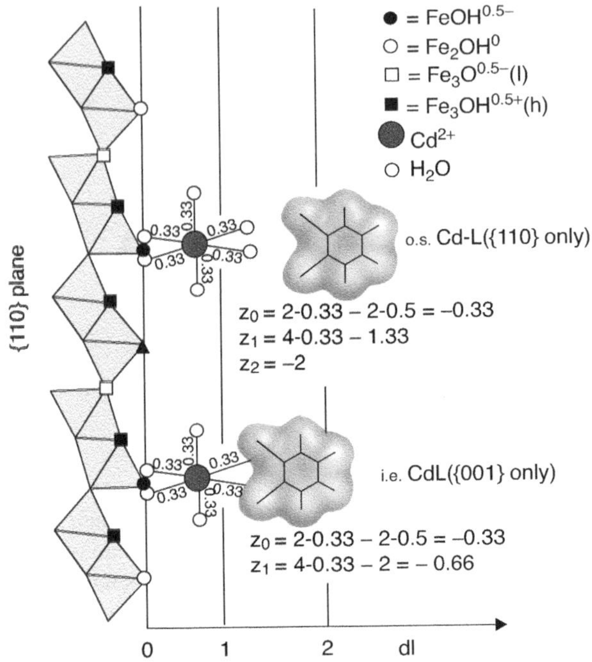

FIGURE 8.20 A representation of the formation of goethite-Cd-phthalate TSCs using the triple layer model approach. In the upper part of the figure, a Type A TSC forms in which Cd is innerspherically bound to goethite and phthalate is outerspherically bound to the adsorbed Cd. In the lower part of the figure, a Type A TSC forms in which Cd is innerspherically bound to boethite, but phthalate is instead inner-spherically bound to the adsorbed Cd. In this figure, the Cd complex is in the 0 plane, outerspherically bound phthalate in the 2 plane, innerspherically-bound phthalate in the 1 plane, and diffuse ions needed to ensure electroneutrality of the interface in the diffuse layer (dl). From Boily et al., 2005 (with permission).

The *vapor pressure* P° of a compound is defined as the pressure of the vapor in equilibrium with its nonvapor (condensed) phase(s). The SI unit for pressure is the pascal (Pa), which is equal to one Newton per square meter ($N{\cdot}m^{-2}$ or $kg{\cdot}m^{-1}{\cdot}s^{-2}$) or 9.87×10^{-6} atm. The vapor pressure of organic compounds can be highly sensitive to temperature, even under the range of ambient conditions. A substance that has a high vapor pressure in the general vicinity of 25°C is often referred to as *volatile*. Figure 8.21 shows the trends of P° values for some important classes of organic compounds. Among members of a homologous series, such as the linear aliphatic hydrocarbon n-alkanes, lower molecular weight compounds tend to be more volatile. Replacing an H atom bound to a C by a halogen such as Cl also tends to decrease equilibrium vapor pressure. Other polar functional groups, which include those that can form hydrogen bonds, also tend to decrease equilibrium vapor pressure, hence volatility.

Raoult's Law states that the vapor pressure of an ideal solution is dependent on the vapor pressure (P_i°) of each chemical component i and the mole fraction (x_i) of the component present in the solution. Raoult's Law can be used to describe the equilibrium partial pressure of a volatile organic compound (P_{org}) in the atmosphere above an ideal solvent

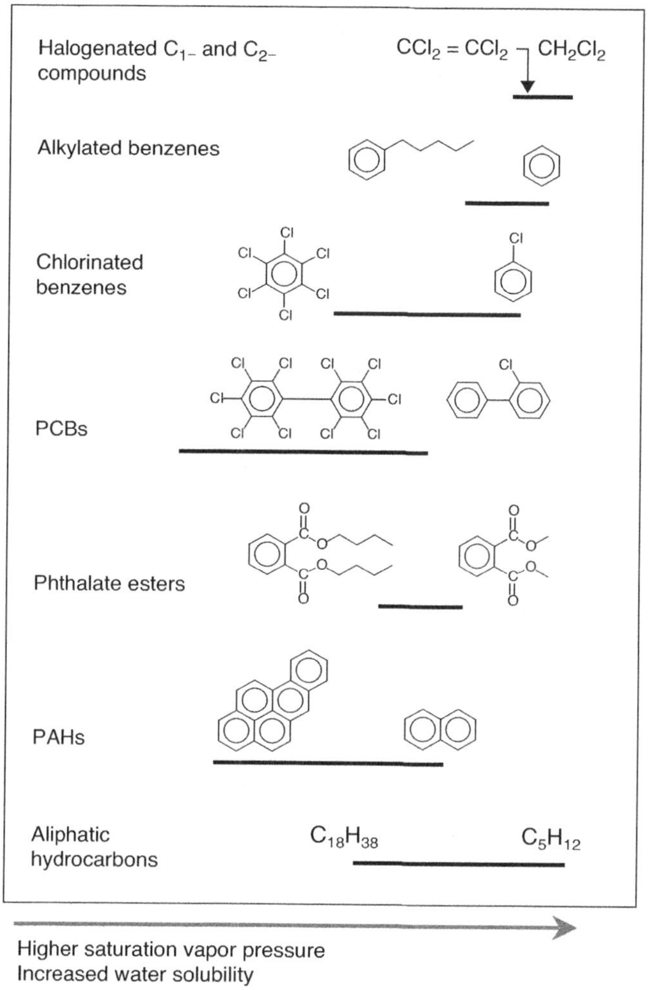

FIGURE 8.21 How saturation vapor pressure (P°) and water solubility (C_w^{sat}) trend for some classes of organic compounds. PCB = polychlorinated biphenyls. Revised from Schwarzenbach et al., 1993.

(such as benzene)

$$P_{org} = x_{org}\, P_{org}^{\circ}. \tag{8.6}$$

For situations in which nonpolar organic compounds are present in polar solvents such as water, the system cannot be considered to be ideal. In such cases, the concept of fugacity is used, where *fugacity* is the "fleeing tendency" of the gas from the solution, and fugacity of a component i in a liquid mixture can be related to its fugacity in a pure liquid by (Schwarzenbach et al., 1993)

$$f_i = \gamma_i x_i f_{i\ \text{pure liquid}} \tag{8.7}$$

$$= \gamma_i x_i P_{i\ \text{pure liquid}}^{\circ}, \tag{8.8}$$

where γ_i accounts for nonideal behavior, as we encountered for activity coefficients in Chapter 1.

Schwarzenbach et al. (1993) define the *aqueous solubility* of an organic compound as "the abundance of the chemical per unit volume in the aqueous phase when the solution is in equilibrium with the pure compound in its actual aggregation state (gas, liquid, solid) at a specified temperature and pressure (e.g., 25°C, 1 atm)." Organic compounds of interest in environmental surface and interface chemistry span many orders of magnitude in aqueous solubility (in mol L^{-1}). Trends for some example compound types are shown in Figure 8.21. For the classes of compounds considered in Figure 8.21, the more volatile compounds also tend to be more water soluble. Temperature also plays an important role in solubility.

The density of an organic compound can also play an important role in its behavior in a surface or subsurface spill. The densities of some organic contaminants are shown in Table 8.2. Also included is a designation as to whether the (insoluble) compound is likely to be a "sinker" or "floater" in aqueous solutions such as typical ground-water environments. Soluble organics dissolve in water and are termed "mixers."

The Octanol-Water Partition Constant, K_{ow}

The solubility of a nonpolar organic compound in a nonpolar solvent can be much greater than its solubility in aqueous solution. Nonpolar organic compounds often partition into

TABLE 8.2 Densities, octanol-water partition coefficients, and aqueous solubilities of some organic contaminants.

Compound	Density, 20°C	sink/float	log K_{ow}	$-\log C_w^{sat}$
Atrazine $C_8H_{14}ClN_5$	1.187[1]	sinker[1]	2.61	3.86[2]
Carbon tetrachloride, CCl_4	1.5940	sinker	1.97	2.20
Trichloroethene TCE, $ClCHCCl_2$	1.4642	sinker	un	2.07
Chlorobenzene C_6H_5Cl	1.1058	sinker	2.84	2.35
1,2,4 Trichlorobenzene $C_6H_3Cl_3$	1.4542	sinker	4.02	3.65
PCBs $C_{12}H_{10-x}Cl_x$	>1	sinker	un	~4.5–10
DDT $C_{14}H_9Cl_5$	>1	sinker	6.36	7.85
Benzene C_6H_6	0.87865	floater	2.13	1.64
Nitrobenzene $C_6H_5NO_2$	1.199[1]	sinker[1]	1.85	1.77
Toluene C_7H_8	0.8669	floater	2.73	2.25
n-Pentane $CH_3(CH_2)_3CH_3$	0.6262	floater	3.39	3.25
n-Octane	0.7025	floater	5.15	5.20
n-Nonane $CH_3(CH_2)_7CH_3$	0.7176	floater	5.42	5.94
n-Decane $CH_3(CH_2)_8CH_3$	0.7300	floater	~6	6.75
Naphthalene $C_{10}H_8$	1.4003	sinker	3.35	3.61
Acetone CH_3COCH_3	0.7899	mixer	−0.24	−1.13
Ethanol C_2H_5OH	0.7893	mixer	−0.30	fully miscible[2]
Formaldehyde HCHO	0.815	mixer	0.35	~1.5[2]
Tetrahydrofuran C_4H_8O	0.888	mixer	0.46	miscible[2]

[1]Density from NIST; sinker designation P. Maurice.
[2]Data from product materials safety data sheets.
[3]Density data and designation as "sinker," "floater," or "mixed" compiled by Pitchford et al. (1988). Log K_{ow} data (at 25°C) from the Sangster Database (logkow.cisti.nrc.ca/logkow; June, 2008). un = data unavailable or more than one compound. Density in g cm^{-3}. Solubility data as compiled by Schwarzenbach et al. (1993), at 25°C, in $-\log C_{sat}^{w}$ with C in mol L^{-1}.

natural organic matter or absorb into the interlayers of swelling clays. The partitioning of nonpolar organic compounds into an organic phase such as an organic solvent can be described by a dimensionless constant (Schwarzenbach et al., 1993)

$$K_{sw} = C_s/C_w \ (\text{mol } L_s^{-1} \text{ mol}^{-1} \text{ } L_w), \qquad (8.9)$$

where C_s and C_w are the concentrations of the nonpolar organic molecule of interest in the organic solvent and water, respectively. The organic solvent most commonly used is n-octanol, and the partition constant is known as the *octanol-water partition constant*, designated K_{ow}. Some values of log K_{ow} are given in Table 8.2.

Organic Fuel and Solvent Leaks: Volatilization, Solubility, Density, and K_{ow}

One environmental scenario in which organic volatilization, solubility, and density can be important is in the spill of an organic contaminant into a water-table aquifer. As shown in Figure 8.22, a surface or subsurface spill of crude oil or other nonalcohol fuel, such as has occurred over many years near Bemidji, Minnesota, involves a variety of processes. Because the density of the fuel is less than that of water, the fuel floats at the top of the water table. Some fuel remains trapped within the unsaturated zone as coatings on soil particles. Some fuel is dissolved into the groundwater and migrates as a solute through the porous media. Other components volatilize and diffuse upward through the soil zone and eventually into the atmosphere. Over time, (bio)geochemical processes such as sorption and biodegradation, affect the contaminant fate and transport.

In the leak of a solvent such as TCE (sometimes called trichloroethylene) into a similar sand and gravel aquifer (Figure 8.23), the fact that TCE is denser than water plays an important role in the subsurface behavior. Some solvent may become trapped at the base of the aquifer, along an aquitard, and continue to contaminate the aquifer for a prolonged period of time. TCE is highly insoluble, so it does not dissolve into the aquifer water. However, it is volatile and evaporates into the soil zone, eventually reaching the atmosphere. Although not shown in Figure 8.23, processes such as sorption and biodegradation also can be important for solvents in the subsurface. In both the fuel and solvent cases, bioavailability of the

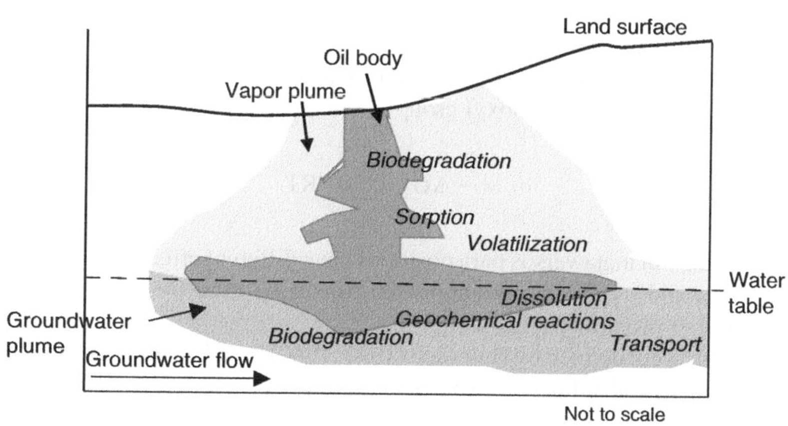

FIGURE 8.22 A schematic illustration of (bio)geochemical processes involved in the fate and transport of hydrocarbons in the subsurface at the *Bemidji* site. After Delin et al., 1998.

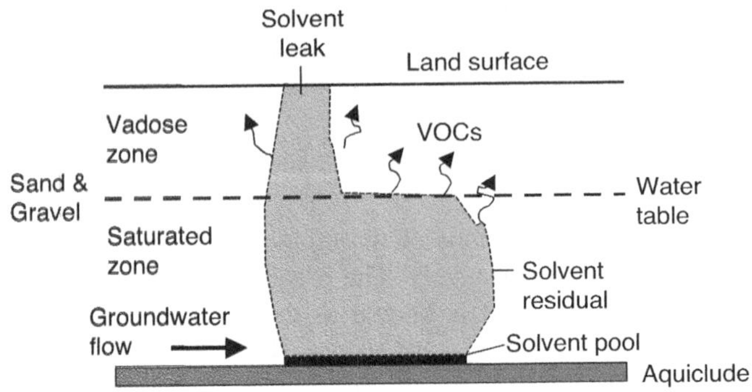

FIGURE 8.23 Schematic of a solvent (TCE) leak over a sand and gravel aquifer. Because TCE is denser than the aquifer water, it sinks to the aquitard at the base of the aquifer and continues to contaminate the aquifer over time. After Pitchford et al., 1988.

organic compound(s) can be important to potential toxicity as well as to bioremediation of the spill.

The Hammett Constant σ for Substituted Aromatic Acids Based on the Benzene Ring

Consider benzoic acid, which consists of a benzene ring with a carboxyl group. Other substituents at different positions on the ring might be expected to alter the acid dissociation constant for the carboxyl group, through their electron donating or withdrawing effects. Louis Plack Hammett (1937, 1940) recognized an empirical relationship between the free energy of dissociation of the carboxyl group of the unsubstituted compound ($\Delta G^\circ_{\mathrm{H}}$) and that of the substituted compound (ΔG°), taking into consideration effects of substituents i in meta- and parapositions

$$\Delta G^\circ = \Delta G^\circ_{\mathrm{H}} + \sum\nolimits_i \Delta G^\circ_i \tag{8.10}$$

(Schwarzenbach et al., 1993).

A constant now known as the *Hammett constant* σ_i is used to express the effect of substituent i on the $\mathrm{p}K_{\mathrm{a}}$ of the carboxyl group

$$\sigma_i = -\Delta G^\circ_{\mathrm{i}}/2.303\ \mathrm{RT}. \tag{8.11}$$

Because substituents in meta versus para positions have different effects, they have different values of σ_i, σ_{m}, and σ_{p}. This same approach cannot be used for substituents in ortho positions because of proximity effects.

The values of σ_i were defined by Hammett (1937, 1940) based on the ionization constants of benzoic acid versus substituted benzoic acids, as

$$\sigma_i = \log\ K_i - \log\ K_{\mathrm{H}}, \tag{8.12}$$

where K_H is the ionization constant for benzoic acid in water (or for some insoluble acids a mixture of water and ethanol) at 25°C and K_i is the constant for the corresponding m- or p-substituted benzoic acid (Hansch et al, 1991).

The resulting and subsequently refined Hammett constants show how the pK_a of the substituted benzoic acid differs from that of the unsubstituted benzoic acid. When there is one added substituent,

$$pK_a = pK_{aH} - \sigma_i. \tag{8.13}$$

For multiple substituents (m- or p-), the Hammett constants are summed and subtracted from the pK_{aH}.

The approach of using Hammett constants for various applications in organic, physical, and environmental chemistry has been criticized because it is based on empirical observations rather than theoretical considerations. Yet, it has proven to be enormously useful in many aspects of environmental chemistry, including investigations of sorption of substituted benzoic acids. Hammett constants for some common substituents of various compounds are shown in Table 8.3. Substituents with Hammett constants that are negative are electron withdrawing, and those that are positive are electron donating. The effects can change depending on the location of the substituent; for example, —OH as a substituent to benzoic acid has $\sigma_m = 0.12$ and $\sigma_p = -0.37$.

Case Study: Sorption of SNBs

Sorption of HOCs to Fe oxides and nonswelling clays such as kaolinite is generally minimal except when the minerals are coated with NOM. HOCs may absorb in the interlayers of swelling clays (e.g., Lee et al., 1990), and/or partition into the NOM (humic and fulvic acids) components of soils and sediments (e.g., Chiou et al., 1979, 2000). Interlayer absorption may involve (1) hydrophobic interactions between HOC benzene rings and the siloxane rings on the surfaces of clay interlayers, (2) hydrophobic interactions between the hydrophobic

TABLE 8.3 Values of the Hammett constant for some substituted benzoic acids.

Substituent	σ_m	σ_p
H	0.00	0.00
COOH	0.37	0.45
CH_3	−0.07	−0.17
OCH_3	0.12	−0.27
OH	0.12	−0.37
Br	0.39	0.23
Cl	0.37	0.23
F	0.34	0.06
I	0.35	0.18
NO_2	0.71	0.78
NH_2	−0.16	−0.66
NH_3^+	0.86	0.60
CN	0.56	0.66

Data from Hansch et al., 1991.

TABLE 8.4 Physicochemical properties of substituted nitrobenzenes.

substitutent	C_w^{sat}	K_f	σ_i
p-NH_2	723	0.0499	−0.66
p-OH	11600	0.182	−0.36
p-OCH_3	590	0.452	−0.12
m-OCH_3	428	0.244	0.10
p-CHO	2320	1.05	0.22
p-$OCOCH_3$	—	0.533	0.31
m-CHO	—	1.02	0.36
m-$COCH_3$	750	2.51	0.36
p-$COOCH_3$	539	3.28	0.44
p-$COCH_3$	1300	2.62	0.52
m-CN	1710	1.03	0.62
p-CN	1650	1.16	0.63
m-NO_2	533	2.87	0.71
p-NO_2	69	2.87	0.78

K_f = Freundlich sorption coefficient, C_w^{sat} = water solubility, in mg L^{-1}, and σ_i = Hammett constant.
Data from Boyd et al., 2001.

components of molecules sorbing next to one another, and/or (3) electrostatic interactions between different functional groups and constituents on the HOCs and hydrated cations, such as K^+, present in the clay interlayers.

Boyd et al. (2001) investigated the sorption, most likely dominated by absorption in the interlayer, of 14 different SNBs (Table 8.4) by smectite clays. Their research used a combination of batch experiments, FTIR, and molecular dynamics modeling. They used both m- and p-substituted SNBs and explored how solubility might influence sorption.

Some example isotherms for the SNB sorption onto smectite from water solution are shown in Figure 8.24. The isotherms were fit to a modification of the Freundlich equation

$$q = \Gamma = K_f C^n, \tag{8.14}$$

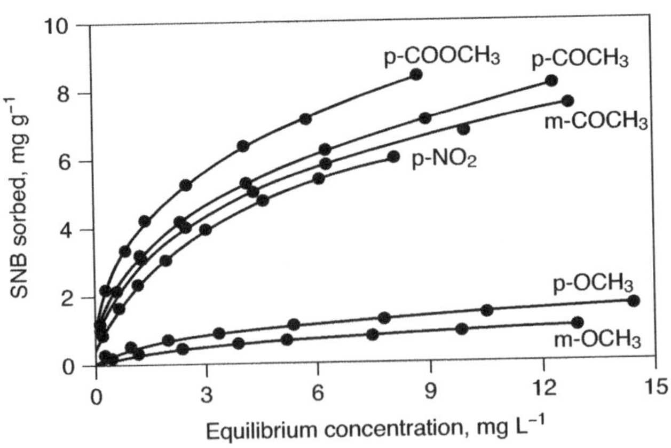

FIGURE 8.24 Sorption isotherms of some SNBs to K-smectite. Revised from Boyd et al., 2001 (with permission).

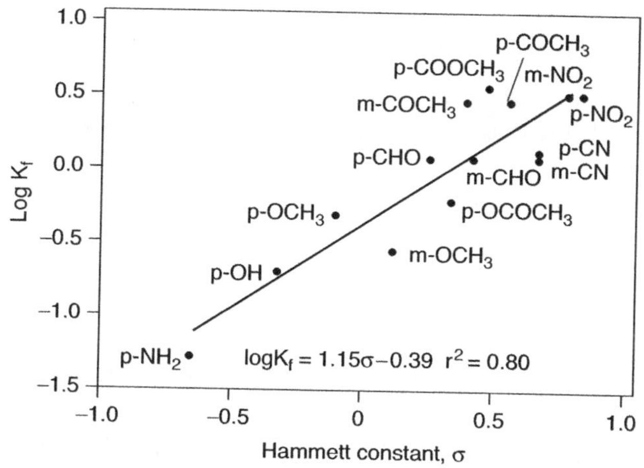

FIGURE 8.25 Plot of log Freundlich coefficients (log K_f) against Hammett constants (σ_i) of the various SNBs. Revised from Boyd et al., 2001 (with permission).

where

q or Γ = concentration of SNB sorbed in mg g^{-1}
C = equilibrium concentration of SNB remaining in solution after sorption in mg L^{-1}
n = constant expressing isotherm curvature
K_f = isotherm coefficient.

The Freundlich coefficients (K_f)s are plotted against Hammett constant σ_i values for SNBs in Figure 8.25. A higher K_f suggests a greater sorption affinity, and a more positive σ indicates more electron withdrawal. Sorption affinities were greater when the substituted groups were strong electron withdrawing (-NO_2, -CN, -$COCH_3$, -$COOCH_3$) and lesser when the substituted groups were strong electron releasing (-NH_2, -OH).

FTIR and modeling results indicated that when the SNBs absorb in the interlayer, laying flat, the NO_2 groups complex the interlayer K^+ either directly or through associated water, with interactions also through the substituted constituents. The ability of different SNBs to complex K^+ seems to play an important role in the observed trends in absorption. The absorption also appears to be affected by the aqueous solubilities of the various SNBs (Table 8.4).

Molecular Dynamics (MD) Modeling of Atrazine Absorption

Atrazine is a widely used herbicide whose fate and transport is of considerable environmental concern. Chapelle et al. (2005) investigated the absorption of atrazine

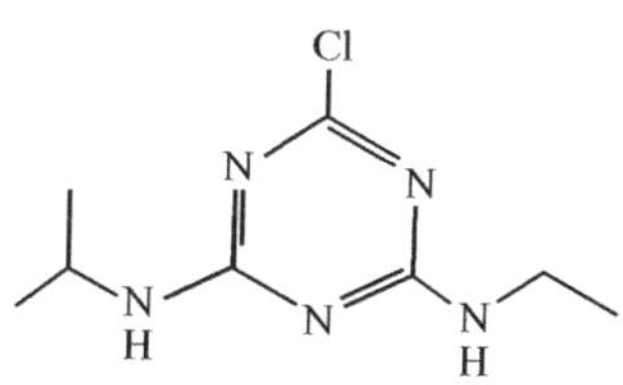

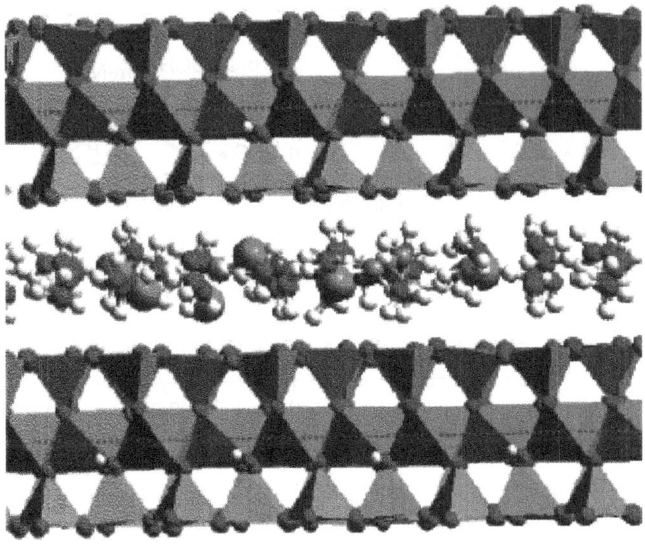

FIGURE 8.26 A molecular model of the herbicide atrazine absorbed in a K-saturated smectite interlayer containing one layer of water. In the presence of one interlayer of water, atrazine molecules are oriented parallel to the basal-plane surfaces exposed in the interlayer and interact directly with the hydrophobic sites on the two surfaces on either side of the interlayer spacing. From Chappell et al., 2005.

in the interlayers of swelling clays, and applied MD modeling. The MD modeling suggested that the lone pair electrons of the ring N atoms on atrazine can form hydrogen bonds with water molecules, particularly the solvation waters around interlayer cations such as K^+. Additionally, the hydrophobic alkyl tails of atrazine may interact with the hydrophobic basal-plane surfaces exposed in the interlayer. In the presence of one interlayer of water, the atrazine molecules are oriented parallel to the basal-plane surfaces exposed in the interlayer and interact directly with the hydrophobic sites on the two surfaces on either side of the interlayer spacing (Figure 8.26). The orientation of atrazine in the interlayer can change when not just one but three layers of water are present in the interlayer (Chapelle et al., 2005). Hence, the hydration state of the clay can play an important role in absorption behavior.

THE K_d APPROACH TO HYDROPHOBIC ORGANIC COMPOUND TRANSPORT IN POROUS MEDIA

One simple approach that has been used widely to incorporate sorption into hydrologic transport models is the linear distribution coefficient, or linear K_d. This approach is based on a simple equation for "partitioning" of a pollutant between dissolved and adsorbed phases

$$C_{i(s)} = K_d \, C_{i(aq)}, \tag{8.15}$$

where

$C_{i(s)}$ = molar concentration of the adsorbate on the surface
$C_{i(aq)}$ = molar concentration of the adsorptive in solution

K_d = distribution coefficient in units of volume solvent over weight adsorptive (L g^{-1} or $cm^3\ g^{-1}$).

The K_d is found by plotting $C_{i(s)}$ on the x-axis and $C_{i(aq)}$ on the y-axis; the slope is the linear K_d. A somewhat better approach uses activities rather than concentrations. The K_d value for a given constituent can then be incorporated into the equation for diffusion and advection in flow, such as

$$dC/dt = \frac{D\delta^2 C}{\delta x^2} - v\frac{\delta C}{\delta x} \pm \text{adsorption term}, \tag{8.16}$$

where

C = concentration
t = time
x = distance
v = velocity
D = diffusion coefficient.

The adsorption term, related to K_d, needs to be incorporated into the advection and diffusion portions (see Drever, 1997).

The K_d approach works best when applied to the sorption of hydrophobic organic compounds, which tend to partition into NOM coatings on mineral grains. The K_d for HOC partitioning to organic-rich sediments has been expressed as

$$K_d = K_{oc} f_{oc}, \tag{8.17}$$

where K_{oc} is a partition coefficient (in $cm^3\ g^{-1}$) for the hydrophobic organic compound between organic matter and water, and f_{oc} is the weight fraction of organic carbon in the porous medium (grams organic carbon per gram aquifer solid, g_{oc}/g_s).

K_{oc} can often be estimated from the octanol-water partition coefficient of a hydrophobic organic compound K_{ow}. K_{ow} is a measure of the relative affinity of the organic compound for hydrophobic octanol versus water

$$K_{ow} = [\text{solute}]_{\text{octanol}}/[\text{solute}]_{\text{water}}. \tag{8.18}$$

If K_{ow} for a given organic compound is known, then K_{oc} can be calculated from one of a variety of equations that have been proposed, such as that of Schwarzenbach and Westall (1981)

$$\log\ K_{oc} = 0.72\ \log\ K_{ow} + \log\ f_{oc} + 0.49, \tag{8.19}$$

which is appropriate for sediments with log $f_{oc} > 0.001$ (with > 0.1% organic matter).

If a sediment contains expanding clays, then their absorption abilities need to be considered, as well. Some values of K_{ow} are given in Table 8.2.

ACTIVATED CARBON AND SORPTION OF VOCs

Thus far, we have discussed surface and interface interactions primarily in aqueous solution. However, solid–vapor or solid–liquid–vapor interactions can be important for many environmental processes. Airborne VOCs can be toxic, and many can also play a role in the formation of ozone-containing "photochemical smog," according to

$$\text{VOC} + \text{NO}_x \underset{h\nu}{\rightarrow} \text{O}_3 \text{ containing "photochemical smog".} \tag{8.20}$$

Although O_3 is essential in the stratosphere for blocking UV radiation, ground-level O_3 can be damaging to human life, animals, and plants such as trees.

One common method for removing VOCs from air and water is by using *activated carbon*, which is also sometimes called *activated charcoal*. Activated carbon is usually formed by processing charcoal to make it extremely porous, with a high internal surface area for sorption. A schematic of the activated carbon structure is shown in Figure 8.27. Activated carbon typically has a Brunauer, Emmett, and Teller (BET) surface area of $\sim$500 m^2g^{-1}. Although the high surface area alone can remove VOCs simply by molecular (physical) sorption, many processes have been developed to enhance sorption or to modify the activated carbon surface to catalyze VOC destruction.

VOC sorption is primarily to the pores within the activated carbon structure. VOC molecules tend to sorb most strongly in areas where the pore diameter of the sorbent is close to the molecular diameter of the VOC compound (Shepherd, 2001). Although activated carbon can sorb most volatile organic compounds, it is most effective for higher molecular weight and higher boiling point compounds. Activated carbon removes VOCs by physical adsorption, and the removal depends on the organic compound characteristics, the temperature of the gas stream that is being processed, and the concentration of the contaminant in the gas stream (Shepherd, 2001). Shepherd (2001) reported that the adsorption capacity of an activated charcol for trichloroethylene varied from 20–65% a concentration increased from 10–10000 ppm in a gas stream.

The amount of VOC that can be adsorbed per unit weight of activated carbon consumed under the particular treatment condition is known as the *adsorption capacity* or *carbon*

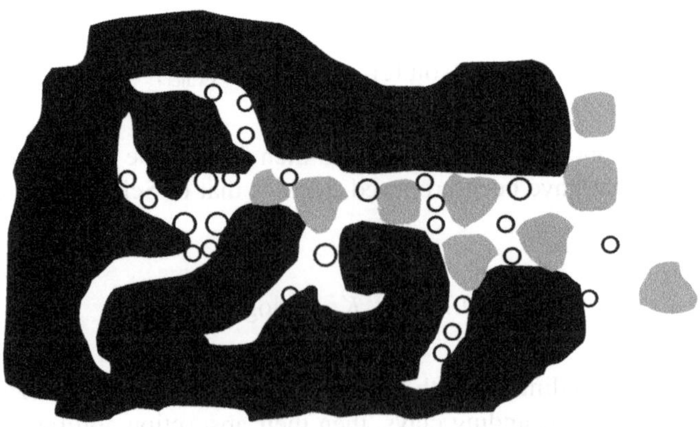

FIGURE 8.27 A schematic representation of VOC sorption to the activated carbon structure. After Shepherd, 2001.

TABLE 8.5 Relative sorption rate to an activated carbon sample of various organic compounds, measured as carbon capacity percent.

Compound	Molecular weight	Boiling point	Carbon capacity %
Nitrobenzene	123	211	51
Tetrachloroethane	166	147	40
Tetrachloroethylene	165	121	35
Styrene	104	145	25
Xylene	106	138	21
Toluene	92	111	20
Benzene	78	80	12
MTBE	88	55	12
Hexane	86	68	7
Methylene chloride	84	40	2
Acetone	58	56	0.8
Vinylchloride	62	−14	0.7
Chloroethane	64	12	0.5
Methane	16	−161	0.0003

Values of molecular weight and boiling point (C) are also given. Data from Shepherd, 2001.

capacity percent. Typical adsorption capacities for so-called "moderately adsorbed" compounds are from 5–30 % of the weight of the activated carbon.

The relative absorption rates (as "weaker" to "stronger") for various VOCs are given in the Table 8.5.

QUESTIONS FOR FURTHER THOUGHT

1. This problem follows from Problem 5 in Chapter 7. A student conducts an experiment to determine the effects of the organic ligand DFO-B on Cd adsorption.
   (a) Plot Cd adsorption density in the presence of DFO-B on the y-axis, with pH on the x-axis.
   (b) In comparing to Chapter 7 Problem 5, how does DFO-B affect Cd adsorption to kaolinite as a function of pH?
   (c) Be sure to include error bars on data points and discuss how they were determined.

   For each experiment, 0.0516 g of kaolinite with 12.1 m^2g^{-1} surface area were used in each batch reaction vessel except for the control. The experiments were run with total solution volume = 25 mL.

Cd adsorption data in the presence of DFO-B.

Sample	pH	[Cd] rep 1	[Cd] rep 2	[Cd] rep 3
Control	5.21	1.20	1.24	1.23
1	4.38	1.22	1.22	1.24
2	5.02	1.20	1.22	1.22
3	5.61	1.21	1.22	1.22
4	5.98	1.19	1.18	1.18
5	6.43	1.19	1.20	1.19

6	6.92	1.15	1.15	1.15
7	7.44	1.05	1.22	1.11
8	7.86	1.00	1.06	1.07
9	8.39	1.02	1.06	1.06
10	8.89	1.03	1.08	1.08

*Note: [Cd] in controls without kaolinite did not vary significantly with pH, [Cd] values in the presence of DFO-B. [Cd] in mgCd L^{-1}.

2. Below are data from an experiment comparing adsorption of an organic ligand to hematite and goethite. For simplicity, only one sample analysis is provided, rather than replicates. All data at pH 4. Solid concentration = 41.1 m^2 L^{-1} for kaolinite and 48.0 m^2 L^{-1} for goethite. The BET surface area of kaolinite is 12.1 m^2 g^{-1} and for goethite it is 58.0 m^2 g^{-1}. Volumes of suspensions = 25 mL.
   (a) Plot as mg C sorbed per m^2 surface versus equilibrium [dissolved organic carbon (DOC)], then
   (b) as % sorbed versus initial [DOC].
   (c) Compare adsorption to kaolinite versus to goethite and give a plausible explanation of the results.
   (d) Fit Langmuir and Freundlich isotherms to the isotherm data, and compare isotherm parameters.
   (e) Next, plot as mg C per g of solid (kaolinite and goethite) versus equilibrium [DOC].
   (f) What difference, does any, does it make to calculate sorption in terms of surface area of solid rather than weight of solid when comparing different minerals?

[DOC] control mgC L^{-1}	[DOC] kaolinite mgC L^{-1}	[DOC] goethite mgC L^{-1}
0.0	0.0	0.0
4.5	1.9	0.0
9.1	6.0	0.0
13.5	10.2	3.1
18.2	14.7	6.9
22.7	19.2	10.8
27.3	23.6	15.5
31.7	28.9	20.3
36.6	33.0	24.7
41.0	37.6	28.9

3. (a) Does adsorption of NOM (humic and fulvic acids) fit the assumptions of the Langmuir equation?
   (b) Why or why not?
4. (a) Would you expect the K_d approach to work well for helping to describe NOM transport through a sand packed column?

(b) Why or why not?

(c) Explain your reasoning.

5. For the four compounds: atrazine, dichloro-diphenyl-trichloroethane (DDT), nitrobenzene, and n-octane,

   (a) calculate K_d values in $cm^3\ g^{-1}$ for an organic-rich sediment with $f_{oc} = 0.01$.

   (b) Which compound would tend to partition most to the sediment?

   (c) Which least?

   (d) Explain your reasoning.

6. (a) Would you expect the adsorption capacity of activated carbon for a VOC such as trichloroethylene to increase or decrease as temperature of the gas stream is increased?

   (b) Explain your reasoning.

7. (a) How might you expect a spill of chlorobenzene on the surface of a sandy aquifer to behave similarly or differently from a spill of n-octane?

   (b) Explain your reasoning.

   (c) What factors should be considered?

8. For the data in Table 8.2,

   (a) plot log K_{ow} versus –log $C^{sat}{}_{w}$.

   (b) Is there any correlation?

   (c) Why or why not would you expect these values to correlate?

   You might use as an approximation a value of $-\log C_w^{sat}$ for miscible compounds = 0.

9. For the data in Table 8.5,

   (a) Plot molecular weight versus boiling point, molecular weight versus carbon capacity%, and boiling point versus carbon capacity %.

   (b) Are there any correlations?

   (c) Would you expect there to be?

   (d) Explain why or why not.

   (e) What, if anything, does this tell you about the adsorption of VOCs to activated charcoal?

10. A student runs a batch SNB sorption experiment of m-NO_2 and p-NO_2 on montmorillonite. The concentration of montmorillonite in suspension in each sample vessel is 1 g L^{-1}. The table below gives the initial and equilibrium SNB concentrations.

    (a) Plot the data for each of the two SNBs and fit to Langmuir, Freundlich, and Frumkin equations.

    (b) Give values of parameters: for Langmuir, K_L and q_m; for Freundlich, K_f and n; for Frumkin B and α.

    (c) Which isotherm equation seems to be most appropriate for these SNBs?

    (d) How do K_f and K_L each compare with the Hammett constant σ values for the corresponding SNBs? For this problem, use isotherm equations provided in Chapter 6.

p-NO_2 initial conc. mg L^{-1}	*p*-NO_2 equil. conc. mg L^{-1}	*m*-NO_2 intial conc. mg L^{-1}	*m*-NO_2 equil. conc. mg L^{-1}
1.3	0.5	5.0	2.3
2.5	1.0	8.3	4.5
3.6	1.5	15.8	9.0
4.7	2.1	23.7	15.1
7.0	3.2	33.0	22.5
9.3	4.8	50.4	38.6
11.3	6.3	54.2	42.1
13.8	8.0		

FURTHER READING

Boyd, S.A., Sheng, G., Teppen, B.J., and Johnston, C.T. 2001. Mechanisms for the adsorption of substituted nitrobenzenes by smectite clays. *Environmental Science and Technology* 35, 4227–4234.

Schwarzenbach, R.P., Gschwend, P.M., and Imboden, D.M. 2002. *Environmental Organic Chemistry* 2nd ed. John Wiley & Sons, New York, 1000 p.

Solomons, T.W.G. 2008. *Organic Chemistry* 9th rev. ed. John Wiley and Sons, Inc., New York.

Zhou, Q., Maurice, P.A., and Cabaniss, S.E. 2001. Size fractionation upon adsorption of fulvic acid on goethite: Equilibrium and kinetic studies. *Geochimica et Cosmochimica Acta* 65, 803–812.

9

MINERAL NUCLEATION AND GROWTH

This chapter discusses mineral precipitation, which can be divided into two stages: nucleation and growth. Although the chapter focuses primarily on precipitation from aqueous solution, many basic concepts also apply to growth from the vapor phase. This chapter, along with Chapter 10 on mineral dissolution and weathering, highlights thermodynamic versus kinetic approaches, as discussed in Chapter 1.

SATURATION STATE AND MINERAL NUCLEATION: AN EXAMPLE OF THE CONFLUENCE OF THERMODYNAMICS AND KINETICS

Recall from Chapter 5 that surfaces and interfaces have free energy associated with them, and that it takes energy to create new surface area. When crystals first nucleate from solution, they are so small that the interfacial free energy is a significant, even dominant, component of the total free energy of the mineral. The *free energy of formation* of a single crystal ΔG_n is given by

$$\Delta G_n = \Delta G_{bulk} + \Delta G_{interf}. \tag{9.1}$$

This equation takes into consideration the bulk change in free energy (ΔG_{bulk}) and the change caused by formation of a new interface (ΔG_{interf}). For crystals larger than about 2 μm, $\Delta G_{interf} << \Delta G_{bulk}$ and the interface component can be ignored (Berner, 1980).

The free energy of formation of a crystal from a supersaturated solution can be related to saturation state (Ω) and surface free energy γ. For a spherical particle,

$$\Delta G_n = -(4/V)\pi r^3 k_B T \ln(\Omega) + 4\pi r^2\gamma, \tag{9.2}$$

Environmental Surfaces and Interfaces from the Nanoscale to the Global Scale, by Patricia A. Maurice

where

V = molecular volume of the precipitate
r = radius of the nucleus
k_B = Boltzmann's constant
Ω = saturation ratio
T = temperature (K)
γ = surface free energy per unit surface area of the nuclei (Burda et al., 2005).

When the solution is supersaturated, $\Omega > 1$, and ΔG_n has a maximum at the critical size of the nucleating crystal r*.

The energy necessary to reach this maximum and to form the *critical nucleus* (r = r*) is known as the *free energy of nucleation* ΔG_n^* and is the activation energy (E_a) barrier that must be overcome for nucleation. Once the critical nucleus is formed, any additional growth of the crystal results in a decrease in free energy. At this point, the process goes from nucleation to crystal growth. When the particles are small, they are called *embryos* (or sometimes nuclei); as they grow, they are called crystals. An increase of Ω causes a decrease in both the free energy of nucleation and the size of the critical nucleus, as reflected by a smaller value of r*.

The size of the critical radius r* can be calculated by setting $\Delta G/dr = 0$ so that

$$r^* = \frac{2V\gamma}{3k_B T \ln(\Omega)}. \qquad (9.3)$$

At any given value of saturation ratio Ω, particles with $r > r^*$ will tend to grow, whereas particles with $r < r^*$ will tend to dissolve. The greater the Ω, the smaller the r*; hence, smaller particles are stable at higher Ω. A schematic of the change in free energy as a function of particle size is provided in Figure 9.2 based on the work of Lower et al. (1998b). In this case, the symbol n is used for the more general size of (or number of atoms in) the crystal, rather than using radius r specifically for a spherical particle; however, the treatments are similar.

To understand more of how the saturation state affects nucleation and crystal growth, it is important to consider whether nucleation is occurring homogeneously in solution or heterogeneously on a preexisting particle surface. In *homogeneous nucleation*, a new crystal nucleates directly from solution, as an individual nucleus not associated with an already existing surface. In *heterogeneous nucleation*, the new crystal nucleates on top of an already existing surface (Figure 9.1). When heterogeneous nucleation occurs, ΔG_{interf} is decreased, which decreases the free energy of nucleation.

Homogeneous nucleation can occur spontaneously in solution when the saturation state of the solution exceeds a certain critical value (Nielsen, 1964). In a *batch* system, such as a beaker or a flask at high supersaturation, many small nuclei will form quickly, and they will only grow a small amount before the solution species are depleted. Thus, many small crystals will result. At somewhat lesser supersaturation, fewer nuclei will form, and they will grow into fewer, larger crystals. In only slightly supersaturated solutions, nucleation will need to take advantage of any surfaces present, and heterogeneous nucleation will occur. This surface for heterogeneous nucleation may be a mineral phase or a bacterium, or even a scratch on a beaker surface. At small values of supersaturation, a precipitation reaction may be thermodynamically favored but not occur for very long times because of the need to overcome the activation energy and kinetic hindrance, especially if no surface for potential heterogeneous nucleation is available. Thus, mineral nucleation shows a link between kinetics and thermodynamics.

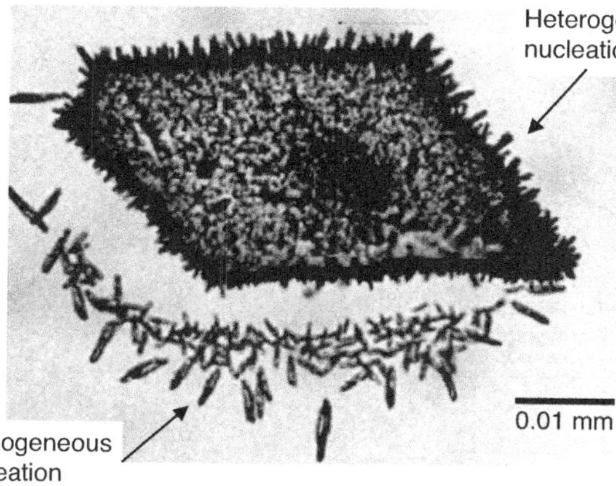

FIGURE 9.1 An example of homogeneous versus heterogeneous nucleation. In this optical microscope image of a (rhombohedral) apatite crystal in Pb-brearing solution, the Pb-phosphate mineral pyromorphite (needle-shaped crystals) has nucleated both heterogeneously on the surface of the apatite and homogeneously in solution (to bottom and left). Adapted from Manecki et al., 2000.

Hydroxypyromorphite Nucleation

Nucleation theory is illustrated in the work by Lower et al. (1998a, b), on nucleation and growth of the Pb-phosphate mineral hydroxypyromorphite (PY) in the presence of the Ca-phosphate mineral hydroxyapatite (AP). The reason why this process was chosen for study is because apatite amendments offer a potential means of remediating Pb-contaminated soils. Batch experiments were conducted in which AP was added as a source of PO_4^{-3}. Rapid dissolution of AP released PO_4^{-3} to solution, and the PO_4^{-3} reacted quickly with dissolved lead (Pb_{aq}) to form PY

$$Ca_5(PO_4)_3OH + 7H^+ \overset{\text{diss}}{\rightleftarrows} 5Ca^{2+} + 3H_2PO_4^- + H_2O \tag{9.4}$$

$$5Pb^{2+} + 3H_2PO_4^- + H_2O \overset{\text{ppt}}{\rightleftarrows} Pb_5(PO_4)_3OH + 7H^+. \tag{9.5}$$

After the reaction, precipitates were imaged by atomic force microscopy (AFM), scanning electron microscopy (SEM), and transmission electron microscopy (TEM) and were analyzed by X-ray diffraction (XRD) and energy-dispersive X-ray electron spectroscopy (EDS).

In Lower et al.'s (1998a, b) work at high supersaturation, many small PY crystals formed (Figure 9.2). At lesser supersaturation, fewer PY nuclei formed, but they grew to larger sizes before the solution species were depleted and the solution no longer was supersaturated. At low degrees of supersaturation, PY nucleation occurred heterogeneously on AP. The results, therefore, were consistent with nucleation theory.

Heterogeneous Nucleation and Epitaxial Growth

The presence of a surface can greatly affect nucleation because the interfacial energy between the crystal nucleus or embryo and a solid substrate is often lower than that between

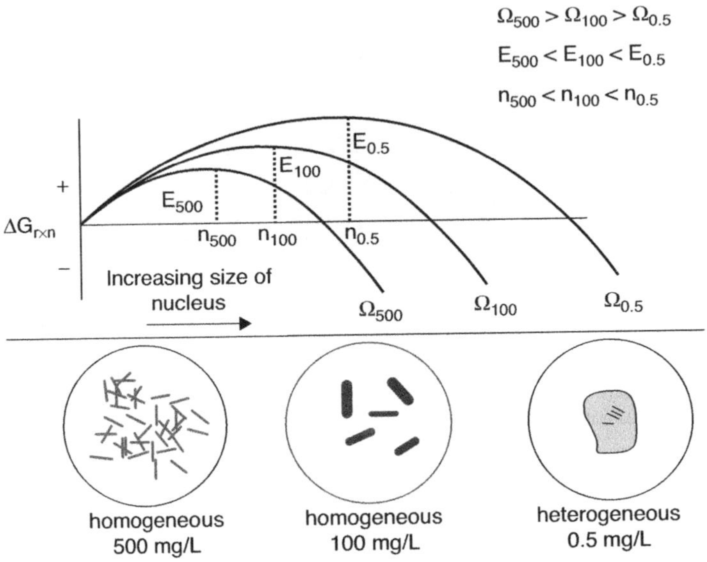

FIGURE 9.2 How the saturation state affects nucleation of pyromorphite in the presence of apatite. Ω = saturation state, E = activation energy at different saturation states, and n = crystal size at different saturation states. From Lower et al., 1998b, after the general construct of Neilsen, 1964.

the nucleus and the solution. This in turn is because the component molecules or ions of the crystal can form bonds with the surface substrate that are stronger than the bonds involved in solvation by solution. The formation of these stronger bonds leads to smaller interfacial free energy.

When one considers heterogeneous nucleation, it is clear that the nature of the surface will be important for determining the interfacial free energy. When the atomic-scale structure of the substrate surface closely matches that of the precipitating mineral, the lattice strain is minimized and nucleation is energetically more favorable. *Oriented overgrowth* of one crystal on another can occur when there is a well-defined minimum in interfacial free energy for some particular orientational relationship of the two crystals, relative to the boundary between them and their crystallographic structures. Such oriented heterogenenous nucleation and crystal growth is often termed *epitaxial*.

Epitaxy is more likely when there is good agreement between the lattice structures, and it is less likely when there is *lattice misfit*. Lattice misfit (d) describes the mismatch between arrays of atoms at the interface between the substrate crystal and the overgrowth crystal (Van der Merve and Ball 1975). The lattice misfit (d_m) is calculated from

$$d_m = D_a / a, \tag{9.6}$$

where

a = lattice parameter of the stress-free substrate crystal
D_a = difference in lattice parameters between the substrate and the overgrowth.

Misfits of <10–15% are generally thought to permit epitaxial growth (Liu et al., 1996). The misfit between the growing crystal and the underlying substrate can lead to

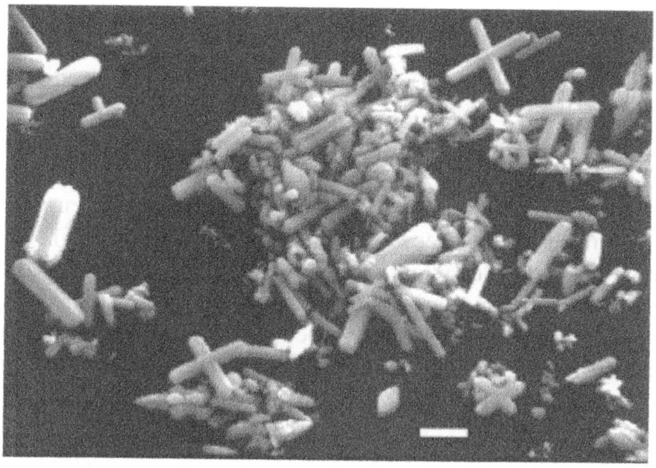

FIGURE 9.3 An SEM image of euhedral pyromorphite crystals formed by homogeneous precipitation from solution. Some penetration twins and parallel intergrowths are apparent. Scale bar = 3.33 μm. From Manecki et al., 2000.

imperfections in the epitaxially growing crystal. Thus, the epitaxially grown crystals typically exhibit higher densities of structural defects than do crystals that nucleate and grow homogeneously from solution (Stowell, 1975).

Manecki et al. (2000) observed homogeneous nucleation at fairly high supersaturation (Figure 9.3), but that under some conditions, PY grew epitaxially on AP crystals (including hydroxyapatite, chlorapatite, and fluorapatite) (Figure 9.4). The basal plane value of lattice misfit, $d = 6.06\%$, 4.04%, and 6.27% for pyromorphite and hydroxyapatite, chlorapatite,

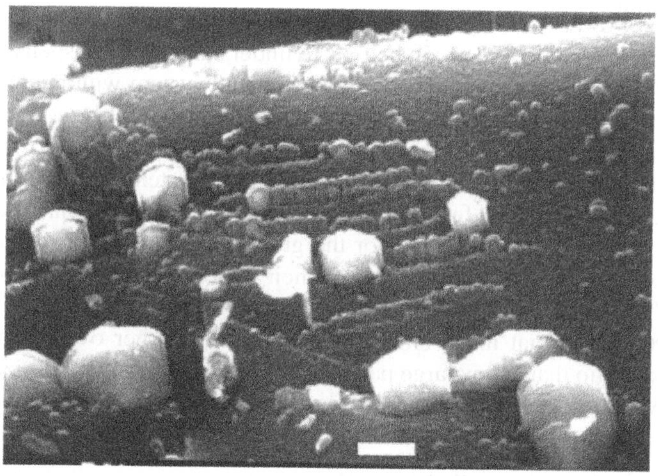

FIGURE 9.4 SEM image of the products of CAP-Pbaq reaction when dry apatite is added to $PbCl_2$ solution. Heterogeneously precipitated PY is apparent, nucleated on the surface of chlorapatite. PY crystals (white and light-gray hexagonal crystals parallel to each other) form preferentially along steps and fractures. The parallel orientation of precipitating PY crystals suggests epitaxial growth on the apatite surface. Scale bar = 1.67 μm. From Manecki et al., 2000.

and fluorapatite, respectively, which indicates a close fit between their lattice structures. Also, the minerals in question are isostructural. Misfit is higher along the crystallographic c-axis, and the basal-plane surface of apatite was found to be preferred for heterogeneous nucleation of PY.

For heterogeneous nucleation, not only is the surface structure important but also the microtopography. Manecki et al. (2000) observed that the heterogeneous precipitation of pyromorphite on chlorapatite and fluorapatite occurred preferentially along surface features such as steps and microcracks (Figure 9.4). As discussed in Chapter 5, atoms and molecules along features such as step edges have different coordination; hence, different energies are associated with them than atoms along flat terraces. The kinetics of detachment and attachment processes at a mineral surface are determined by the energy barriers encountered by the molecules, and these energy barriers are different for different types of surface sites. Growth often can occur preferentially along step edges and particularly at kink sites.

FROM NUCLEATION TO GROWTH

Once the crystal begins to grow beyond the embryo, the following processes need to be considered: (1) transport of species to the surface of the crystal; (2) various surface reactions such as adsorption and surface diffusion, that result in incorporation of the species into the crystal lattice; and (3) removal of any reaction products that are not becoming part of the crystal. An important component of (2) is desolvation or removal of water molecules from the attaching material. As the crystal grows, the relative increase in surface area is smaller, and the contribution of surface free energy to the total energetics of the system becomes less.

Ostwald Ripening

Surface free energy considerations can help us to understand how a suspension of particles of a particular mineral evolves over time. Suppose that a system with many small crystals formed from a highly supersaturated solution. These crystals will have a fairly high total surface area. The formation of many small crystals is kinetically favored. However, the high surface area and high interfacial free energy means that the system is not thermodynamically favored relative to a system with fewer large crystals and lower interfacial free energy. Over time, the small crystals will act as fuel for the growth of big crystals; some of the small crystals will dissolve, whereas others grow at their expense. This spontaneous process is called *Ostwald Ripening*, which is named after Wilhelm Ostwald who first described the process in 1896. Given that many small particles have a larger overall surface-area-to-volume (or mass) ratio than a few large particles, the evolution is to minimize the total free energy of the system.

Transport and Surface Controlled Growth

As discussed in Chapter 1, the rates of crystal growth of different minerals may be controlled by transport processes, surface process(es), or an intermediate case of both. Generally, *transport-controlled* crystal growth is considered to be sensitive to flow rate because flow rate controls the transport of species to and from the surface—up to a point when flow rate is so fast that there is no longer additional sensitivity. *Surface-controlled* crystal growth is thought to be insensitive to flow rate (or stir rate) because the surface process is slow relative to transport. Surface-controlled crystal growth is generally slower than transport-controlled crystal growth, and it has greater temperature sensitivity associated with higher activation energy. Factors such as solution pH, saturation state, temperature, presence of impurities, and particle size of a mineral can influence whether a mineral grows by surface, transport, or mixed mechanisms in any particular setting.

Recall from Chapter 1 that when the growth rate is surface controlled, the concentrations of reactant(s) near the surface will be uniform. When growth rate is transport controlled, there will be a gradient in reactant concentration near the surface, with the reactant most depleted close to the surface. In mixed control growth, there will be some depletion of reactant(s) near the surface but less than for transport-controlled growth.

For surface-controlled crystal growth, surface microtopography can be important to the rate because sites associated with features such as steps, kinks, and dislocations have different coordinations and different energies. Different rates of reaction may be associated with different surface features. The slowest step in surface-controlled dissolution is often desolvation of the reactant as it attaches to the surface, and this desolvation is often the rate-controlling process (Sunagawa, 2005). Surface-controlled growth generally leads to crystals with distinct morphologies, whereas transport-controlled growth generally leads to rounded crystals (Sunagawa, 2005).

The Special Importance of Kink Sites

Kink sites are particularly important for crystal growth from aqueous solutions. One way of understanding the role of different sites is to consider the (100) face of a crystal with a simple cubic lattice known as a *Kossel crystal*, which contains a single monatomic step, meaning a step one atom high. We can consider the following bonding arrangements of atoms associated with different surface sites (Figure 9.5): (1) embedded on a terrace removed from the step edge, (2) embedded along the step edge, (3) at a kink site, (4) adsorbed to the step edge, and (5) adsorbed to a terrace site. Each of these five atoms has a unique bonding arrangement.

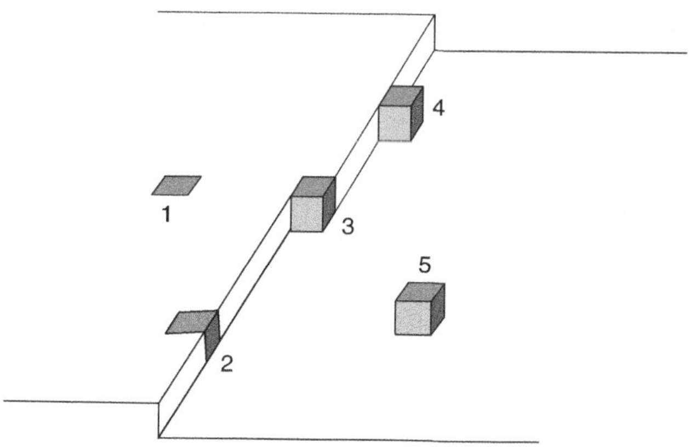

FIGURE 9.5 Various sites that an atom can occupy on a surface: (1) atom embedded on a terrace, (2) atom embedded into the edge of a step, (3) atom in a kink position, (4) atom absorbed at a step edge, and (5) atom adsorbed on a terrace (adatom). Based on Markov, 1995.

For example, (5) has one saturated bond and five dangling bonds. Removal of (5) from the surface would lead to a change in the ratio of saturated versus dangling bonds, which results in a change in surface free enegy. The kink site (3) is the only site that has an equal number of saturated versus dangling bonds, three of each. Removal of the atom from the kink site thus does not change the ratio of saturated to dangling bonds. Removal of atom (3) creates another equivalent kink site so that the process can be repeated until the edge of the crystal is met. In this way, the kink sites are regenerating. The process can be viewed in reverse, as attachment (growth or adsorption) rather than as detachment (dissolution or desorption).

BCF Theory

For the surface of a single crystal, the solution saturation state plays an important role in determining the rate and mode of crystal growth from supersaturated solutions. Concepts of mineral growth of "perfect" crystals were put forward as early as the late 1800s (Gibbs, 1878). At very high supersaturation, a new crystal surface may nucleate at many locations across the substrate surface, leading to surface roughening. Only at very high supersaturation is the new surface free energy associated with this rapid nucleation offset by formation of bulk mineral. At somewhat lesser supersaturation, two-dimensional (2-D) nuclei may form on the flat substrate surface and spread out in a 2-D fashion, in a process sometimes known as "birth and spread" (Figure 9.6).

Burton, Cabrera, and Frank, in a series of papers that culminated in their landmark work titled "The growth of crystals and the equilibrium structure of their surfaces" (Burton, et al., 1951), developed the theoretical framework for crystal growth in less highly supersaturated solutions (and vapors). Because many natural waters are not highly supersaturated, this work is of relevance for environmental surface chemistry and has led to what is currently called *BCF theory*. These authors recognized the importance of (1) highly reactive *kink sites* that occur along step edges, and (2) *screw dislocations*, which can be a source of continuously propagating steps. Because of these structures, mineral surfaces can undergo nucleation and growth from only moderately supersaturated conditions.

Screw dislocations (Chapter 5) have particular importance in crystal nucleation and growth. One can imagine that as a mineral grew from a kink site along a straight step, eventually the step would migrate all the way to the edge of the crystal. However, when screw dislocations intersect with crystal surfaces (Figure 9.6), they can generate sources of

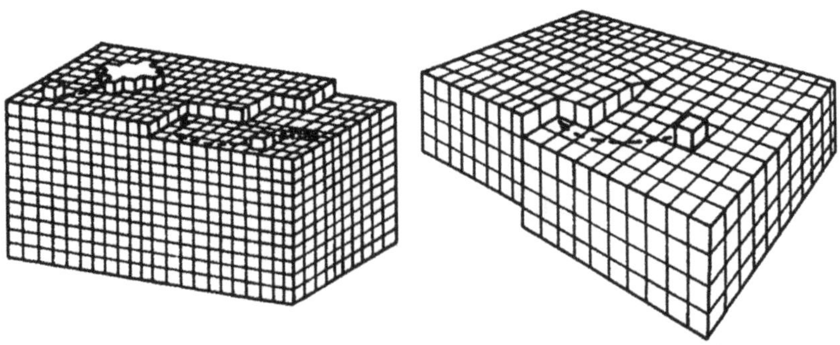

FIGURE 9.6 Schematic illustration of surface nucleation-controlled growth and screw dislocation-controlled growth. From Hochella, 1990 (with permission).

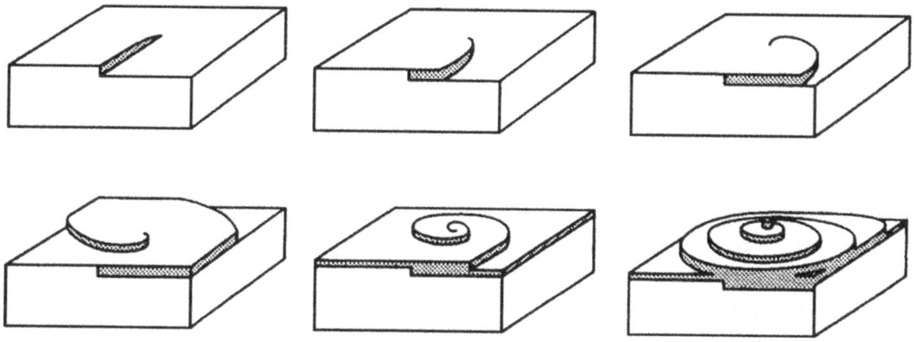

FIGURE 9.7 The mechanism of spiral growth from a screw dislocation (upper left) beginning with a single spiral and eventually forming multiple spirals (lower right). From Sunagawa, 2005 (with permission).

steps continuously (Figure 9.7). Figure 9.8 shows spiral growth hillocks that occur when growth nucleates at a dislocation, and a series of steps is generated.

In the hypothetical case of a crystal face that is atomically smooth, the rate of growth on the surface is controlled by the formation of steps and the lateral movement of the steps across the surface (Markov, 1995). These two processes are independent of one another. On a hypothetical atomically flat surface, the rate is controlled by the frequency

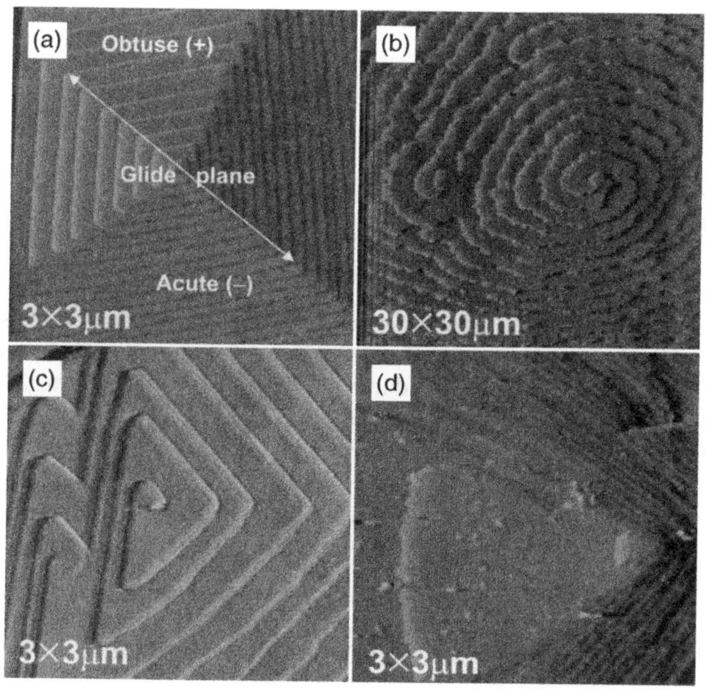

FIGURE 9.8 AFM images of growth hillocks that formed by spiral growth on various surfaces: (a) calcite, (b) canavalin, (c) brushite, and (d) calcium oxalate monohydrate. From DeYoreo and Vekilov, 2003 (with permission).

of formation of 2-D nuclei on the surface, and the nuclei edges are steps. Forming 2-D nuclei requires overcoming the energy barrier associated with creating new surface area and is restricted to saturation states above a certain critical supersaturation. However, when screw dislocations intersect the surface, they provide initial attachment sites and serve as a continuous source of steps so that the growth rate is no longer limited by the formation of new nuclei but rather by the rate of step advance. The rate of step advance, and hence of crystal growth, is dependent on factors such as (1) the step height and structure, (2) the rate of surface diffusion, (3) the interaction between steps, (4) the density of crystal defects at the surface, and (5) the concentrations of impurities that may, for example, serve to poison step movement (Markov, 1995).

For spiral growth, in which a growth hillock is formed, the distance between successive turns of the spiral plays an important role in determining the rate of growth of the crystal face. The interstep distance is inversely proportional to the degree of supersaturation, so that more highly supersaturated solutions result in steeper-sided growth hillocks generated from screw dislocations. Pairs of dislocations can produce growth spirals that intersect and interact in complex ways depending on factors such as whether the dislocations are of like or opposite sign, the distance between the dislocations, and the degree of supersaturation (Burton, et al., 1951).

Growth Mode and Driving Force

The discussion above can be summarized and restated in terms of *driving force* considerations. The driving force associated with crystal growth is related to the change in *chemical potential* $\Delta\mu$, and hence to *supersaturation state* (σ), according to

$$\text{Driving force} = \Delta\mu / k_B\mathrm{T} = \sigma, \tag{9.7}$$

where

k_B = Boltzmann constant
T = absolute temperature.

At high σ, many nuclei can form on a surface and grow simultaneously before a single layer is formed across the surface. In this case, the surface can become significantly roughened even at relatively low temperature when crystal surfaces tend to be fairly smooth. This form of growth is often termed "adhesive" type growth (Sunagawa, 2005) and the process leads to "kinetic roughness" (Markov, 1995).

The rate of growth (R_g) is related to the *driving force* according to

$$R_g = \mathrm{A}(\Delta\mu / k_B\mathrm{T}), \tag{9.8}$$

where A is a constant.

At somewhat lesser values of supersaturation σ, growth on a flat surface occurs by 2-D nucleation and growth, often termed "birth and spread." In this case,

$$R_g = \mathrm{A}e^{-\mathrm{B}(\Delta\mu / k_B\mathrm{T})}, \tag{9.9}$$

where A and B are constants.

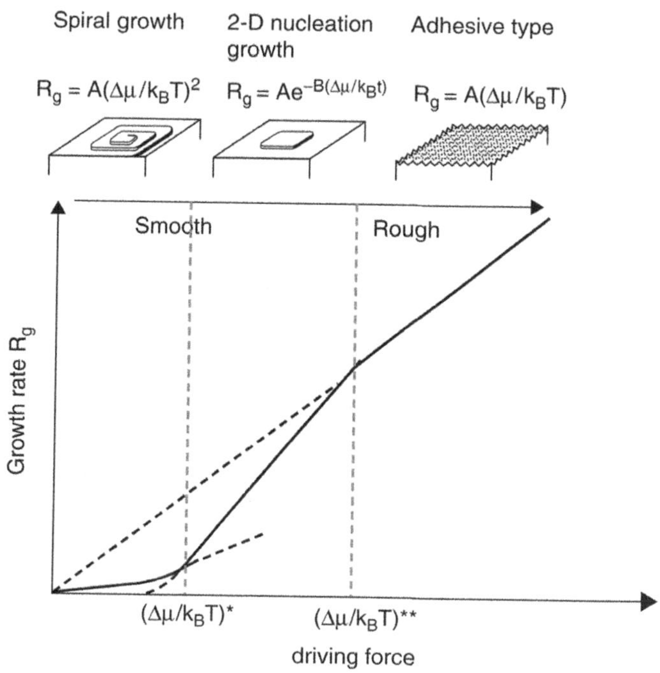

FIGURE 9.9 Different growth modes that lead to rough and smooth interfaces. Growth rate R_g increases with driving force $\Delta\mu$/kT and hence saturation at a given temperature. After Sunagawa, 2005 (with permission).

At still lower supersaturation, growth will occur at screw dislocations following the spiral mechanism, and

$$R_g = A(\Delta\mu/k_B T)^2 \tag{9.10}$$

(Sunagawa, 2005). The relationship between driving force and growth rate for a given crystallographic face and growth direction is shown in Figure 9.9.

Case Study: Calcite Birth and Spread versus Spiral Growth: BCF Theory

Carbonate minerals have particular importance in environmental surface chemisty because carbonates are widely distributed, the carbonate system plays a key role in controlling the pH of many waters, and many organisms contain carbonate shells or other components (Chapters 2 and 3). Thus, both the thermodynamics and the kinetics of carbonate systems have been studied extensively. Teng et al. (2000) used in-solution AFM to image the calcite surface and to measure spiral growth rates at different values of supersaturation. Supersaturation σ (or driving force) was calculated as

$$\sigma = \frac{\Delta\mu}{k_B T} = \ln\left(\frac{a_{Ca^{2+}} a_{CO_3^{2-}}}{K_{sp}}\right), \tag{9.11}$$

where

$\Delta\mu$ = difference in chemical potential difference between $CaCO_3$ molecules in the aqueous phase versus solid calcite
k_B = Boltzmann constant
a_i = measured activity of the given species i
K_{sp} = solubility product of calcite at ionic strength = 0 (Teng et al., 2000).

They found that when σ was < 0.8, growth occurred by step migration at screw dislocations and other surface defects. However, at $\sigma > 0.8$, 2-D surface nucleation became important, in which islands formed and grew on the surface ("birth and spread"), many eventually merging. At $\sigma \sim 1$, both mechanisms were observed. Examples of AFM images of the different nucleation mechanisms are shown in Figure 9.10.

This saturation state dependence of growth—at low supersaturation, growth only occurs at screw dislocations sites, but at higher saturation, growth can occur by 2-D birth and spread—is in agreement with surface free energy considerations and with the basic concepts of crystal growth, as outlined in the section on *BCF Theory*. Many research groups are using techniques such as SEM, in-solution AFM, vertical scanning interferometry (VSI), and synchrotron-based analysis, along with modeling, to investigate how factors such as saturation state, temperature, presence of impurities (including bacteria and biofilms), and crystallographic indices of surfaces affect the growth and dissolution of minerals such as calcite, dolomite, and barite (e.g., Gratz et al., 1993; Reeder, 1996; Lower et al., 1998a, b; Lüttge et al., 2005; Buss et al., 2007; Weaver et al., 2007).

Rates of Step Advancement

As described above, crystals often grow by advancement of steps that contain kinks across a mineral surface. The rate of advancement of a step R_s is given by the difference between the attachment rate, R_a and the detachment rate, R_d (values in molecules per unit time) (DeYoreo and Vekilov, 2003):

$$R_s = R_a - R_d. \tag{9.12}$$

The step speed is

$$v_s = R_s b, \tag{9.13}$$

where b is the depth of the kink.

For a monatomic step, the rate of step propagation is a product of the molecular size, the kink density on the mineral surface, and the net flux of reactants into the kink (DeYoreo and Vekilov, 2003). The flux is proportional to the reactant concencentration.

When a crystal grows, the reactants are consumed in the local area, leading to a dilute region. Solutes will tend to diffuse from the bulk solution toward the growing surface in a process known as *volume* or *bulk diffusion* (Sunagawa, 2005). Markov (1995) describes in detail the volume diffusion field around an isolated monatomic straight-edged step edge of height a (Figure 9.11) that is far removed from other steps. A brief description is given here. If C is the concentration at a given location, C_{st} is the concentration in the vicinity of the step, and C_∞ is the bulk solution concentration, then the radius vector r for boundary

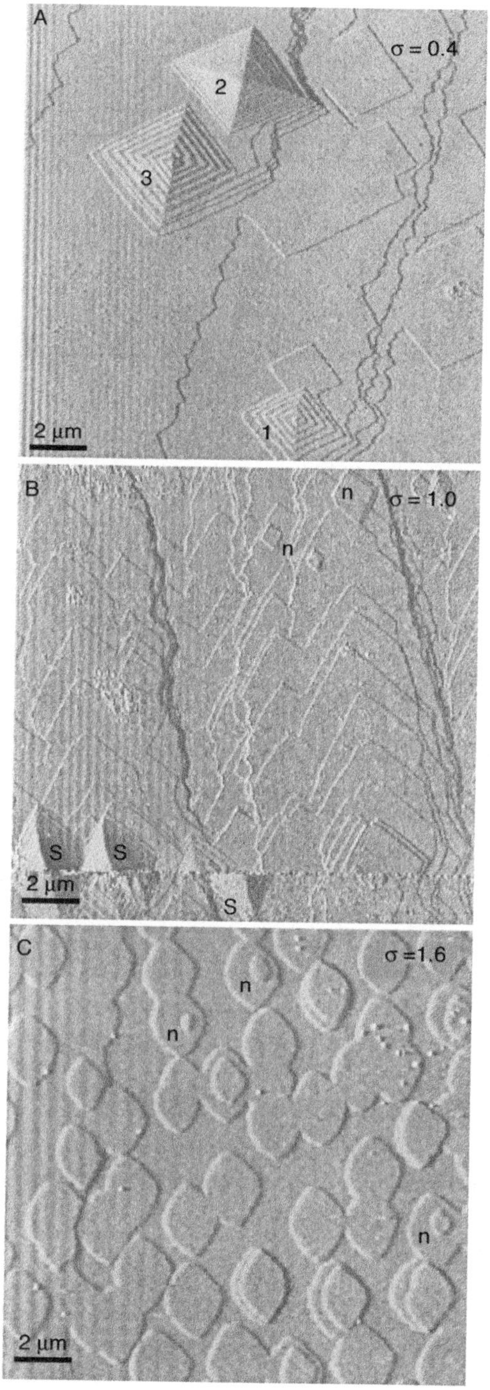

FIGURE 9.10 In-solution AFM images of calcite growth at different saturation states. (a) Spiral growth at $\sigma \sim 0.4$. (b) Spiral growth and surface nucleation growth coexisting at $\sigma \sim 1.0$. (c) Primarily 2-D surface nucleation and growth at $\sigma \sim 1.6$. From Teng et al., 2000 (with permission).

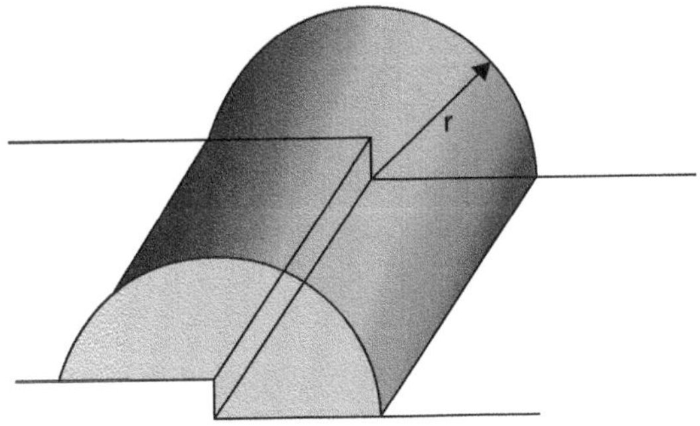

FIGURE 9.11 The volume diffusion field around an isolated step of height a has cylindrical symmetry and radius vector r. Based on Markov, 1995.

conditions of the volume diffusion field is given by

$$r = a/\pi \quad \text{for} \quad C = C_{st}, \tag{9.14}$$

meaning that a step with height a is approximated by a cylindrical radius $r = a/\pi$, and

$$r = d \quad \text{for} \quad C = C_{\infty}, \tag{9.15}$$

meaning that at some radius d distance from the step edge, the concentration within the volume diffusion field attains that of the bulk solution. If D is the bulk diffusion coefficient, then $D\pi/a$ is the rate of diffusion.

The equation for the rate of step advance across a surface in solution is different depending on whether the rate of diffusion (transport) or the rate of surface reaction is slowest. For an isolated straight-edged monatomic step, the rate of step migration ν_{∞} in a kinetic or surface-reaction regime for which the rate of diffusion ($D\pi/a$) is much greater than the rate of crystallization (given by the kinetic coefficient, β_{st}) is given by (Markov, 1995)

$$\nu_{\infty} = \beta_{st} C_0 \nu_c \sigma \quad \text{when} \quad (D\pi/a >> \beta_{st}), \tag{9.16}$$

where

β_{st} = rate of crystallization or the "kinetic coefficient" of the step
C_0 = equilibrium concentration of the solute reactant at the given temperature
ν_c = volume of a growth unit in the crystal
σ = supersaturation.

The value of β_{st} for calcite is 0.3 cm s^{-1} (DeYoreo and Vekilov, 2003).

In the other end-member regime of diffusion limitation,

$$\nu_{\infty} = \frac{\pi D C_0 \nu_c \sigma}{a \, \ln(\pi d/a)} \quad \text{when} \quad (\pi D/a << \beta_{st}) \tag{9.17}$$

Markov (2005).

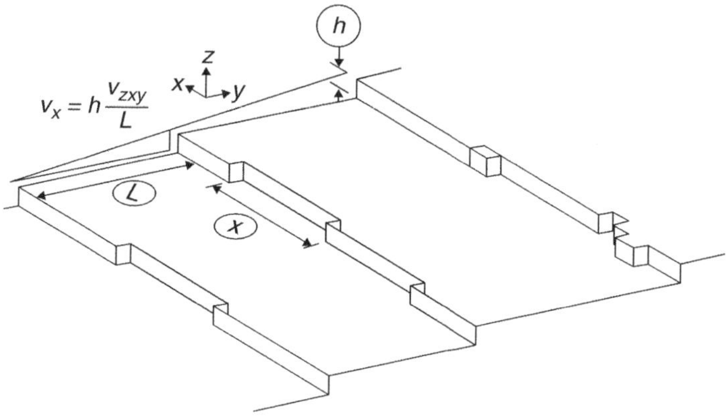

FIGURE 9.12 Schematic illustration of the three parameters that relate crystal growth rate at steps on a surface with bulk growth rate. These include surface step height (h), kink spacing (x), and interstep spacing or terrace width (L). From Morse et al., 2007 (with permission).

A monatomic step will migrate across a surface faster than a polyatomic step, because a polyatomic step requires a faster flux of atoms to move at the same rate as a monatomic step. When monatomic steps are migrating across a surface, local fluctuations in supersaturation or the presence of an impurity can cause the steps to merge. The resulting diatomic step will move more slowly, and another monatomic step will catch up with it, then another, and so on, leading to a polyatomic step. As this process continues, thick macrosteps can occur on surfaces, eventually forming ledges. Equations for the rate of advance of steps in a train of steps are given by Markov (2005).

Figure 9.12 illustrates the following three parameters that relate crystal growth (and dissolution) from steps on a surface with bulk growth: (1) the surface step height (h), (2) the kink spacing (x), and (3) the interstep spacing or terrace width (L). Many studies have shown that step movement on cleavage faces of carbonates can be related to overall rate (growth or dissolution) as measured in bulk solution (Morse et al., 2007, and references therein).

IMPURITIES AND GROWTH AT STEPS

During crystal growth at step edges, various processes occur by which impurities present in the solution can associate with the surface and cause the growth to be modified. Modification mechanisms include (1) *incorporation*, in which the impurity is incorporated to become part of the growing crystal; (2) *step pinning*, in which impurities block the attachment of molecules to step edges; (3) *kink blocking*, in which impurities adsorb to kink sites for short times, which reduces the kink site density; and (4) *step edge adsorption*, in which the adsorbing impurity (such as a surfactant) lowers interfacial energy by adsorbing (DeYoreo and Vekilov, 2003). The extent of inhibition can be affected not only by the concentration of the impurity but also by other factors, such as temperature, saturation state, and the concentration and identity of the background electrolyte. One frequently encountered ‘poison’ is phosphate, which adsorbs strongly and can affect the growth rates of many oxide, silicate, and carbonate minerals.

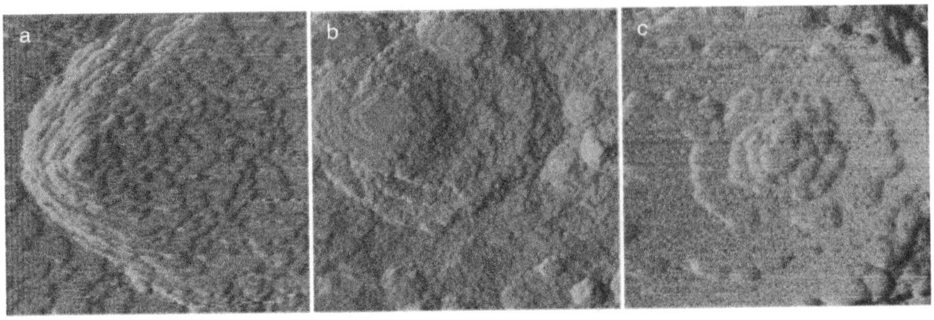

FIGURE 9.13 AFM images of growth hillocks on calcium oxalate monohydrate grown in the presence of citrate. (a) [citrate] = 2 μM and $\sigma = 0.70$, image 1.5 μm across; (b) [citrate] = 20 μM and $\sigma = 0.68$, 4.5 μm across; (c) higher resolution of (b), 2 μm across. From Weaver et al., 2007 (with permission).

The ability of various impurities to inhibit crystal growth is important for a broad range of plants and animals in which certain types of mineralization may be pathologic. One key example involves kidney stones, which are composed of calcium oxalate monohydrate (COM) and whose mineralization is inhibited by citrate that is present in the urinary tract or used in therapy. Weaver et al. (2007) showed that citrate inhibition of COM formation involves a combination of kink blocking and step pinning. The inhibition could be described by a classic step-pinning model developed by Cabrera and Vermilyea (1958), which is known as the C–V model. Figure 9.13 shows growth hillocks formed on COM in the presence of different concentrations of citrate. When no citrate was present, the growth hillocks had smooth edges (not shown here). At relatively low concentrations, the growth hillocks had the same shape but serrated edges (Figure 9.13 a). At higher citrate concentration, the shape of the growth hillocks changed and the edges became very rough (Figures 9.13 b–c).

MONTE CARLO SIMULATIONS OF CRYSTAL GROWTH

Monte Carlo simulations are one means by which the kinetics of mineral growth (and dissolution) have been investigated in a manner that allows comparison between theory and experimental observations. The Monte Carlo approach models surface structure and dynamic processes on surfaces based on statistical sampling with random numbers (Lasaga, 1990; Philpot, 2004). Molecular structures and configurations are optimized based on analysis of many randomly generated trial configurations (Cygan, 2001).

The processes of attachment of an ion or a molecule from solution, surface diffusion, and detachment back to solution dominate the dynamics of adsorption, growth, and dissolution (Lasaga, 1990). The rate constants for these three processes and their variations with local surface microtopography can be determined. The rates can then be converted to probabilities, so that surface dynamics can be treated stochastically.

Monte Carlo simulation techniques have been applied successfully for many years to study crystal growth (e.g., Gilmer, 1976, 1980) and dissolution (e.g. Blum and Lasaga, 1987), as well as other processes such as the configurations of water and cations in clay interlayers (Sposito et al., 1999) and microbial attachment to mineral surfaces (Lüttge et al., 2005). Figure 9.14 shows a Monte Carlo simulation for spiral growth that originated at a

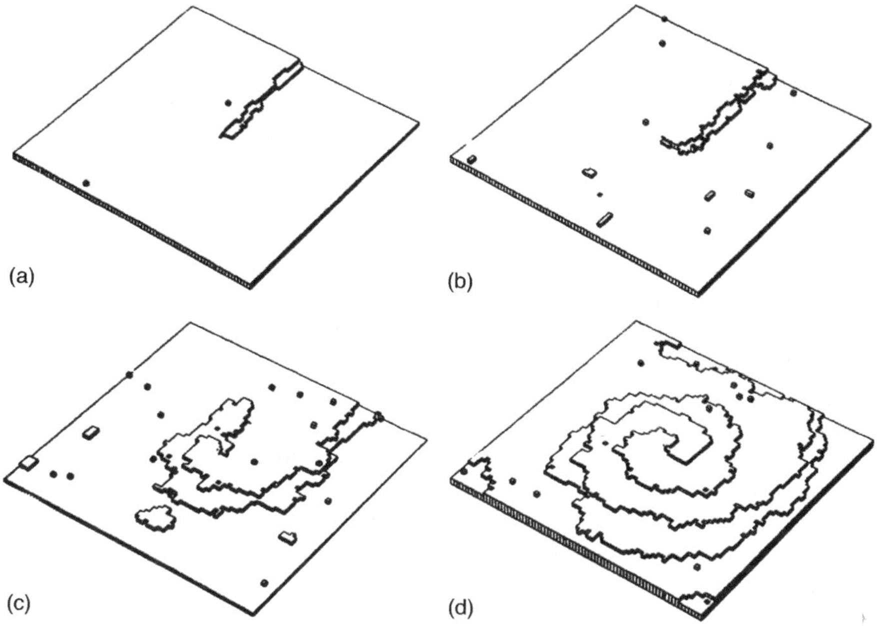

FIGURE 9.14 Monte Carlo simulation of spiral growth from a dislocation on a simple cubic (100) face. From Gilmer, 1980 (with permission).

screw dislocation (Gilmer, 1980). Two parallel steps start out at the dislocation, but as the steps advance across the surface, a complex array of step sites results. This simulation shows the formation of multiple spirals from a single screw dislocation.

One advantage of simulations such as Monte Carlo for mineral growth (and dissolution) studies is that because of rapid advances in computer technologies, the dynamic motions of atoms or clusters, the movement of steps, the growth of islands on surfaces, and so on can be visualized and measured. The simulations can be compared with observations, for example by in-solution AFM or VSI. Increasing simulations to larger areas of surfaces is important for greater comparability with such experimental techniques. Current Monte Carlo simulations can encompass hundreds of thousands to millions of molecules (Lüttge et al., 2005).

BIOMINERALIZATION

According to traditional definitions, a mineral is a naturally occurring solid substance that is formed through geological processes and that has a characteristic chemical composition, a highly ordered atomic structure, and specific physical properties (Hurlbut and Klein, 1985). Skinner (2005) argues that the definition of a mineral should be expanded to include formation by "biogeochemical processes," which "reflects a greater appreciation, although an incomplete understanding, of the processes of mineral formation by living forms."

This refined definition recognizes that the nucleation and growth of minerals may be influenced by the presence of organisms. A *biomineral* is a mineral produced by one or more organisms, and most biominerals contain both mineral and organic compounds (Weiner and

Dove, 2003; Ferris, 2005). Biominerals often have morphologic, microtopographic, and compositional properties, such as trace-element and isotopic compositions, that are distinct from their abiotically produced counterparts.

Two principal modes of biomineralization include *biologically induced* mineralization and *biologically controlled* mineralization (Weiner and Dove, 2003). In biologically induced mineralization, the cell promotes mineralization by altering solution conditions and/or providing surfaces for heterogeneous nucleation, but it does not directly control the mineral type or habit. In biologically controlled mineralization, the cellular activities of the organism play a direct role in controlling nucleation and growth, along with the morphology and depositional location of the mineral formed. Biologically controlled biomineralization may be extracellular, intercellular, or intracellular. Extracellular is taken to mean on the outer wall of the cell, within the cell wall, or in the tissue areas immediate surrounding the cell wall (Skinner and Jahren, 2003). Intercellular means between adjacent cells. Intracellular refers to biomineralization within the cell (Skinner and Jahren, 2003). When biomineralization is intracellular, the mineral may stay within the cell or be secreted to extracellular locations (Weiner and Dove, 2003).

The calcium carbonate ($CaCO_3$) polymorph minerals calcite, aragonite, and vaterite are often formed by biomineralization. Some examples of $CaCO_3$ biomineralizing organisms include corals, whose skeletons are mostly nucleated as aragonite, often containing impurities such as Sr (Cohen and McConnaughey, 2003); mollusks, with calcite and aragonite precipitation by bivalves and vaterite precipitation by freshwater snails (in snail egg capsules) (Skinner and Jahren, 2003); coccolithophores, which are single-celled oceanic plankton whose calcareous shells (known as coccoliths) have formed vast chalk formations (Young and Henricksen, 2003); and cyanobacteria, which were likely some of the earliest biomineralizing organisms and which may precipitate calcite or aragonite as described in Chapter 12. An example of a coccolith is shown in Figure 9.15. Research has shown that

FIGURE 9.15 SEM image of a coccolith, as an example of biologically controlled biomineralization of calcite. Courtesy of Richard Lampitt and Jeremy Young, The Natural History Museum, London.

the calcite forms in intracellular vesicles in coccolithophores, which provides an example of biologically controlled biomineralization (Young and Henricksen, 2003).

Silica, primarily opal; apatite; and iron and manganese (hydr)oxides also commonly form as biominerals. A variety of sulfur-containing minerals, which includes elemental sulfur (S^0), may also be formed by biomineralization. For example, purple sulfur bacteria accumulate S^0 intracelluarly, whereas green sulfur bacteria accumulate S^0 extracelluarly (Skinner and Jahren, 2003). A wide range of other minerals may occur as biominerals, along with a variety of organic crystals (Skinner, 2005).

Biomineralization of apatite is of great interest in biomedicine, because apatite minerals make up the bones and teeth of vertebrates, including humans (e.g., Pasteris et al., 2008). Pasteris et al. (2004) showed using laser Raman microprobe spectroscopy that tooth enamel is made of hydroxylapatite. However, contrary to common belief, bone apatite is not hydroxylated and therefore not *hydroxyl*apatite. These authors suggested that the human body biochemically controls the formation of hydroxylapatite versus (nonhydroxyl) apatite.

Examples of biomineralization by plants are described in Chapter 11. Examples of biomineralization by microorganisms, particularly bacteria, are described in Chapter 12.

CARBONATE PRECIPITATION IN THE MARINE ENVIRONMENT

Carbonate mineral precipitation in the marine environment is an example of the many complexities associated with mineral nucleation and growth. It is such a complex topic that we can only touch on some key points here. The complementary reviews of Morse et al. (2007) and Millero et al. (2007) summarize many of these key points as well as numerous other considerations.

Many carbonate minerals of importance are found in the oceans, including primarily calcite, aragonite, and magnesian carbonate. The shallow oceans are saturated with respect to calcite and aragonite, whereas deeper ocean waters are undersaturated (Berner and Berner, 1987). As noted in Chapter 1, carbonate growth and dissolution can be controlled by surface, transport, or mixed processes. Carbonates can form phases at the surface that are distinct from the bulk mineral, and these phases can affect both precipitation and dissolution in complex ways that are not yet fully understood (Morse et al., 2007).

As discussed above, many organisms can biomineralize carbonates, and we are only beginning to understand the processes of biomineralization. Some examples in the oceans include coccolithophorid and mollusc formation of calcite, coral, and mollusk formation of aragonite, and coralline algae and echinoderm formation of Mg-calcite.

The high salinity of the oceans plays a role in both the thermodynamics and kinetics of carbonate precipitation. Mucci (1983) showed that the solubilities of calcite and aragonite increase with increasing ionic strength, which can be anticipated from our discussion in Chapter 1. Calcite precipitation rate tends to increase with ionic strength within the realm of ocean waters, although the exact reasons for this increase are not fully understood (Morse et al., 2007, and references therein). As also discussed in Chapter 1, the solubility of calcite tends to decrease with increasing temperature; the same is true for aragonite.

Seawater contains several constituents that may be taken up into carbonates and/or act as poisons to block precipitation. The uptake of Mg to form magnesian calcite increases with increasing temperature (Mucci, 1987). Sulfate also can be taken up through coprecipitation, more into calcite than into aragonite (Morse et al., 2007, and references therein). Mn and Sr also are commonly taken up into carbonates. Inorganic phosphate strongly adsorbs to calcite and aragonite and strongly inhibits carbonate precipitation. Gratz and Hillner (1993) showed that phosphate binds to steps and poisons step movement, changing step shape and surface morphology. Dissolved organic matter can have complex effects, often serving as an inhibitor.

Models suggest that because of the increase in atmospheric P_{CO_2} caused by the burning of fossil fuels (Chapter 14), the shallow surface ocean waters are likely to experience a decrease in pH and to become undersaturated with aragonite within the next 100 to 200 years (Millero, 2007, and references therein; Figure 9.16). Because kinetics and saturation state are linked, this could potentially have profound implications for biomineralization and ocean biology.

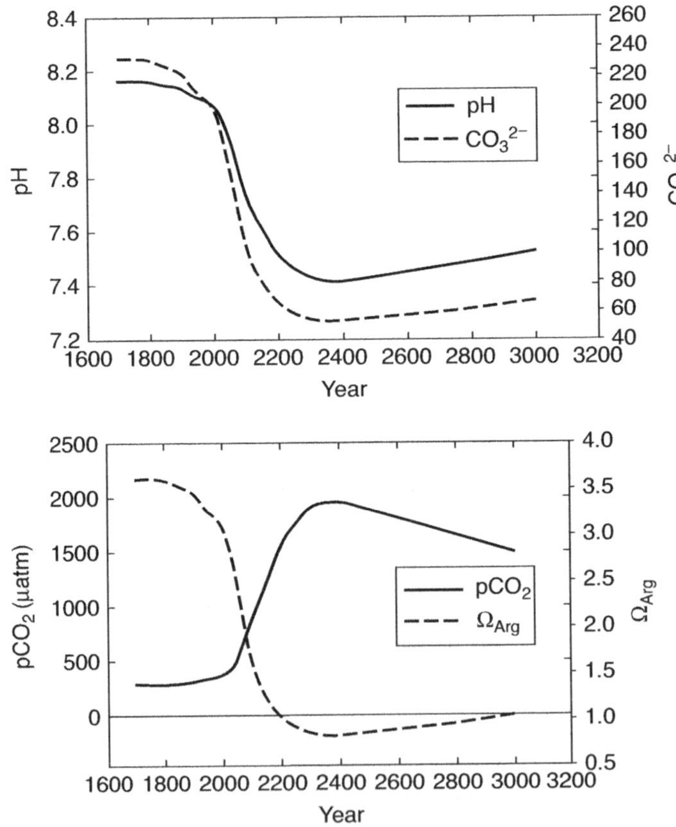

FIGURE 9.16 Measured and projected changes in pH, $[CO_3^{2-}]$ (in μmol kg^{-1}), pCO_2 (atm), and the saturation state of aragonite over time. From Millero, 2007 (with permission).

QUESTIONS FOR FURTHER THOUGHT

1. Read one of the following papers. Summarize it from the standpoint of mineral–water interface and environmental surface chemistry, with a particular emphasis on crystal nucleation and growth. What new thing(s) did you learn from the paper? Write a 2–3 page essay.

   De Yoreo, J.J. and Dove, P.M. 2004. Shaping crystals with biomolecules, *Science* 306, 1301–1302.

   Davis, K.J., Dove, P.M., De Yoreo, J.J., The role of Mg^{2+} as an impurity in calcite growth. *Science* 290, 1134–1137.

2. You have been given the assignment of determining how the presence of phosphate might affect the nucleation and growth of hematite from a supersaturated solution. Design an experiment dealing with this question, describing the reasoning behind your approach and the methods that you would use. Assume that any methods you needed would be available.
3. (a) How does one define a surface and an interface for the purposes of investigating mineral growth?
   (b) What are some factors that would need to be considered?
4. Adsorption and mineral growth both involve attachment to a surface. Comment on the similarities and differences between these two processes.
5. Synthesizing Fe (hydr)oxides often involves starting with a supersaturated solution and then placing these solutions in bottles in an incubator at a set temperature for a given amount of time.
   (a) Explain why temperature and duration of incubation might affect properties such as particle size and surface area.
   (b) In one set of experiments, researchers found that a different product in terms of particle size and surface area resulted if the solutions were mixed up and incubated in the same room versus if they were mixed up, carried down the hall and downstairs, and then incubated at the same temperature for the same period of time. What could have caused the differences?
   (c) What might be the advantages of growing mineral particles such as goethite in a plastic vessel rather than a glass vessel?
6. Explain, based on concepts and equations presented in this chapter and Chapter 1, why crystal growth may be related to or affected by:
   (a) temperature,
   (b) pH,
   (c) reactant concentration,
   (d) solution ionic strength, and
   (e) presence of impurities.
7. How well would you expect each of the following mineral surfaces to serve as a substrate for heterogeneous nucleation of gibbsite? Explain your reasoning, including what factors need to be considered:
   (a) kaolinite,
   (b) hematite,
   (c) goethite, and
   (d) calcite.

FURTHER READING

Berner, R.A. 1980. Early Diagenesis: A Theoretical Approach. Princeton Series in Geochemistry. Princeton University Press, Princeton, NJ, 241 p.

DeYoreo, J.J., and Vekilov, P.G. 2003. Principles of crystal nucleation and growth. In Dove, P.M., DeYoreo, J.J., and Weiner, S., eds. *Biomineralization Reviews in Mineralogy and Geochemistry*, vol. 54. *Mineralogy Society of America*, Chantilly, VA. pp. 56–93.

Lower, S.K., *et al.* 1998. Aqueous Pb sorption by hydroxylapatite: Applications of atomic force microscopy to dissolution, nucleation, and growth studies. *American Mineralogist* 83, 147–158.

Manecki, M., Maurice, P.A., and Traina, S.J. 2000. Uptake of aqueous Pb by Cl-, F-, and OH- apatites: Mineralogic evidence for nucleation mechanisms. *American Mineralogist* 85, 932–942.

Markov, I.V. 1995. Crystal Growth for Beginners: Fundamentals of Nucleation, Growth, and Epitaxy. World Scientific, Singapore, 422 p.

Sunagawa, I. 2005. Crystals: Growth, Morphology, and Perfection. Cambridge University Press, Cambridge, UK, 295 p.

10

MINERAL WEATHERING AND DISSOLUTION

The overall process by which rocks are broken down, sometimes accompanied by formation of new minerals, is termed *weathering*. One component of weathering is mineral dissolution. Mineral dissolution is in effect the opposite of crystal growth, and many of the same types of factors and processes need to be considered. Dissolution and weathering have been studied from both field and laboratory perspectives. This chapter focuses primarily on abiotic mineral dissolution and weathering, considering both thermodynamic and kinetic approaches. Plant and microbial effects on mineral weathering are discussed in Chapters 11 and 12.

CHEMICAL, PHYSICAL, AND BIOLOGICAL WEATHERING

In *chemical weathering*, minerals are dissolved and sometimes new minerals are formed through chemical reactions. The chemical weathering of rocks results in partitioning of the components of the rocks into the following three main classes of weathering products (Figure 10.1):

1. *Dissolved ions* in solution
2. *Secondary weathering products* such as kaolinite formed from feldspar dissolution (see below)
3. *Incompletely weathered residuum*, which consists of highly stable minerals not easily broken down such as quartz, titanium oxide, and aluminum and iron oxides.

Physical weathering refers to the breakdown of rocks through primarily non chemical processes, such as wind, ice wedging, or abrasion caused by currents or wave action. *Biological weathering* refers to rock breakdown by biological processes, which may be

Environmental Surfaces and Interfaces from the Nanoscale to the Global Scale, by Patricia A. Maurice

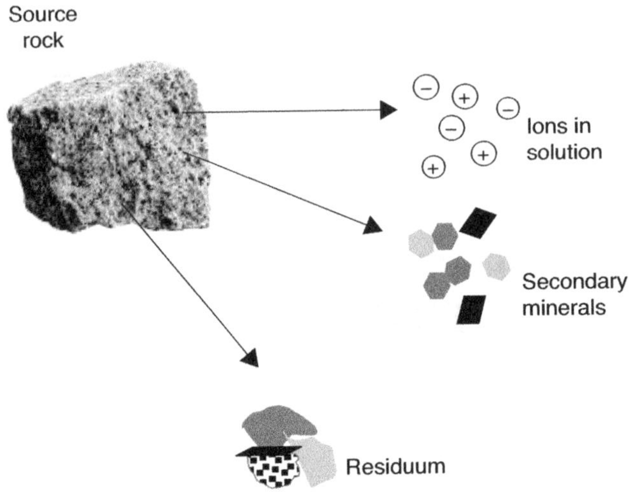

FIGURE 10.1 The results of chemical weathering.

physical and/or chemical. This process includes, for example, rocks being physically torn apart by roots and at the same time dissolved by root exudates.

(Bio)chemical weathering rates depend on factors that are *intrinsic* to a mineral, such as mineral composition, surface area, and defect densities, along with *extrinsic* factors, such as climate, solution composition, and biological activity (White, 2004). Weathering is a complex function of factors such as rainfall and runoff, rock composition (lithology), temperature (including cyclic freeze/thaw), topography, relief, vegetation, and plant and microbial activity. The relative roles of different controlling factors has been the subject of much debate, and these roles vary both spatially and temporally.

In general, silicate minerals that crystallize at highest temperatures from igneous melts tend to be more susceptible to weathering processes than those that form at lower temperatures in the crystallization process (Goldich, 1938). This means that among the mafic minerals, olivine tends to be highly weatherable, followed by pyroxene, amphibole, and finally biotite as least weatherable. Among the felsic minerals, Ca-Na plagioclase tends to be most weatherable, followed by Na-Ca plagioclase, K-feldspar, and muscovite as least weatherable. However, the complexities of weathering mean that these broad generalizations are not observed in many instances in the field. Gislason et al. (1996) showed that the weathering rates of volcanic glasses are about one order of magnitude faster than those of crystalline minerals of comparable composition.

Chemical weathering involves reactions between water and rock that result in the breakdown of minerals and sometimes in the formation of new minerals. Examples of two important types of weathering reactions are *carbonate dissolution* and *hydrolysis*.

An example of a carbonate dissolution reaction is the dissolution of calcite in water open to exchange with atmospheric CO_2

$$\underset{\text{Calcite}}{CaCO_3(s)} + CO_2(g) + H_2O(l) \rightarrow Ca^{2+}(aq) + 2HCO_3^-(aq). \tag{10.1}$$

Most dissolution reactions that involve oxides and silicates are through the process of hydrolysis. An example of an hydrolysis reaction is the weathering of K-feldspar to form the clay mineral kaolinite

$$\underset{\text{K - feldspar}}{2KAlSi_3O_8(s)} + 2H^+(aq) + 9H_2O(l) \rightarrow \underset{\text{kaolinite}}{Al_2Si_2O_5(OH)_4(s)} + 2K^+(aq) + 4H_4SiO_4(aq). \quad (10.2)$$

Sometimes, feldspar *phenocrysts* can be completely altered to kaolinite, with the original shape of the phenocrysts remaining intact.

Reaction 10.2, which combines feldspar with protons and water to yield kaolinite clay and solution species, is an example of *incongruent dissolution*. When a mineral dissolves incongruently, the release of constituents to solution does not reflect the original stoichiometric relationships in the mineral phase. In this case, a secondary mineral (kaolinite) is formed. Incongruent dissolution also may occur if certain constituents are selectively leached from the surface, potentially leaving behind a *leached layer*. Much research has focused on the formation of leached layers and whether a leached layer "armors" the underlying mineral against any more weathering. An amorphous layer at the surface of a natural field sample of K-feldspar is shown in Figure 10.2. Although the genesis of this amorphous material is not known for certain, it may potentially represent a leached layer.

Congruent dissolution occurs when the mineral dissolves stoichiometrically, with no secondary-phase formation. Under acidic conditions, kaolinite can be weathered through the hydrolysis reaction

$$\underset{\text{kaolinite}}{Al_2Si_2O_5(OH)_4(s)} + 6H^+ \rightarrow 2\,Al^{3+}(aq) + 2H_4SiO_4(aq) + H_2O\,(l). \quad (10.3)$$

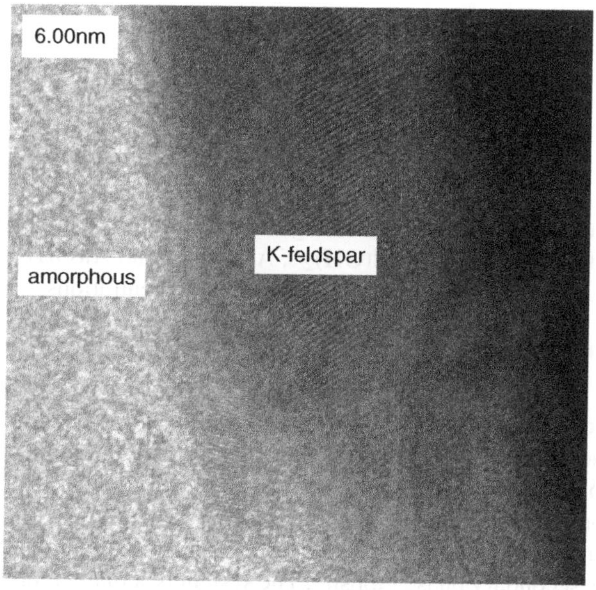

FIGURE 10.2 A transmission electron spectroscopy (TEM) image of K-feldspar from the Navajo Sandstone Aquifer (Arizona) with an amorphous layer at the surface, from Zhu et al., 2006. It is possible that this amorphous layer might represent a "leached layer" (with permission).

This reaction is an example of congruent dissolution, where the stoichiometry of components released to solution matches that of the original mineral. The carbonate dissolution reaction in Equation 10.1 is another example of congruent dissolution.

(Bio)chemical weathering may be studied using a variety of different approaches, including analysis of (1) bulk compositional changes in the regolith, (2) changes in mineral surface chemistry and microtopography, and (3) analysis of solution chemistry. Studies may be conducted in the field, the laboratory, or a combination of both. They may include analysis of major components and trace elements, along with isotopic studies that involve either solid and/or solution.

THERMODYNAMICS OF MINERAL WEATHERING

One way of considering mineral weathering is to take a thermodynamic approach. Even though many natural systems at the surface of the Earth are not at equilibrium, a thermodynamic approach allows us to predict what the system would be like—for example, what minerals would be present and the distributions of species in solution—it if were at equilibrium. Second, we can compare field observations with predictions from thermodynamics to determine whether the assemblage seems to be at equilibrium, and if not, then in which direction the system deviates from equilibrium. If a system is supersaturated with respect to some phase such as quartz or amorphous silica, then one should ask why the supersaturation exists.

Two of the most important elements to consider are Al and Si, as they are constituents of many (hydr)oxide and aluminosilicate minerals. When considering the controls on Al concentrations in solution, we often use an activity diagram to plot how the concentration of Al in solution in equilibrium with the mineral gibbsite $Al(OH)_3$ varies as a function of pH.

Under acidic pH conditions, gibbsite dissolves according to the following overall reaction:

$$\overset{\text{gibbsite}}{Al(OH)_3(s)} + 3H^+(aq) \rightarrow Al^{3+}(aq) + 3H_2O\,(l). \qquad (10.4)$$

In solutions without other ligands (such as organic acids and F) to bind Al, the Al solution species will include Al^{3+}, $AlOH^{2+}$, $Al(OH)_2^+$, $Al(OH)_3^0$, and $Al(OH)_4^-$, so that the total dissolved Al concentration is the sum of these species concentrations

$$Al_{tot} = [Al^{3+}] + [AlOH^{2+}] + [Al(OH)_2^{\ +}] + [Al(OH)_3^{\ 0}] + [Al(OH)_4^{\ -}]. \qquad (10.5)$$

Figure 10.3 is constructed based on the equilibrium reaction equations that relate various species to one another and use relevant thermodynamic data. This figure shows that gibbsite is only sparingly soluble at circumneutral pH, but it increases in solubility as pH either decreases or increases. This factor impacts the weathering of aluminosilicate minerals, because Al released by weathering of aluminosilicates at circumneutral pH tends to reprecipitate as the secondary phase gibbsite. The system is actually more complicated than this; amorphous aluminum hydroxide or some other phase may precipitate, depending on the composition of the solution. In systems that are impacted by acid rain, such as in the Adirondack Mountains of New York, the decrease in pH of soil pore waters and surface

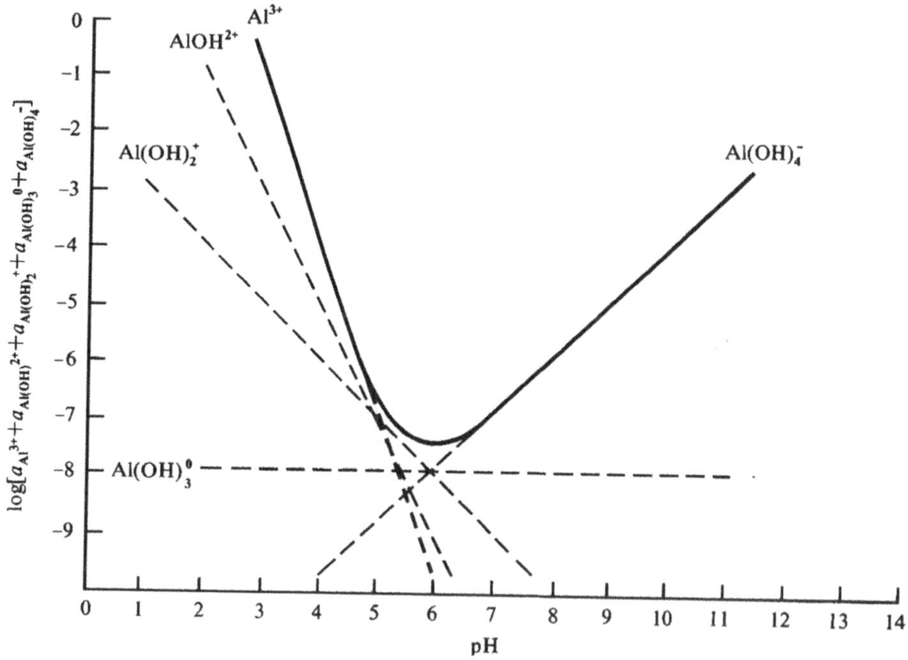

FIGURE 10.3 Gibbsite solubility as a function of pH at 25°C. From Nordstrom and Muñoz, 1986 (with permission).

waters can lead to increases in the concentrations of inorganic Al species that are toxic to aquatic life.

For Si, we generally consider quartz and amorphous SiO_2. At pH < 9, quartz dissolves according to the reaction

$$\underset{\text{quartz}}{SiO_2(s)} + 2H_2O(l) \rightleftarrows H_4SiO_4(aq) \tag{10.6}$$

$$Keq = a_{H_4SIO_4}$$
$$= 1 \times 10^{-4} \text{ at } 25°C; \tag{10.7}$$

amorphous silica has a solubility constant of 2×10^{-3} at 25°C (Drever, 1997).

In many environments, the aqueous silica concentration is controlled not by quartz solubility but by the solubility of amorphous silica. If we were to run a computer program that calculated species concentrations at equilibrium and found that the system was supersaturated with respect to quartz, then we would have to consider potential control by amorphous silica.

At high pH, H_4SiO_4 dissociates to form $H_3SiO_4^-$ and $H_2SiO_4^{2-}$. The total dissolved silica concentration will be the sum of the concentrations of the three species

$$SiO_{2\,tot} = [H_4SiO_4] + [H_3SiO_4^-] + [H_2SiO_4^{2-}]. \tag{10.8}$$

Thus, the total dissolved silica concentration in equilibrium with quartz or amorphous silica increases at high pH values (Figure 10.4).

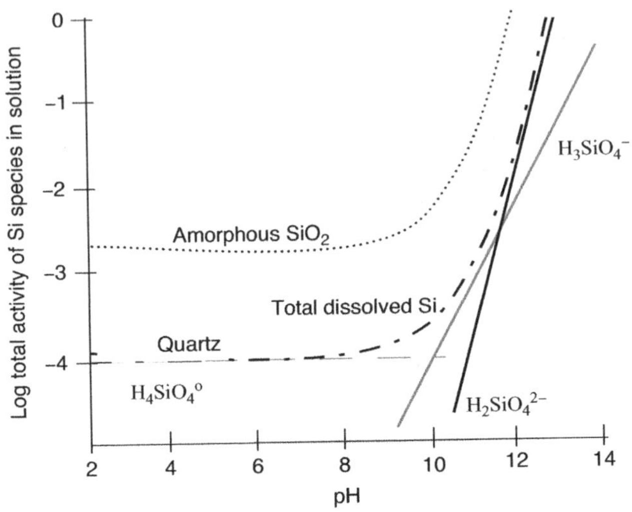

FIGURE 10.4 The activities of dissolved silica species in equilibrium with quartz and amorphous silica at 25°C.

A similar thermodynamic approach can be used for minerals that contains cations in addition to aluminum and silicon. We can consider reactions between pairs of minerals in the system K_2O-Al_2O_3-SiO_2-H_2O

muscovite ⇄ kaolinite

$$2KAl_3Si_3O_{10}(OH)_2 + 2H^+ + 3H_2O \rightleftarrows 3Al_2Si_2O_5(OH)_4 + 2K^+, \tag{10.9}$$

K-feldspar ⇄ kaolinite

$$2KAlSi_3O_8 + 2H^+ + 9H_2O \rightleftarrows Al_2Si_2O_5(OH)_4 + 2K^+ + 4H_4SiO_4, \tag{10.10}$$

K-feldspar ⇄ muscovite

$$3KAlSi_3O_8 + 2H^+ + 12H_2O \rightleftarrows KAl_3Si_3O_{10}(OH)_2 + 2K^+ + 6H_4SiO, \tag{10.11}$$

muscovite ⇄ gibbsite

$$KAl_3Si_3O_{10}(OH)_2 + H^+ + 9H_2O \rightleftarrows 3Al(OH)_3 + K^+ + 3H_4SiO_4, \tag{10.12}$$

and

pyrophyllite ⇄ kaolinite

$$Al_2Si_4O_{10}(OH)_2 + 5\,H_2O \rightleftarrows Al_2Si_2O_5(OH)_4 + 2H_4SiO_4. \tag{10.13}$$

All the equilibrium expressions can be displayed as straight lines on a graph with log (a_{K+}/a_{H+}) and log a_{H4SiO4} as axes. The slopes of the lines are determined by reaction stoichiometry. The positions of the lines are determined by equilibrium constants, which are calculated from free energies of the reactions. As an example, consider the muscovite/kaolinite boundary. The expression for the equilibrium constant for Equation 10.9 is:

$$K_{eq} = [K^+]^2/[H^+]^2 = 10^{8.0}, \tag{10.14}$$

which on taking the logarithm of both sides and then dividing both sides by 2 gives

$$\log[K^+]/[H^+] = 4.0, \tag{10.15}$$

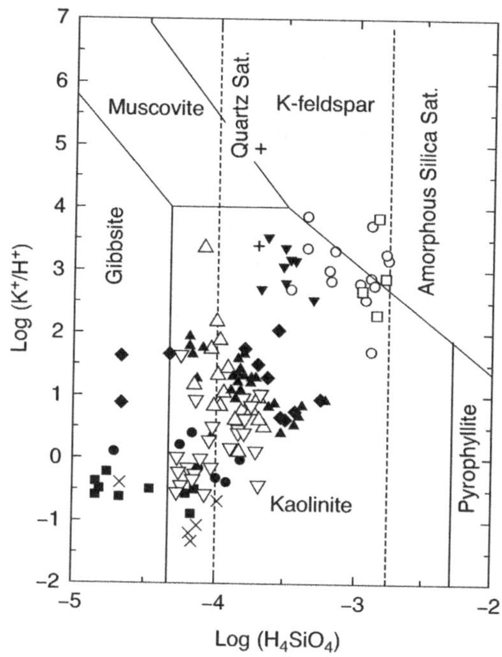

FIGURE 10.5 Stability fields in the K_2O-Al_2O_3-SiO_2-H_2O system, and soil water data plotted from a wide variety of sites (see White, 1995, for site locations and sources for different points). From White, 1995 (with permission).

so that the boundary plots as a straight horizontal line (slope = 0) intersecting the y-axis at 4.0. Such a plot is shown in Figure 10.5. The dashed vertical lines in Figure 10.5 are for quartz and amorphous silica.

White (1995) plotted solute concentrations from a wide range of soils on mineral stability field diagrams for the K_2O-Al_2O_3-SiO_2-H_2O system. As shown in Figure 10.5, the soil solutions are either supersaturated or undersaturated with respect to quartz, but none of the soils is supersaturated with respect to amorphous silica. Most samples plot in the kaolinite stability field. White (1995) points out that kaolinite is the most common secondary mineral phase that forms in association with common silicate rocks such as granite. Other samples plot in the gibbsite stability field, and gibbsite can also be a product of aluminosilicate weathering. This diagram only accounts for a small fraction of minerals. An alternative diagram of log ($[Na^+]/[H^+]$) versus log $[H_4SiO_4]$ presented by White (1995), for example, showed that some soil solutions were supersaturated with respect to Na-smectite, which is not considered in the K-based plot.

Thermodynamically based computer models (e.g., HYDRAQL, Papelis et al., 1988; PHREEQC, Parkhurst and Appelo, 1999) can calculate saturation indices for a vast array of different minerals and amorphous solids. It is important to realize that just because model output shows that a solution may be supersaturated or undersaturated with respect to a mineral (e.g., albite) does not mean that the mineral is precipitating or dissolving, respectively, in a given system. The actual mineralogy at the field site—not just the major minerals, but also the minor phases—needs to be analyzed carefully. Kinetic constraints may occur that are inhibiting precipitation or dissolution, or a mineral phase may not be available to dissolve. Although thermodynamic models do not describe time-dependent behavior, they help us to understand constraints on possible processes.

KINETICS OF MINERAL DISSOLUTION

As discussed in Chapter 1, kinetic studies focus on the rates and mechanisms of reactions. When a mineral dissolves, the following successive elementary steps may be involved (Stumm, 1992):

1. Mass transport of dissolved reactants from bulk solution to the mineral surface
2. Adsorption of solutes
3. Interlattice transfer of reacting species
4. Chemical reactions
5. Detachment of reactants from the surface and
6. Mass transport into the bulk solution.

Calcareous and evaporite minerals tend to be soluble and tend to dissolve rapidly and congruently. Dissolution of aluminosilicate minerals can be complex and often incongruent, with formation of secondary phases or leached layers.

As described in Chapter 1, mineral dissolution is generally either *transport* or *surface controlled*. In many cases, both mechanisms are important. We shall focus primarily on surface-controlled dissolution here.

In a surface-controlled dissolution batch experiment, one plots the concentration of product on the y-axis and time on the x-axis. The initial portions of such curves are often parapolic because of such factors as (1) the initial solution is far from equilibrium and (2) the initial surfaces may have a high density of reactive surface sites, sometimes including fine particles and defects left over from sample preparation. Once steady state is achieved, the plot becomes linear and the slope gives the rate of dissolution. An example is shown in Figure 10.6.

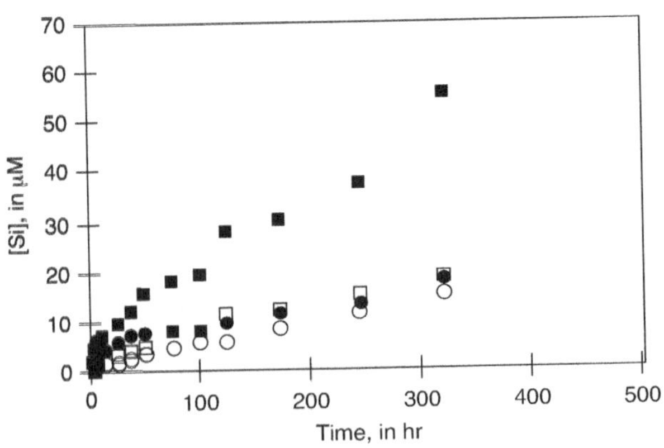

FIGURE 10.6 Silica release from kaolinite during dissolution in oxalate (filled) or nitric acid (open) in a batch system. Squares are kaolinite sample KGa-2, and circles are kaolinite sample KGa-1b. KGa-1b is a more highly ordered kaolinite sample, and KGa-2 is a less well-ordered kaolinite sample. Redrafted from Sutheimer et al., 1999.

For surface-controlled dissolution, the concentrations of solutes adjacent to the surface will be the same as in bulk solution (see Chapter 1).

Just like in crystal growth, mineral surface structure and microtopography play an important role in surface-controlled dissolution. Surface features such as steps, kinks, dislocations, and pits establish sites with different activation energies and different rates of reaction

$$\text{Fast reaction} \quad \text{More active sites} \rightarrow \text{Me(aq)}, \tag{10.16}$$

$$\text{Slower reaction} \quad \text{Less active sites} \rightarrow \text{Me(aq)}, \tag{10.17}$$

where Me(aq) represents metal ion released to solution.

Because these reactions are in parallel, the surface features that dissolve most quickly dominate the overall rate, until or unless all are consumed by reaction. Generally, pit formation or step migration is taken to be evidence for surface-controlled dissolution at particular sites.

Etch Pit Formation

When the dissolution rate is surface controlled, dissolution shows a preference for high-energy surface sites. Similar to mineral nucleation and growth, dissolution can occur by migration of steps across the surface but with steps retreating rather than growing. *Etch pits* (Figures 10.7 and 10.8) may be created when dissolution occurs at certain other types

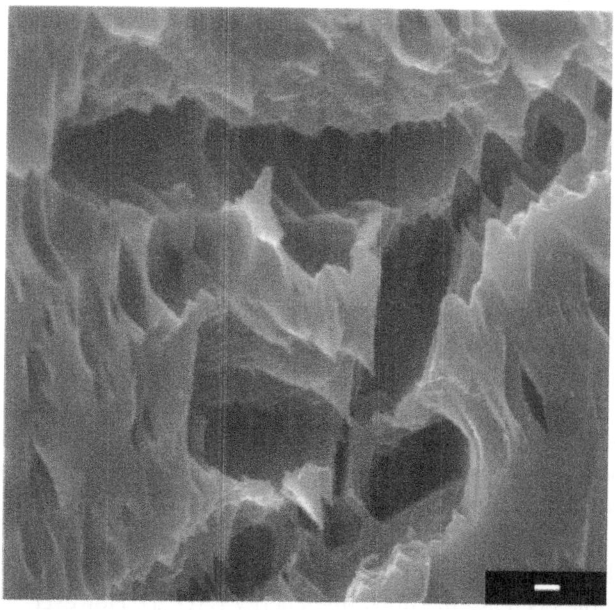

FIGURE 10.7 An SEM image of an apatite surface showing etch pits that formed upon laboratory dissolution in peat water for 6 months. Scale bar = 1 μm. From Bennett et al., 2001 (with permission).

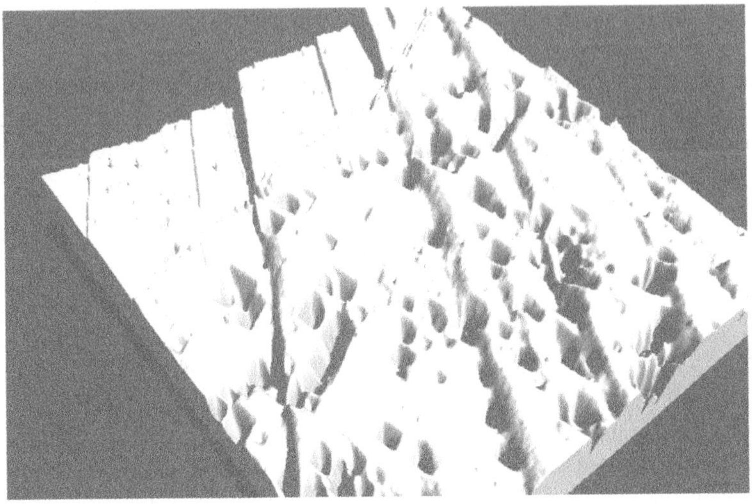

FIGURE 10.8 A vertical scanning interferometry (VSI) height map of a calcite surface with extensive crystallographically controlled flat-bottomed etch pits. From Lüttge et al., 2005.

of imperfections. Etch pits can increase mineral surface area and can present new surfaces, including internal pore space, for sorption. Sometimes, etch features may occur because of compositional heterogeneities. For example, in peristerite feldspars, which have alternating albite and oligoclase lamellae, deep grooves may form by preferential dissolution of one of the feldspar components (e.g., Nugent et al., 1998a, b).

Tiny, microscopic etch pits may nucleate at impurities such as point defects. Dislocations may result in sizable etch pits (see Chapter 5 for a discussion of dislocations). Dislocations have associated elastic strain, and dissolution in the region of the dislocation will thus be energetically favorable relative to dissolution at a flat terrace site. The strain energy density u(r) at a screw dislocation is given by (Blum and Lasaga, 1987)

$$u(r) = \frac{\mu_s b^2/(8\pi^2\kappa)}{r^2{}_h + r^2}, \tag{10.18}$$

where

μ_s = shear modulus (in ergs cm^{-3})
b = magnitude of the Burger's vector (in Å)
r_h = radius of the core region (in Å)
r = distance from the dislocation line (in Å)
κ = constant that describes the relative contribution of screw and edge components in the dislocation strain field.
$\kappa = 1$ for a pure screw dislocation and $\kappa = 1 - \nu$ for an edge dislocation, where ν is the Poisson ratio (generally ~0.25) (Blum and Lasaga, 1987).

Dissolution of a given volume of crystal structure under conditions of undersaturation results in a favorable decrease in Gibbs free energy (ΔG_{vol}). However, when a pit forms, this favorable effect is countered by the formation of a new surface, which results in an increase in surface free energy (ΔG_{surf}). The elastic strain energy (ΔG_{str}) that is associated with a dislocation is available to help offset the energy associated with formation of a new surface after nucleation of a pit. When an etch pit nucleates at a dislocation, the overall free energy change after dissolution at the dislocation (ΔG_{dis}) will be

$$\Delta G_{\text{dis}} = -\Delta G_{\text{vol}} + \Delta G_{\text{surf}} - \Delta G_{\text{str}} \tag{10.19}$$

(MacInnis and Brantley, 1992).

The greatest strain energy associated with a dislocation is at the dislocation's core, and it decreases considerably by about 10 Å from the dislocation line (Blum et al., 1990). When a pit nucleates and begins to grow at a dislocation, the unfavorable formation of new surface area ($+\Delta G_{\text{surf}}$) will be countered by the favorable lowering of energy associated with overall dissolution in unsaturated conditions ($-\Delta G_{\text{vol}}$) and the release of energy related to strain ($-\Delta G_{\text{str}}$). As the pit continues to grow, the incremental increase in new surface area decreases. Thus, $+\Delta G_{\text{surf}}$ tends to decrease simultaneously with the pit growing beyond the region of highest strain. For highly undersaturated solutions, once a small pit nucleates at a dislocation, dissolution will continue to propagate outward across the unstressed portion of the crystal surface, forming a sizable etch pit. However, closer to saturation, a microscopic hole may form in the region of the dislocation, but dissolution will be restricted so that a macroscopic etch pit will not form (Lasaga and Blum, 1987).

The conditions under which etch pits form and the thermodynamic controls on etch pit formation remain the subjects of much study. For example, Teng et al. (2004) showed that dissolution of calcite could occur at pits or step edges or both, depending on the degree of undersaturation of the system with respect to calcite.

The formation of etch pits is an important topic of corrosion research in materials science and engineering. On the one hand, finding and preventing corrosion is crucial to fields ranging from aerospace engineering to computer science. On the other hand, purposeful and controlled corrosion can be a useful tool in emerging fields such as nanotechnology (Chapter 13). Very small cavities or spaces formed by etching can be used for nanofabrication or as tiny reaction chambers for molecular-level reactions.

Oxalate Promoted Dissolution of Hematite

As an example of mineral dissolution mechanisms, we shall discuss the dissolution of hematite (α-Fe_2O_3) by the simple organic ligand oxalate, which is commonly found in soils. Much of the seminal work on organic ligand-promoted dissolution of Fe(III)(hydroxides) was undertaken by Swiss chemist Professor Werner Stumm and his students and colleagues.

For the laboratory dissolution of simple oxides at a constant temperature, fairly simple and straightforward rate equations may be developed. Hematite may undergo proton-promoted dissolution, in which an activated complex forms when surface-protonation and Fe(III) is released to solution. The rate law for proton-promoted dissolution is

$$R_{\text{H}} = k_{\text{H}}[\text{H}_{\text{ads}}]^{\text{n}}, \tag{10.20}$$

where R_{H} is the rate, $[\text{H}_{\text{ads}}]$ is the activity of adsorbed protons in excess of the protonation state at the point of zero charge, k_{H} is the rate constant (s^{-1}), and n is the reaction order.

Hematite dissolution by oxalate can occur through a direct surface-controlled, ligand-promoted dissolution mechanism, in which oxalate binds directly to the oxide surface to form a surface chelate, and then removes the metal ion to solution (e.g., Stumm and Furrer,

1987). If the surface speciation remains constant over the experimental conditions, then the dissolution rate R_L is proportional to the concentration of surface-bound ligands, [>ML]

$$R_L = k_L[>\mathrm{ML}], \tag{10.21}$$

where k_L is the rate constant and >M is the Fe site on the hematite surface. The overall dissolution rate in an acidic solution of oxalate is then

$$R_{net} = (R_H + R_L) f \Delta G, \tag{10.22}$$

where $f\Delta G$ is a function of the solution saturation state with respect to hematite (Kraemer, 2004, and references therein).

Chapter 8 shows some examples of the correlation between dissolution rate and ligand concentration for a variety of simple organic ligands. Maurice et al. (1995) observed pit formation on the hematite surface in the presence of oxalic acid, using in-solution atomic force microscopy (AFM), consistent with surface-controlled dissolution.

Photochemical reductive dissolution of hematite can occur in the presence of oxalate and light, resulting in the formation of dissolved Fe(II) and oxidized oxalate. Fe(II)-catalyzed "thermal" dissolution of hematite can occur in which electrons are transferred from dissolved Fe(II) through the oxalate ligand to the surface Fe(III). As the surface Fe(III) is reduced, the resulting Fe(II) detaches easily from the surface. A reaction sequence proposed by Suter et al. (1991) for Fe(II)-catalyzed thermal dissolution of goethite is shown in Figure 10.9.

Many other examples of organic- and microbially mediated reductive dissolution of hematite and other Fe(hydr)oxides are available. See, for example, sections on microbial–mineral interactions (Chapter 12) and acid mine drainage (Chapter 14).

COMPARISON OF LABORATORY- AND FIELD-BASED DISSOLUTION RATES

Two basic approaches to studies of mineral weathering and dissolution are laboratory and field based. Laboratory experiments allow for a more controlled environment where individual processes may be probed. However, they do not account for the complexity of the "real world," in which a variety of different processes and mechanisms may occur over space and time. Field-based observations can provide field-applicable information more directly, but they are often more expensive and time-consuming, and sometimes it is difficult to estimate how long a mineral has been weathering. A fundamental problem for both approaches is that weathering operates over geological time scales that can be difficult to account for correctly in either field or laboratory experiments.

Because of their abundance and importance to many hydrobiogeochemical processes, feldspars have been the subjects of many field, laboratory, and comparative studies. Feldspar dissolution rates tend to vary in a complex fashion with pH. A commonly used equation for feldspar hydrobiogeochemical dissolution rate (R_d) is

$$R_d = k_H a_{H^+}^n + k_{H_2O} + k_{OH} a_{OH^-}^m, \tag{10.23}$$

from Brantley, 2003, where k_H, k_{H_2O}, and k_{OH} are the rate constants at acidic, near neutral, and basic pH, respectively; a_i is the activity of species i; and n and m are reaction orders.

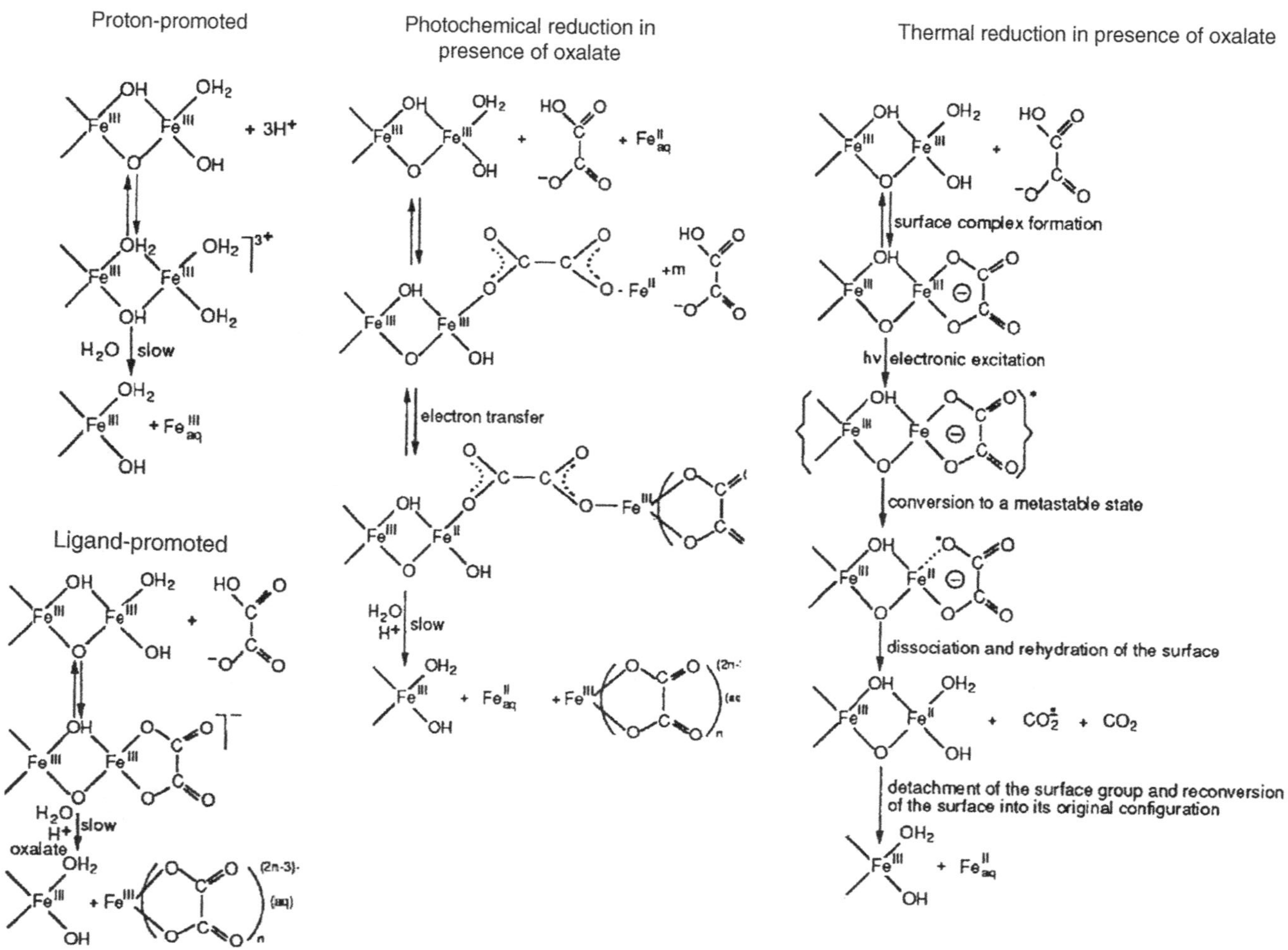

FIGURE 10.9 Some modes of hematite dissolution in the presence of oxalic acid. Revised from Stumm, 1992 (with permission).

This equation reflects a V-shaped curve of log dissolution rate versus pH, with very slow dissolution rates at near neutral pH and greater rates at acidic and basic pH.

The weathering rates of feldspars have often been found to be up to 10,000 times slower in the field than determined experimentally in the laboratory (Blum and Stillings, 1995). Factors that may contribute to this discrepancy include (Blum and Stillings, 1995) the following:

(1) *Time*, because laboratory studies are generally last from days to months, but field weathering can go on for millions of years.
(2) *Surface areas*, because we need to determine how *reactive* surface area relates to *geometric* or Brunauer, Emmett, and Teller (*BET)* surface area.
(3) *Defects* in the crystal structure and especially at the mineral surface. Cutting, grinding, polishing, and other preparation techniques for laboratory experiments may induce defects that affect dissolution rate.
(4) *Unsaturated hydrology*, because the degree of saturation can affect the reactive surface area exposed to water, cyclic evaporation may cause precipitation of certain phases, and micropores may have more static flow than macropores and the pore solutions may be less unsaturated.
(5) Effects of *solution chemistry, including saturation state and local pH*.
(6) The presence of *surface coatings*, such as Fe and Al oxides, organic coatings, and biofilms.
(7) *Biological effects*, as geobiological interactions in the field are poorly understood.

For studies in which the mineral substrates are the same in the laboratory versus the field, so that intrinsic factors are normalized and only extrinsic factors vary, the discrepancy between field and laboratory rates is about two orders of magnitude (Brantley, 2003). In many instances, both field and laboratory approaches are combined in a single study. One way of merging laboratory and field methods is the in situ microcosm approach, in which a reaction vessel that contains a mineral surface is buried in a field site and then exhumed for analysis. This approach is also discussed in Chapter 13.

Nugent et al. (1998a, b) applied a combination of scanning electron microscopy (SEM). AFM, secondary ion mass spectrometry (SIMS), X-ray photoelectron spectroscopy (XPS), and transmission electron spectroscopy (TEM) with energy-dispersive X-ray electron spectroscopy (EDS) to analyze the surfaces of feldspar samples that were emplaced in acidic soils in in situ reactors for up to several years. As shown in Figure 10.10, AFM revealed that fine particles accumulated on the feldspar surface. The presence of these particles had to be taken into consideration to interpret XPS and SIMS analyses correctly. Such particles could potentially affect the dissolution rate of the underlying surface.

Reactive Surface Area and Feldspar Dissolution

The weathering rate (R_d) in mol m^{-2} s^{-1} of feldspars and other silicates is generally inversely related to surface area, S (m^2) and time, t (in seconds), according to

$$R_d = Q/(St), \tag{10.24}$$

where Q (mol) is the moles of material reacted (White, 2004). Calculation of feldspar dissolution rates requires determining what fraction of the *total surface area* is reactive.

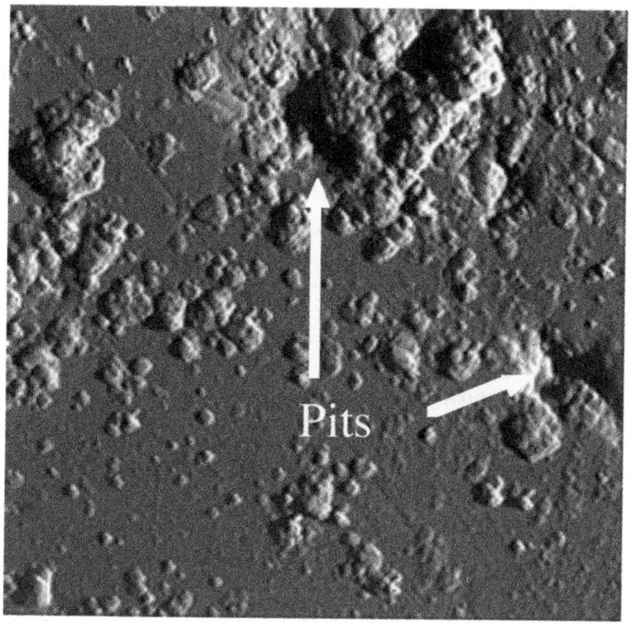

FIGURE 10.10 Tapping mode AFM (TMAFM) amplitude image of weathered, pitted persiterite feldspar, partially covered with a natural coating. Image is 10 μm on a side. From Nugent et al., 1998a.

White and Brantley (2003) conducted column experiments for more than 6 years in which the starting materials were fresh and field-weathered granites from Panola Mountain, Georgia. The surface areas of the fresh granites increased by a factor of 3.5, whereas those of field-weathered granites increased by only a factor of 1.2. The experimental plagioclase weathering rate for the weathered Panola granite sample was 2.2×10^{-15} mol m^{-2} s^{-1}, which was more than two orders of magnitude slower than that of the fresh granite sample, 7.0×10^{-14} mol m^{-2} s^{-1}, but more than one order of magnitude faster than that of the field determined plagioclase dissolution rate at Panola Mountain, 2.8×10^{-16} mol m^{-2} s^{-1}.

Using field and laboratory data, White and Brantley (2003) showed that silicate surface roughness λ (the ratio of BET surface area divided by a geometric estimate of surface area) increases over time during weathering according to the relationship

$$\lambda = 13.6\,t^{0.2}. \tag{10.25}$$

This equation was used to account for changes in surface area in previously reported felspar dissolution studies, when not directly measured. By plotting plagioclase weathering rates from field and laboratory studies against weathering duration (Figure 10.11), they showed that plagioclase weathering rate (R_{dp}) decreases over time according to the equation

$$R_{dp} = 3.47 \times 10^{-13}\,t^{-0.56}, \tag{10.26}$$

where R_{dp} is in moles m^{-2} s^{-1}.

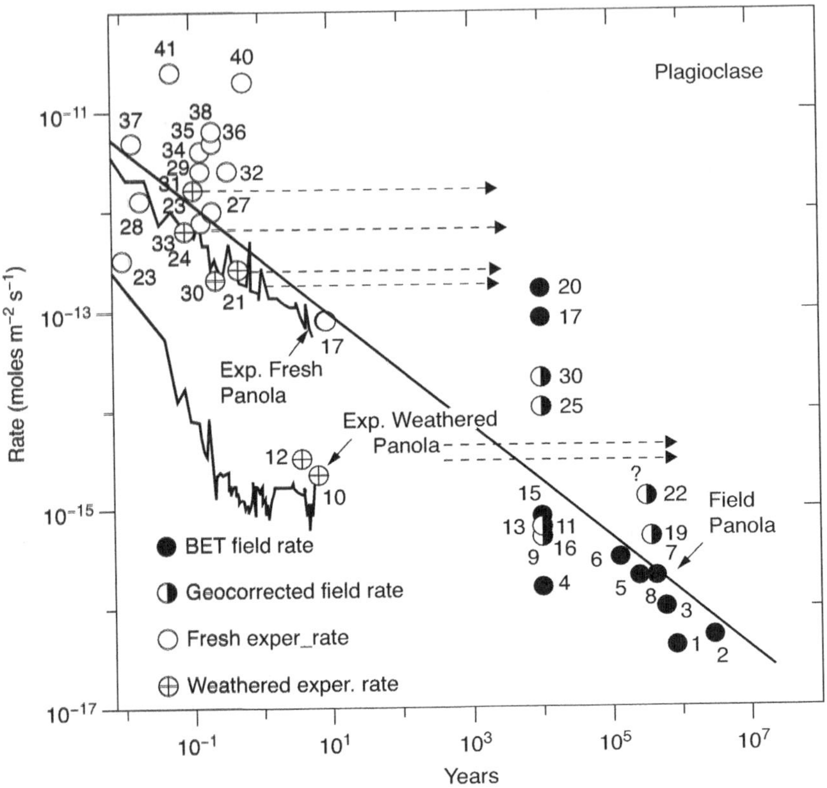

FIGURE 10.11 Experimental and field-based weathering rates of plagioclase feldspar as a function of time. The arrows extend to the approximate ages of naturally weathered samples that were subsequently weathered experimentally (in the laboratory). Field rates are based on the estimated age of the weathering environment. The symbols refer to rates based on surface areas determined by BET measurements or geometrically corrected. Figure from White and Brantley, 2003, who provide a key to the location/sample numbers (with permission).

Brantley et al. (1999) showed that fractal analysis (see Chapter 5) was helpful for understanding how specific surface area A depends on feldspar particle size r,

$$A \sim r^{D-3}, \tag{10.27}$$

where D is the fractal dimensionality.

Because laboratory studies are conducted over short periods of time, part of the discrepancy in field versus laboratory rates of feldspars is caused by the duration of weathering. Longer weathering times result in depletion of reactive surface sites, accumulation of leached layers and secondary precipitates, and other factors such as pore waters becoming more saturated. Dissolution of feldspars also can lead to development of fine-scale porosity that increases surface area but decreases the dissolution rate on a surface-area basis because the solution in the fine pores comes to equilibrium more quickly than the surrounding solution. These different processes and factors continue to be topics of study and debate.

RAINFALL AND WEATHERING: AN EXAMPLE FROM THE HAWAIIAN ISLANDS

The Hawaiian Islands offer a glimpse at the role of climate in chemical, physical, and biological weathering. As the remotest island chain on Earth, the Hawaiian Islands have formed as the Pacific Plate rides over a hot spot. The rock type (basalt) is similar from island to island. But the ages of rocks decrease from ~4 million years old on the northwestern island of Kauai to newly formed basalt from ongoing eruptions on the southeastern island of Hawaii. Pronounced gradients are evident in rainfall and temperature with elevation and/or leeward versus windward side. As the prevailing trade winds encounter mountainous terrain, orographic rainfall occurs most often on the windward side of the mountains but in some instances at the summit and/or the leeward side of the summit as well. On the island of Kauai, for example, the windward side receives some of the highest rainfall rates on Earth, often $>10{,}000$ mm y^{-1}, which results in erosion and steep-sided canyons (Figure 10.12), whereas the leeward side is dry. Individual, dated lava flows can be investigated across climate gradients (Figure 10.12).

Many studies of weathering and ecology in Hawaii have compared rocks of different ages, different climate scenarios, transects along single lava flows or across climate regimes,

FIGURE 10.12 Clockwise from upper left: Steep-sided, highly weathered canyons in the rainy section of Kauai; layered flows at ancient archeological site on Hawaii; basalt remaining from a recent lava flow that cross a street on Hawaii; highly weathered and vegetation portion of Maui. Steep gradients in rainfall and temperature, a rich diversity of microclimates, and dated lava flows make the Hawaiian Islands an excellent location for studying rates and mechanisms of weathering of basalt.

and other factors. Porder et al. (2007) studied total mass loss and chemical weathering along three Hawaiian lava flows of ~10 ka, 170 ka, and 350 ka age (ka = thousand years). Each flow crosses a precipitation gradient from ~600 to ~2500 mm y^{-1}. Taking into consideration wind-blown sand coming from Asia that accumulates in Hawaiian soils and assuming that similar rainfall gradients have occurred in the past, these authors found that weathering generally increases with increasing rainfall. A threshold occurs when rainfall exceeds evapotranspiration. Beyond this threshold, the weathering rates increase dramatically, but increasing rainfall still does not result in substantially faster weathering. Weathering rates are fastest on younger flows that receive high rainfall. As flows age, the weathering rates decrease in agreement with the general observation that silicate dissolution rates tend to decrease over time.

Case Study: Weathering in the Antarctic Dry Valleys

One philosophy for choosing a field site is to select an "end-member site" that highlights one or more processes of interest. The Antarctic Dry Valleys represent an end-member site because they contain no land plants, are one of the coldest locations on Earth, and water flow is restricted to a few weeks during the Austral summer when glacial meltwater flows through stream channels and permeates the hyporheic zone (Figure 10.13). Generally, the rates of mineral dissolution and weathering tend to be faster in warmer and wetter locations than in colder and/or drier locations. But the Dry Valleys highlight some unique processes.

Research by U.S. Geological Survey geochemist Alex Blum and Professors Diane McKnight (University of Colorado), W. Berry Lyons (Ohio State University), and their colleagues has focused on determining the rates of chemical weathering in the Antarctic Dry Valleys (Figure 10.13). Because of the extreme cold, lack of water, and absence of land plants, one would anticipate very little chemical weathering or dissolution of minerals. However, during the Antarctic summer, which only lasts a few weeks, glaciers melt and the meltwater flows down streambeds. The meltwater can permeate into the *hyporheic zone*, which is the zone of sediments directly underlying the stream.

Weathering occurs in the hyporheic zone in the Antarctic Dry Valleys despite the cold climate. Solute concentrations of streams tend to increase downstream (Chin, 1993), which suggests as exchange with hyporheic zone water where weathering may occur. Hydrogeochemical modeling by Lyons et al. (1997) coupled with analysis of K:Cl versus Ca:Cl ratios in lakes suggest a continental source of these cations rather than a marine-salt source. Lyons et al. (1997) showed that $^{87}Sr/^{86}Sr$ isotopic ratios of stream waters are consistent with weathering of intrusive igneous rocks. *Gibbs plots* based on stream composition suggest that both chemical weathering and atmospheric precipitation contribute to the dissolved loads of streams.

To investigate weathering in the hyporheic zone, an in situ microcosm approach was used in which pieces of muscovite mica were freshly cleaved and buried for the 3-month Antarctic summer in reaction vessels that consisted of holey bottles in the hyporheic zone of Green Creek, which is a McMurdo Dry Valley stream. Despite the short exposure time, AFM of the basal-plane surfaces revealed small, anhedral ~10-Å deep etch pits covering ~4% of the surfaces, along with colonization by bacteria. Based on the etch-pit geometries and geometric surface area, an approximate basal-plane dissolution rate of 8.13×10^{-18} mol muscovite cm^{-2} s^{-1} was calculated. If dissolution were stoichiometric, then the Si

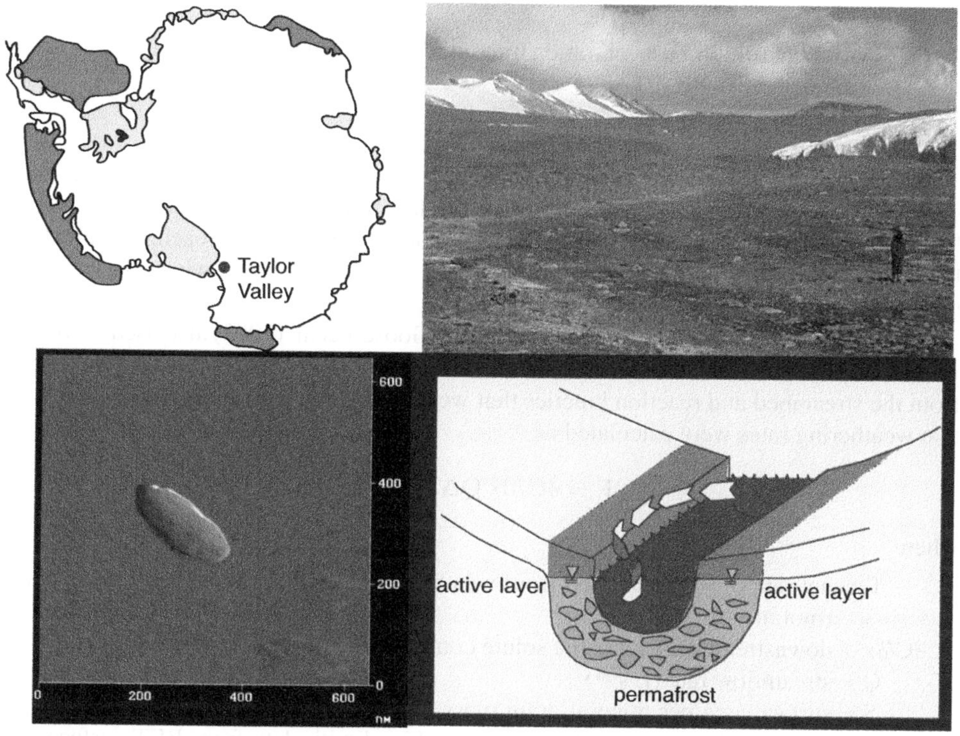

FIGURE 10.13 Chemical weathering in the hyporheic zone of an Antarctic Dry Valley Stream. Clockwise from upper left: Location of Taylor Valley, Antactica; a photo of Dry Valley sediments and glacial meltwaters; a model of the hyporheic zone and hyporheic exchange in the Dry Valleys; and an AFM image of a bacterium on the surface of a mica sample that had been buried in the hyporheic zone. AFM image from Maurice et al., 2002. Other subfigures courtesy of D. McKnight (University of Colorado at Boulder).

dissolution rate would be 2.5×10^{-17} mol cm^{-2} s^{-1}. This rate is within an order of magnitude of some laboratory-based muscovite dissolution rates (e.g., Nickel, 1973 in Kalinowski and Schweda, 1996, and Lin and Clemency, 1981a,b). Bacteria (α-proteobacteria were determined from analysis of sediment samples) were attached to the surfaces, and bacteria might have enhanced dissolution to obtain nutrients such as Fe and K. XPS analysis suggested a small amount of Fe depletion on the mica samples, along with accumulation of N and C consistent with biofilm formation. Alternatively, or in addition, the bacteria might have been using the mica surfaces simply as solid substrates.

Gooseff et al. (2002) investigated chemical weathering at a nearby stream based on analysis of streambed sediment mineralogic composition (by XRD) and downstream changes in Si and K concentrations. First, they used a static mineral surface approach to calculate the *land surface normalized weathering flux* (LWF). In this approach, data on chemical compositions of stream waters at specific points were used to calculate LWF (in moles m^{-2} y^{-1}) values according to the formula

$$LWF = C_{avg}Q_{ann}/A_w, \quad (10.28)$$

where

C_{avg} = average solution concentration measured at a given stream sampling location (M)
Q_{ann} = annual streamflow ($L\ y^{-1}$)
A_w = watershed area (m^2).

Given that there is no precipitation contribution to streamflow, only meltwater from glaciers, the A_w was taken to be the area of the hyporheic zone, estimated from stream length and active stream channel/hyporheic zone width. Using this approach, the calculated LWF values ranged from 1.1×10^{-2} to 2.4×10^{-2} mol Si $m^{-2}\ y^{-1}$, where the m^2 refers to land surface area.

To obtain rates relative to mineral surface area, Gooseff et al. (2002) also used a *constant weathering contribution* (CRC) model that assumed constant flux of solute to the stream from the streambed and reaction kinetics that were slow relative to the transport rate, such that weathering rates were calculated as

$$\Gamma = \delta C/dx\ Q\Delta x/m_{sed}S, \tag{10.29}$$

where

Γ = net rate of solute derived from weathering normalized to mineral surface area (mol $m^{-2}\ s^{-1}$)
$\delta C/\delta x$ = downstream change in the solute concentration (mol $L^{-1}\ m^{-1}$)
Q = streamflow rate ($L\ s^{-1}$)
S = surface area per mineral grain mass ($m^2\ g^{-1}$), where geometric surface areas of minerals were calculated and combined with data from BET surface area measurements
Δx = finite different spatial step (m)
m_{sed} = mass of the sediment, calculated from

$$m_{sed} = (1 - \phi)\rho_x \Delta x A_c, \tag{10.30}$$

where ϕ is sediment porosity ($cm^3\ cm^{-3}$) and A_c is the cross-sectional area of active hyporheic zone sediments exposed to weathering and water exchange with the stream (m^2) (Gooseff et al., 2002).

Finally, *hydrologic exchange* (HE) modeling was conducted based on the OTIS model, which includes transient storage and chemical weathering. The details of this approach (Gooseff, 2002) are beyond the scope of this text. However, the generalized one-dimensional transport with inflow and storage (OTIS) model is described in Chapter 2.

Gooseff et al. (2002) calculated that the average rate of Si supplied by hyporheic zone weathering over a 5.2-km stream length was 6.1×10^{-9} mol Si $L^{-1}\ m^{-1}$. Using the CRC and HE methods, they determined Si dissolution rates as a function of mineral surface area ranging from 0.6×10^{-12} mol Si $m^{-2}\ s^{-1}$ to 4.5×10^{-12} mol Si $m^{-2}\ s^{-1}$. It is not suprising that this rate is faster than that calculated from the muscovite in situ microcosm experiment, because muscovite dissolution is generally slow relative to that of many other aluminosilicates.

The field-observed rate of Gooseff et al. (2002) is fast, even relative to Si dissolution rates observed at study sites with much wetter and warmer climates. According to White and Brantley (2003), the rates of plagioclase weathering based on field studies in the United States and Europe are generally in the range of 10^{-16} to 10^{-12} mol $m^{-2}\ s^{-1}$. The relatively fast Dry Valley rate is counter to expectations based on the cold temperature and lack of plants, but it is perhaps caused by continuous saturation and rapid flushing of hyporheic sediments during the Austral summer (Gooseff et al., 2002). Where liquid water is available and flowing in the Antartic Dry Valleys, chemical weathering can be rapid most likely because of hydrologic factors that involve interaction between the hyporheic zone and the stream.

REACTORS FOR DISSOLUTION EXPERIMENTS

Many laboratory studies of mineral dissolution are conducted in which a mineral is reacted with a given solution, and solution concentrations of various reactants and/or products are monitored over the course of the reaction. The experiments can be conducted in either *batch* or *flow-through* reaction vessels.

Batch reactors are the simplest types of reactors, and they consist of a vessel such as a beaker, a flask, or a bottle that is either shaken, stirred (Figure 10.14), or in the case of a capped bottle or test tube, mixed on a rotary shaker. The experiments may be set up in a temperature-controlled room or incubator; alternatively, the reaction vessel may be double jacketed to allow circulation of water from a temperature-controlled bath. In a typical experiment, a suspension is created, the pH is adjusted, and the system is allowed to react toward equilibrium. For biotic experiments, the solution and mineral components may be sterilized, and aseptic techniques may be used. Microorganisms can be added to the vessels, along with nutrients appropriate for microbial growth. Batch reactors may be open or closed to the atmosphere, and they may be continuously purged using nitrogen or another inert gas, or kept under specific conditions of O_2 and/or CO_2.

Reaction progress is monitored either by removing small aliquots or by sacrificing replicate samples at different time intervals. In the former approach, the change in sample volume must be taken into consideration. For studying adsorption, one set of kinetic experiments generally is conducted to establish equilibration time. Then, additional experiments

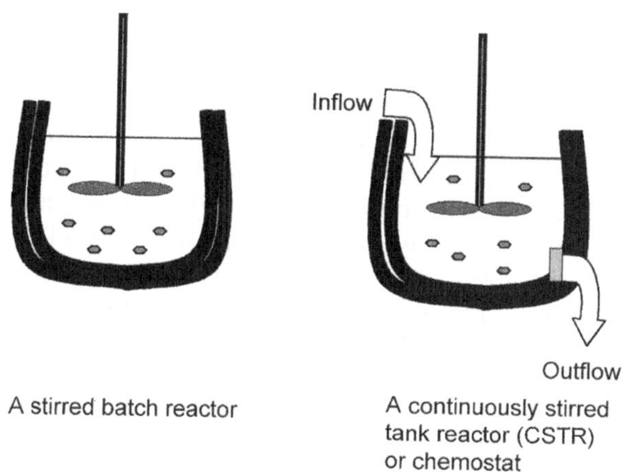

FIGURE 10.14 Comparison of a simple stirred batch reactor versus a somewhat more complex continuously stirred tank reactor.

are run at that equilibration time. At the end, the mineral is separated from the solution by filtration or centrifugation (both of which can impose potential artifacts), and the amount adsorbed is calculated by difference of the initial versus reacted solution conditions.

The dissolution rate from batch experiments can be calculated from the linear portion of the dissolution curve using

$$R = \frac{dn/dt}{\nu A_s m}, \tag{10.31}$$

where

n = moles of element (such as Fe, Si, Al) released to solution
ν = stoichiometric coefficient of the element in the mineral
t = time
A_s = specific surface area of mineral in $m^2\ g^{-1}$
m = mineral mass in g at $t = 0$ (Brantley et al., 2006).

To calculate the change in moles over time *dn/dt* when small aliquots are removed at different time intervals, the following equation can be used (Brantley et al., 2006) for *n* moles at different time periods *t*:

$$n_t = n_{t-1} + ([i]_t - [i]_{t-1})V_{t-1}, \tag{10.32}$$

where

n_t = moles of an element (e.g., Fe, Al, and Si) at time t
$n_{t\text{-}1}$ = moles of the element at time $t - 1$
$[i]$ = concentration in the medium in the flask in moles per liter at time t or $t - 1$
V_{t-1} = volume of medium in the flask in liters at time $t - 1$.

Measuring dissolution rate in a batch system is not optimal because the system is changing over time both in terms of the solution and the mineral surface. The surface of the mineral may change in structure, composition, and surface area. As the system approaches overall equilibrium, the data can become less representative of dissolution rate because reaction products accumulate and back reactions become more pronounced. Secondary precipitates may form. Batch systems (Figure 10.14) are generally used for adsorption experiments, for many microbial experiments, and for some dissolution experiments when rate and mechanism are not the primary focus.

In many instances, flow-through systems (Figure 10.14) provide more mechanistically useful data. Flow-through reactors allow the experimentalist to maintain constant solution conditions, which facilitates data interpretation especially for dissolution experiments.

The *mixed-flow reactor* (or *continuously stirred tank reactor*, CSTR; Figure 10.14) was developed for mineral dissolution studies by Rimstidt and Dove (1986). The mineral of interest is placed in a stirred reaction vessel of volume V_o through which a solution is pumped at a constant flow rate of Q. The input solution is kept to a constant concentration of reactants and a set pH, ionic strength, and so on, and the system is temperature controlled. The effluent is sampled continuously or at various sampling intervals to determine reaction rate. The rate of the reaction, R (mol $m^{-2}\ s^{-1}$) is calculated from:

$$R = \frac{Q([i]_o - [i]_i)}{A_s m}, \tag{10.33}$$

where

Q = flow rate
$[i]_o$ = outlet or effluent concentration
$[i]_i$ = inlet concentration
A_s = specific surface area of the mineral
m = mass of the mineral (Brantley et al., 2006).

Steady-state dissolution is defined when the outlet concentration becomes constant. Changes in pH or mineral surface area must be taken into account.

A type of CSTR used in (geo)microbiology is a *chemostat* (from the chemical environment is static). A chemostat consists of a nutrient reservoir and a growth chamber. By controlling the concentration of a limiting nutrient, the bacterial culture can be maintained growing under carefully controlled conditions over long periods of time at a specific growth phase. All the microbial parameters, such as the volume of the growth chamber, the cell density, the concentration of dissolved oxygen, the pH, and the nutrient concentrations remain constant throughout the experiment. By operating under flow conditions, the metabolic wastes are removed so that the cells do not become poisoned by accumulation of waste products over time. Temperature control is an important component that helps to maintain the culture in a particular metabolic state and constant cell numbers over time. A photograph of a chemostat from Pacific Northwest National Laboratory is shown in Figure 10.15.

FIGURE 10.15 A chemostat. Courtesy of E. Hill, Microbial Cell Dynamics Laboratory, Pacific Northwest National Laboratory.

A third type of reactor that has been used less frequently for dissolution studies is the *plug flow* or *column reactor*. Particles (generally the size of sand) are packed into a column and the solution is introduced generally by gravity flow. These reactors may more closely resemble actual porous media through which water and solutes are flowing. The equation used for a CSTR can also be used for columns; however, (1) solution conditions change with distance in the column and (2) effluent concentrations represent only the conditions passing out of the column, not through the entire column over time. These problems can be diminished by running the columns over long periods of time so that the gradients within the column are minimized. Over long periods of time, it is often possible to buffer pH by controlling CO_2, depending on the system and the target pH.

Field and laboratory studies can be brought together, to some extent, with the use of in situ *microcosms* in which a reaction vessel with a mineral sample of interest is embedded at a field site. An in situ microcosm can be a simple holey bottle buried in soil or sediment or a more complex vessel that pumps solution from a given source such as a river or a hydrothermal lake through a packed column. Some examples of in situ microcosms are shown in Chapter 12.

An important component of dissolution studies is incorporation of biological processes, which is the subject of Chapters 11 and 12.

THE USE OF RADIOGENIC ISOTOPES IN WEATHERING STUDIES

The use of radiogenic isotope analysis for interpreting rates and mechanisms of weathering is an evolving discipline, and thus it is only briefly discussed here. One of the most commonly used isotopic ratios is $^{87}Sr/^{86}Sr$. ^{86}Sr is a natural stable isotope that is not radiogenic. ^{87}Sr is stable but is radiogenic, forming from decay of ^{87}Rb, which has a $t_{1/2}$ of 4.88×10^{10} years (see Chapter 1). Although Sr has sometimes been considered to be a surrogate for Ca and thus indicative of Ca weathering from minerals, White et al. (1999) showed that Sr release during weathering is nonstoichiometric relative to major elements and that the nonstoichiometry may be sensitive to climate. Brantley et al. (1998) showed that Sr release from feldspars may be related to defects or inclusions, and thus again it is not necessarily indicative of overall feldspar weathering, at least during early stages of weathering.

Erel et al. (2004) suggested that because changes in the isotopic compositions of Sr and Pb released by weathering of granitoid rocks reflects release from different mineral components that have different isotopic ratios, isotopic ratios could be used to map different weathering stages. In a study of granitoid weathering, they found that changes in $^{87}Sr/^{86}Sr$ release over time reflected changing release from interlayer sites in biotite mica, from traces of calcite/aragonite associated with the rock, and from plagioclase feldspar weathering. Overall, the use of radiogenic isotope ratios in chemical weathering studies highlights the complexities of weathering processes and demonstrates that many additional years of research are needed.

QUESTIONS FOR FURTHER THOUGHT

1. Why is it important to discuss mineral dissolution rates in terms of surface area (per meters squared) rather than in terms of weight (grams) of mineral?
2. You have been given the assignment of determining how the presence of phosphate affects goethite dissolution by oxalic acid. Write a 4–6-page (double-spaced) summary

of experimental design with methodology. Assume that any methods needed are available.

3. What are the advantages and disadvantages of a flow-through reactor for mineral dissolution experiments versus a batch reactor?
4. Read one of the following papers on dissolution and summarize it in a 2–3-page (double-spaced) essay. Be sure to describe the various methods that were used and how each method contributed to understanding.

   Nugent, M., Brantley, S., Pantano, C., and Maurice, P. 1998. The influence of natural mineral coatings on feldspar dissolution. *Nature* 395, 588–591.

   Rufe, E., and Hochella, M.F., Jr. 1999. Quantitative assessment of reactive surface area of phlogopite during acid dissolution. *Science* 285, 874–876.

   The following paper is in the field of corrosion in materials science and gives a different perspective, but it is equally interesting.

   Punckt, C., Bolscher, M., Rotermund, H.H., Mikhailov, A.S., Organic, L., Budiansky, N., Scully, J.R., and Hudson, J.L. 2004. Sudden onset of pitting corrosion on stainless steel as a critical phenomenon. *Science* 305, 1133–1136.
5. Over time, stability constants in the literature change as new data are collected.
   (a) Construct a diagram showing the stability relations among K-feldspar, mica, kaolinite, and gibbsite at 25°C and 1 atm, as functions of $\log[K^+]/[H^+]$ and $\log[H_4SiO_4]$. Note that pyrophyllite is neglected in this exercise. Use the reactions and constants given below.

   (b) Next, compare the stability fields in your diagram with that of White (1995).

Kspar-muscovite

$$3KAlSi_3O_8 + 2H^+ + 12H_2O \rightleftarrows KAl_3Si_3O_{10}(OH)_2 + 6H_4SiO_4 + 2K^+$$
$$\log K = -13.4$$

muscovite/kaolinite

$$2KAl_3Si_3O_{10}(OH)_2 + 2H^+ + 3H_2O \rightleftarrows 3Al_2Si_2O_5(OH)_4 + 2K^+$$
$$\log K = 10.0$$

kspar-kaolinite

$$2KAlSi_3O_8 + 2H^+ + 9H_2O \rightleftarrows Al_2Si_2O_5(OH)_4 + 4H_4SiO_4 + 2K^+$$
$$\log K = -5.6$$

muscovite-gibbsite

$$KAl_3Si_3O_{10}(OH)_2 + H^+ + 9H_2O \rightleftarrows 3Al(OH)_3 + 3H_4SiO_4 + K^+$$
$$\log K = -10.8$$

kaolinite-gibbsite

$$Al_2Si_2O_5(OH)_4 + 5H_2O \rightleftarrows 2Al(OH)_3 + 2H_4SiO_4$$
$$\log K = -10.5.$$

   Note also that a solution will be saturated with respect to amorphous silica at $\log[H_4SiO_4] = -2.7$ and with respect to quartz at $\log[H_4SiO_4] = -4$.

   (c) What implications do the differences in thermodynamic values used in your calculations versus by White (1995) have for geochemists studying waters in equilibrium with aluminosilicate minerals?

   (d) What implications are there of not considering all solid species and reactions?

   (e) Why is it important to use up-to-date databases of thermodynamic values?

6. Why should the base sodium hydroxide (NaOH) never be stored in a glass bottle?
7. A geochemist studying the dissolution rate of kaolinite conducts experiments using two different samples (from different source localities) of Clay Minerals Society kaolinite standards. The dissolution rates of the two samples seem to differ by almost an order of magnitude.
   (a) What factors might contribute to the different dissolution rates?
   (b) Suggest a set of experiments to determine which factors are most important, and to what degree?
8. A geochemist runs batch dissolution experiments of kaolinite in 0.001 M oxalic acid at pH 3. The reaction vessels are centrifuge tubes that are capped and placed on a rotating shaker at 22°C. Separate centrifuge tubes are used and sacrificed at each reaction time period: 1, 2, 6, 12, 24, and 48 hours (rather than aliquots being removed from a single reaction vessel). The kaolinite used has a BET surface area of 22.2 m^2 g^{-1}, 0.022 g of kaolinite and 25 mL of solution are present in each reaction vessel. Calculate the Si-based kaolinite dissolution rate R_{Si} based on the linear portion of the plot of Si concentration in solution over time (i.e., when steady state is achieved), in moles Si cm^{-2} s^{-1}. (Hint, use Equation 10.31).

Time (hours)	Si (ug L^{-1})
1	26.20
2	34.26
6	64.77
12	81.11
24	97.99
48	126.31

9. A student runs a dissolution experiment using K feldspar in a continuously stirred tank reactor. The influent concentration of Si is 0.28 ug L^{-1} and the effluent concentration is 5.6 ug L^{-1} at steady state. The flow rate is 0.01 mL s^{-1}. Ten grams of feldspar are used, with a surface area of 2 m^2 g^{-1}.
   (a) Neglecting change in surface area and weight of feldspar, what is the feldspar dissolution rate in moles Si m^{-2} s^{-1}?
   (b) Comment on how one must consider the flow rate versus the mass (weight) of feldspar used in an experiment of this type, to ensure effluent Si concentrations detectable by current methods. Note: You may need to do some research on methods of Si analysis and pump speeds to answer the second part of this question.
10. Plot how feldspar surface roughness and dissolution rate can be expected to change from 1,000 to 10,000,000 years of weathering duration. (Hint: You will need to consider which format is best to use for the plot axes).
11. The small temperate zone watershed (6-km^2 surface area) introduced in Chapter 2, McDonalds Branch basin, has an average streamflow of 6.4×10^4 cm^3 s^{-1} and the stream outflow had an average Si concencentration of 3.9 mgSi L^{-1} over a 10-year time period.
   (a) Calculate the LWF in mol Si m^{-2} y^{-1}.
   (b) How does this LWF compare with that calculated by Gooseff et al. for an Antarctic Dry Valley stream?

FURTHER READING

Brantley, S.L., Ruebush, S., Jang, J.-H., and Tien, M. 2006. Analysis of (bio)geochemical kinetics of Fe(III) oxides. In Maurice, P.A., and Warren, L.A. (eds), Methods for Study of Microbe-Mineral Interactions. vol. 14, CMS Workshops Lectures. pp. 79–116.

Drever, J.I. (ed). 2005. Surface and Ground Water, Weathering, and Soils, vol. 5: Treatise on Geochemistry. Elsevier, New York, 644 p.

White, A.F., and Brantley, S.L. (eds.) 1995. Chemical Weathering Rates of Silicate Minerals. Reviews in Mineralogy, vol. 31. Mineralogical Society of America, Chantilly, VA, 583 p.

11

PLANTS AS ENVIRONMENTAL SURFACES

Plants represent an important component of the biosphere within the critical zone. They both affect and are strongly affected by the hydrologic cycle, which is the basis for much of the modern field of ecohydrology. In this chapter, we also discuss some ways in which plants can affect processes such as mineral weathering, development of cave formations, and remediation of inorganic and organic pollutants. Many important hydrobiogeochemical interface interactions occur within the *rhizosphere*. This region contains soil, roots, and microorganisms and extends to a few millimeters from the root surface.

There are currently 12 divisions of plants, which are arranged into the following four groups:

1. Bryophytes, which are the nonvascular plants
2. Seedless vascular plants, which reproduce by spores
3. Gymnosperms, which are "naked seed" plants—cone-bearing trees are gymnosperms
4. Angiosperms, which are the flowering plants (see Vaccari et al., 2006).

This chapter focuses primarily on the *angiosperms*, as they have been the dominant land plants for ~100 million years.

ECOHYDROLOGY AND SOIL MOISTURE BALANCE

Plants link the atmosphere, the biosphere, and the hydrosphere, both affecting and affected by climate and water cycling. Thus, an important element of the modern field of *ecohydrology* is plant–climate–soil dynamics.

As water passes through canopy vegetation, some of it is intercepted by branch and leaf surfaces, and it is either evaporated back to the atmosphere, taken up by certain epiphytes (in

Environmental Surfaces and Interfaces from the Nanoscale to the Global Scale, by Patricia A. Maurice

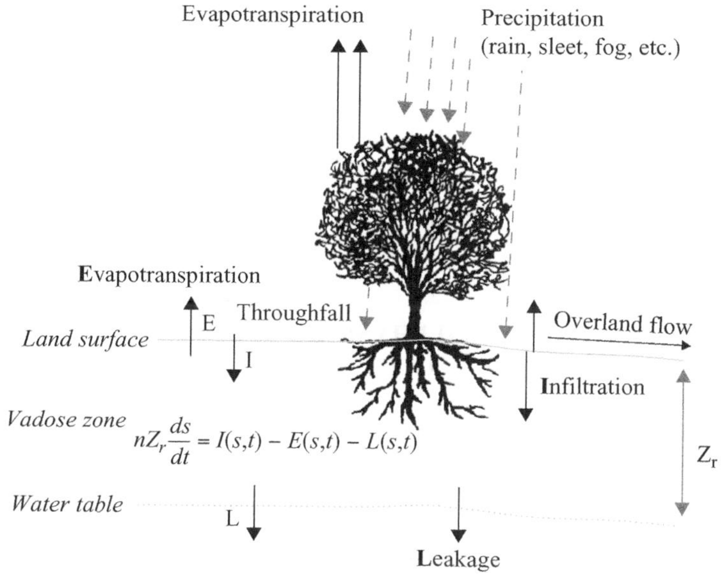

FIGURE 11.1 Land plants, water, and the soil water balance: some fundamental components of ecohydrology.

regions where they grow), or passed through as "throughfall" (Figure 11.1). Together with direct precipitation, throughfall contributes moisture to the underlying soils. A key area of study in hydrology relates to the *soil moisture balance equation*, which in the absence of pronounced topography is written as (Rodriguez-Iturbe, 2000)

$$n_p Z_r \frac{ds}{dt} = I(s,t) - E(s,t) - L(s,t), \tag{11.1}$$

where

n_p = soil porosity
Z_r = soil depth
s = relative soil moisture content
t = time
$I(s, t)$ = infiltration into the soil (primarily of water from precipitation)
$E(s, t)$ = evapotranspiration
$L(s, t)$ = leakage, or loss of water from the soil into the underlying environment.

This equation relates the soil moisture changes over time for a given soil porosity and depth to the input of water from precipitation minus the output of water through evapotranspiration and leakage into the underlying subsurface. Although this equation may seem fairly simple, it can be highly complex to solve because I, E, and L are all functions of the relative soil moisture content. For example, the amount of precipitation that infiltrates into the soil versus running off over the land surface varies depending on how saturated the soil is.

The soil moisture equation is fundamental to the remainder of this chapter because plant physiology is linked directly to the availability of water, primarily in soils. In areas with ample rainfall, such as tropical or temperate rainforests, plant growth is not limited by water and is dependent on the availability of nutrients. In more arid regions, however, water availability can be a major limiting factor. Plant types and distributions can be linked closely to hydrologic conditions (especially in the long term), and vegetation is often used as a component of remote sensing in hydrology. Under conditions of water stress, plants tend to decrease both transpiration and photosynthesis, thus impacting climate and carbon dioxide flux. Soil moisture can play an important role in many surface/interface processes that range from mineral weathering (Chapter 11) to soil microbial community structure (Chapter 13).

SOME NOTES ON ANGIOSPERM PHYSIOLOGY

Plant roots provide an anchor, may be used in storage of sugars, and take up water and minerals. In angiosperms, water, minerals, and food are transported through the *xylem* and the *phloem*. Leaves are the primary location of photosynthesis in plants, with the net reaction:

$$6CO_2 + 6H_2O \xrightarrow[h\nu]{} C_6H_{12}O_6 + 6O_2. \tag{11.2}$$

The epidermis of leaves contains pores known as *stomata* that admit CO_2. The stomata are also the location of much of the transpiration of water from the plant.

THE NUTRIENT NEEDS OF PLANTS

Because of photosynthesis, plants do not require an organic food source; in other words, they are *autotrophic*. C, H, and O, which are incorporated during photosynthesis, are not generally considered to be mineral nutrients, although they are major components of plants. Plants need to take up 13 *essential nutrients* from the soil pore waters for use in plant growth, development, or reproduction: N, P, K, Ca, Mg, Si, Fe, Cl, Zn, Mn, Cu, B, and Mo (ITRC, 2001). Some plant species also need Na, Co, I, and V (Lucas, 2001). An element is considered to be "essential" if the plant will fail to complete its life cycle if the element is not present at a sufficient bioavailable concentration. Other nonessential inorganics, such as Pb, Cd, As, and Cl, can be taken up as well. Although some elements such as Al can potentially be toxic, they can be found in plants because of their abundance in soil solutions under many conditions (Lucas, 2001). Table 11.1 shows the typical composition of plants as compiled by Lucas (2001).

A few uses of key components and some of their uses are described here. As a component of *adenosine triphosphate* (ATP), *phosphorus* is essential for the conversion of light energy to chemical energy during photosynthesis. Phosphorus can also be used to modify the activity of various enzymes and can be used in cell signaling. Phosphorus is essential for the formation of many plant biomolecules. *Nitrogen* is an essential component of DNA and of all proteins. *Potassium* plays a key role in water regulation in plants by controlling the opening and closing of the stoma via a potassium ion pump mechanism. *Magnesium* takes part in ATP production and is a component of chlorophyll. *Iron* is essential for plant photosynthesis and occurs as an enzyme cofactor in plants. Iron deficiency (as well as

TABLE 11.1 The typical elemental composition of plants.

Element	Dry matter
C	~440 mg g^{-1}
O	~440 mg g^{-1}
H	~60 mg g^{-1}
K	5–80 mg g^{-1}
Ca	<60 mg g^{-1}
Mg	<50 mg g^{-1}
P	<50 mg g^{-1}
S	<15 mg g^{-1}
Na	<10 mg g^{-1}
Cl	<8 mg g^{-1}
Si	<100 mg g^{-1}
Al	<10 mg g^{-1}
Fe	20–600 ppm
B	2–800 ppm
Mn	10–600 ppm
Zn	5–250 ppm
Cu	5–50 ppm
Ni	0.05–5 ppm
Co	0.05–10 ppm
Mo	0.1–10 ppm

After Lucas, 2001.

deficiencies in some other nutrients or water) can result in *chlorosis*, which is a condition that occurs when leaves produce insufficient chlorophyll. It can be of considerable concern for agriculture, particularly in limestone terrains, because of the tendency of Fe(hydr)oxides to be insoluble at moderate to somewhat alkaline pH. Chlorosis can also occur in some acid-soil regions, because of formation of insoluble Fe-P compounds.

Uptake may be either passive through the transpiration mechanism or active through root transport membranes. Many constituents are taken from the soil porewaters, although some trace constituents are believed to be taken up directly from leaf surfaces. Once dissolved constituents enter the plant through the root, they can be transported throughout the plant for use or sequestered, for example in cell vacuoles, if potentially toxic. Plants can also act to prevent toxins from entering their systems through sequestration mechanisms on the plant root surfaces or in the soil.

Plant uptake of nutrients may show seasonal variations. For example, in a study near Santa Cruz (CA), White et al. (2006) showed that plant biomass is lowest in the late fall; it increases during the winter rainy season and then decreases over the dry summer. This pattern results in seasonal changes in biomass concentrations of major cations, as shown in Figure 11.2. The effect is most pronounced for K and Ca.

Because different plants have different tolerances in terms of sunlight, hydrologic characteristics, salinity, acidic or basic conditions, and nutrient levels, as well as different abilities to cope with potential toxins, plant community structure is tied closely to a wide variety of surface and interface interactions.

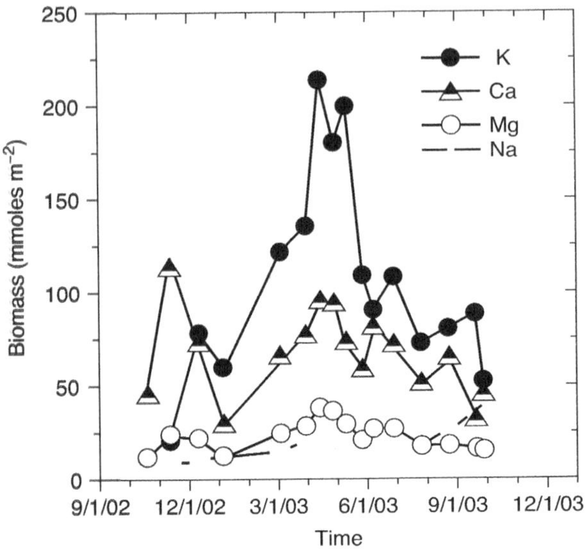

FIGURE 11.2 Changes in major cations in above-ground vegetation normalized to 1 m^2 of land surface for the 1-year growth cycle from November 2002 to October 2003. From White et al., 2006 (with permission).

EFFECTS OF PLANTS ON MINERAL DISSOLUTION AND WEATHERING

Elements such as C and N are present as gas phases in the atmosphere, and they enter terrestrial ecosystems either as wet or dry deposition (see Chapter 14) or through such biological processes as photosynthesis and nitrogen fixation. However, plants generally need to obtain many nutrients such as Ca, Mg, K, PO_4, Fe, and others listed above from the soil minerals, so that enhancement of weathering is essential for plant growth (e.g., Lucas, 2001; Berner et al., 2003; Likens, 2004). Reviews of plant effects on weathering are provided by Lucas (2001) and Berner et al. (2003). Plants also can derive some components, such as Ca and Mg, from aerosols including sea salts (Chadwick et al., 1999). Many of the plant interactions that affect weathering and the chemical composition of soil waters are shown in Figure 11.3.

Plants and the microorganisms associated with plant roots can increase mineral dissolution rates by exuding low-molecular-weight organic acids that lower the pH of the soil solution and that can form complexes with Fe and Al. Roots also can excrete protons or hydroxide ions that can alter the pH of the soil solution by as much as 2.5 pH units (Lucas, 2001). Roots can take up base cations (e.g., Ca^{2+}, Mg^{2+}, Na^+, and K^+) and excrete acid anions (e.g., SO_4^{2-} and NO_3^-). Respiration by plant roots and associated microorganisms releases CO_2 that increases the carbonic acid content of soil pore waters. As rainfall passes through the canopy, becoming *throughfall*, it can collect different constituents such as organic acids, SO_4^{2-} and K^+ depending on the vegetation type, the rainfall chemistry, and the composition of dry deposition (e.g., aerosols, particles, sea salts). The complexation ability of organic acids can increase weathering directly through surface-controlled, ligand-promoted dissolution as described in Chapter 10, or indirectly by altering saturation states

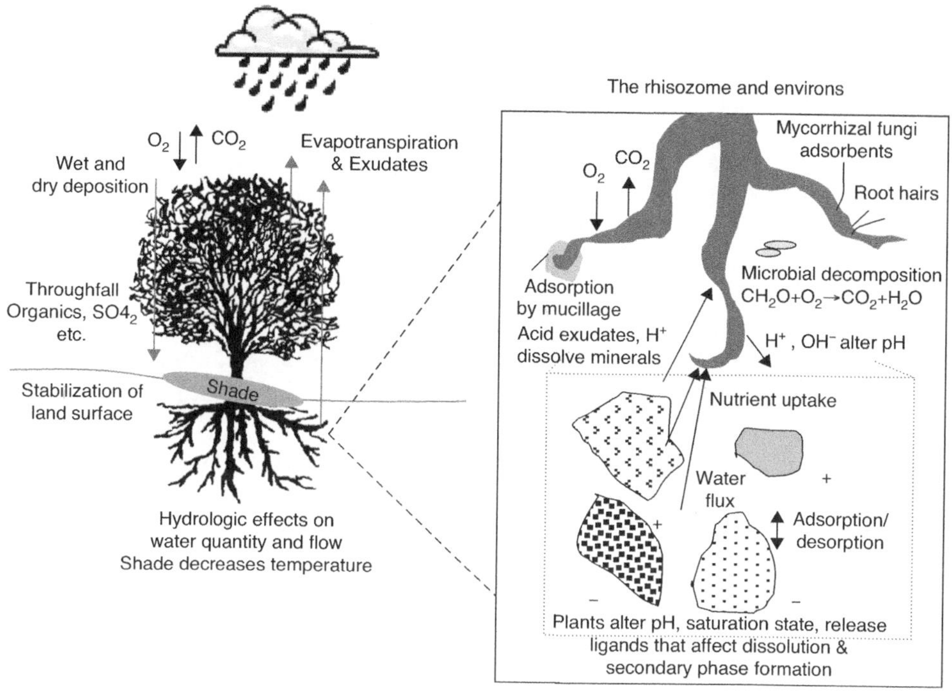

FIGURE 11.3 Some effects of plants on weathering and soil porewater composition.

and preventing formation of secondary weathering products such as aluminum hydroxides and clays (Berner et al., 2003).

Microbial breakdown of organic debris from plants in the uppermost soil horizons leads to CO_2 production that forms carbonic acid, which in turn accelerates mineral dissolution, especially of carbonates. Microbial degradation also releases natural organic matter (humic and fulvic acids; Chapter 8) that can complex metals and alter soil aggregation and fertility. Plants also have hydrologic effects, such as removal of water from soils by evapotranspiration, enhancing local precipitation by holding water on leaf surfaces that evaporates and leads to new rainfall, intercepting fog, and decreasing surface runoff. Shading by trees can affect soil temperatures and also weathering rates. Plants also can affect physical weathering, for example by decreasing erosion on slopes, altering stormflow, or exposing new mineral surfaces to chemical weathering. Plants can affect the thickness of soil cover, which can affect water–rock interactions and rates of weathering.

According to Berner et al. (2003), the net effect of plants is to increase aluminosilicate weathering rates by a factor of ~2 to 10. Angiosperms and gymnosperms appear to have similar effects. Berner et al. (2003) argue that lichens and mosses do not accelerate weathering to anywhere near the same extent as large vascular plants such as trees because (1) the contact area between small organims such as lichens and minerals is much less than that between massive root hairs and minerals, (2) the rate of growth of vascular plants is much greater and the need for nutrients thus much larger, and (3) lichen–metal complexing agents are rare and not very effective in weathering for a variety of reasons.

Processes such as fire and clear cutting by humans' hurricanes or volcanic explosions can further demonstrate the importance of plants. For example, human clear-cutting of

FIGURE 11.4 Erosion after deforestation of the island of Madagascar. Courtesy of R. Butler, wildmadagascar. org.

a portion of the Hubbard Brook Experimental Forest, New Hampshire (Likens, 2004) resulted in hydrologic changes including greatly decreased evapotranspiration and increased storm runoff accompanied by an increase in decomposition and nitrification. Effects of clear-cutting decreased over time as the forest floor was eventually revegetated, but changes to nutrient concentrations in streams were observed to last decades. Increased erosion and sediment transport because of deforestation is a well-known phenomenon, sometimes leading to such massive soil loss that the deforested area cannot recover for thousands of years or more. Erosion that occurs after deforestation on the island of Madagascar is of enormous proportions (see Figure 11.4).

The uptake of nutrients and non-nutrient components by plants needs to be considered in studies of the input/output budgets of chemical constituents from terrestrial ecosystems, especially forests. If a forest is at steady state, meaning that it is not growing or shrinking overall, the amount of new growth will be balanced by death and decay so that there will be no net effect on input/output budgets of nutrients. An actively increasing forest can take up nutrients; a decimated and decaying forest can lead to spikes in nutrients and sediments being released to streams through increased decay and erosion. In tropical forests, soils can be greatly depleted such that the bulk of nutrients is tied up in the biota. When tropical forests undergo slash and burn, the remaining nutrient-poor soils often can only support agriculture for a short time.

One generalized model for the evolution of terrestrial ecosystems and of the geological materials on which they are formed links duration of weathering with nutrient status and how nutrients are derived (Chadwick et al., 1999). Newly formed rocks, such as from basalt flows currently forming on the island of Hawaii, tend to be rich in mineral nutrients but relatively poor in atmospherically derived elements so that these systems generally are nitrogen limited. As the system ages, the soils are weathered, and they become depleted in elements such as Ca, Mg, K, and P, whereas atmospherically derived components are supplied continuously (Chadwick et al., 1999). In a study of ecosystem dynamics in the Hawaiian Islands, which are located in the Pacific Ocean far removed from continents, Chadwick et al. (1999) showed that plants growing on weathered soils can obtain cations

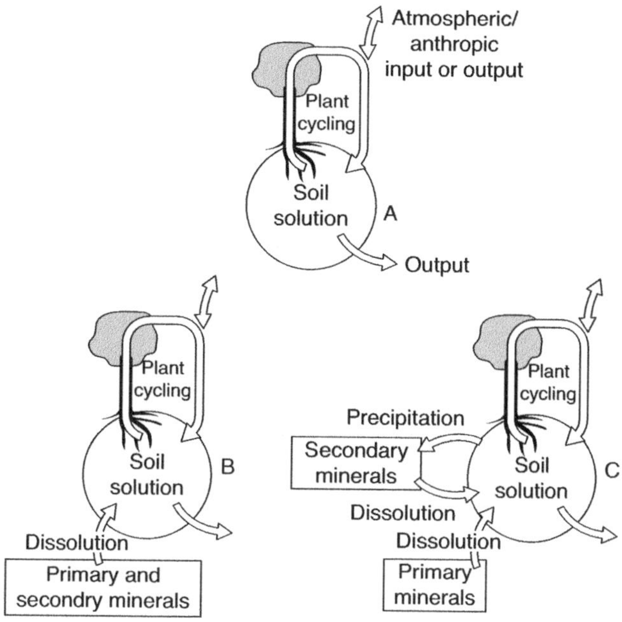

FIGURE 11.5 Three types of plant cycling interactions with soil minerals. See explanations of types A, B, and C in the text. From Lucas, 2001 (with permission).

from marine aerosols and P from wind-blown dust particles that originate thousands of miles away. These authors conclude that "the dynamics of long-term soil and ecosystem development cannot be evaluated in isolation: *nowhere on Earth is that isolated*" [italics added].

MODES OF PLANT ELEMENTAL CYCLING

Lucas (2001) describes three main types of plant cycling of elements, as shown in Figure 11.5. Type A cycling occurs when an element is not present in the soil minerals, and thus only involves the atmosphere, soil solution, and plant. This type of cycling can occur often for N and S, and for Ca, Mg, and K in highly leached soils. Type B cycling occurs when the element is a constituent of the soil minerals, but its concentration in the soil pore waters is below saturation with respect to precipitation of secondary phases, as often encountered for Ca, Mg, and K. Type C cycling occurs when the element is a constituent of the soil minerals, and its concentration/speciation is at or above saturation with respect to secondary phase formation, as often encountered for Si, Al, or Fe. Soil mineral dissolution and secondary phase formation can clearly be influenced by the presence of plants.

PLANTS AND BIOMINERALIZATION: PHYTOLITHS

Plants widely practice biomineralization, which is the precipitation of a mineral (or an amorphous solid) as a result of the metabolic functioning of a living organism (Skinner and

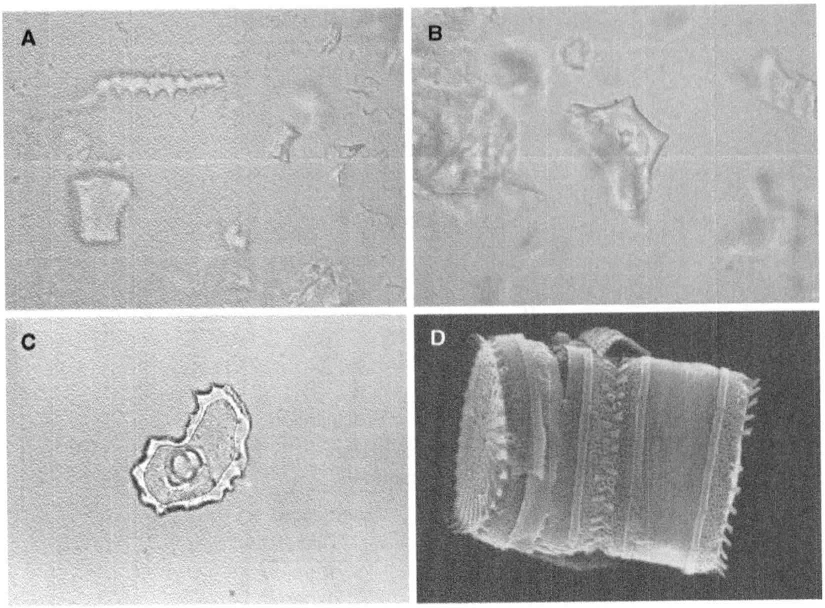

FIGURE 11.6 Pictures of siliceous phytoliths (a–c) and a diatom (d). From Conley, 2002 (with permission).

Jahren, 2004). *Phytoliths* ("plant stones") are small mineral precipitates formed in many plants, at least in some cases to provide structural stability in leafs and stalks. Opaline silica, calcium oxalate, and calcium carbonate are produced by many plant species. Oxides of magnesium, iron, aluminum, and manganese have also been found in a small number of plant species (Skinner and Jahren, 2003).

Si phytoliths are amorphous and are sometimes known as opal phytoliths or as *biogenic silica (BSi)* (Figure 11.6). The morphology of the Si phytolith is indicative of plant taxon. Phytoliths can remain stable in soils for thousands of years (Lucas, 2001), and they can be used in archaeology, paleontology, or studies of *pedogenesis* (Pearsall and Piperno, 1993). They can also be useful in climate change studies as indicators of vegetation changes. Si in plants can (1) affect the bioavailability and uptake of a variety of elements, (2) decrease transpiration, (3) increase strength, (3) keep leaves erect and improve light interception, and (4) perform either directly or indirectly a variety of other functions. One of the main roles of Si is analogous to that of lignin, providing strength and structure, but Si incorporation is energetically much cheaper than lignin production (Conley, 2002). Terrestrial plants can contain 1–10% or even more Si (Conley, 2002). Hodson et al. showed that ferns, gymnosperms, and angiosperms generally accumulate less Si in their shoots than do nonvascular plants (Hodson et al., 2005). Bamboo can contain up to 70% silica by dry weight (Jones et al., 1966), and rice husks at harvest can contain up to 20% silica (Garrity et al., 1984).

Phytolith formation, dissolution, and transport can play an important role in the terrestrial Si cycle and may be a major source of Si to the oceans. The export of dissolved silica can be small relative to a forest's internal biogeochemical Si cycle. Alexandre et al. (1997) showed that 74% of the dissolved silica in soil solution in a Congo rain forest originated from the dissolution of phytoliths, although the Si in phytoliths ultimately came from mineral weathering. The contribution of Si phytoliths to soils can play an important role in the stability of secondary minerals such as kaolinite (Lucas, 2001). The storage of BSi in plants as phytoliths needs to be considered in determining rates of mineral weathering, yet it is often overlooked. Conley (2002) calculated that the annual production of BSi in plants as phytoliths ranges from 60–200 Tmole Si/year, of similar magnitude to the oceanic fixation of BSi by diatoms (240 Tmole Si/year). However, additional study is needed to help better constrain the phytolith estimate.

Calcium oxalate is precipitated in plants as the monohydrate mineral whewellite ($CaC_2O_4{\cdot}H_2O$) and the dihydrate mineral weddellite ($CaC_2O_4{\cdot}2H_2O$) (Skinner and Jahren, 2003). Calcite precipitation inside or outside plant tissue in the root-rhizosphere system can occur in periodic wet/dry climates such as around the Mediterranean (Jaillard, 1987; Jaillard et al., 1991). Precipitation outside root tissues is often associated with fungal activity. Cystoliths are small carbonate concretions that can form in plant leaves such as that of *Ficus elastica*, which is the Indian rubber plant. It is believed that lichens, which are a symbiotic association between a fungus and an alga, may also have the ability to biomineralize calcium oxalate or carbonate (Skinner and Jahren, 2003).

PLANTS AND FORMATIONS IN LIMESTONE CAVES

The breakdown of organic debris in soils plays an important role in geochemical weathering and ultimately in development of carbonate formations such as stalactites, stalagmites, and other speleothems in limestone caves. This process is an excellent example of carbonate equilibria in systems closed and open to the atmosphere. Caves form when underground water dissolves limestone to form the cave passages. When the cave moves above the water table (or the water table falls), subsequent interactions can foster growth of carbonate cave formations such as *stalactites, stalagmites*, and other *speleothems*.

When water passes through soils, the breakdown of organic matter (represented as CH_2O) by microorganisms in the soil profile can result in a build up of CO_2 to levels above atmospheric

$$CH_2O + O_2 \rightleftarrows CO_2(g) + H_2O. \tag{11.3}$$

The CO_2 then combines with water to form carbonic acid, which partly dissociates to H^+ and the bicarbonate ion, HCO_3^-.

The net result is increased proton concentration, hence increased acidity

$$CO_2(g) + H_2O \rightarrow H_2CO_3 \tag{11.4}$$
$$H_2CO_3 \rightarrow H^+ + HCO_3^-. \tag{11.5}$$

This acidity can then cause weathering of limestone, given by the overall generalized reactions

$$H_2CO_3 + CaCO_3 \rightarrow Ca^{2+} + 2HCO_3^- \tag{11.6}$$
$$\text{or } H^+ + CaCO_3 \rightarrow Ca^{2+} + HCO_3^-. \tag{11.7}$$

The resulting concentrations of Ca^{2+} and HCO_3^- can build up to high values.

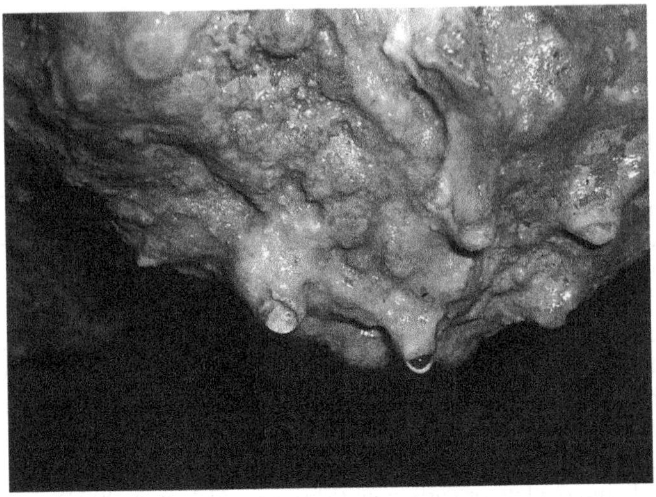

FIGURE 11.7 Drops forming on stalactites, illustrating the process by which stalactites are formed. Photo taken by Pedro Sanchez at Lol-Tun caves at Yucatan, December 31, 2005. Courtesy of P. Sanchez.

When the water percolates through the cave walls and into an air-filled cave, it has to come into equilibrium with the partial pressure of carbon dioxide, P_{CO_2} inside the cave. The result is often simply atmospheric P_{CO_2}. The water will then degas to reequilibrate with the atmosphere in the cave, which causes the reactions to go in the other direction and the cave formations to precipitate out of solution. This process is illustrated in the cave photo showing drops forming at the base of stalactites; see Figure 11.7 and the schematic in Figure 11.8.

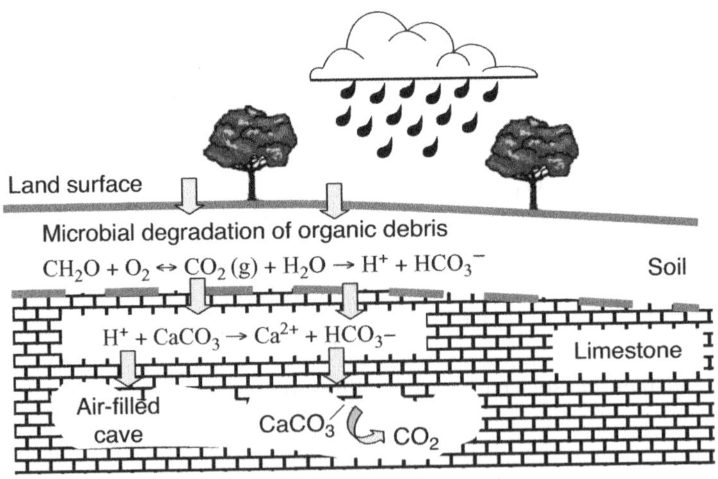

FIGURE 11.8 The breakdown of organic debris in soils leads to the generation of acidity. As this acidic water passes through limestone, it becomes enriched in dissolved carbonate components. When the resulting water percolates into a limestone cave, degassing of CO_2 is accompanied by precipitation of carbonate formations such as stalactites, stalagmites, and other carbonate speleothems.

Although we do not generally think of plants when we think of cave formations, it is the breakdown of plant and other organic debris in soils that leads to the carbonic acid accumulation, dissolution of carbonates, and eventually precipitation of carbonate cave speleothems, such as stalactites, stalagmites, columns, and "bacon rinds." It should be noted that not all speleothems form in this way; for example, microbial mediation of sulfuric acid speleogensis and associated formations has been the topic of ongoing research (e.g., Engel et al., 2004).

PHYTOREMEDIATION AS AN EXAMPLE OF PLANT-MINERAL-CONTAMINANT INTERACTIONS

Plants can also affect the mobility and bioavailability of toxic metals, radionuclides, and organic pollutants such as pesticides, petroleum products, and explosives. The ability for soils to take up or transform organic and inorganic constituents is being taken advantage of in the growing field of *phytoremediation* (e.g., ITRC, 2001; Pulford and Watson, 2003; Burken and Schnoor, 1997; Schnoor, 2000; Dietz and Schnoor, 2001). Phytoremediation approaches have the potential to be more cost effective than wholesale digging up and removal of contaminated soils, which then need to be dealt with, or than application of various pump and wash procedures, which often leave a percentage of the contaminants behind on mineral surfaces. Phytoremediation can sometimes offer advantages over microbial bioremediation, especially in systems where microorganisms are difficult to maintain in the proper metabolic states and community structures.

Some of the primary means of phytoremediation include (Pulford and Watson, 2003; ITRC, 2001; Vaccari et al., 2006) (Table 11.2; Figure 11.9) the following:

1. *Phytostabilization*, in which plants decrease pollutant mobility and bioavailability in the soils or sediments either by accumulation in the roots, adsorption onto the roots, or precipitation or sorption associated with soils in the root zone. An example is by promoting the reduction of highly mobile and bioavailable Cr(VI) to less mobile Cr(III). Phytostabilization involves proteins and enzymes either exuded into the rhyzosphere or associated with the root cell walls.
2. *Rhizodegradation* is the breakdown of organic contaminants in soil through rhizosphere bioactivity associated with plant exudates or soil organisms such as yeasts, bacteria, and fungi. Rhizodegradation involves symbiosis between plants and soil microorganisms.
3. *Phytoaccumulation* refers to the concentration of inorganic contaminants in the roots, shoots, or leaves. This process can affect the concentrations and distributions of metals, such as Cd, Ni, Zn, As, Se, Fe, and Mn, and radionuclides, such as U and Cs in soils, and ultimately water–rock interactions such as precipitation, dissolution, and sorption. Many phytoaccumulation mechanisms have been evolved by plants to detoxify contaminants effectively. Phytoaccumulation can make plants inedible by humans or animals. Remediation schemes that rely on phytoaccumulation require a strategy to dispose of toxins when plants are harvested.
4. *Phytodegradation* (or *phytotransformation*) is a process in which plants and/or their microbial associates take up and degrade or transform organic pollutants such as chlorinated solvents, phenols, pesticides, and explosives. An example is

TABLE 11.2 A summary of phytotechnology mechanisms.

Mechanism	Plant Types	Media	Typical contaminants
Phytostabilization	herbaceous species grasses, trees wetland species	soils sediments sludges	As, Cd, Cr, Cu, Pb, Zn
Rhizodegradation	herbaceous species grasses, trees wetland species	soils, sediments sludges groundwater	organic compounds e.g., PAHs, PCBs, pesticides
Phytoaccumulation	herbaceous species grasses, trees wetland species	soils, sediments sludges	metals and radionuclides Ag, Au, Cd, Co, Cr, Cu, Hg, Mn Mo, Ni, Pb, Zn, Sr, Cs, Pu, U
Phyrodegradation	algae, trees herbaceous species wetland species	soils, sediments sludes, surface & groundwater	organic compounds e.g., solvents, munitiions
Phytovolatilization	herbaceous species trees wetland species	soils, sediments sludges groundwater	chlorinated solvents MTBE Se, Hg, As
Evapotranspiration	herbaceous species grasses, trees wetland species	groundwater surface water storm water	water soluble organics & inorganics
Rhizofiltration	sunflowers trees wetland plants	sludges wetlands	Pb, Cu, Zn, Ni, Sr, Cd U(VI), Mn, Cr(VI)

After ITRC, 2001, with revisions.
PAH = polycyclic aromatic hydrocarbon, PCB = polychlorinated biphenyl, MTBE = methyl tert-butyl ether.

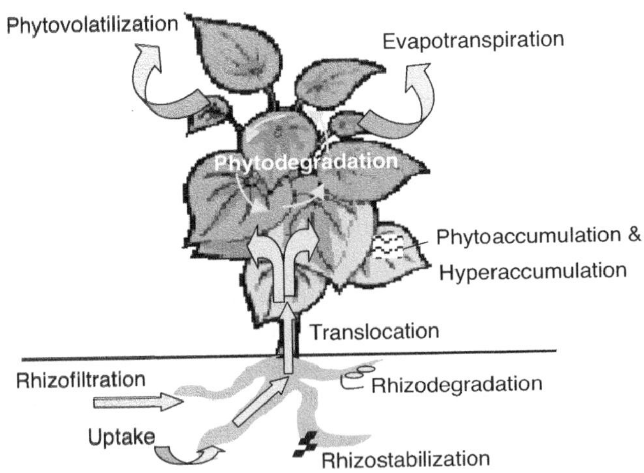

FIGURE 11.9 Some mechanisms of phytoremediation.

the production of nitroreductase enzymes that help to break down nitroaromatic compounds.

5. *Phytovolatilization*, in which organic and inorganic pollutants such as chlorinated solvents and mercury are volatized into the atmosphere through the uptake, translocation (from roots to shoots or leaves), and transpiration action of plants. In some instances, plant enzymes may transform mercury into a volatile form.
6. *Evapotranspiration*, which is the combined evaporation and transpiration processes associated with plants, can alter soil hydrology and hydraulics, thus altering pollutant mobility.
7. *Rhizofiltration*, in which plants are used to filter metals and radionuclides from wastewaters through biochemical processes. This process is particularly useful for metals and radionuclides present in water at relatively low concentrations and when large amounts of water need to be remediated (Cooney, 1996). Over time, the roots become saturated with contaminants and the plants are removed. Because only the roots are involved in the accumulation, less overall contaminated biomass is produced. Although this process removes metals and radionuclides from the soils and waters, the contaminants then need to be disposed of.

Hyperaccumulating plants can take up at least 1000 mg/kg dry weight of a metal or metalloid (ITRC 2001). Examples include *Berkheya codii*, which hyperaccumulates Ni; *Astragalus racemosus*, which hyperaccumulates Se; and *Pteris vittata*, which hyperaccumulates As (Pulford and Watson, 2003). Genetic engineering is being used to promote hyperaccumulation. The large uptake of metals by hyperaccumulators can be expected to have significant effects on sorption, dissolution, and precipitation. Many heavy metals, radionuclides, and organic contaminants can be amenable to phytoaccumulation by common plants. Poplar can be used to remediate chlorinated solvents and nitrates in groundwater and heavy metals in soil, and sunflower is being studied for remediating radionuclides in groundwater.

Case Study: Phytoremediation of Atrazine

Phytoremediation offers the potential to degrade pollutant organics such as atrazine into less toxic compounds. Atrazine (2-chloro-4-(ethylamino)-6-isopropylamino-1,3,5-triazine) is a widely used herbicide that is an endocrine disruptor and thus a danger to mammals, amphibians, fish, and ultimately entire ecosytems. Atrazine is hydrophobic and sorbs to soils primarily through partitioning into soil organic matter, along with absorption into the interlayers of swelling clays (Lee et al., 1990; See Chapter 8). Atrazine can be broken down by microbial processes in soils, eventually leading the organic compound to be "mineralized" or transformed to $CO_2 + H_2O$. The purpose of phytoremediation of atrazine is to use plants to break it down into (hopefully) less toxic compounds.

Figure 11.10 shows how poplar trees can be used to phytoremediate atrazine. One advantage of using poplar for phytoremediation is that although poplars usually have fairly shallow roots, they can be induced to grow deeper root systems through certain irrigation processes. Poplars also tend to be fast growers and to take up and use water rapidly.

When atrazine is taken up into poplar roots, it is translocated so that atrazine metabolism occurs in the roots, stems, and leaves, continuing over time (Burken and Schnoor, 1997). Remaining atrazine and degradation products are translocated into the stem and leaves, where additional transformations can occur. Metabolic products from atrazine metabolism

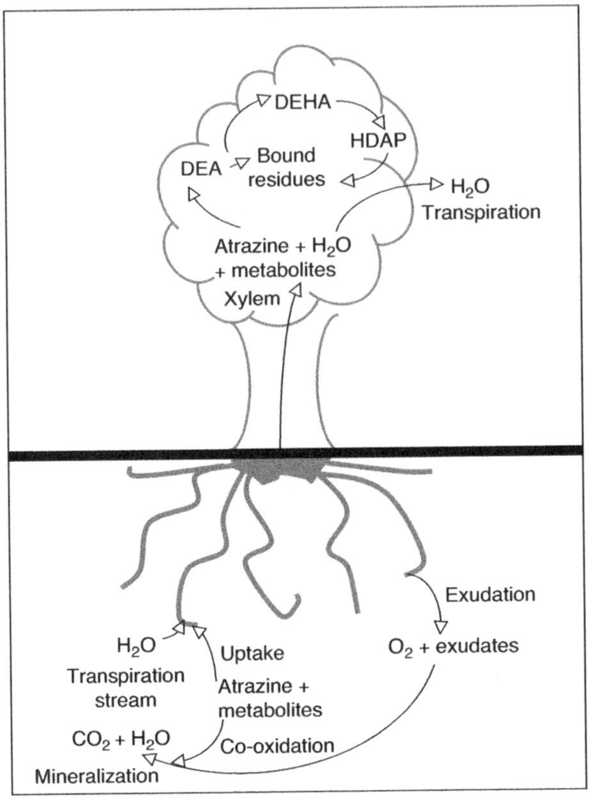

FIGURE 11.10 Atrazine uptake, metabolism, and bound residues in poplar trees. After Burken and Schnoor, 1997.

in poplar trees include: hydroxyatrazine (HA), deethylatrazine (DEA), deisopropylatrazine (DIA), deethylhydroxyatrazine (DEHA), didealkylatedatrazine (DDA), and hydroxylated dealkylated products (HDAP). The overall process can be complex kinetically, involving both simultaneous and sequential reactions.

The ability of plants to compete with mineral surfaces for pollutants or even to mobilize and then sequester pollutants from soil particle surfaces is a key component of phytoremediation. The growing field of phytoremediation argues for the importance of investigating environmental interface interactions in the context of biological processes, which include not only microbiologic but also macrobiologic processes. Phytoremediation also reveals that other environmental surfaces beyond minerals, such as those of plant roots, can be important to natural environmental processes and to pollutant remediation strategies.

QUESTIONS FOR FURTHER THOUGHT

1. What are some ways in which clear cutting of forests might impact water–rock interactions and environmental surface chemistry in a given ecosystem?
2. Compare the molar ratios of the various elements in plants to those of the average oceans and mean river water.

3. (a) How might one account for plant cycling of elements in calculating the input/output budget for a small watershed?
   (b) What might some challenges be to such a calculation?
4. A former military base has significant contamination of trinitrotoluene, Cr, and Pb.
   (a) Which phytoremediation techniques might be employed?
   (b) What properties of the site might need to be considered in development of a phytoremediation strategy?
   (c) What are some potential pitfalls?
5. Stormwater trapping basins are often mandated for construction sites, as a means of slowing the flow of surface runoff and causing suspended sediments to settle out.
   (a) Why might it be important to vegetate the walls of the trapping basin quickly after construction?
   (b) What might be some potential benefits or pitfalls associated with using water-loving plants such as cattails to help trap sediments in the basin?
6. Why might applications of natural humic substances derived from soils be used to help remediate chlorosis?

FURTHER READING

Berner, E.K., Berner, R.A., and Moulton, K.L. 2004. Plants and mineral weathering: Present and past. In Drever, J.I. (ed.), Treatise on Geochemistry vol. 5, Surface and Groundwater, Weathering and Soils. Elsevier, New York, pp. 169–188.

Lucas, Y. 2001. The role of plants in controlling rates and products of weathering: Importance of biological pumping. *Annual Review of Earth and Planetary Science*, 29, 135–163.

Rodriguez-Iturbe, J. 2007. Ecohydrology of Water-Controlled Ecosystems. Cambridge University Press, Cambridge, UK.

Schnoor, J.L. 2000. Degradation by Plants – Phytoremediation, Biotechnology vol. IIB, Wiley-VCH, Weinheim, Germany.

Skinner, H.C.W., and Jahren, A.H. 2003. Biomineralization. In Schlesinger W.H. (ed), Biogeochemistry, Treatise on Geochemistry vol. 8. Elsevier Science, New York, pp. 117–184.

12

MICROORGANISMS AS ENVIRONMENTAL SURFACES

Microorganisms are ubiquitous in the Earth's surface environment known as the critical zone. They can even live under extreme conditions, such as hot springs in Yellowstone National Park (Figure 12.1), deep ocean vents, acid rock drainage environments, the Arctic, and the Antarctic. They can exist as single cells or in colonies. In nature, they generally exist in complex community structures.

Figure 12.2 shows the phylogenetic "tree of life" based on ribosomal RNA (rRNA) analysis. Woese et al. (1990) proposed that this molecular-based comparison indicated that life divides into three domains: archaea, bacteria, and eucarya. Microorganisms encompass archaea (formerly known as archaebacteria) and "true" bacteria (formerly known as eubacteria). Eucarya include some microorganisms as well as higher life forms. Viruses are also generally considered to be microorganisms, although they are noncellular and are not shown in Figure 12.2. The bacteria differ from the archaea in terms of their metabolic abilities, the composition of their membranes, and the structure of their ribosomes.

Of the various microorganisms, the bacteria have been the focus of the bulk of research on interface reactions. Fungi have also been the subject of a good deal of study, particularly in soil science. We shall focus primarily but not exclusively on bacteria here.

Bacteria have been shown to affect mineral growth and dissolution, mineral particle aggregation, and sorption phenomena. Many surfaces in nature, in laboratories, and even in human bodies (e.g., teeth, orifices, and prosthetic devices) are coated with bacteria and/or bacterial exudates. The presence of bacteria can affect the chemical composition of water, the mobilities of nutrients and toxic metals, the degradation of natural and contaminant organic compounds, and the surface properties of minerals. In many instances, what we think of as the mineral–water interface is in reality a mineral–bacteria–water interface.

Over the past few decades, an explosion of new biochemical and genetic approaches has been developed to investigate bacterial community structure, diversity, and controls on

Environmental Surfaces and Interfaces from the Nanoscale to the Global Scale, by Patricia A. Maurice

FIGURE 12.1 Microorganisms are ubiquitious and can even flourish in and around hot springs. Shield Spring is in the Upper Geyser Basin, Yellowstone National Park. Photo by P. Maurice.

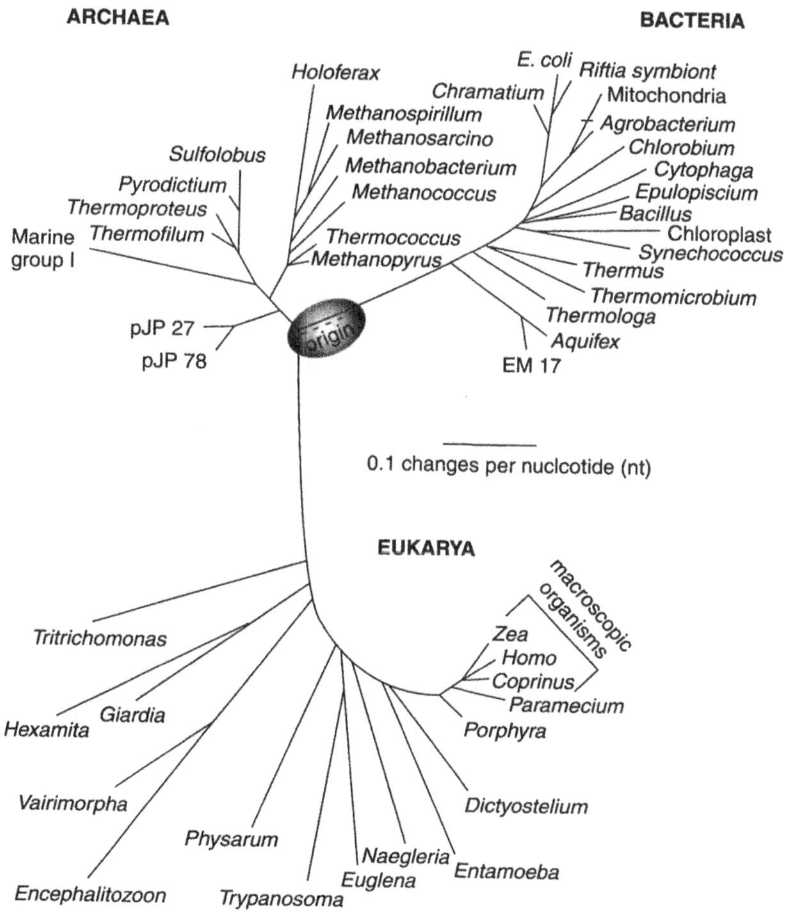

FIGURE 12.2 The phylogenetic tree of life. From Atlas, 1997 (with permission).

activity. No attempt is made here to discuss these fields, as they are developing rapidly along with our understanding of proper application to studies of microbial–mineral, –metal–organic contaminant, and other interactions.

HOW MICROORGANISMS "MAKE A LIVING"

One general means of differentiating microorganisms is based on the varied metabolic processes that may be used (Figure 12.3). *Autotrophs* fix their own C from CO_2, whereas *heterotrophs* use only organic carbon from other sources. *Photolithotrophs* obtain energy directly from sunlight and assimilate carbon as CO_2, bicarbonate, or carbonate. *Photo-heterotrophs* derive all or part of their energy from sunlight but assimilate organic C as the C source. Most bacteria are *chemotrophs*, which means that they access energy that is stored in chemical components. *Chemoorganotrophic* bacteria obtain energy from organic molecules. *Chemolithotrophs* obtain their energy from inorganic molecules such as iron, sulfur, and hydrogen sulfide, and they assimilate carbon as CO_2, bicarbonate, or carbonate. The chemolithotrophs can be especially effective in enhancing dissolution of minerals such as Fe and Mn oxides. Most chemolithotrophs only use O_2 as an electron acceptor, with some notable exceptions such as *Thiobacillus denitrificans* (e.g., Beller et al., 2006). *Mixotrophs* may derive energy from a mixture of processes as shown in Figure 12.3.

Metabolic Pathways

Table 12.1 summarizes environmentally relevant metabolic pathways in prokaryotes and eukaryotes. Some of the most common examples are discussed in Table 12.1.

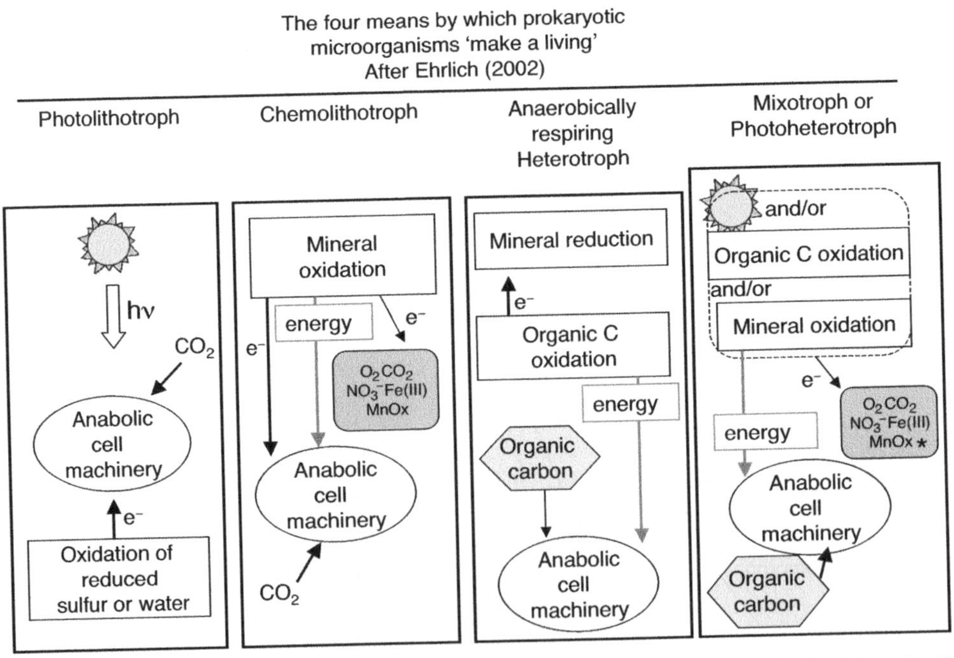

FIGURE 12.3 How prokaryotic microorganisms make a living. (∗) This mechanism is for mixotrophs but not photoheterotrophs. Based on Ehrlich, 2002.

TABLE 12.1 Environmentally relevant metabolic pathways and their relative occurrence in prokaryotes and eukaryotes.

Metabolic Pathway		Prokaryotes	Eukaryotes
		Occurrence √ = yes X = no	
Autotrophy (ability to grow with CO_2 as carbon source)			
Phototrophy	Oxygenic	√	√
	Anoxygenic	√	X
	Electron donor		
	H_2	√	X
	H_2S, S^0	√	X
	Organic C	√	X
	Fe^{2+}	√	X
Lithotrophy	Energy source		
	H_2	√	X
	HS^-, S^0, $S_2O_3^{2-}$	√	X
	NH_4^+, NO_3^-	√	X
	Fe^{2+}	√	X
	Mn^{2+}	√	X
Heterotrophy (requires organic C for growth)			
	Fermentation	√	X[1]
	Aerobic respiration	√	√
	Anaerobic respiration		
	NO_3^-, NO_2^-	√	X
	S°, $S_2O_3^{2-}$, SO_4^{2-}	√	X
	Fe^{3+}, Mn^{4+}	√	X
	CO_2	√	X
	Fumarate (organic C)	√	X

[1]Only a few yeasts among the eukaryotes can grow by fermentation, although fermentation may occur under stress.

Adapted from Nealson and Stahl, 1997.

The microbial world is driven by and encompasses a wide variety of redox processes. Organic compounds often serve to transfer electrons to an inorganic electron acceptor. This process of coupling oxidation of organic compounds with a reduction of external electron acceptors is termed *respiration*. The most common external electron acceptors used by microorganisms in natural environments are (Chapelle, 2001) O_2, NO_3^-, Fe(III), SO_4^{2-}, and CO_2. Additional electron acceptors used by microorganisms include managanate, selenate, and uranyl ions. In considering microbial redox processes, it is important to identify both the electron donor and the electron acceptor.

Adenosine triphosphate (ATP) is an important and widely used energy storage compound. *Aerobic respiration* requires O_2 to generate ATP. Energetically, this process is favorable, so that aerobic organisms tend to dominate the microbial community wherever O_2 is present. A *facultative anaerobic* bacterium makes ATP by aerobic respiration if oxygen is present, but also it can switch to fermentation. In contrast, *obligate anaerobes* cannot survive in the presence of oxygen.

Fermentation is the process by which energy is derived from the oxidation of organic compounds such as carbohydrates using an endogenous electron acceptor, usually another organic compound, which is reduced. Fermentation can be used by some bacteria, algae, fungi, and a few protozoa, and it can occur occasionally in higher organisms including

Microbial denitrification

Nitrogen species	Nitrogen oxidation state	Catalyzing enzyme
Nitrate NO_3^-	+5	
⇩		Nitrate reductase
Nitrite NO_2^-	+3	
⇩		Nitrite reductase
Nitric Oxide NO	+2	
⇩		Nitric oxide reductase
Nitrous oxide N_2O	+1	
⇩		Nitrous oxide reductase
Dinitrogen N_2	0	

FIGURE 12.4 The microbial denitrification pathway.

humans during intense exercise. The reaction varies with the organic compound being degraded. A typical fermentation reaction in which yeast break down glucose sugar to produce ethanol is

$$C_6H_{12}O_6 + O_2 \rightarrow 2\,CH_3CH_2OH + 2\,CO_2 + 2\,\text{ATP}. \tag{12.1}$$

In anaerobic environments, many bacteria carry out respiration by using nitrate as the terminal electron acceptor. This process is not as energetically favorable as is aerobic respiration. Many organisms reduce nitrate (NO_3^-) to nitrite (NO_2^-); a few organisms can reduce NO_2^- to ammonia (NH_3). Some microorganisms can reduce NO_3^- and NO_2^- ultimately to N_2 gas, in a process called *denitrification* (Figure 12.4). A generalized reaction for the oxidation of a carbohydrate through denitrification is

$$\tfrac{5}{4}\,CH_2O + NO_3^- \rightarrow \tfrac{5}{4}\,CO_2 + \tfrac{1}{2}\,N_2 + \tfrac{3}{4}\,H_2O + OH^- \tag{12.2}$$

(Vaccari et al., 2006). Some microorganisms can oxidize ammonia to nitrite and nitrite to nitrate, in the overall process of *nitrification*. However, N_2 gas seems not to be oxidized microbially.

Methanogens are obligate anaerobic archaea that carry out *methanogenesis* or the production of methane (CH_4). An example reaction is the anaerobic respiration in which CO_2 is reduced to CH_4 and H_2 is oxidized

$$4H_2 + CO_2 \rightarrow CH_4 + 2\,H_2O. \tag{12.3}$$

Another example reaction produces not only methane but also CO_2

$$CH_3COOH \rightarrow CH_4 + CO_2. \tag{12.4}$$

Methanogenesis is a far less energetically favorable process than is aerobic respiration.

A vast array of microbial processes involves sulfur (S), such as dissimilatory sulfate and S reduction, assimilatory sulfate reduction, sulfur oxidation, and S disproportionation (in which some S compounds are oxidized and others are reduced). An example of sulfide-oxidizing bacteria are members of the genera *Thiobacillus* and *Acidithiobacillus*, which are chemolithotrophic obligate aerobes that are gram negative (see below) and rod shaped. Examples of sulfide oxidation reactions include

$$2H_2S + O_2 \rightarrow 2H_2O + 2S \qquad (12.5)$$

and

$$2S + 3O_2 + 2H_2O \rightarrow 2H_2SO_4. \qquad (12.6)$$

Many *Thiobacilli* grow best at circumneutral pH, which can present difficulties when sulfuric acid is produced. However, the acidiphilic (note: the genus name for the acidiphilic *Thiobacillus* was changed to *Acidithiobacillus* by Kelly and Wood, 2000) *Acidithiobacillus thiooxidans* thrives under acidic conditions as low as pH ~ 2 (Chapelle, 2001). *Desulfovibrio* are sulfate reducers that depend on fermentative bacteria to supply simple organic compounds.

Various microorganisms can use Fe or Mn redox processes. Fe(III)oxyhydroxide reduction by *Geobacter* and *Shewanella* is discussed in some detail below. *Acidithiobacillus ferrooxidans* is an acidiphilic bacterium that can derive energy from oxidizing ferrous iron, and it can also oxidize reduced froms of sulfur. It plays a key role in pyrite oxidation in acid mine drainage environments. The ferrous-iron–oxidizing *Gallionella* forms highly insoluble Fe(III)oxyhydroxides that can clog wells (Chapelle, 2001). According to Tebo et al. (1997), a wide array of microorganisms can accelerate the rate of Mn(II) oxidation by up to five orders of magnitude in comparison with abiotic processes. One of the most highly studied is the β-proteobacterium *Leptothrix discophora*. A variety of fungi is also known to possess Mn(II)-oxidizing enzymes (Barr and Aust, 1994). From this discussion, it is clear that microorganisms can play an important role in the distributions of S-, Mn-, and Fe-bearing minerals.

Taken as a whole, the microbial world exhibits a tremendous range of biogeochemical processes that can affect the chemical cycling of the elements. Table 12.2 is a summary of the microbially mediated transformations among CHONSP atoms, ions, and molecules, giving an indication of this variety and complexity. Much work has been done on the processes by which microorganisms affect nutrient and metal cycling. Figure 12.5 shows major microbial processes that affect the C cycle with depth in a column of sediments, which illustrates how the presence of light and dissolved oxygen affect C cycling.

A single bacterial strain can be flexible in its use of different mechanisms to obtain energy, such as using Fe(III), Mn(IV), or U(VI) as a terminal electron acceptor. A variety of different mechanisms can be used by a given strain to acquire nutrients as well. For example, the aerobic bacterium *Pseudomonas mendocina* requires Fe as a basic nutrient and can acquire Fe by altering pH or releasing Fe(III) chelating agents and reductants. Fe can also be acquired through cell-wall–associated reduction (Hersman et al., 2001). *Shewanella putrifaciens* can respire on a wide variety of terminal electron acceptors, including (but not necessarily limited to) oxygen, nitrate (NO_3^-), nitrite (NO_2^-), trimethylamine-*N*-oxide (TMAO), sulfite (SO_3^{2-}), thiosulfate ($S_2O_3^{2-}$), elemental sulfur [S(0)], fumarate, uranyl carbonate (U(VI)), selenite (Se(IV)), Fe(III), and Mn(IV) (DiChristina et al., 2002).

TABLE 12.2 Microbially mediated transformations amongst CNOSP.

	Species reduced				
Species oxidized	O	C	N	S	H
O		Photosynthesis $CO_2 \rightarrow$ Organic C $H_2O \rightarrow O_2$			
C	Aerobic respiration $O_2 \rightarrow H_2O$ organic C $\rightarrow CO_2$		Denitrification $NO_3^- \rightarrow N_2$ organic C $\rightarrow CO_2$	Sulfate reduction $SO_4^- \rightarrow H_2S$ organic C$\rightarrow CO_2$	Fermentation organic H $\rightarrow H_2$ organic C $\rightarrow CO_2\rightarrow$
N	Ammonia oxidation $O_2 \rightarrow H_2O$ $NH_4^+ \rightarrow NO_2$	Nitrification $CO_2 \rightarrow$ organic C $NH_4^+ \rightarrow NO_3^-$			
S	Sulfide oxidation $O_2 \rightarrow H_2O$ $HS^- \rightarrow S^0 \rightarrow H_2SO_4$	Chemoautotrophy $CO_2 \rightarrow$ organic C $H_2S \rightarrow S^0 \rightarrow H_2SO_4$	Denitrification/S oxid. $NO_3^- \rightarrow N_2$ $S^0 \rightarrow SO_4^{2-}$		
H		Methanogenesis $CO_2 \rightarrow CH_4$ organic C $\rightarrow CH_4$ $H_2 \rightarrow H_2O$	Denitrification$\rightarrow$ $NO_3^- \rightarrow N_2$ $H_2 \rightarrow H_2O$		

After Kaplan and Bartley, 2000.

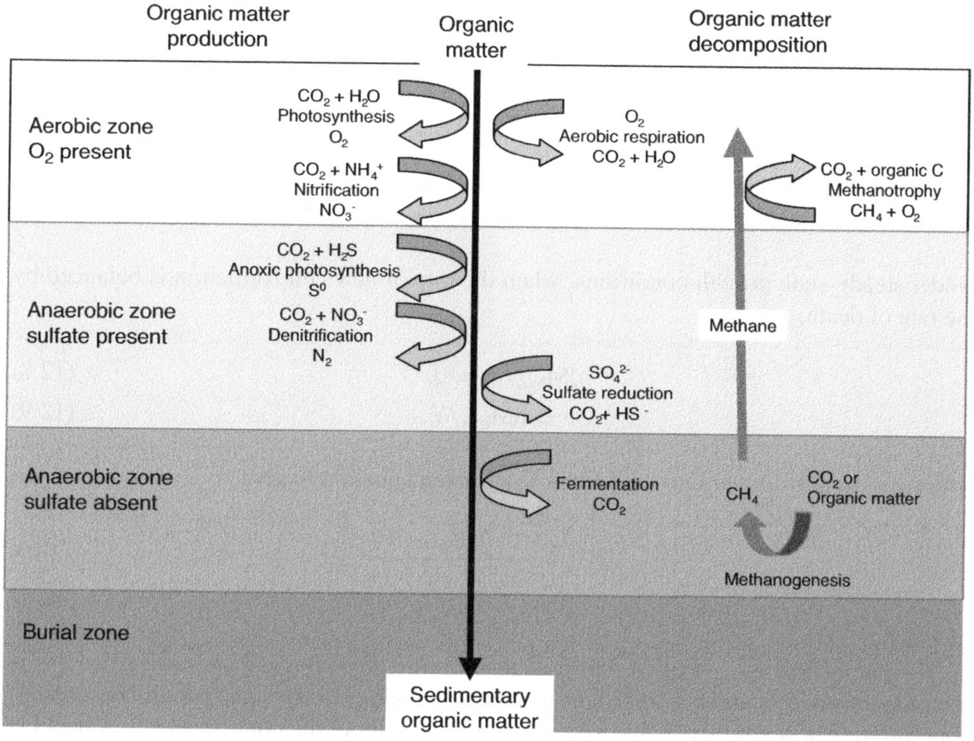

FIGURE 12.5 The microbial carbon cycle in a sediment sequence from aerobic (and with light during the day) at the top to anaerobic (and dark) at depth. Below the aerobic zone, the anaerobic organisms show zonation determined by the availability of electron acceptors. Based on Kaplan and Bartley, 2000.

Microbial Redox Reactions and Michaelis-Menten Kinetics

Microbial respiration using H_2 as a substrate can be described by Michaelis-Menten kinetics, with implications for the concentrations of H_2 in sediments containing bacteria that use different metabolic pathways (Chapelle, 2001; Lovley et al., 2004). Recall from Chapter 1, Equation 1.83, that the Michaelis-Menten equation has the form

$$\nu = \frac{\nu_{max}[E_T][S]}{K_m + [S]}, \tag{1.83}$$

where

ν = rate of substrate uptake (in the example above, H_2 consumption in moles Per unit time)

ν_{max} = maximum rate of substrate uptake when uptake rate is not limited by substrate availability

K_m = Michaelis-Menten rate constant = $(k_{-1}+k_2)/k_1$ and when $[S] = K_m$, $\nu = \frac{1}{2}\nu_{max}$.

If the substrate S is H_2, and the total enzyme concentration is set equal to the number of grams of hydrogen-consuming microorganisms M_{hyd}, then the rate of growth of the microorganisms (grams of cells per unit time) on H_2 can be described by

$$dM_{hyd}/dt = (\nu Y) - (bM_{hyd}), \tag{12.7}$$

where

Y = yield coefficient (in grams of cells formed per mole of hydrogen consumed)
b = cell death coefficient (in units of time^{-1}).

Under steady-state growth conditions, when the rate of new cell formation is balanced by the rate of death,

$$dM_{hyd}/dt = 0, \tag{12.8}$$

$$\nu = bM_{hyd}/Y, \tag{12.9}$$

which can be substituted into the Michaelis-Menten equation to give

$$[H_2] = \frac{K_m}{(\nu_{max} Y/b - 1)} \tag{12.10}$$

(Chapelle, 2001).

The parameters K_m, ν_{max}, Y, and b depend on the physiological characteristics of the microorganism. Equation 12.10 thus means that under steady-state conditions, the H_2 concentration depends on the physiological characteristics of the microorganism (Chapelle, 2001). This concentration would likely be appropriate for many groundwater systems, where H_2 respiration would be occurring. A similar construct would likely not be as valuable for other microbial processes in shallow systems with greater fluxes.

Chapelle (2001) tabulates the potential energy yield characteristic of a variety of anaerobic oxidation processes (Table 12.3), and from this result, the author calculates the characteristic hydrogen concentration that would occur when each process was predominant. Note that the lowest characteristic hydrogen concentrations are for sediments undergoing nitrogen reduction and the highest concentrations are for sediments undergoing methanogenesis. The different energy yields can lead to sedimentary sequences or zones of activity. The data in Table 12.3 can be used to help predict which processes will tend to dominate over others under various conditions and provide insight into microbial competition. For example, if Fe(III)minerals are highly abundant in a sediment, then the H_2 concentrations will be so low

TABLE 12.3 The potential energy yields and steady-state H_2 concentrations that are characteristic of different aerobic oxidation processes

Reaction	Characteristic $[H_2]$, in nM	Potential energy yield kJ per H_2	
$2NO_3^+ + 5H_2 + 2H^+ \rightarrow N_2 + 6H_2O$	0.01–0.05	224	↑
$MnO_2 + H_2 \rightarrow Mn(OH)_2$	0.1–0.3	163	greater
$2Fe(OH)_3 \rightarrow 2Fe(OH)_2 + 2H_2O$	0.2–0.8	50	energy
$SO_4^{2-} + 4H_2 + H^+ \rightarrow HS^- + 4H_2O$	1.0–4.0	38	yield
$HCO_3^- + 4H_2 + H^+ \rightarrow CH_4 + 3H_2O$	5.0–5.0	34	

After Chapelle, 2001.

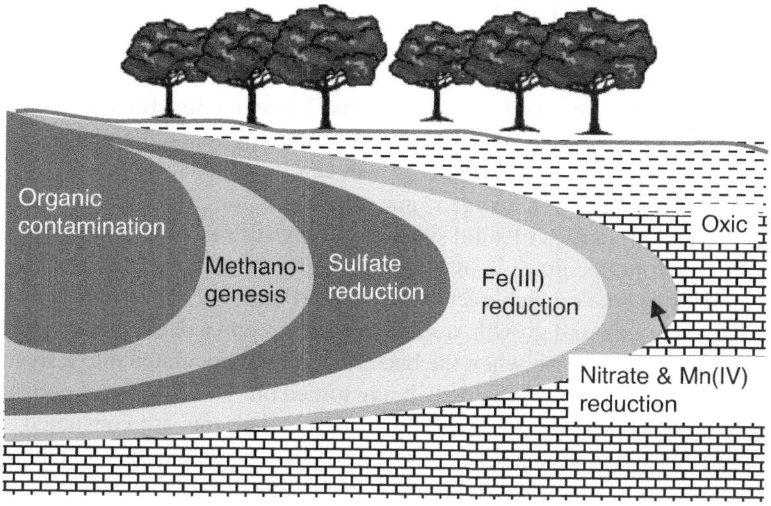

FIGURE 12.6 Zonation of redox processes for an organic-contaminated, electron-donor-rich aquifer. After Chapelle, 2001 (with permission).

that sulfate-reducing and methanogenic bacteria will be out-competed (Chapelle, 2001). Different zones of microbial community composition and activity may occur because of the interactions. Whether a subsurface environment is electron-donor-poor as many pristine aquifers or electron-donor-rich, as at the organic-contaminated *Bemidji* (MN) site can affect the zonation (Chapelle, 2001; Lovley et al., 1994 (Figure 12.6).

Microbial Temperature Ranges and Extremophiles

Microorganisms have been found in a wide range of temperature environments (e.g., Brock, 1985). *Mesophiles* grow within the temperature range from 25 to ~45°C. *Thermophiles* grow at higher temperatures. *Thermus aquaticus*, which was discovered in hot springs of Yellowstone National Park, can inhabit waters ranging from 50 to 80°C (122 to 176°F) (Brock, 1994). *Hyperthermophilic* bacteria have been identified growing at temperatures around the boiling point of water. For example, Kashefi and Lovley (2004) reported that an Fe-reducing bacterium from a black smoker hydrothermal vent along the Juan de Fuca Ridge grew on an Fe(III) substrate at temperatures of 85 to 121°C. The water is not actually boiling in this environment because of the high pressure. *Psychrophiles* can grow at low temperatures as associated with the polar regions. For example, bacteria were observed in the hyporheic zone of an Antarctic Dry Valley stream (Maurice et al., 2002) in a study described in Chapter 10.

Halophilic thermophilic bacteria such as *Alterococcus agarolyticus* (e.g., Shieh and Jean, 1998) can tolerate both high temperatures and high salinity. *Thermoacidophiles* such as *Sulfolobus acidocaldarius* (member of archaea) can tolerate high temperatures and high acidity. *S. acidocaldarius* are hyperthermophiles and can oxidize sulfur to sulfuric acid (Ehrlich, 2002). *S. acidocaldarius* have been found in mud pots and acidic hot springs at Yellowstone National Park (Brock, 1994). Bacteria that can live in extreme environments in terms of temperature, pH, and/or salinity are known as *extremophiles*.

MICROBIAL GROWTH CURVES

A major factor differentiating microbial systems from abiotic systems is that microorganisms are living organisms and thus can reproduce under favorable conditions. Most laboratory studies of microbial growth are done in batch reactors, although chemostats are becoming more commonplace. Microbial growth can be affected by factors such as temperature, water availability, pH, oxygen, macronutrients, micronutrients, light/radiation, toxins, presence and properties of solid substrates, flow/stir rate, and grazing by protozoa.

When a bacterial monoculture is introduced into fresh medium in a batch system, the growth curve tends to take on a characteristic shape (Figure 12.7), including an initial lag phase, a period of exponential growth, a stationary phase, and a death phase. The lag phase is a period of little or no growth when the bacteria are first inoculated into a fresh medium and various metabolic activities are slowly being turned on. The "log" or exponential phase is the subsequent period of rapid bacterial growth. As nutrients are consumed and waste products accumulate, rates of bacterial growth and death become equal so that the overall bacterial population numbers stay approximately the same during the stationary phase. The final death phase occurs when the cells die, potentially accompanied by cell lysis (Madigan et al., 1997), as toxic bacterial by-products accumulate and nutrients are exhausted.

For the portion of the growth curve that corresponds to exponential growth, the change in population size as cell numbers (N) over time (t) is a function of N

$$dN/dt = rN, \qquad (12.11)$$

where r is the rate constant, with units of time^{-1}.

If the initial number of cells is N_0, then

$$N = N_0e^{rt}. \qquad (12.12)$$

The amount of time from one generation to the next is known as the generation time t_g. Because bacteria split or double at each new generation, t_g for bacteria is equal to the

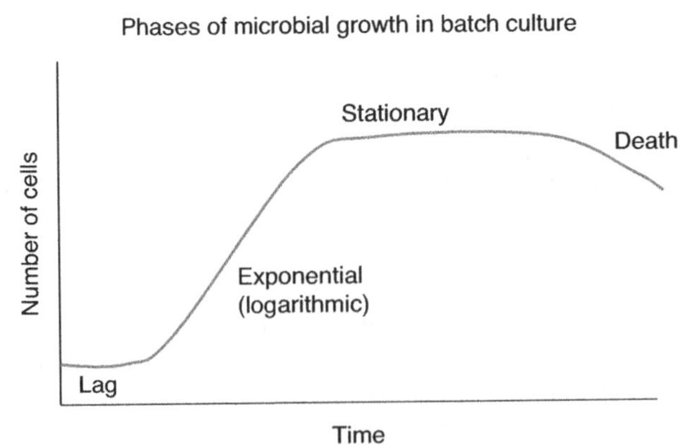

FIGURE 12.7 Schematic of a typical microbial growth curve in a batch reactor.

doubling time t_d and can be found from

$$t_d = \ln 2/r. \qquad (12.13)$$

BACTERIAL GROUPS

At the current time, some disagreement exists among biologists regarding how to classify members of the domain bacteria; some divide the domain into kingdoms and others into phyla. However they are grouped, three of the most important groups are the cyanobacteria, proteobacteria, and firmicutes (Vaccari et al., 2006).

The cyanobacteria, which are also known as blue-green bacteria, are mainly oxygen-producing aerobic photoautotrophs whose evolution probably played an important role in transforming the early oxygen-poor atmosphere into the oxygen-rich atmosphere that can support higher life forms. Most use CO_2 as a carbon source to produce organic carbon compounds; a few can use organic carbon compounds (Vaccari et al., 2006). The cyanobacteria were previously mistaken for algae and thus were formerly known as blue-green algae (Ehrlich, 2002).

The proteobacteria are gram negative (see below) and have been divided into the following five classes or subdivisions based on 16S rRNA analysis: α, β, γ, δ, and ε. Mitochondria appear likely to have evolved from early proteobacteria (Vacarri et al., 2006). The *Pseudomonads* are aerobic proteobacteria that are common in soils and water (e.g., *P. mendocina* and *P. putida*) but can also be human pathogens (e.g., *P. aeruginosa*). Members of the *Geobacter* and *Shewanella* genera are proteobacteria that are highly studied for their metal (or S) reducing capabilities and for use in bioremediation strategies and fuel cells. *Acidithiobacillus thiooxidans* is a member of the γ-proteobacteria that transforms sulfide into sulfuric acid. *Escherichia* such as *E. coli* are also members of the γ-proteobacteria. These organisms are perhaps the best studied of the bacteria because *E. coli* is present in the human intestinal tract and because some *Escherischia* can be pathogenic. They are so heavily studied that much is known about their behavior relative to a wide range of metals and other environmentally important compounds (e.g., Ehrlich, 2002; Nies, 1999).

The firmicutes are mostly (but not all) gram-positive bacteria that are found in soil environments but also within the human body, some playing supportive roles and others pathogenic. Bacilli such as *Bacillus subtilis* are members of the firmicutes.

BACTERIAL CELL WALLS

The bacteria have been categorized into two broad groups, gram positive and gram negative, depending on the characteristics of their cell walls. This designation is based on a positive or negative reaction to the gram stain, which was developed in 1884 by Dutch bacteriologist Hans Christian Joachim Gram (Vaccari et al., 2006). The differences in staining—blue-violet for gram-positive and red for gram-negative as viewed by light microscopy—is related to differences in cell wall characteristics (Beveridge, 1988).

Both gram-negative and gram-positive bacterial cell walls contain components that tend to provide an electronegative charge. This characteristic makes the cell walls of many bacteria tend to adsorb metal cations (Beveridge and Doyle, 1989; Beveridge and

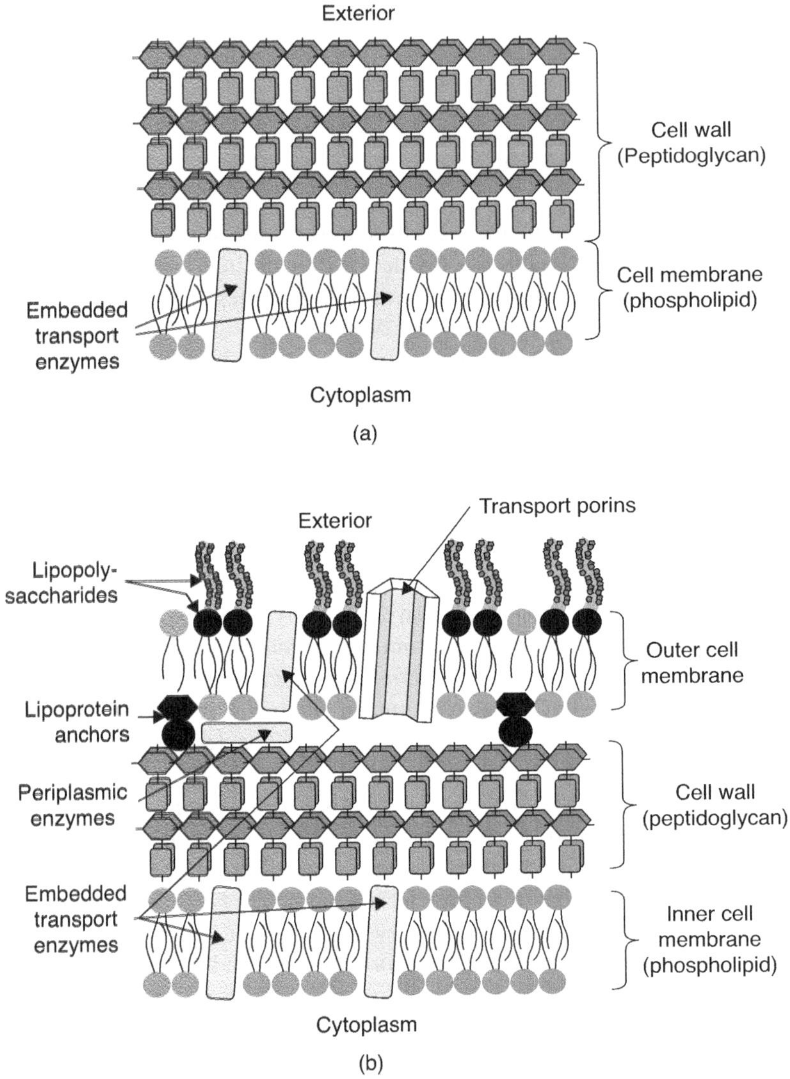

FIGURE 12.8 Comparison of gram-positive (a) and gram-negative (b) bacterial cell walls. From Vaccari et al., 2006 (with permission).

Koval, 1981) and to adsorb to positively charged mineral surfaces. Gram-positive cell walls (Figure 12.8) are dominated by a layer of peptidoglycan (PG) that is rich in carboxylate groups. Secondary polymers such as teichoic or teichuronic acids that are also present in the peptidoglycan material can add additional carboxyl or phosphoryl groups (Beveridge, 1981; Fortin et al., 1997). As shown in Figure 12.8, the gram-negative cell walls have more complex structure and composition, which contain a thin layer of PG between two lipid/protein bilayers known as the plasma membrane (PM) and the outer membrane (OM). The outer membrane contains negatively charged lipopolysaccharide (LPS), such that gram-negative bacterial surfaces also tend to be electronegative (Fortin et al., 1997).

However, there have been reports of net positively charged bacterial surfaces at pH <5, which could facilitate their attachment to silica at low pH (e.g., Amores and Warren, 2007). This development of positive surface charge is likely related to the presence of amino groups (Claessens et al., 2005).

It is important to recognize that bacterial and biofilm surfaces may be far more complex than the surfaces of abiotic mineral phases. This is because bacteria are living organisms, and biofilm environments are dynamic both spatially and temporally. Bacteria can mediate pH through metabolic processes (e.g., Claessens et al., 2005). Dead cells and materials released to solution when cells break apart or lyse can also play biogeochemical roles.

Even when not actively metabolizing, bacterial surfaces are complex. The charge that is present on bacterial surfaces is not concentrated in a single plane, which presents challenges to modeling bacterial surface site protonation/deprotonation and metal adsorption behavior (Turner and Fein, 2006). Surface titrations have been used for determining the pK_a values of sites on viable but nonmetabolizing bacterial surfaces (van der Wal et al., 1997; Turner and Fein, 2006).

BACTERIAL ADHESION AND BIOFILMS

Many surfaces in nature are covered with microbial biofilms. The presence of microbial biofilms can alter the charge on mineral surfaces, the aggregation of soil and other particles, and even the chemical composition (pH, Eh, concentrations of metals and other constituents) of solution at and near the mineral surface.

Microbial attachment to both natural and man-made surfaces, from clays to concrete to hip implants, has been the topic of much study. Indeed, much of the corrosion of metals and concrete is driven by microbial biofilms. Marshall et al. (1971) suggested that bacteria initially adsorb reversibly to a solid surface, largely because of electrostatic interactions, and later adsorb irreversibly as extracellular fibrils are developed that attach the cell firmly to the surface (Marshall and Cruickshank, 1973). Ams et al. (2004) showed that viable but nonmetabolizing gram-positive and gram-negative bacteria adsorbed considerably more to Fe-oxide-coated quartz sand than to uncoated quartz sand. Under the conditions of the experiment, the quartz surface would be negatively charged, whereas the coatings would impart a positive charge. The results of Ams et al. (2004) agreed with the observation that initial attachment is generally reversible and likely caused by electrostatic interactions.

The irreversible phase of attachment involves production of *exopolysaccharides* (EPS) during increased cell division, forming biofilms. Over time in the environment, other organisms join the biofilm to form a complex community structure. Eventually, microorganisms also can leave the biofilm to form colonies elsewhere. Biofilms are microbial accumulations complete with various attachment features including exopolysaccharide (EPS) "glue" and including minerals and sometimes other components (such as blood cells in biofilms inhabiting prosthetic devices in humans). A schematic model for the formation of a biofilm is show in Figure 12.9.

Bacterial biofilms have complex, heterogeneous structures (Costerton et al., 1995). Clusters of microorganisms are separated by interstitial voids so that nutrients are allowed to flow through the system (Lewandowski et al., 1995). Figure 12.10 shows an epifluorescence image of biofilm formed by the aerobic bacterium *P. mendocina* cultured in the presence of Fe(hydr)oxide aggregates. The biofilm is an open network with bacteria and mineral

FIGURE 12.9 Schematic model showing the stages in biofilm formation and development. Revised from D. Davis (http://biology.plosjournals.org/perlserv/?request=slideshow&type=figure&doi=10.1371/journal.pbio.0050307&id=89595); Creative commons.

aggregates embedded in an EPS network. The detailed structure of the EPS, which includes both fibrous attachment features and less-structured organic material, is shown in the atomic force microscopy (AFM) close up of a biofilm from *P. mendocina*, in Figure 4.12. Microorganisms within biofilms may create environments that are different from the bulk in terms of pH, dissolved oxygen, nutrients, metals, and other constituents (Little et al., 1997). Biofilms can thus serve as sites of enhanced mineral corrosion or precipitation (Figure 12.11).

The open biofilm structure is crucial because bacteria rely on transport of nutrients and wastes through the biofilm. Biofilms can range in thickness from a few tens of microns to as much as several centimeters in thickness. Even for a single strain of bacteria grown in the laboratory, biofilm architecture can vary greatly from more open to more clumped depending on such conditions as C source, stir rate, dissolved oxygen distribution, and temperature. The fungus *Exophiala lecanii* changes from more compact and yeast like to to more open and filamentous biofilm structures in response to N limitation (Figure 12.12)

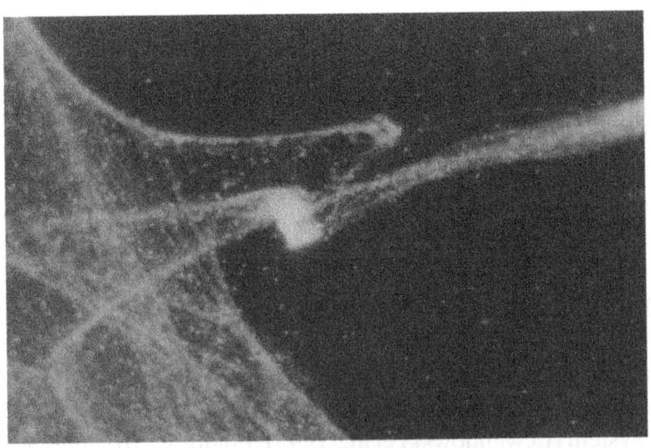

FIGURE 12.10 Epifluorescence image. Acridine Orange stained, 40× *P. mendocina* and Fe oxides. The small specs are bacteria. From Forsythe et al., 1998 (with permission).

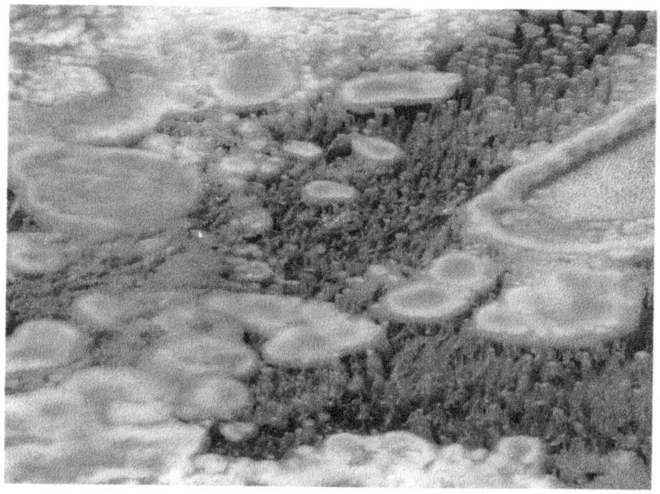

FIGURE 12.11 A microbial mat from a thermal environment at Yellowstone National Park. (photo by P. Maurice). Shown in this image are conical, columnar, and lilypad stromatolites shaped by cyanobacteria (*Leptolyngbya*) that form in some of the quiet hydrothermal pools. Silicate precipitates are present, in association with the biofilm. Cyanobacteria aggregate into cones and columns even in the absence of silica precipitation, but the precipitation of silica preserves the shapes. Conical structures that are found in the deepest parts of the pools are thought to be modern analogs of ancient conical stromatolites that formed in the Precambrian (personal communication, Tanja Bosak, Massachusetts Institute of Technology, March 2008).

(Woertz and Kinney, 2002). The more open structure would permit faster diffusion and more exposure of microbial surfaces to solution, hence improving access to low levels of N.

Biofilms of *P. aeruginosa* grown in the laboratory have different architectures depending on whether C is supplied as glucose, glutamate, or succinate (Figure 12.13).

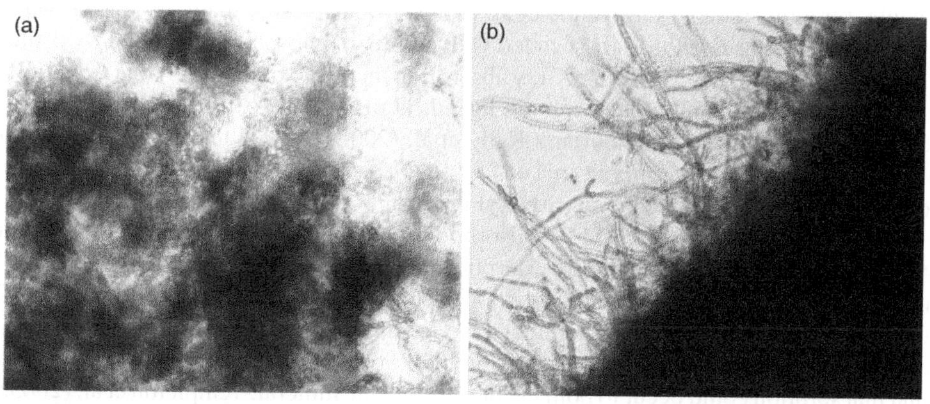

FIGURE 12.12 Change in biofilm architecture of the dimorphic fungus *Exophiala lecanii* from (a) yeast like to (b) filamentous in response to N limitation. The more open structure in (b) would allow for greater access to small concentrations of N. Courtesy J. Woertz, Univ. of Notre Dame.

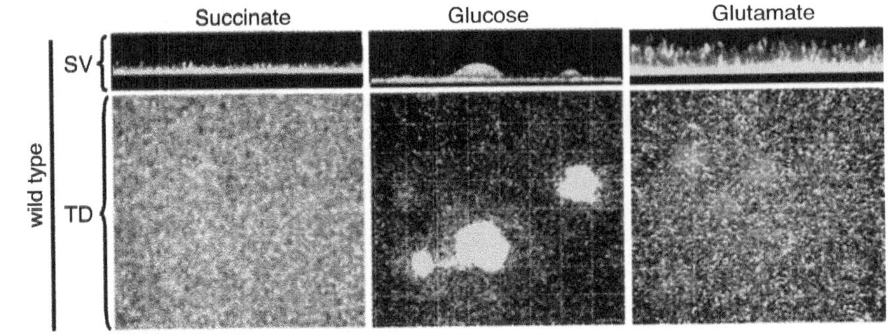

FIGURE 12.13 Confocal scanning laser microscopy (CSLM) images of *P. aeruginosa* biofilms grown in different media. Top is side view; bottom is top view. TD = top display; SV = side view. From Shrout et al., 2006 (with permission).

The formation of biofims is controlled at least in part by a process known as *quorum sensing*. Bacteria can coordinate either within a given species or among species by producing signaling molecules that allow cell–cell communication. The local density of bacteria and the release of signaling molecules can influence the gene expression, which thereby alters behavior such as EPS production, swarming motility, and biofilm formation.

In the field, a biofilm may be organized with different organisms inhabiting different layers and providing various nutrients or otherwise altering chemical parameters to make the system more or less habitable to other microorganisms.

Bacterial–Metal Interactions

Bacteria and bacterial biofilms can also be important locations of metal and radionuclide adsorption. Metal adsorption to live, viable, but nonmetabolizing bacteria generally increases with increasing pH (e.g., Borrok et al., 2004).

Metal sorption to bacteria and biofilms has been the topic of recent investigations that combine bechtop experiments with a wide array of synchtron-based techniques. For example, titration studies by Boyanov et al. (2003) revealed three likely groups on viable but nonmetabolizing *B. subtilis*, associated with carboxyl, phosphoryl, and hydroxyl sites. Boyanov et al. (2003) showed using X-ray absorption fine structure (XAFS) that Cd adsorbs to *B. subtilis* predominantly to phosphoryl ligands at pH < 4.4. As pH increases, carboxyl groups become increasingly important. Kelly et al. (2002) showed that UO_2^{2+} adsorbs to phosphoryl groups on *B. subtilis* at a pH of 1.67. As pH is increased to 4.8, binding becomes progressively more associated with carboxyl groups.

Toner et al. (2005) investigated Zn adsorption to biofilms of *P. putida* using extended XAFS (EXAFS). They found that Zn adsorption at a pH of 9 was primarily associated with phosphoryl groups, with a smaller contribution of carbonyl-type complexes. Templeton et al. (2003a) used a synchtron-based standing-wave technique to investigate Pb(II) sorption at the interface between *Burkholderia cepacia* biofilms and hematite or corundum surfaces and found that sorption could occur to both the biofilm and the mineral. Templeton et al. (2003b) used synchrotron-based grazing-incidence spectroscopy combined with X-ray absorption near-edge structure (XANES) to investigate selenium sorption to *B. cepacia* biofilms and associated α-Al_2O_3. Partitioning could be to both the mineral surface and the biofilm.

Because Se is redox active, changes in the Se partitioning behavior over time were influenced by microbially induced reduction of Se(VI) and Se(IV) to Se(0). Kemner et al. (2004) used synchrotron-based hard X-ray imaging and elemental mapping at 150-nm spatial resolution coupled with XANES to investigate Cr interactions with *Pseudomonas fluorescens*. They showed that Cr associated with phosphoryl groups. Several groups have used AFM to measure forces between bacteria and surfaces, as discussed in Chapter 4.

Metals do not just attach to bacterial surfaces. In metabolizing bacteria, metals can be taken up by the cell. Some metals are essential, some are toxic, and many are essential but toxic at high concentrations. Two general types of uptake systems are fast and unspecific versus slow and specific (Nies, 1999). The fast systems are generally driven just by the chemiosmotic gradient across the bacteria's cytoplasmic membrane. The slower system has high substrate specificity, often uses ATP hydrolysis as the energy source, and only tends to be used when the cell is starving or in need (Nies, 1999). Three possible mechanisms for microbial resistance to heavy metals are as follows (Nies, 1999): (1) efflux or active extrusion of some heavy-metal ions from the cell, (2) segregation of certain metals (primarily "sulfur lovers") by thiol-containing compounds, or (3) reduction of metals to a less toxic state, for metals whose toxicity is strongly redox sensitive. Sometimes a combination of mechanisms is used.

BACTERIAL-PROMOTED MINERAL DISSOLUTION

Bacteria can affect rates of mineral dissolution through a variety of processes, such as exuding organic ligands, some of which may have metal chelation or reductive effects; altering pH; altering chemical equilibria and saturation state; metabolizing organic compounds; or attaching directly to mineral surfaces and enhancing dissolution by cell-wall–associated processes, most notably reduction. Much has been written on these topics. A few examples are provided here.

Dissolution of Fe(III)(hydr)oxides by DIRB

The dissolution of Fe(hydr)oxides by dissimilatory Fe-reducing bacteria (DIRB) has been the topic of much study because Fe(hydr)oxides have an ability to bind metals and radionuclides strongly and thus to affect mobility through porous media. Reductive dissolution of Fe(hydr)oxides could remobilize adsorbed contaminants. Two of the most studied genera of DIRB are *Geobacter*, which are strict anaerobes, and *Shewanella*, which are facultative anaerobes. *Geobacter* and *Shewanella* strains have the ability to couple organic matter or in some cases H_2 oxidation with Fe(hydr)oxide reductive dissolution as a means of gaining energy. The latter occurs by

$$0.5\ H_2 + 2H^+ + FeOOH \rightarrow Fe^{2+} + 2H_2O. \qquad (12.14)$$

The ferric iron either serves as the dominant terminal electron acceptor for bacterial respiration or accompanies fermentation as a supplementary terminal electron acceptor. Both of these are *dissimilatory* processes, in which the Fe is not being incorporated into specific cellular components. If Fe were incorporated directly into cell components, the process would be called *assimilatory*. The reduction of Fe(III) coupled with the degradation of acetate to allow population growth of *Geobacter* is illustrated in Figure 12.14.

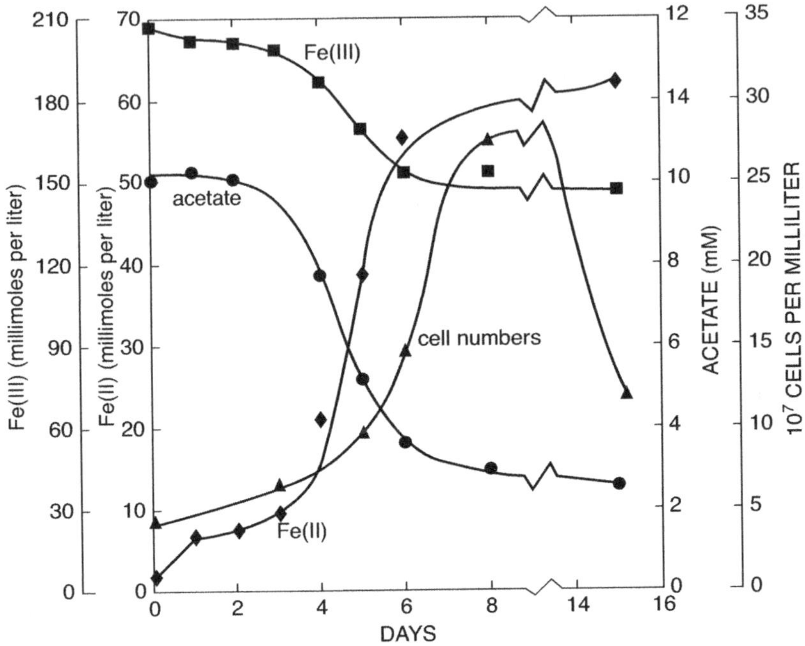

FIGURE 12.14 Reduction of Fe(III) by acetate mediated by *Geobacter metallireducens* (GS-15). From Lovley and Phillips, 1988 (with permission).

One key question that has been asked is whether *Geobacter* and *Shewanella* only dissolve a solid Fe(III)(hydr)oxide [or a Mn(IV) oxide] at contact points between the bacterial cell and the solid surface or whether the bacteria can effect dissolution at a distance from the location of the cell itself. The DIRB have developed several strategies, which include (1) transferring electrons directly to the mineral surface via multiheme cytochromes associated with the outer membrane of the cell, which requires direct cell-mineral contact; (2) using conductive pili or "nanowires" produced by and radiating out of the cell to extend the region in which the activity can occur; or using electron-shuttling compounds either produced by (3) the cell itself or (4) present in the medium (such as humic acids) to transfer electrons at a distance removed from the cell (Gorby et al., 2006). Some studies and results are highlighted below, but this remains a topic of much research and discussion.

Nevin and Lovley (2000) showed that *Geobacter metallireducens* is incapable of producing and exuding extracellular electron-shuttling compounds that would be needed for process (3). Dissolution is generally limited to the contact area at the bacteria–mineral interface, and often through process (1). *Geobacter* can also produce nanowires to conduct electrons as in processs (2). Nonlocal dissolution can occur if an electron shuttling component such as a humic acid is present to transfer electrons between *Geobacter* and the mineral surface (4) (Lovley et al., 1996). This process is illustrated in Figure 12.15 and includes oxidation of acetate from the growth medium, transfer of the electrons through *Geobacter* to the humic, and transfer of the electrons from the humic to the mineral surface to reduce Fe(III) to Fe(II). Fe(II) is then released to solution.

Shewanella can use several means of reducing Fe oxides, and all four processes have been observed for various *Shewanella* strains. Decaheme cytochromes OmcA and MtrC

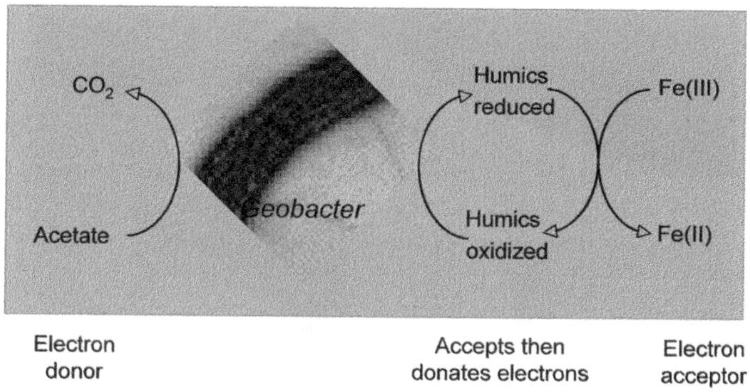

FIGURE 12.15 The process of humic shuttling of electrons used in *Geobacter* reduction of Fe(III). After Lovley, 1996, including a photomicrograph of *Geobacter* courtesy of D. Lovley (University of Massachusetts).

associated with the cell outer membrane of *Shewanella oneidensis* are believed to act as terminal metal-reducing enzymes when in direct contact or close proximity with the mineral surface (e.g., Shi et al., 2006). In contrast to *Geobacter*, *Shewanella* seems to be able to synthesize and exude electron shuttling compounds that allow dissolution to occur at some distance removed from the actual sites of cell attachment to Fe(hydr)oxides, even in the absence of a compound such as a humic acid. The shuttle molecules, likely quinones, receive electrons from the outer-membrane cytochromes and transfer them to the mineral surface (Newman and Kolter, 2000). *Shewanella* also can produce nanowires. An AFM image of hematite particles reacted with *Shewanella* shows small etch pits that do not mimic the shape of the bacteria (Figure 12.16). This suggests that dissolution either occurred at some distance from the cells via a shuttling compound and/or by the activity of nanowires (Rosso et al., 2003; Gorby et al., 2006).

Dissimilatory Metal-Reducing Bacteria

Many bacteria can reduce a wide array of metals and radionuclides and are known more broadly as dissimilatory metal-reducing bacteria (DMRB). For example, Lovley et al. (1991) showed that *Geobacter* GS-15 could couple acetate oxidation with reduction of more mobile U(VI) to less mobile U(IV) according to the generalized reaction:

$$CH_3COO^- + 4U(VI) + 4H_2O \rightarrow 4U(IV) + 2HCO_3^-\ 9H^+. \qquad (12.15)$$

Cell growth was coupled with reduction, as shown in Figure 12.17. The bacteria can obtain energy for population growth from this electron-transfer mechanism.

Microbial Effects on Carbonate Dissolution

Microorganisms may influence rates of carbonate growth and dissolution and incorporation of contaminants. Lüttge and Conrad (2004) and Davis et al. (2007) used AFM and vertical-scanning inferometry (VSI) to study the effects of surface colonization by *S. oneidensis* MR-1 on rates of calcite dissolution. Both AFM and VSI can operate on surfaces immersed in growth medium, allowing for three-dimensional (3-D) topographic measurements over

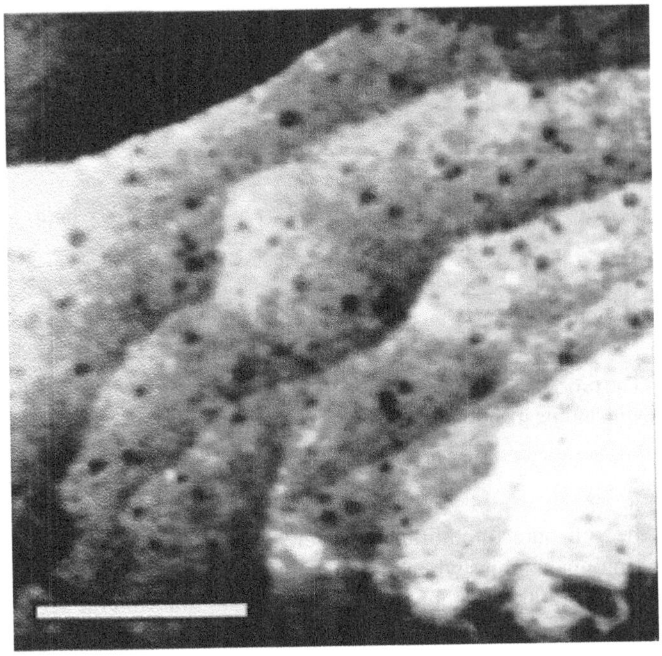

FIGURE 12.16 AFM image of small etch pits formed on the surfaces of hematite particles in the presence of *Shewanella*. Because these pits do not mimic the shape of the bacteria, it was proposed that they represented dissolution by an electron shuttling compound. It is also possible that they could have formed from nanowires produced from the bacteria. The etch pit shape suggests dissolution at point defects. From Rosso et al., 2003 (with permission). Scale bar = 1 μm.

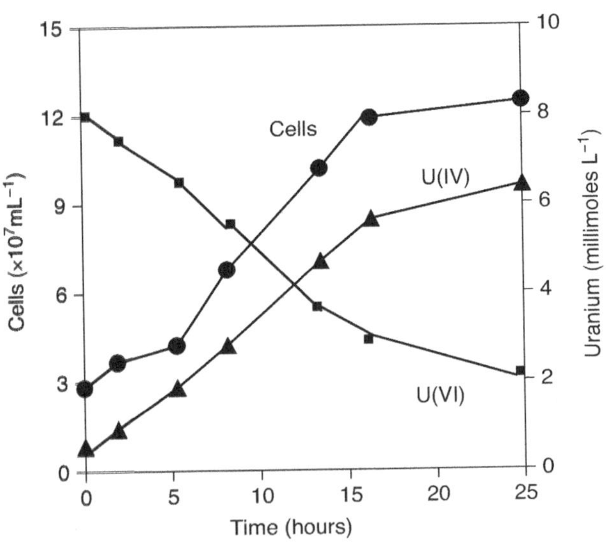

FIGURE 12.17 *Geobacter* growth using uranium reduction. Redrafted for clarity from Lovley et al., 1991 (with permission).

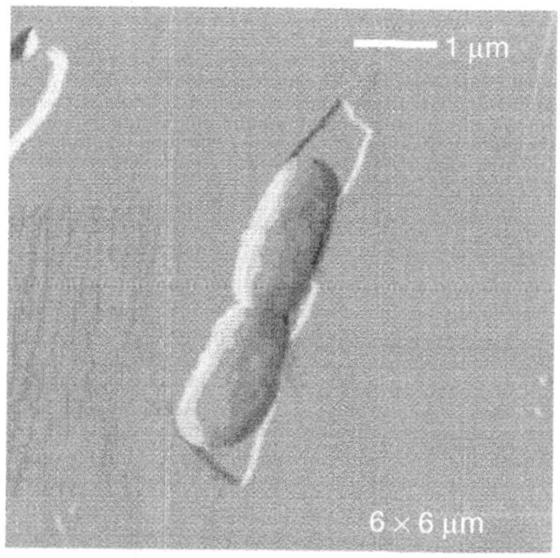

FIGURE 12.18 Atomic force microscope images of calcite with *S. oneidensis* attached to a dissolution feature. From Davis et al., 2007 (with permission).

the course of reaction. AFM has higher lateral resolution (potentially Å to nm scale) but can be destructive; VSI is nondestructive but has lesser lateral resolution (hundreds of nm to mm scale). Hence, the two techniques are complementary.

Under aerobic conditions, *S. oneidensis* attached to previously nucleated etch features on the calcite surface and inhibited calcite dissolution by interfering with etch pit development. The inhibition process resulted in a 40–70% decrease in the calcite surface normalized dissolution rate in the presence of the bacteria. Figure 12.18a shows an AFM image of *S. oneidensis* attached to nascent etch features on the calcite surface. In the absence of the bacteria, dissolution features would grow into substantial etch pits. But in the presence of the bacteria, etch pit growth was blocked. *S. oneidensis* enhanced dolomite [$CaMg(CO_3)_2$] dissolution at the site of the cell attachment but inhibited additional dissolution in the area of attachment and biofilm formation.

The Importance of Field-Based Studies

Many studies of microbially mediated mineral dissolution or biomineralization are conducted in the laboratory on a single strain. However, what we most want to know is how bacteria interact with minerals and nutrients in the field, where different processes may occur and where natural consortia are active rather than a single strain.

Conducting field-based studies is important because natural microbial consortia remain difficult or impossible to maintain in the laboratory for substantial periods of time and because consortia are difficult to characterize. Many microbial strains are resistant to cultivation in the laboratory. Thus, numerous laboratory dissolution studies have been conducted on a few key strains of bacteria. New methods are extending cultivation to far more strains of bacteria, but it is still difficult or impossible in the laboratory to approach the complex biogeochemical processes of bacterial consortia in the field. One means of coupling field and laboratory methods is the *in situ microcosm* approach.

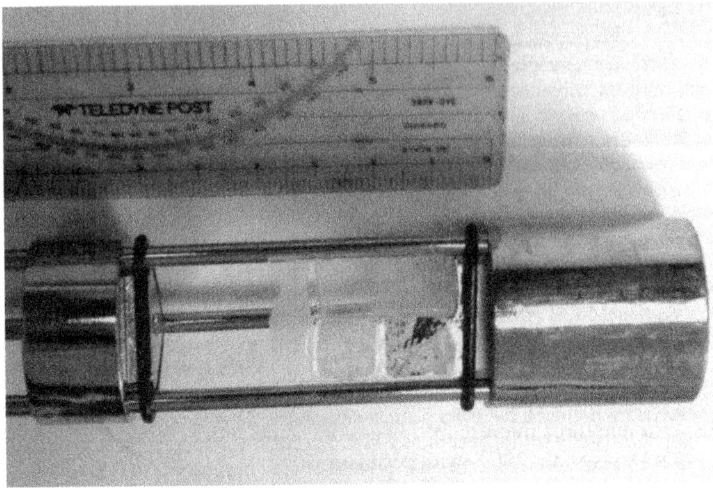

FIGURE 12.19 In situ microcosm for emplacement in a well. From Bennett, 2006 (with permission).

Case Study: The In Situ Microcosm Approach

Bennett et al. (2001) suggested that microorganisms may acquire P, which can be a limiting nutrient, from apatite inclusions in feldspar. Bennett's group developed an *in situ* microcosm approach in which clean, sterile mineral chips were placed in permeable polyethylene chambers that were then buried or inserted into wells (Figure 12.19). The field site was the oil-contaminated Bemidji Aquifer, where microbial processes and aluminosilicate weathering are highlighted. After 3 months of exposure in the reactors, feldspar samples with no apatite inclusions showed little or no evidence of bacterial colonization or weathering. In contrast, mineral samples with significant P, including feldspars and basalt, all showed heavy colonization and weathering, which suggested that the bacteria were acquiring P from these minerals (Figure 12.20 a, b). The authors pointed out that differences in surface charge and other properties might also explain the results, at least in part.

To control surface composition systematically while minimizing potential confounding factors, Rogers and Bennett (2004) used manufactured glasses with different amounts of apatite (for P) or goethite (for Fe) inclusions. The silicate glass experiments showed little or no colonization of samples without the nutrients, but heavy colonization and dissolution of P-containing samples most likely also affected by the presence of Fe (Figure 12.21). The observations of enhanced colonization of glasses with greater nutrient content supported the hypothesis that the bacteria can benefit from dissolution of nutrient-bearing substrates, which include minerals.

Coupling In Situ Microcosms with Community Analysis

The in situ microcosm approach can be coupled with methods to determine the community structure of microorganisms attached to mineral surfaces. For example, Boyd et al., (2007) used genetic analysis to show that the detailed structure and composition of microbial

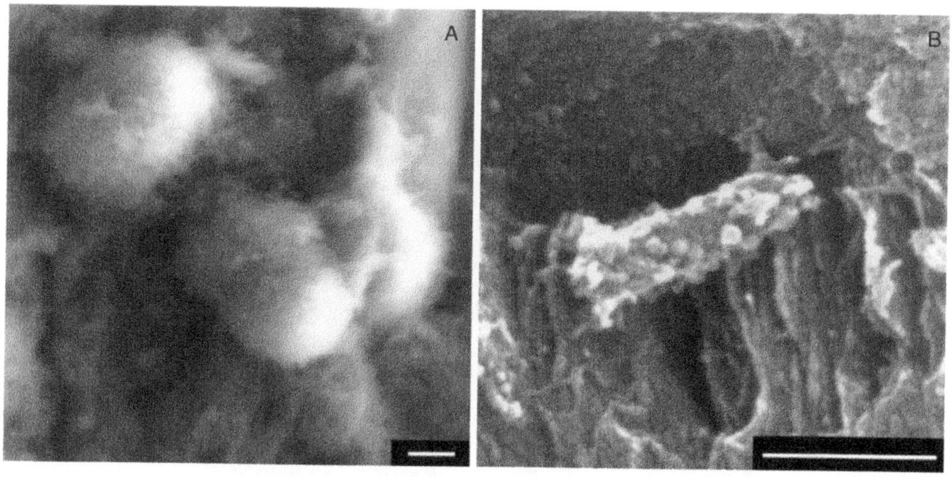

FIGURE 12.20 (a) An environmental SEM image of anorthoclase feldspar after 1 year in an in situ microcosm. The surface is weathered and covered by spherical cocci and diplococci bacteria, which are coated in a secondary clay precipitate. (b) A confocal-SEM image of basalt after 3 months of extensive weathering. Precipitates are present on the bacterium near the center of the image. Scale bar = 1 μm in both images. From Bennett et al., 2001 (with permission).

communities associated with different mineral subtrates from in situ microcosms at a field site in Oak Ridge Tennesee were distinct. This study thus demonstrated that the biofilm community structure can depend on small-scale mineralogic heterogeneity.

Professor Gill Geesey and his group at Montana State University have developed in situ continuous flow reactors (Figure 12.22) that can be emplaced in surface-water environments such as geothermal springs at Yellowstone National Park and allowed to incubate for extended periods of time. By coupling such an in situ method with laboratory

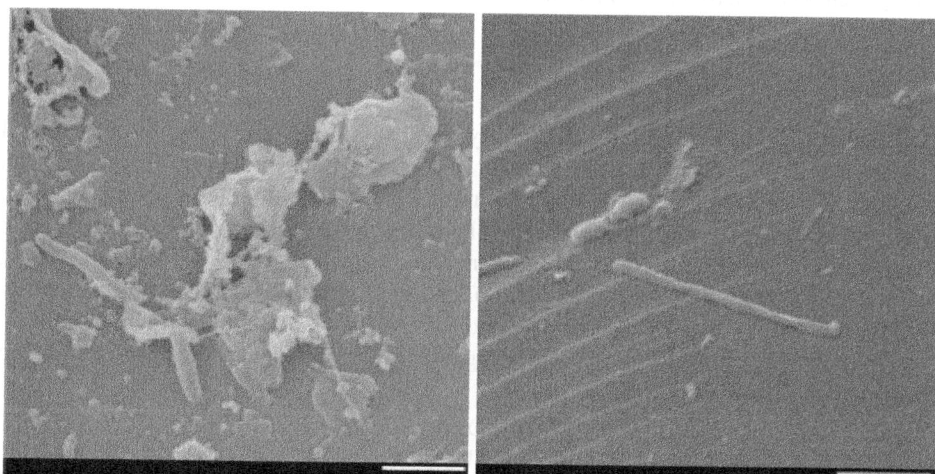

FIGURE 12.21 SEM photomicrographs of glass samples that were in groundwater for 9 months. Enhanced microbial colonization of nutrient-bearing glass (left) versus non-nutrient glass (right) as observed by Rogers and Bennett, 2004 (with permission).

FIGURE 12.22 In situ flow-through reactors used in biogeochemical investigations at a hot spring in Yellowstone National Park. Photo courtesy of E. Boyd and G. Geesey, Montana State University.

analysis, the field-based microbial community composition as well as the physiological and biogeochemical activity can be determined.

Siderophores

Fe(III)(hydr)oxide minerals tend to have very low solubilities in aerobic, circum-neutral environments. This presents a challenge for plants and microorganisms living in such environments, because Fe is an important nutrient. One strategy that has evolved for some aerobic microorganisms and graminaceous plants for the acquisition of Fe is production of low-molecular-weight Fe-complexing ligands known as siderophores (Neilands, 1981; Raymond and Dertz, 2004; Kraemer, 2004). Siderophore-Fe(III) 1:1 complex stability constants can be as high as 10^{52} (Raymond and Dertz, 2004; Albrecht-Gary and Crumbliss, 1998). Figure 12.23 shows some common ligands found in siderophores, which include catechol and hydroxamate.

Figure 12.24 illustrates the siderophore cycling process involved in aerobic bacterial acquisition of Fe. Fe may be removed from the siderophore at the cell wall and the siderophore may be recycled or the siderophore-Fe complex may be taken up into the cell and the siderophore may be broken down intracellularly.

Siderophores may enhance Fe(hydr)oxide dissolution either indirectly through their pronounced effects on solubility or directly by attaching to the surface for ligand-promoted,

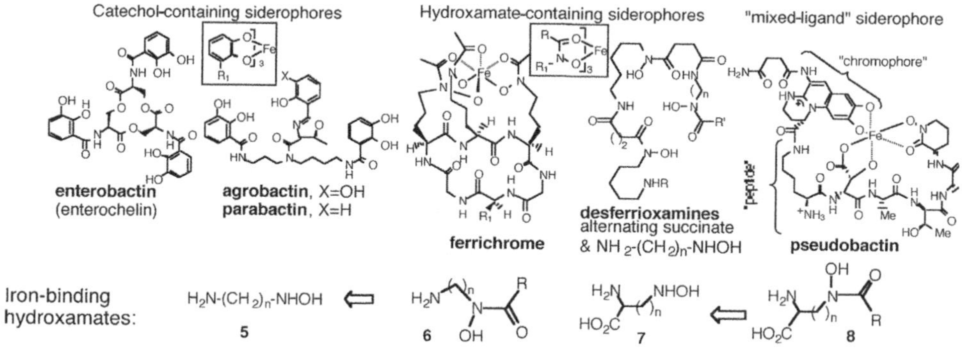

FIGURE 12.23 Some common siderophore structures. Courtesy of M. Miller, University of Notre Dame.

surface-controlled dissolution process (Kraemer et al., 1999; Kraemer, 2004). Siderophores may also allow bacteria to acquire Fe from Fe-bearing silicate minerals (e.g., Liermann et al., 2000; Ams et al., 2002; Buss et al., 2006).

One of the most highly studied siderophores from an environmental standpoint is the trihydroxamate siderophore desferrioxamine B (DFO-B), which is a naturally occurring and environmentally relevant siderophore that is also commercially available. DFO-B was first isolated from cultures of the soil bacterium *Streptomyces pilosus* (Bickel et al., 1960). DFO-B is present as a monovalent cation over a wide pH range (Duckworth and Sposito, 2005). Holmén et al. (1997) showed that the small monohydroxamate ligand acetohydroxamic acid (aHA) forms a bidentate mononuclear surface complex with surface Fe^{3+} through both oxygen atoms of the hydroxamate functional group (Holmén et al., 1997). DFO-B may form at least binuclear complexes with surface Fe sites (Kalinowski et al. 2000; Kendall et al., 2005). DFO-B has been shown to enhance dissolution of a variety of minerals, such

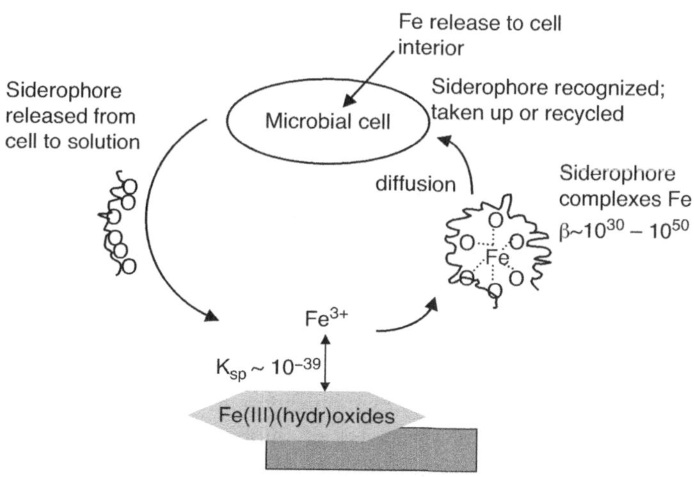

FIGURE 12.24 A schematic representation of siderophore-mediated acquisition of Fe from the environment by microorganisms.

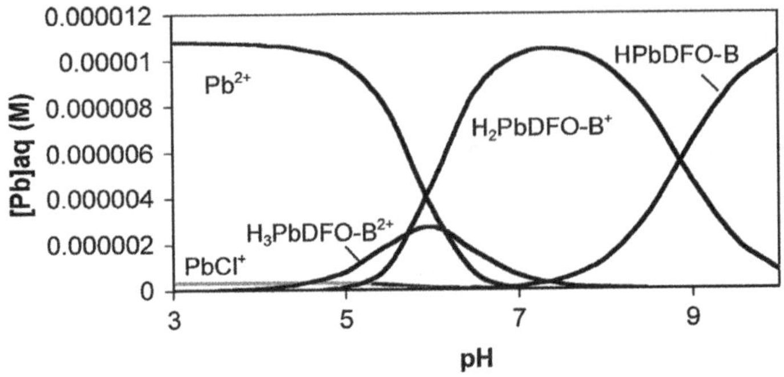

FIGURE 12.25 The distribution of dissolved Pb species in the presence of the siderophore DFO-B. From Hepinstall et al., 2005 (with permission).

as goethite, kaolinite, and hornblende (Kraemer et al., 1999; Maurice et al., 2001; Buss et al., 2007).

DFO-B binds not only Fe(III) but also other metals like Pb and Cd strongly, which suggests that siderophores may affect metal mobility in porous media. Speciation of Pb in the presence of DFO-B is shown in Figure 12.25. DFO-B forms complexes with Pb at pH > ~5 (for the particular conditions). Adsorption experiments by Hepinstall et al. (2005) showed that DFO-B enhanced Pb adsorption to kaolinite slightly around a pH of 6 but inhibited adsorption at a pH above 6. EXAFS analysis by Mishra et al. (2007) showed that DFO-B–Pb-kaolinite ternary surface complexes occurred at pH 6–9. Haack et al. (2008) showed that DFO-B absorbs in the interlayers of montmorillonite to form bilayers at high absorption density and enhances Fe dissolution from montmorillonite.

MICROBIAL BIOMINERALIZATION

Biomineralization is the process by which living organisms catalyze or otherwise mediate the formation of minerals (Chapter 9). Bacteria may affect mineral precipitation either directly by altering solution saturation state with respect to a given mineral phase or more indirectly by providing solid surfaces for heterogeneous nucleation. In some cases, biominerals may actually form inside of bacteria. Bacteria may affect the precipitation of a wide range of minerals, such as carbonates; phosphates; sulfates and sulfides; iron, manganese, and uranium (hydr)oxides; and many silicates. Amorphous silica and metallic S may also be formed through bacterial biomineralization. The role of microorganisms in mineral nucleation and growth can be profound. For example, microorganisms seem to be responsible for most Mn(II) oxidation in the environment, and hence they play a dominant role in the formation of Mn oxides (Tebo et al., 1997).

Carbonate Precipitation

The role of microorganisms, particularly cyanobacteria, in carbonate mineral precipitation has received a great deal of interest because of implications for the evolution of early life on Earth. Cyanobacteria are believed to be one of the earliest organisms to have biomineralized, and they are one of the most abundant life forms on Earth (Skinner and Jahren, 2003). Metabolic uptake and fixation of dissolved CO_2 by cyanobacteria increases saturation state, brings about carbonate precipitation (e.g., Thompson and Ferris, 1990), and releases O_2 to the atmosphere. Cyanobacteria calcification is extracellular, occurring on a mucilaginous and fibrous polysaccharide sheath (Skinner and Jahren, 2003). It is likely that both biologically induced and biologically controlled (through molecular templates) mineralization occur (Skinner and Jahren, 2003).

Stromatolites are layered sedimentary structures often formed by carbonate minerals, although siliceous stromatolites may exist in hydrothermal environments. Stromatolite-like carbonate structures date back as much as ~3.5 million years in the sedimentary record (Schopf et al., 1983). It is not always clear that the older formations were formed by microbial processes, because of lack of clear fossil evidence and because postdepositional diagenetic changes are likely.

Much research is aimed at determining whether various stromatolite-containing formations were biotically formed. Recent research includes work in Australia on the 3.43-billion-year-old Strelley Pool Chert (Allwood et al., 2006) and the 2.724-billion-year-old Tumbiana Formation (Lepot et al., 2008), both seeming to contain microbially produced stromatolites. With the evolution of grazing protozoa, stromatolites decreased in abundance and distribution. Modern carbonate stromatolites are found in hypersaline locations such as Shark Bay, Australia and freshwater sites such as Cuatro Ciénegas, Mexico.

Algae are the most abundant calcium-carbonate-depositing organisms, and they account for most of the oceanic carbonate sedimentation and worldwide chalk deposits (Lowenstam and Weiner, 1989, in Skinner and Jahren, 2003). Bacterially mediated carbonate precipitation also occurs and has implications for contaminant fate and transport and potential deep CO_2 sequestration to help mitigate global warming (see Chapter 14). Dr. Grant Ferris (University of Toronto) and his colleagues have investigated microbially induced Ca-carbonate precipitation as a potential tool for bioremediating radionuclide and trace metal contaminants such as Sr in groundwater systems. One area of focus is the role of the bacterial degradation of urea by the bacterial enzyme urease to derive ammonia. This process increases solution pH and alkalinity such that it can promote carbonate precipitation, according to

$$CO(NH_2)_2 \text{ (urea)} + 2H_2O + H^+ \rightleftarrows 2NH_4^+ + HCO_3^- \qquad (12.16)$$

$$2HCO_3^- + Ca^{2+} \rightleftarrows CaCO_3(s) + CO_2 + H_2O \qquad (12.17)$$

(Warren et al., 2001).

Warren et al. (2001) observed carbonate precipitation associated with the surfaces of *Bacillus pasteurii* suggesting heterogeneous nucleation; some evidence for additional homogeneous nucleation was also observed. Scanning electron microscopy (SEM) and AFM images showed a wide variety of precipitate morphologies, including dumbbell shapes, some of which were composed of crystalline rhombohedra (Figure 12.26 a, b). Sr could be taken up, forming solid solution with Ca carbonate. UO_2 was also sequestered but not as effectively. In the presence of Sr and UO_2, not only calcite but also vaterite were detected by a combination of SEM and X-ray diffraction (XRD) (Warren et al., 2001).

Fe(III)(hydr)oxide Precipitaton: BIOS

Ferris (2005) has defined biogenic iron oxides (BIOS) as "composite materials that consist of intact and partly degraded remains of bacterial cells intermixed with variable amounts of poorly ordered hydrous ferric oxide (HFO) minerals." BIOS are often nanoparticulate,

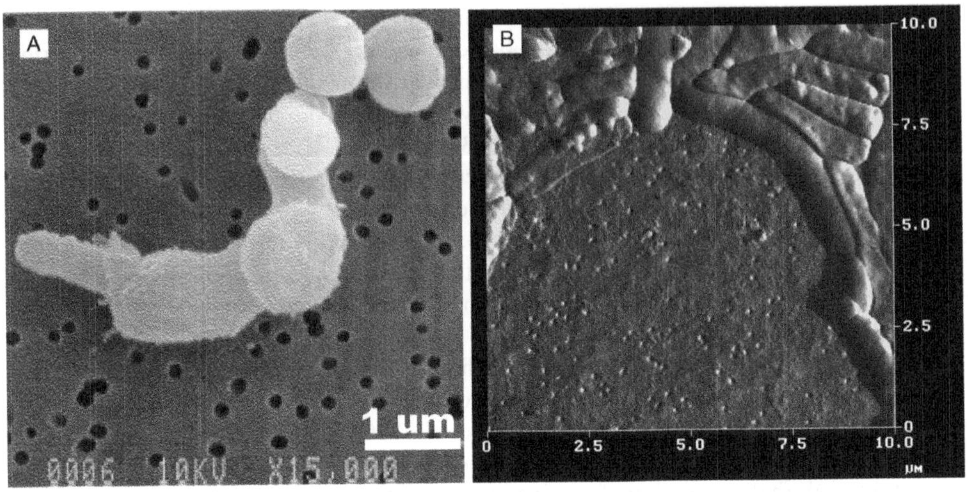

FIGURE 12.26 (a) Carbonate mineral nuclei on *Bacillus pasteurii* imaged by SEM. Sr was present in this experiment. (b) *Bacillus pasteurii* with carbonate precipitates imaged by AFM; compare with SEM image. This sample was in the absence of Sr. From Warren et al., 2001 (with permission).

and they play important roles in adsorption of metals, radionuclides, and phosphate. BIOS form when bacteria oxidize Fe(II) to Fe(III), which then undergoes hydrolysis to precipitate in close association with the cells. The HFO minerals in BIOS often form first by adsorption of Fe(III) on mineral surfaces, then mineral precipitation (Warren and Ferris, 1998). In circumneutral environments, BIOS formation is generally dominated by two-line ferrihydrite, whereas in acidic environments, a range of minerals can form, such as ferrihydite, goethite, jarosite $[(H,Na,K)Fe_3(OH)_6(SO_4)_2]$ (generally at a pH of <3.0), and Schwertmannite $[Fe_8O_8(OH)_6SO_4]$ (generally at a pH of 3.0–4.5) (Ferris, 2005).

Sometimes, Fe oxide precipitation can occur intracellularly. An example of amorphous Fe-rich nanoparticles (with some evidence of crystalline structure) precipitated within the cytoplasm of *S. putrifaciens* is shown in Figure 12.27. The bacteria were allowed to react with HFO, particles of which are present on the external surface of the bacterium. As the reaction progressed, internal nanoparticles formed over time.

Magnetotactic bacteria take up complexed ferric iron and form magnetite (Fe_3O_4) deposits in membrane-bound structures called magnetosomes (Figure 12.28). The magnetite crystals are single-domain magnets that allow the motile bacteria to align with the Earth's magnetic field. Bazylinski and Frankel (2004) describe microbial magnetosome formation as a "complex process that involves several discrete steps, including magnetosome vesicle formation, iron uptake by the cell, iron transport into the magnetosome vesicle and controlled Fe_3O_4 (or Fe_3S_4) biomineralization within the magnetosome vesicle." A transmission electron microscopy (TEM) micrograph of magnetosomes from microbiologist T. Beveridge is shown in Figure 12.28.

Banded Iron Formations (BIF)

The ability of microorganisms to influence the precipitation of Fe(hydr)oxides has led to much discussion of their potential role(s) in the formation of the Precambrian *banded iron formations* (BIF). These formations are laminated chemical sediments that consist of millimeter-thick layers of hematite or magnetite intermbedded with quartz or chert (Cornell

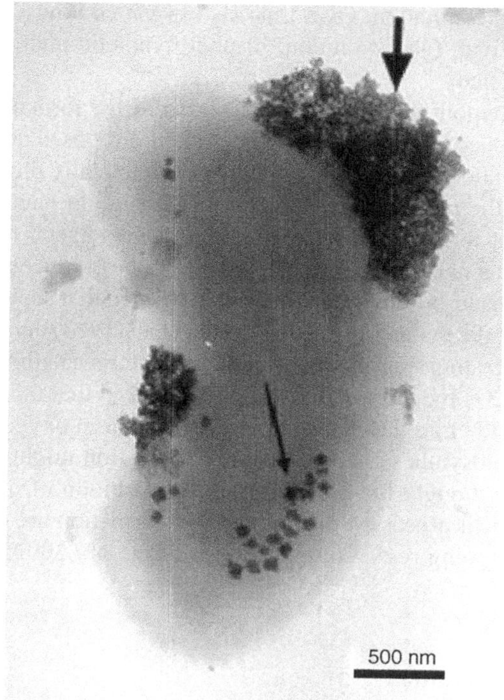

FIGURE 12.27 A TEM micrograph of a cell of *Shewanella putrfaciens* CN32. The large arrow shows extracellular ferrihydrite. Anaerobic respiration involving the ferrihydrite resulted in intracellular iron-rich minerals (small arrow). From Glasauer et al., 2002 (with permission).

and Schwertmann, 2003, and references therein) that represent a major source of iron ore and that have been taken to provide important evidence for the evolution of the Earth's atmosphere. They formed between 3.8 and 0.8 Ga (Ga = thousand million years ago) and have an average iron oxidation state of +2.4 (Klein and Beukes, 1992). Conventionally, it has been thought that the Fe-mineral layers formed in sea water as oxygen released by photosynthetic cyanobacteria combined with dissolved iron to form insoluble iron oxides, including magnetite, which precipitated out (Cloud, 1965). The early photosynthesizers most likely would be poisoned by oxygen production, but oxygen would have been

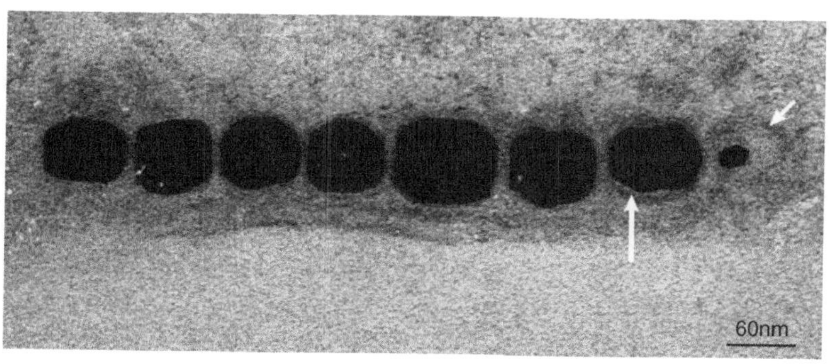

FIGURE 12.28 A TEM miocrograph of a thin section of a magnetosome chain in a cell of the marine spirillum strain MV-4. Image by T. Beveridge as published in Bazylinski and Frankel, 2004 (with permission).

removed by Fe oxide precipitation. Over time, as oxygen continued to accumulate in the oceans through the activity (and evolution) of photosynthetic microorganisms, dissolved Fe eventually was depleted.

A variety of other hypotheses have been put forward for formation of BIF, including abiotic, microbially induced and microbially controlled. Proposed abiotic mechanisms include ultraviolet (UV) photooxidation or oxidation by abiotically produced O_2. Perry et al. (1973) suggested that the magnetite component of BIFs might have formed through biogenic processes similar to those observed today in magnetotactic bacteria. Lovley et al. (1987) demonstrated that nanoparticulate magnetite may be produced by DIRB from amorphous ferric oxide in anaerobic environments and suggested that such a process might have formed at least some of the magnetite in BIF. Widdel et al. (1993) demonstrated that anoxygenic (not oxygen generating) purple photosynthetic bacteria can directly oxidize Fe(II) to Fe(III), coupled with CO_2 reduction. This finding suggested that the oxidized iron in BIFs might have come about before the evolution of cyanobacterial oxygen production through oxygenic two-step photosynthesis. If so, then BIF formation might not be as compelling evidence as previously thought for the timing of the evolution of oxygenic photosynthesis. The (bio)geochemical processes involved in BIF formation are hotly debated and are the subject of much ongoing research (e.g., Konhauser et al., 2002; Kappler et al., 2005; Trouwborst et al., 2007).

(Alumino)silicate Precipitation

Microorganisms play an important role in the Si cycle and in both the precipitation and dissolution of silicate minerals and amorphous solids. Silica is taken up by a variety of bacteria, fungi, and other microorganisms. Diatoms and radiolarians have been studied extensively because they are eukaryotic microorganisms that form siliceous shells, which consitute an important part of the fossil record.

Formation of siliceous deposits in association with microbial mats in hydrothermal environments such as Yellowstone National Park, the North Island of New Zealand, and Iceland has also been studied for many years (e.g., Walter et al., 1976; Konhauser et al., 2001; Jones et al., 2001, 2002; Amores and Warren, 2007). Much controversy has raged over whether the hot springs silicification is purely abiotic, biologically induced, or biologically controlled. Work by Amores and Warren (2008) suggests that biosilicification may occur under acidic conditions at moderate temperatures (35°C), when SiO_2 polymerization is favored. Not only extracellular (Jones et al. 2002) (Figure 12.29) but also intracellular silicification

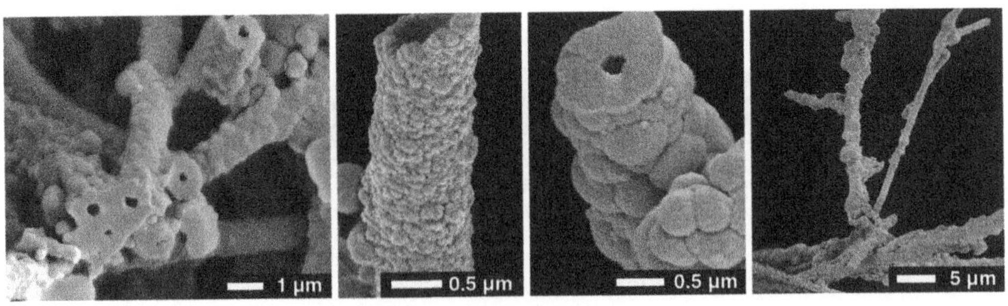

FIGURE 12.29 SEM photomicrographs of some silicified microorganisms from hydrothermal localities in New Zealand. From Jones et al., 2002.

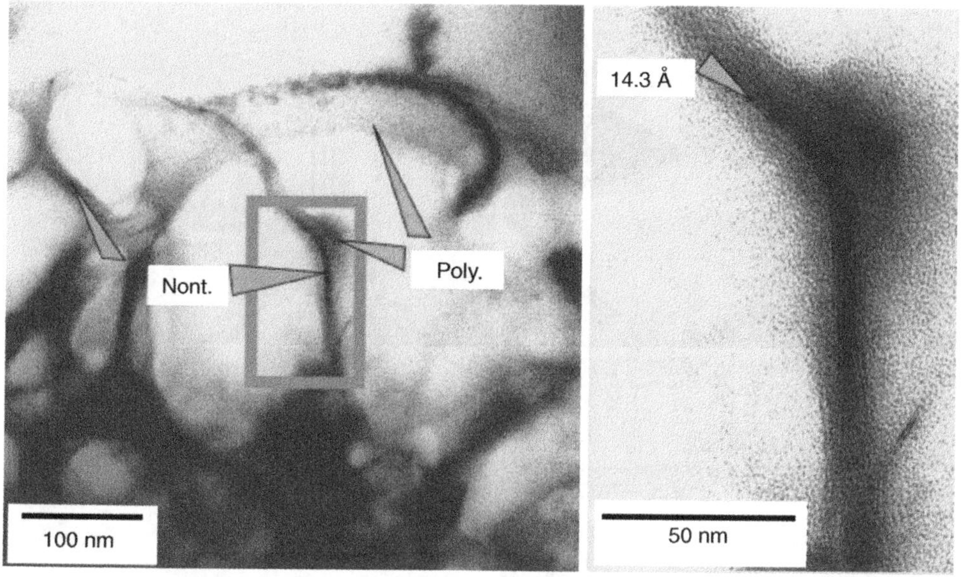

FIGURE 12.30 TEM images of bacterial exopolymers (Ru-red stained TEM), which suggest enhancement of nontronite nucleation as described by Ueshima and Tazaki, 2001. Courtesy of M. Ueshima, University of Notre Dame.

(Konhauser et al., 2001) have been observed, which suggests that the biosilicification may at times be biologically controlled rather than only biologically induced.

In some cases, bacterial exopolysaccharides could potentially help to template aluminosilicate mineral nucleation. For example, Ueshima and Tazaki (2001) observed the Fe-rich smectite mineral nontronite associated with bacteria and biofilm materials in samples from surface layers of deep-sea sediments from Iheya Basin, Okinawa Trough, Japan, using TEM. An experiment was then conducted in which a solution of polysaccharides (dextrin and pectin) was mixed with ferrosiliceous groundwater. Fe-rich layered silicates formed in association with the polysaccharides (Figure 12.30). Although additional study is needed, this research suggested that the exopolysaccharides might play a role in the formation of layered silicates.

Case Study: Bioremediation of U at the Oak Ridge National Laboratory SITE

The use of microorganisms to biodegrade toxic organic compounds into (hopefully) less toxic products has become relatively well established and used at a wide array of Environmental Protection Agency (EPA) sites. The ability of microorganisms to change the redox speciation of metals and radionuclides and hence to affect their adsorption and/or precipitation is well established from laboratory studies (e.g., Lovley et al., 1991; see above). But, application to actual field sites requires a better understanding of the microbial processes and the long-term effectiveness of this immobilization strategy.

Since 2003, the Oak Ridge National Laboratory Field Research Center has been the subject of an intense study of subsurface U bioremediation. Like many department of energy (DOE) sites, Oak Ridge has a legacy of subsurface U contamination from the

FIGURE 12.31 Researchers at Oak Ridge National Laboratory and collaborators from a variety of universities are using biostimulation to transform mobile U(VI) to immobile U(IV). Figure of leaking pits (upper left; from the Department of Energy) and equipment at the site. Courtesy of C. Criddle, Stanford University.

Manhattan Project and the Cold War. U concentrations of up to 800 mg kg^{-1} have been observed in the containant plume (Phillips et al., 2006).

The fundamental premise of this work is that U can be immobilized or sequestered in the subsurface through microbial bioreduction of more mobile U(VI) to less mobile U(IV) (Figure 12.31). U(VI) bioreduction may be accomplished by pure or mixed cultures of a wide range of DIRB including *Geobacter spp.*, sulfate reducers such as *Desulfovibrio*, and denitrifying bacteria such as *Acidovorax* spp (Wall and Krumholz, 2006; Cardenas et al., 2008). To biostimulate microbial activity, ethanol is injected into the subsurface as an electron donor. Techniques from surface science, geochemistry, hydrology, and (geo)microbiology are being used to help understand the rates and mechanisms of reactions and the microbially mediated processes.

Kelly et al. (2008) found that U(VI) was partially reduced to U(IV) during biostimulation, with a change in U primarily bound to phosphate or organic ligands (the latter likely mobile) to primarily adsorbed to Fe/Mn minerals. During subsequent sample storage, nanoparticles of uraninite formed, which suggests that uraninite might eventually also form in the field. Cardenas et al. (2008) showed that hydrologic factors played a role in controling microbial community structure associated with U(VI) reduction, most likely by influencing ethanol distribution.

MICROBIAL FUEL CELLS

The ability of some microorganisms to transfer electrons has led to an interest in microbial fuel cells (e.g., Logan et al., 2006; Lovley, 2006; Du et al., 2007). The potential for production of electricity by microorganisms was recognized ~100 years ago (Potter, M.C., 1910, 1911 referenced in Lovley, 2006). Many bacteria can transfer electrons and drive fuel cells. In addition to interest in electricity generation, other potential applications of microbial fuel cells include (1) to degrade organic components of wastewater, using the electrical energy generated to decrease the cost of wastewater treatment; (2) to produce H_2 biologically; and (3) as sensors for biological oxygen demand (BOD) (Du et al., 2007).

A fuel cell converts chemical energy into electrical energy. In the case of microbial fuel cells, organic matter is degraded by microorganisms in a dissimilatory process in which organics are oxidized and electrons are released. The electrons are transferred to electron carriers that are electrochemically active. To develop a microbial fuel cell, microorganisms must be cultured in such a way that anaerobic metabolism occurs at (or near) the anode (negative terminal), and the anode serves as the electron acceptor. The electrons flow to the cathode (positive terminal). In typical microbial fuel cells, the electrons at the cathode combine with protons that diffuse through a separating device [a proton exchange membrane (PEM), or a salt bridge] and oxygen from the air to produce water.

Many fuel cells rely on microbial fermentation. If the fermenting bacteria oxidize acetate, which is a common constituent of domestic wastewater, then the half-cell reaction at the anode can be represented as

$$CH_3COO^- + 4H_2O \rightarrow 2HCO_3^- + 9H^+ + 8e^- \quad \text{Anodic reaction,} \tag{12.18}$$

and oxygen is used as the electron acceptor

$$O_2 + H^+ + 4e^- \rightarrow 2H_2O \quad \text{Cathodic reaction} \tag{12.19}$$

(Logan et al., 2007).

Although CO_2 ultimately is generated, the process is CO_2 neutral overall because the carbon used was originally generated by photosynthesis.

The cell electromotive force (emf) is calculated from:

$$E_{\text{emf}} = E_{\text{cat}} - E_{\text{an}}, \tag{12.20}$$

where

E_{cat} = electromotive force for a specific reaction taking place at the cathode
E_{an} = electromotive force for a specific reaction taking place at the anode (Logan et al., 2006).

The emf is a thermodynamic property, which provides an upper limit for the cell voltage. But, the actual microbial fuel cells will be subject to a variety of potential losses (Logan et al., 2006).

Many different processes and likely reactions could be used in a microbial fuel cell. Engineers have long sought to convert sugars to electricity in microbial fuel cells. *Rhodoferax ferrireducens* is able to operate on sugars without a fermenter or an electron-shuttling mediator (Chaudhuri and Lovley, 2003). *R. ferrireducens* is a dissimilatory Fe(III)-reducing microorganism that oxidizes glucose and reduces Fe(III) according to the reaction

$$C_6H_{12}O_6 + 6\ H_2O + 24\ Fe(III) \rightarrow 6\ CO_2 + 24\ Fe(II) + 24\ H^+. \qquad (12.21)$$

This reaction is an example of anaerobic respiration. *R. ferrireducens* can oxidize sugars such as glucose, fructose, and lactose to CO_2 with over 80% of the electrons derived from the sugar recovered as electricity (Lovley, 2006).

Microorganisms within fuel cells have been proposed to operate many different ways, depending on the type of microorganism(s) (Lovley, 2006) (Figure 12.32). The example organic compound for the proposed processes shown in Figure 12.32 is glucose. In an *indirect* microbial fuel cell, a fermenting organism converts glucose to hydrogen, which reacts directly with the anode to produce electrons and protons. In this process, organic products accumulate in the anode chamber, so that not all possible electrons are available and the anode tends to foul. A *mediator-driven* microbial fuel cell relies on an electron-shuttling mediator compound that accepts electrons from the microbial cell and transfers them to the anode. The mediator can go through multiple reduction/oxidation cycles. This

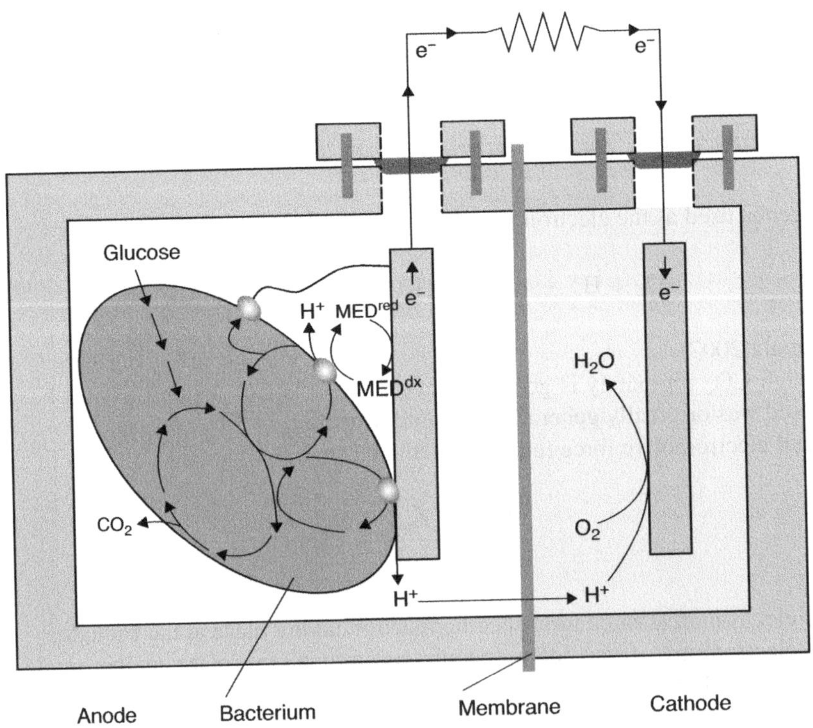

FIGURE 12.32 A schematic of microbial fuel cells showing different means of electron transfer (indirect, mediator driven, and by transfer through cell membrane or nanowires). This fuel cell has a PEM membrane that allows diffusion of H^+ from the anode chamber to the cathode chamber. Electrons flow from the anode through an external resistance to the cathode where they react with O_2, which is the final electron acceptor, and H^+ to form H_2O. From Logan et al., 2006 (with permission).

type of cell often does not use the glucose fully and many mediator compounds are toxic. Some microorganisms produce their own shuttle compounds. In a third type of fuel cell, the organic compound is taken into the microbial cell and oxidized to carbon dioxide. The electrons produced are transferred through the cell membranes or along nanowires to the anode (Lovley, 2006).

A good deal of research has been published using *Geobacterae*, which can oxidize organic compounds with electron transfer to insoluble Fe(III)oxides. Other electron acceptors such as humic substances and Mn(IV) oxides also can be reduced (Lovley, 2006). *Geobacter* can oxidize acetate to CO_2, but in the presence of fermenting organisms, other organic compounds can be used by the consortium. This coupling of fermentation with a Fe^{3+}-reducing microorganism can result in more complete oxidation of the organic compound than for fermentation alone (Lovley, 2006). *Geobacter* must be attached to the eletrode surface, and a shuttling compound is not needed. Reguera et al. (2006) showed that the transfer of electrons by *Geobacter sulfurreducens* through a biofilm in a fuel cell involved electrically conductive pili. These authors suggest that "electron transfer via intertwined pili... might establish a 'nano power grid'" through which the bacteria can transfer electrons through a biofilm tens of micrometers in thickness. The role of pili and

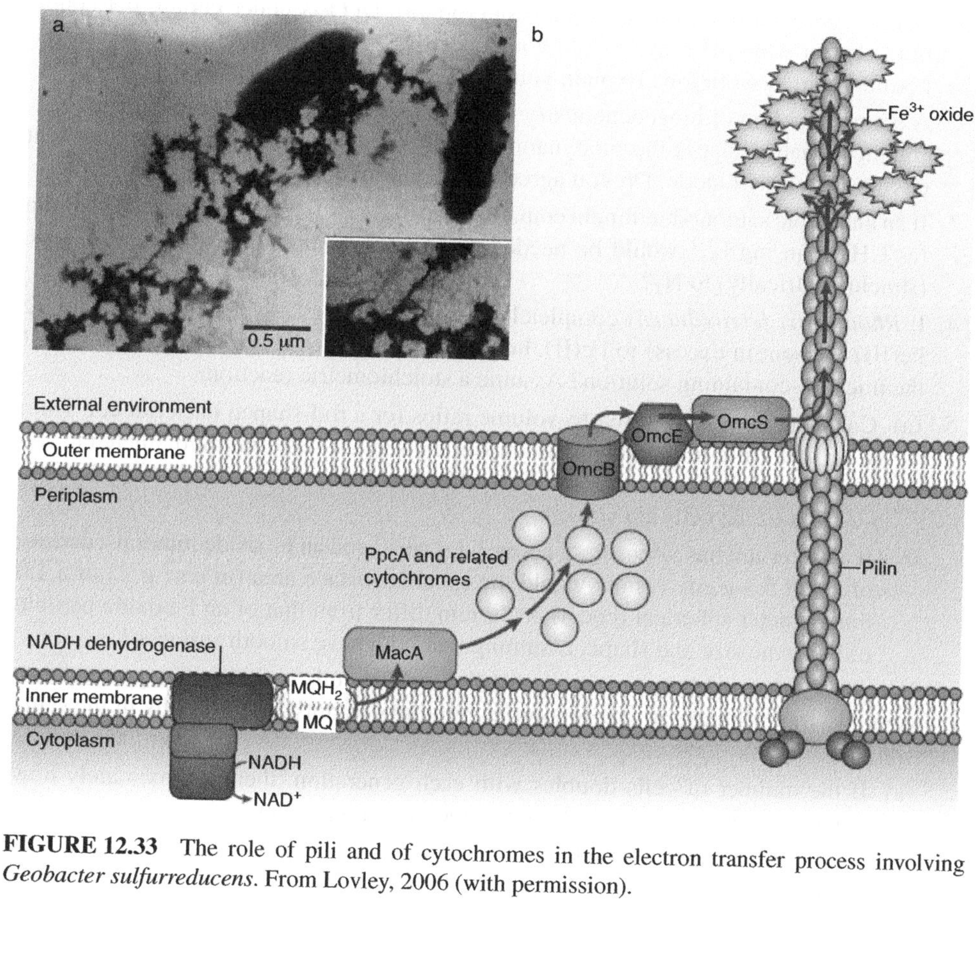

FIGURE 12.33 The role of pili and of cytochromes in the electron transfer process involving *Geobacter sulfurreducens*. From Lovley, 2006 (with permission).

of cytochromes in the electron transfer process involving *G. sulfurreducens* is shown in Figure 12.33.

Many other microorganisms have been explored (Du et al., 2007). Microbial fuel cells operated with mixed cultures tend to achieve greater power densities than those that rely on pure cultures. The mechanism(s) of electron transfer for a wide variety of different microorganisms is an area of active research.

In this chapter, we have explored how microorganisms, with an emphasis here on bacteria, represent important environmental surfaces. They can also interact with other environmental surfaces that range from Fe and Mn oxides and clay minerals to bones and teeth. This field of research is growing rapidly with implications for the environment, for human health, and even for industrial applications such as development of biofilm-based adhesives and fuel cell technologies.

QUESTIONS FOR FURTHER THOUGHT

1. The Pb-phosphate mineral pyromorphite has always been considered to be highly insoluble and nonbioavailable. However, recent research has shown that pyromorphite can be dissolved quite readily by the siderophore DFO-B under certain pH conditions. Under what pH range would you expect DFO-B to be most likely to promote pyromorphite dissolution? Explain your reasoning.
2. A common axiom of biogeochemistry is that bacteria can only accomplish a biogeochemical change that is thermodynamically possible and that their biogeochemical effects are purely kinetic. Do you agree? Why or why not?
3. If an anaerobic saturated sediment contains 30 mg L^{-1} NO_3^-, how much carbohydrate (as CH_2O) in mg L^{-1} would be needed to denitrify completely (to N_2) the nitrate (stoichiometrically) to N_2?
4. If *Rhodoferax ferrireducens* completely oxidizes 100 mg L^{-1} glucose in converting Fe(III) (present in excess) to Fe(II), how many mg of Fe(II) are produced in 1 L of the microbe-containing solution? Assume a stoichiometric reaction.
5. (a) Calculate the surface-area-to-volume ratios for a rod-shaped bacterial cell 1 um wide and 2 um in length versus a spherical (cocci) cell 1 um in diameteter. You may choose to estimate the rod-shaped bacterium by a cylinder. Assume that the surfaces of the cells are smooth.
   (b) If a bacterium has a density of about 1.1 g cm^{-3} and an Fe oxide mineral a density of about 5.3 g cm^{-3}, how would the specific surface area (in cm^2 g^{-1}) of a 1.0 um diameter spherical (cocci) bacterium differ from that of an Fe oxide particle of the same size and shape, assuming that both have smooth surfaces?
6. You have a culture that contains 1×10^6 bacterial cells in 1 mL that grows exponentially for 24 hours, at which point the culture contains 2×10^9 cells in the same volume.
   (a) If the number of cells doubles with each generation, then approximately how many generations does the 24-hour time period represent?
   (b) What is the doubling time t_d?
   (c) What is the rate constant r in $hour^{-1}$?

(d) If an average generation time t_g for humans is about 2 decades, then how many orders of magnitude difference exists between the generation times of the microorganims in your vessel versus humans?

(e) Let's say that the batch culture continued this same rate of growth for another 48 hours. What would be the final concentration of bacteria?

(f) If a single bacterium has a mass of 10^{-12} g and can be approximated by a sphere 1 μm in radius, then what would be the mass of bacteria by the end of the 48 hour experiment?

(g) The approximate volume of cells?

(h) Does it seem reasonable that the growth continued to be exponential over this time period? Explain your reasoning.

7. What factors might contribute to differences in bacterial cell numbers and community structures (which groups are present and their relative abundance) between a sandy sediment versus a clay lens?

8. Explain why people who have undergone hip replacement surgery and therefore have artificial implants (for example, metal or ceramic) may be particularly susceptible to infection, even years after the actual surgery.

9. (a) If a robot were to be sent to the planet Mars to collect rock samples to search for evidence of microbial life, what sorts of measurements would be needed?

   (b) Suggest a protocol for determining whether microorganisms occur on Mars today or were present in the past.

10. A student conducts a batch adsorption experiment on the adsorption of viable but nonmetabolizing bacteria to quartz sand. The student uses spherical (cocci) bacteria 1 μm in radius at a cell density of 10^6 cells mL^{-1} and 0.1 mg of quartz sand with 0.1-m^2 g^{-1} Brunauer, Emmett, and Teller (BET) surface area in a 1-L reaction vessel. The student measures adsorption by measuring cell density (by absorption at 600-nm wavelength) in the suspension before and after adsorption (removing the sand with any adhered bacteria before the "after adsorption" measurement). The student notes no measured change in the cell density in suspension. The student states that the bacterium does not seem to adsorb appreciably to quartz.

    (a) Does this conclusion seem to be warranted?

    (b) Why or why not?

    (c) What (if any) additional information might be helpful in answering this question?

11. Sewage gas typically contains hydrogen sulfide. What problem might the presence of *Acidithiobacillus thiooxidans* present for managers of sewage systems?

FURTHER READING

Banfield, J.F., and Nealson, K.H. (eds). 1997. Geomicrobiology: Interactions Between Microbes and Minerals. Reviews in Mineralogy vol. 35. Mineralogical Society of America, Chantilly, VA, 448 p.

Dove, P.M., DeYoreo, J.J., and Weiner, S. (eds). 2003. Biomineralization. Reviews in Mineralogy and Geochemistry vol. 54. Mineralogical Society of America and Geochemical Society, Chantilly, VA, 381 p.

Ehrlich, H.L. 2002. Geomicrobiology 4th ed. Marcel Dekker, Inc., New York, 768 p.

Maurice, P.A., and Warren, L.A. (eds). 2006. Methods for Study of Microbe-Mineral Interactions vol. 14. CMS Workshop Lectures. The Clay Minerals Society. Chantilly, VA, 165 p.

Vaccari, D.A., Strom, P.F., and Alleman, J.E. 2006. Environmental Biology for Engineers and Scientists. John Wiley and Sons, Inc., Hoboken, NJ, 931 p.

13

ENVIRONMENTAL NANOSCIENCE AND NANOTECHNOLOGY

nano from Greek "dwarf"

In the environmental sciences, the unique importance of small particles has been recognized for many decades. For example, the Clay Minerals Society was founded in the early 1960s to address the special properties of minerals $<2\ \mu m$ in diameter. But it is only within the last few decades that methods and approaches have been developed to measure, image, and explore the properties of the very smallest particles that are just larger than multinuclear clusters and dissolved species. These particles are the *nanoparticles*, where a *nanometer* (nm) is equal in length to one billionth of a meter, or 10^{-9} meters in scientific notation. As such, 1 nm = 10 Å.

Nanoparticles exist throughout the critical zone and include both a natural and an increasingly broader range of synthetic substances. Much research has been conducted on nanoscale minerals such as ferrihydrite, on expanding clays with Å- to nanoscale interlayer spacing, on minerals such as goethite that may have nanoscale domains, and on natural organic matter that often occurs in nanoscale spheroids and aggregates. A fundamental leap in understanding has occurred over the last few decades because of the realization that within the range of $\sim$1 to 100 nm (10^{-9} to 10^{-7} m), which is often referred to as the nanoscale, materials may display physical properties and behaviors that are unique and different from either simple molecules or larger, bulk materials. We are only just beginning to develop the basic and applied nanosciences, especially in the context of complex environmental phenomena.

WHAT IS A NANOPARTICLE?

One formal definition of *nanoparticle* is a "subclassification of ultrafine particle with lengths in two or three dimensions greater than 0.001 μm (1 nm) and smaller than about

Environmental Surfaces and Interfaces from the Nanoscale to the Global Scale, by Patricia A. Maurice

0.1 μm (100 nm) and which may or may not exhibit a size-related intensive property." (ASTM 2006). Hochella et al. (2008) provide a slightly different definition as particles that "are as small as roughly 1 nm and as large as several tens of nanometers in at least one dimension." However, *nanoparticles* are sometimes defined based on the size at which fundamental properties differ from those of the corresponding bulk material (e.g., Banfield and Zhang, 2001).

Colloids have been defined as particles between 1 nm and 1 μm in size (Buffle, 2006), so that nanoparticles overlap with colloids in size range. Compounds that pass through a 0.45-μm filter have been operationally defined as "dissolved." This definition means that nanoparticles in water samples are often sampled with the "dissolved" fraction, even though they have different behaviors from aqueous dissolved species. In considering air pollution, the term "particulate matter" is commonly used. The U.S. Environmental Protection Agency set standards in 1997 for particulate matter less than 2.5 μm in size, which is designated as the $PM_{2.5}$ standard. The term "ultrafine particle" is generally used for particles <0.1 μm in diameter ($PM_{0.1}$), which overlaps with the nanoparticles. If we consider the electromagnetic spectrum, wavelengths around the ultraviolet (UV) to X-ray range are approximately nanoscale.

The size range of nanoparticles versus other particles of environmental interest is shown in Figure 13.1. Note that viruses and many inorganic colloids are within the nanoparticle range. Bacteria are generally not considered to be nanoparticles but may display nanoscale features and form nanoscale biominerals. Nanoparticles often occur as aggregates, and this aggregation process can affect their mobilities and transport in porous media and their bioavailabilities, as well as other properties and processes.

Nanominerals have been defined as minerals such as ferrihydrite that exist only in the nanoparticle size range or as clays that only exist with at least one dimension in that size range (Hochella et al., 2008). *Mineral nanoparticles* may exist at larger sizes as well.

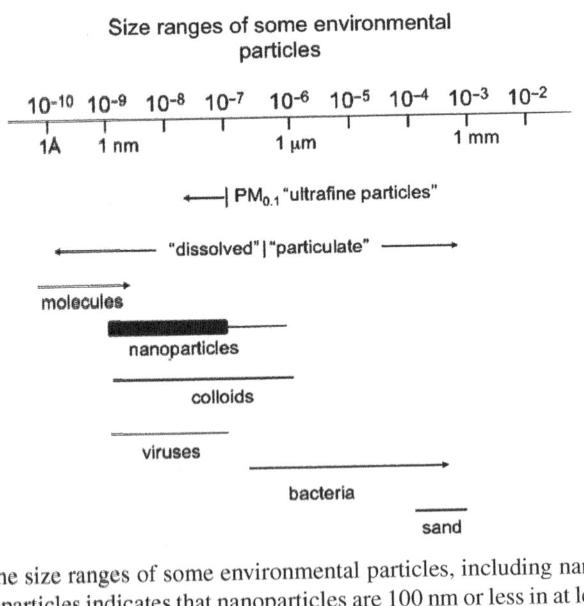

FIGURE 13.1 The size ranges of some environmental particles, including nanoparticles. The box with a line for nanoparticles indicates that nanoparticles are 100 nm or less in at least two dimensions, but they can be longer in the third dimension (such as nanotubes and nanowires).

Both nanominerals and mineral nanoparticles may have properties, including reactivity, that change as a function of size within the nanosize range, even with a fixed composition. This property makes nanominerals and nanoparticles distinct from larger scale minerals.

Small atomic clusters cannot qualify as nanominerals because minerals are by definition ordered crystalline structures (Hochella et al., 2008). An atomic arrangement greater than one unit cell of a mineral is therefore needed for a nanoparticle to qualify as a nanomineral. Nanominerals with the composition of sphalerite (Zn-S) as small as 1.5 nm have been observed. Dissolved Zn-S clusters ($Zn_3S_3(H_2O)_6$; Luther and Rickard, 2005) are 0.7 nm in diameter, and the sphalerite unit cell is 0.54 nm. Therefore, although it is possible that sphalerite nanoparticles of about 1.0 nm in diameter might occur, this measurement is probably at the lower bounds of what could qualify as a sphalerite nanoparticle (Hochella et al., 2008).

NANOPARTICLE OCCURRENCE AND DISTRIBUTION

Nanoparticles may be natural or manufactured. Natural nanoparticles include materials such as humic and fulvic acids (Chapter 8), viruses, soot, clays, and atmospheric sea salts. Examples of manufactured nanoparticles include ZnO and TiO_2 for sunscreens, pigments, and paints; alumina nanoparticles for detergents and antiperspirants; and carbon fullerenes and nanotubes for use in electronics, chemical engineering, and other industries.

Many oxides, hydroxides, and aluminosilicate clays occur as nanoparticles, and they are often aggregated. An example is the nanomineral ferrihydrite, which is used widely in adsorption and microbial Fe reduction experiments because of its high surface area (50–200 $m^2\ g^{-1}$, Waychunas et al., 2005) related to small particle size (generally <10 nm). Ferrihydrate and other Fe(hydr)oxides such as lepidocrocite and magnetite can occur as nanoparticulate biominerals (Figures 13.2a, b, c)

Nanoparticles occur throughout the critical zone (Chapter 2). Nanoparticles are widespread in the atmosphere, which result from processes such as volcanic emissions, forest fires, and sea spray. Nanoparticles of black carbon (see below) may contribute to radiative forcing (Chapter 14). On a mass basis, the average particle size of airborne mineral dust is 2–5 μm, but on the basis of number of particles, the average size is $\sim$100 nm, with many particles <100 nm (Hochella and Madden, 2005; Nowack and Bucheli, 2007). Far less is known about nanoparticles in the oceans, although they are almost certainly widespread. Berquist and Boyle (2007) suggested that nanoparticulate Fe plays an important role in oceanic Fe cycling and can help to explain some differences in Fe profiles in different parts of the ocean. In soil and groundwater environments, nanoparticles represent important reactive surfaces and play a major role in pollutant fate and transport. Weathering within soils produces nanoparticles as secondary weathering products, and natural organic matter present in soils often occurs in nanoparticle structures and aggregates. An example of nanoparticles in surface waters is the formation of Fe and Mn oxides in streams receiving acid mine drainage, as discussed in Chapter 14. Nanoparticles in the Earth's crust can be generated by mechanical grinding associated with movement along faults and can play a role in earthquake processes (Wilson et al., 2005).

Although bacteria probably do not occur as nanoparticles themselves, because of size constraints on the chemical compounds and reaction environment required for life (Maniloff, 1997; Nealson and Stanl, 1997), they may exhibit nanoscale features such as nanowires (Chapter 12). Dissimilatory metal-reducing bacteria have been shown to be

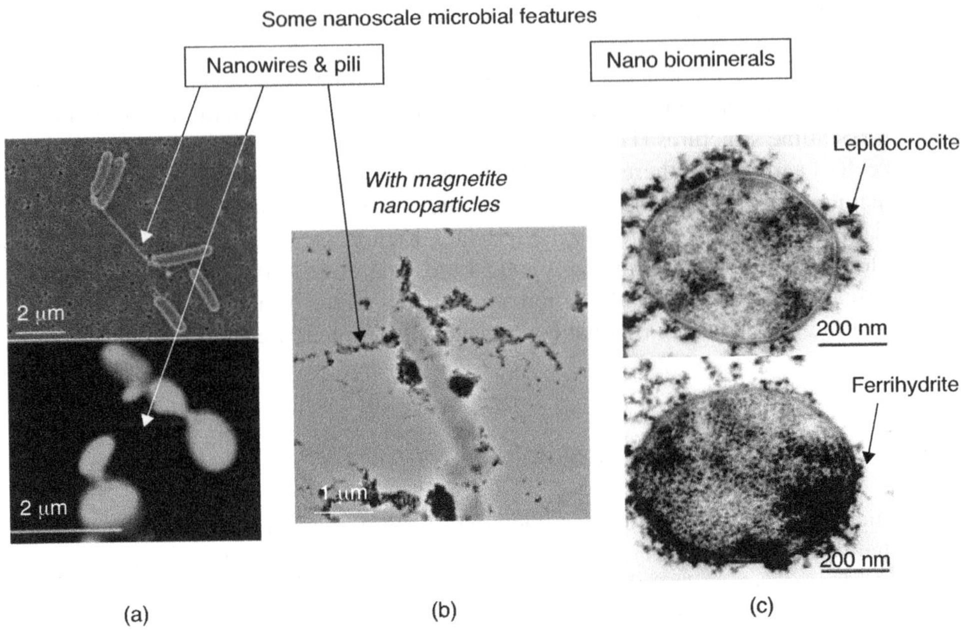

FIGURE 13.2 (a) Pili or "nanowires" on *Shewanella* bacteria are 50 to 150 nm in width and can be microns long (Gorby et al., 2006) (with permission), (b) Nanocrystalline magnetite that has grown along nanowires (Gorby et al., 2006) (with permission), and (c) nanocrystals of lepidocrocite and ferrihydrite on *Bacillus subtilis*. From Fortin and Langley, 2005 (with permission).

able to shuttle electrons through electrically conductive pili or "nanowires" to Fe(III) and Mn(IV) oxyhydroxides (Figure 13.2a) (Reguera et al., 2005; Gorby et al., 2006). Gorby et al. (2006) showed that nanocrystalline magnetite grew along nanowires from the dissimilatory Fe-reducing bacteria (DIRB) *Shewanella oneidensis* strain MR-1 (Figure 13.2b). Bacterial bioprecipitation often results in nanoscale minerals and amorphous solids (e.g., Fortin and Langley, 2005; Figure 13.2b, c). Suzuki et al. (2002) showed that nanoparticles of uraninite (UO_2) generally <2 nm across can be formed by bacterial bioreduction of hexavalent U(VI) to U(IV).

Both *carbon black* and *black carbon* (BC) are nanoparticles formed by the incomplete combustion of fossil fuels. Carbon black is produced industrially and is used in many rubber compounds. BC is released to the atmosphere from the incomplete combustion of fossil fuels or biomass and transported to soils, surface waters, and the surfaces of snow and glaciers (Hedges et al., 2000). BC can occur as nanoparticles or larger particles, and carbon black is specifically produced to be tens to hundreds of nanometers in size, with a high surface-area-to-volume ratio. Because BC is highly unreactive with respect to oxidation to CO_2, it may represent an important sink of C in the global C cycle (Kuhlbusch et al., 1995). Figure 13.3 shows a high-resolution transmission electron microscopy (TEM) image of soil black carbon, along with a sketch helping to identify the different components of the image, which includes both concentric-organized and amorphous carbon black nanoparticles.

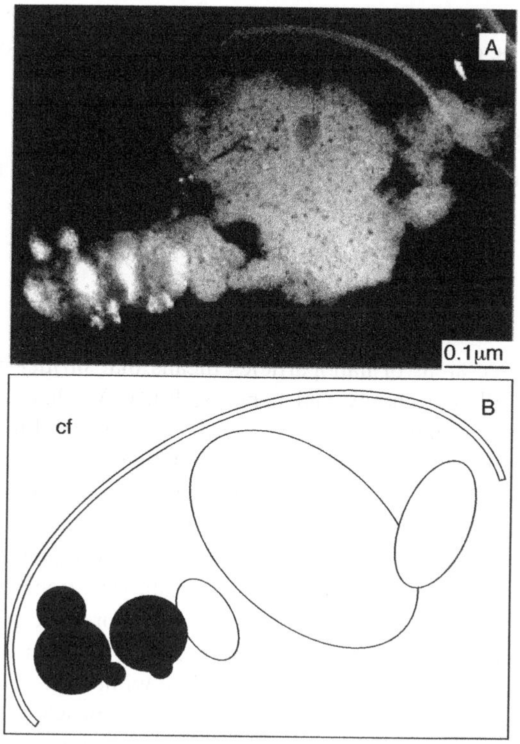

FIGURE 13.3 (a) Transmission electron micrograph (TEM) images of the chemically resistant fraction isolated from a soil showing the presence of at least two types of BC nanoparticles. (b) A schematic drawing of the same image area. In the schematic, dark black circles represent BC nanoparticles with a concentric structure, whereas open ovals represent nanoparticles with a poorly organized structure. cf represents a hole in the TEM grid. From Poirier et al., 2000 (with permission).

WHAT MAKES A NANOPARTICLE DIFFERENT?

We have already examined many equations and concepts throughout this text that point to small particles having distinctly different behaviors. Some examples include the Young-Laplace equation, which shows that the pressure associated with a bubble or a droplet of water increases greatly with decreasing size; Stoke's law, which demonstrates that gravitational settling becomes less and less important as particles shrink to colloid size and below; and crystal nucleation theory, which states that particles below a certain critical radius tend to be unstable. However, many classic approaches and techniques cannot adequately address the special properties, reactivities, and behaviors of nanoparticles. For example, Darjaguin, Landau, Verwey, and Overbeek (DLVO) theory was developed for particles whose radius of curvature is greater than the thickness of the double layer.

Nanoscale size may lead to a different crystalline order or structure, different chemical reactivity, different stability, and unique thermal, electrical, optical, tensile strength, and/or magnetic properties than observed in either molecules or macromaterials. Understanding what makes a nanoparticle different from an atom or a molecule or a macroscale solid requires a detailed understanding of the physics and chemistry of atoms, molecules, and crystals that is beyond the scope of this text. Some basic principles are addressed here, and more information can be obtained from the readings listed at the end of this chapter.

Nanoparticle Surface Area, Stability, and Reactivity

The extremely small size of nanoparticles means that they contain very high specific surface areas compared with larger particles. For a spherical particle, the ratio of surface area ($S = 4\pi r^2$) to volume ($V = 4\pi r^3/3$) is inversely proportional to the particle radius (r), such that

$$S/V = 3/r. \tag{13.1}$$

The high specific surface areas of nanoparticles means that *surface free energy* plays an important role in overall free energy and hence stability. We learned in Chapter 9 that very small particles with a radius less than the *critical radius* tend to be unstable because of their high surface free energies, which is a fundamental tenet of *crystal nucleation theory*. Because of the tendency for very small particles to be unstable, the synthesis of nanoparticles often requires (1) the presence of some agent that attaches to and stabilizes the particle relative to dissolution or (2) immersion in an inert environment (Burda et al., 2005) to achieve products that are stable enough for analysis. Recent research is showing that some materials actually seem to become less soluble or more stable with decreasing size. Although nanoparticles tend to be unstable or metastable because of their high specific surface areas, they can often remain for periods of time without growing or recrystallizing. We now recognize that many factors, such as the medium in which a particle forms, the temperature, the presence of various impurities and defects, and the rate of crystal growth, can affect nanoparticle stability in ways that can be difficult to predict. These factors are discussed in greater detail below.

One of the principal ways in which a nanoparticle differs from a bulk material is that a relatively high proportion of atoms associated with a nanoparticle occurs at the surface. As discussed in Chapter 5, surface structure can be different from bulk structure not only in terms of overall coordination environment but also because of processes such as *surface relaxation* and *reconstruction*. Figure 13.4 shows a molecular dynamics (MD) model of a TiO_2 particle ~5 nm in diameter. The surface and near-surface layers are highly distorted relative to the interior of the nanoparticle, which maintains a bulk-type structure. As nanoparticles become smaller, the surface and near-surface areas contribute an increasingly larger component of the overall energy of the particle. As TiO_2 particle size decreases to ~2 nm, the entire particle has distorted structure and the bulk-structure core is lost (Banfield and Zhang, 2001).

The high proportion of surface atoms can lead to different chemical reactivity with respect to processes such as adsorption, dissolution, redox reactions, and catalysis than for corresponding bulk materials. Because of the tight curvature of nanoparticle surfaces, more step and edge structures and deformation might extend into the crystal from the surface. These factors also can influence reactivity (Waychunas et al., 2005).

Nanoparticles Have a Different Electronic Structure

The electronic behavior of bulk materials often can be described reasonably well by *classic physics*. However, nanoparticle electronic behavior often requires explanations based on *quantum physics*. Taking an example from the electronics industry, the behavior of

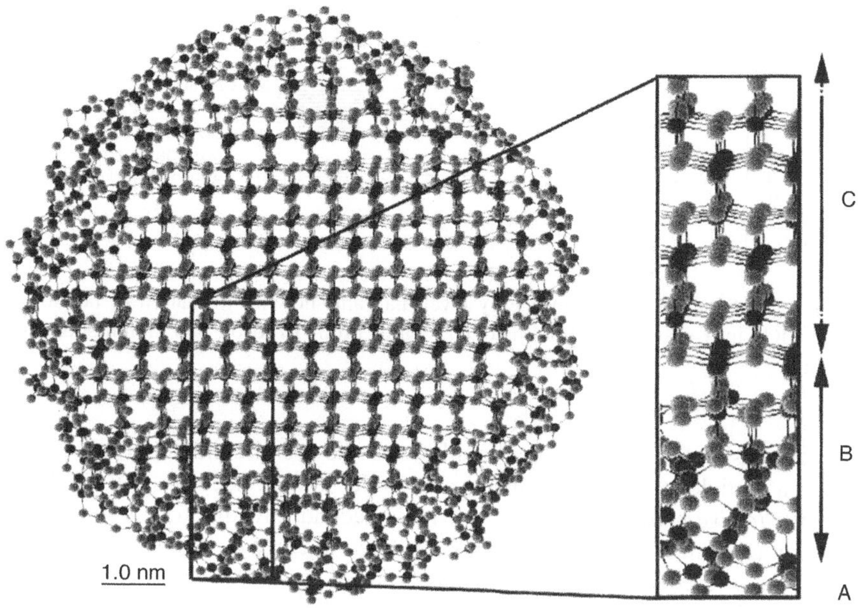

FIGURE 13.4 Cross section through an MD simulation of a TiO_2 particle ~5 nm in diameter. A is the surface layer, B the near-surface layer, and C the interior, which has a typical anatase (bulk) structure. A and B are both distorted because of the presence or proximity of the surface. From Banfield and Zhang, 2001 (with permission).

transistors on the order of 100 nm in size can be described according to classic physics. However, as transistors become smaller and smaller, more purely quantum effects emerge, such as *electron tunneling* (Chapter 4), which can greatly alter the performance of the transistor (Hanson, 2008).

As background for understanding the behavior of electrons in nanoparticles—which is essential to understanding their reactive properties—we can consider how the energy levels of electrons differ from a single atom to a larger, one-dimensional (1-D) solid, to clusters and eventually to three-dimensional (3-D) solids. The behavior of an electron in an isolated atom is generally described using the "particle-in-a-box" construct in which an electron is confined to a defined region of space. For an individual atom, electrons are confined to the region of the atom with quantized electronic states.

Figure 13.5 shows how the outermost *atomic orbitals* (AOs) of atoms overlap in a diatomic molecule to form two *bonding molecular orbitals* as well as *antibonding molecular orbitals* (MOs). As the number of atoms increases to form a one-dimensional solid, in-phase overlaps give molecular orbitals of lower energy known as the bonding molecular orbitals. Out-of-phase overlap results in higher energy antibonding molecular orbitals. Similarly, bonding and antibonding molecular orbitals occur in 3-D solids (not shown in this schematic 1-D figure). In solids, forbidden energy levels known as *bandgaps* develop, which are large for insulators and small for semiconductors. The number and energies of electronic states vary as we increase the size of a system, eventually to the scale of an extended solid material.

In the band structures of semiconducting solids, some electrons with energies just above the bandgap are free, which means that they are not associated with a single atom or bond.

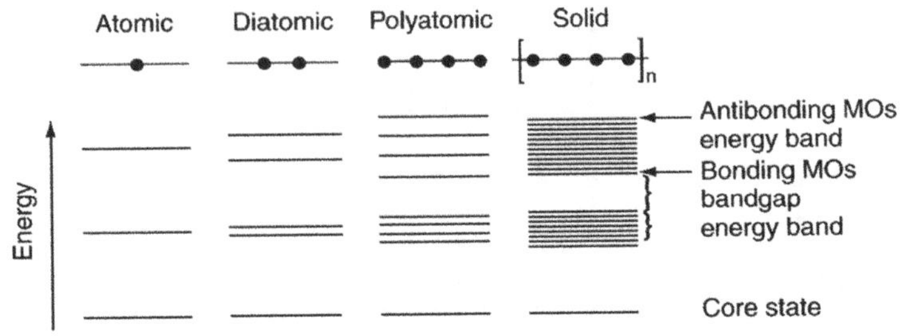

FIGURE 13.5 Schematic electron energy level diagram showing formation of bonding and antibonding MOs in a system from atom to 1-D solid. From Brydson and Hammond, 2005 (with permission).

These electrons in the semiconductors allow for the (semi)conducting nature. Near the surface of a particle, electrons occur that are similar to free electrons in the bulk but have a different *effective mass*. Recall from Newton's Second Law that force equals mass times acceleration (F = ma), so electrons that respond differently to an applied force are often said to have a different effective mass. Given the small size of nanoparticles, these electrons can be coherent over the whole volume of the nanoparticle, which leads to some unique behaviors.

How Electronic Structure Influences Nanoparticle Behavior

The boundaries of nanoparticles can lead to *size quantization* effects. These size quantization effects can result in differences in rates of electron transfer from nanoparticle surfaces to solution or sorbates and, hence, to differences in rates of redox reactions. This effect can increase *quantum yields for* electron transfer processes substantially. Higher quantum yields can lead, for example, to different rates of redox reactions and to differences in catalysis, including photocatalysis. The result for semiconductor nanoparticles is a strong dependence of electronic structure on particle size and shape. Madden and Hochella (2005) showed that the rate of heterogeneous manganese oxidation was from 1 to 1.5 orders of magnitude greater for 7.3-nm diameter hematite particles than for 37-nm hematite particles. The effect was a result of size quantization of the hematite electronic structure and was apparent even when the oxidation rate was normalized to particle surface area.

When noble metals such as gold, silver, platinum, and palladium occur at the nanoscale, a new very strong absorption of light is observed that results from the electrons in the conduction band from one surface of the particle to the other undergoing collective oscillation (Burda et al., 2005). The oscillation is at a frequency that absorbs visible light, which is called surface plasmon absorption. The resulting vivid coloration has been taken advantage of by pottery makers for hundreds of years and by Medieval European craftsman in making stained glass windows, but it has only recently been fully understood. The frequencies of absorption are a strong function of the size and shape of nanoparticles. This phenomenon of vivid coloration can also occur for some semiconducting nanoparticles.

Nanoparticles may have particularly useful catalytic properties because of the high surface areas they expose to the atmosphere or to solution and because of differences in surface structure, quantum yields, and other surface properties. As discussed in Chapter 2, nanoparticles can serve as catalysts for ozone depletion in the atmosphere. Nanoparticles of TiO_2 have been shown to photocatalyze degradation of pollutant organics and to be toxic to bacteria such as *Bacillus subtilis* (e.g., Armaleo et al., 2007), which may result in new techniques for water treatment.

Nanoparticle Disorder and Defect Structures

Nanoparticles often have varying degrees of disorder with different structures than encountered in chemically analogous bulk crystals. Defect structures may have different energies for nanoparticles than for bulk crystals, and the kinetics of nanoparticle growth plays a role in determining their structures, which include defect structures and stabilities. For example, Chernyshova et al. (2007) showed that the structure of hematite nanoparticles from 7 to 130 nm in size, including lattice disorder and inclusions of maghemite-like defects, varied not only with nanoparticle size but also with growth kinetics. Hence, not just the size per se of nanoparticles affects their structure, including defect structures, and stability, but also the conditions under which the particles grew.

Brydson and Hammond (2005) argued that *dislocations* may play a less important role in reactivity of nanocrystals than of bulk crystals. This is because the free energy of dislocations may be overwhelmed by the free energy associated with the high density of surface and interface sites on the nanocrystals, and because the elastic strain energy associated with nanocrystals is greatly decreased relative to that of bulk crystals.

The *core energy* of a dislocation is associated with the region within about three lattice planes around the dislocation core. The elastic strain energy extends from outside the core to the boundaries of the crystal. Additional free energy associated with the dislocation develops from entropy effects (Brydson and Hammond, 2005). For microparticles, the core energy is about one tenth of the elastic strain energy. For nanoparticles about 10 nm in diameter, the core energy is approximately equal to the elastic strain energy, which is only about one third that associated with dislocations in microcrystals (Brydson and Hammond, 2005). Because of the decreased elastic strain energy and other factors related to small particle size, dislocations tend to be relatively immobile in nanoparticles. Gilbert et al. (2004) note that nanocrystals may have inhomogeneities in internal strain as a result of competing relaxations from curved and highly irregular surfaces. This strain can result in structural stiffening, where defects that cause strain are maintained in the crystal.

Ferrihydrite Size, Structure, and Stability

The structure of the nanomineral ferrihydrite, which typically exists as particles <10 nm in diameter, has been the subject of much study and controversy. Recent research by Michel et al. (2007a) suggests that ferrihydrite nanoparticles from 2 to 6 nm in size all have the same composition ($Fe_{10}O_{14}(OH)_2$) and structure described by the hexagonal space group P6(3)mc. The particles show evidence of increasing disorder and distortion of some sites as particle size decreases from 6 nm to 2 nm (Michel et al., 2007a). These authors showed that for 6-nm particles, the hexagonal unit cell dimensions are $\sim$5.95 Å and c$\sim$9.06 Å, and the structure contains 20% tetrahedrally and 80% octahedrally coordinated Fe. But, for 2-nm particles, the hexagonal unit cell dimensions decrease slightly, and some Fe sites have lesser site occupancy or show evidence of distortion. The structural relaxation that occurs with decreasing particle size is likely related at least in part to the high proportion of surface relative to interior sites. Michel et al. (2007b) showed that the surface-adsorbed water content also varied with particle size. The model proposed by Michel et al. (2007a) is new and still being considered and questioned in the normal "vetting" procedure of the scientific community (e.g., Rancourt and Meunier, 2008).

Effects of pH and Adsorbed Ions on Nanoparticle Stabilities

The predominance of surface sites associated with nanoparticles makes the structure, stability, and properties of nanoparticles particularly sensitive to the surrounding environment (Banfield and Zhang, 2001). The stabilities of nanoparticles may be different in air, vacuum, or various liquids. In aqueous solutions, the pH and concentrations of adsorbing ions or molecules can affect stability.

In Chapter 5, we learned that the total free energy of a particle is equal to the sum of the bulk free energy (G_b) plus the surface or interfacial free energy. For nanoparticles in solution, in which the surface component can be significant, it is important to consider not just the surface or interfacial tension (related to van der Waals forces and hydrogen bonding) but also the contribution to surface free energy of the surface electrostatic energy ($\sigma\psi A$). In aqueous solutions, surfaces often are charged, and this surface charge component can be a substantial contribution to the surface/interfacial (G_s) and total free energies (G_t) of nanoparticles in solution

$$G_t = G_s + G_b \tag{13.2}$$

$$G_s = \gamma A + \sigma\psi A, \tag{13.3}$$

where

γ = surface or interfacial tension (Jm^{-2})
A = surface area (m^{-2})
σ = surface charge per unit area in (cm^{-2})
ψ = surface potential (V) (Finnegan et al., 2007).

The relative contributions of γA and $\sigma\psi A$ vary with solution conditions. At pH far away from the mineral point of zero charge (p.z.c.), $\sigma\psi A$ is relatively high, whereas γA tends to be decreased by adsorption of "potential determining ions," according to the Gibbs equation (Finnegan et al., 2007). Finnegan et al. (2007) argue that the stabilities of nanoparticles, as determined by G_t, may be influenced most strongly by γA close to the p.z.c. and by $\sigma\psi A$ further from the p.z.c.

Finnegan et al. (2007) showed that the stabilities of nanoparticles of the TiO_2 phases anatase, rutile, and brookite were pH sensitive. For nanoparticles several tens of nanometers in diameter, rutile is the most stable phase over a wide range of pH values. For small nanoparticles, rutile is more stable than anatase at very low pH, but at high pH, anatase is stabilized relative to rutile and brookite. This size-dependent crossover in stability is shown schematically in Figure 13.6.

For large particles, specifically adsorbed ions represent a negligible component of the overall particle composition. But, for nanoparticles smaller than about 10 nm in diameter, the overall composition can be significantly influenced by specific adsorption. Specific adsorption to the smallest nanoparticles can affect their stabilities by changing the nanoparticle composition and free energy of formation (Fukushi and Sato, 2005). This means that particles may become more or less stable after attachment/detachment of specifically adsorbed ions. Adsorbates are sometimes used in nanotechnology to stabilize nanoparticles of

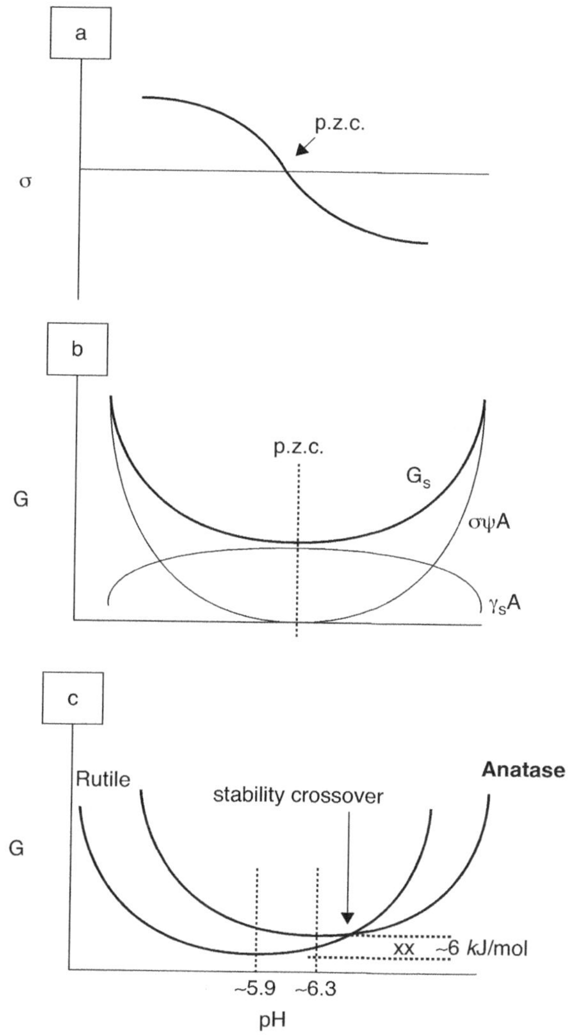

FIGURE 13.6 Stability crossover between anatase and rutile nanoparticles with pH. The points of zero charge for the minerals are ~6.3 and ~5.9, respectively. From Finnegan et al., 2007 (with permission).

a specific size range, structure, or composition. The effects may be thermodynamic and/or kinetic.

Case Study: Fe(hydr)oxide Size and Stability

Oxides of Ti, Al, Fe, and Zr show varying degrees of stability as their particle size changes, and the particles do not always become more stable with increasing size (Navrotsky, 2001; 2003; Navrotsky et al., 2008). Nanoparticulate oxides can exist as different stable polymorphs at different size ranges, as illustrated schematically in Figure 13.7. For the smallest particles, surface energies can dominate and dictate which polymorph will be stable. As a nanoparticle grows, different phases may become more or less stable such that the particle

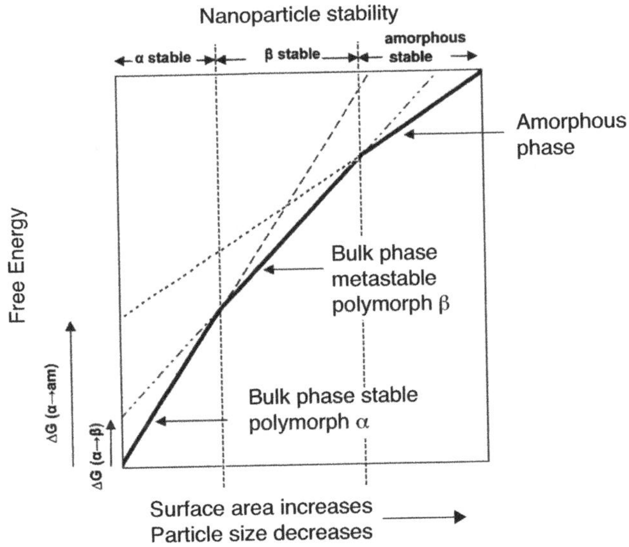

FIGURE 13.7 An example of the crossovers in free energy (ΔG), hence stability, for a nanoparticle system. Free energy decreases (becomes less positive) from polymorph α, which is stable for larger, bulk-space crystals, to polymorph β, which is metastable for larger crystals, to amorphous as surface area increases (and particle size decreases). Adapted from Navrotsky, 2001 (with permission).

may undergo several phase transformations during growth (e.g., Waychunas et al., 2005), which leads to crossovers in stabilities as a function of particle size (Figure 13.6).

Alexandra Navrotsky and her coworkers have focused a great deal of research on Fe(hydr)oxide particle size and stability, which is helping to explain previous apparent discrepancies in stability data. The thermodynamics of the different nanoscale polymorphs depend on the energetics of the bulk polymorphs, the particle size (and surface area), and the extent of hydration. Navrotsky et al. (2008) found that stability crossovers can be different for hydrated versus anhydrous surfaces of the Fe(hydr)oxide system: hematite, goethite, ferrihydrite, lepidocrocite, akageneite, and maghemite (Figure 13.8). Focusing on hematite and goethite, particle size and hydration can play a key role in determining the relative stabilities (Navrotsky et al., 2008). Surface enthalpy and free energy is generally much higher for oxides such as hematite than for hydroxides such as goethite. The lower surface enthalpy/energy for the hydrated goethite allows it to be stabilized relative to hematite at a small particle size. The stabilities of hematite and goethite relative to bulk hematite cross over at ~60-nm particle size or 15-m^2 g^{-1} surface area, with goethite becoming relatively more stable than hematite at the smaller particle sizes and higher surface areas. Although bulk hematite in water tends to be more stable than goethite, small particle size favors goethite stability. The hematite–goethite equilibrium relationships are also sensitive to variations in temperature, the activity of water, and Fe-Al substitution.

Secondary Growth of Nanoparticles

As well as growing according to classic theories (see Chapter 9) such as molecular addition and Ostwald Ripening, nanocrystals may grow by assembly of aggregates in a process

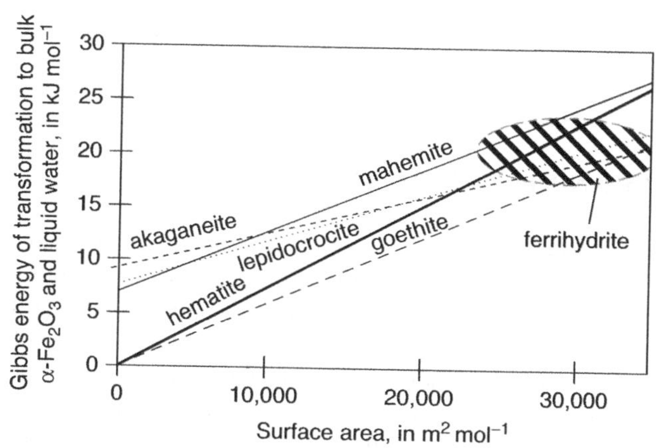

FIGURE 13.8 Free energy (approximated by enthalpy as entropy changes are approximately the same) of nanoparticles relative to coarse hematite plus liquid water at 298 K of various iron oxide and oxyhydroxide polymorphs as a function of surface area per mole $FeO_{1.5}$, FeOOH, or $Fe(OH)_3$. This plot is for hydrated surfaces; anhydrous surfaces show a different pattern (not shown here). Based on Navrotsky et al., 2008.

known as *oriented attachment* (Penn and Banfield, 1998a, 1998b), which is also known as *secondary growth* (Burda et al., 2005). As the oriented nanoparticles aggregate, the removal of the boundary between adjacent particles decreases surface free energy (Waychunas, 2001). Impurities may potentially be trapped as the aggregates form into single crystals, with implications for pullutant mobility, bioavailablity, and remediation strategies (Waychunas et al., 2005). Secondary growth proceeds much faster than the process of molecular addition.

When the oriented attachment is imperfect, screw, edge, or mixed dislocations may result, along with twin boundaries (Figure 13.9) (Penn and Banfield, 1998a). Thus, imperfect oriented attachment is one means of generating dislocations in nanocrystals. For the TiO_2 system, Penn and Banfield (1998b) showed that brookite, which is a polytype of anatase,

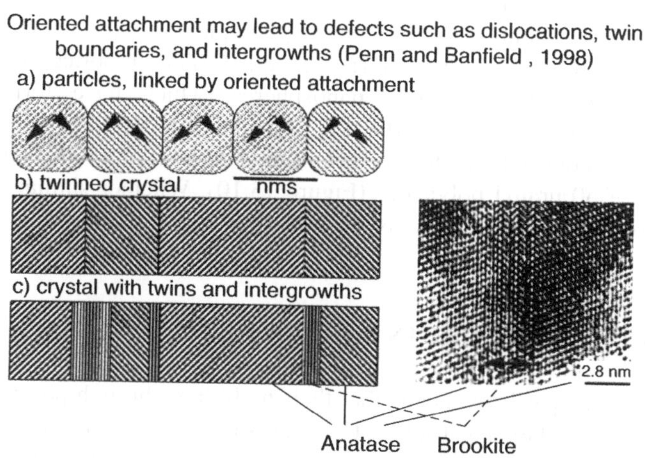

FIGURE 13.9 How oriented attachment may lead to imperfections in nanocrystals. Model and TEM image of anatase and brookite from Penn and Banfield, 1998 (with permission).

can occur at some of the boundaries that formed during oriented attachment of anatase nanoparticles (Figure 13.9).

SELF-ASSEMBLY AND TEMPLATING

Nanoparticles tend to undergo *templating* and *self-assembly* phenomena, and such processes can play a major role in natural biological processes such as biomineralization as well as synthetic technologies such as liquid crystal formation (e.g., Hamley, 2005). Self-assembly of polymers, including amphiphilic surfactants, to form micelles and oriented surface layers has been well documented. The formation of micelles in water is driven by minimization of surface free energy. Liquid crystals are self-assembling materials with a molecular order intermediate between that of a liquid and that of a crystal. *Lyotrophic* liquid crystals are formed by amphiphilic molecules in solution. DNA may form lyotrophic liquid crystal phases in solution (Hamley, 2005). *Thermotrophic* liquid crystals are formed by organic molecules after heating at a low-temperature crystal phase in the absence of solvent (Hamley, 2005). The resulting structure may be rod like or disk like. Self-assembly of natural organic matter on mineral surfaces in solution is demonstrated in Figure 8.18.

Templating can occur naturally, but it is also used in many synthetic, technological applications. Templating is a surface/interface process. *Epitaxy* in mineral heterogeneous nucleation is an example of templating, as is formation of ordered organic monolayers and bilayers within swelling clays such as smectite. Scientists have speculated as to whether templating by clays might have played a role in the origin of life, by helping to form complex organic components. Bacterial surfaces and exudates may potentially template formation of nanoscale biominerals. Self-assembled *surfactants* can be used to template silica, alumina, and titania structures, which result in structures that resemble those of zeolites (Hamley, 2005). Templating of mesoporous silica by a host of organic molecules and ionic liquids has been especially well studied. Nanoscale templating has been a hot area of study in biophysics and medical biochemistry because of the potential to develop artificial bones, soft tissues, artificial blood, and other applications.

Self-assembly is a component of *fullerene* formation (see below) and of the formation of aqueous aluminum polynuclear complexes and nanoclusters such as the polyoxocation (Casey, 2006). Burns et al. (2005) reported newly analyzed structures of self-assembled actinide nanospheres, which have formulas such as $[K_{16}(H_2O)_2(UO_2)(O_2)_2(H_2O)_2\{(UO_2)(O_2)_{1.5}\}_{28}]_{14}$ and $[Li_6(H_2O)_8NpO_2(H_2O)_4\{(NpO_2)(O_2)(OH)\}_{24}]_{20}$ and are stable in alkaline solutions. Forbes et al. (2008) reported a spherical nanocluster of 50 uranyl polyhedra (Figure 13.10). Whether similar structures may form in high-level waste storage tanks is the subject of ongoing research.

NANOPARTICLE TRANSPORT IN POROUS MEDIA

The small size of nanoparticles means that their mobilities through porous media can be different from those of larger particles. The high surface areas of nanoparticles suggest that they may have the potential to adsorb to the surfaces of porous media and/or mobile colloids. Given their small size, nanoparticles may penetrate into micropores in soils and sediments, thus decreasing mobility, or they may remain in the macropores. The tendency

FIGURE 13.10 A 50-uranyl polyhedron nanocluster. Courtesy P. Burns., University of Notre Dame.

of many nanoparticles to form colloidal-sized aggregates can play an important role in fate and transport.

The mobility of a particle in porous media is controlled to a large extent by the deposition of particles onto the porous media surfaces known as the *collectors*. This deposition process involves transport through the solution to the mineral grains—which is controlled by fluid flow along with diffusion related to Brownian motion and sedimentation caused by gravity—and attachment or adhesion to the collector surfaces. For nanoparticles, gravity-induced sedimentation is negligible relative to Brownian motion. Two terms commonly used in colloidal transport and deposition are the frequency at which colloidal particles strike the collector, which is also known as the *single-collector contact efficiency* (η_T), and the sticking efficiency (α). Both of these terms are also being applied to nanoparticles, but the theoretical considerations often cannot be transferred directly and need to be nanoparticle specific.

Saiers and Ryan (2005) investigated how the shape of a smooth collector and the roughness of a spherical collector affected the η_T values for particles of difference sizes that range from nanoparticles to larger colloids. The η_T is equal to the sum of the dimensionless transport rates associated with diffusion (η_D), interception (η_I), and sedimentation (η_S)

$$\eta_T = \eta_D + \eta_I + \eta_s = F_S/F_A, \tag{13.4}$$

where F_S is the rate at which colloids strike the collector and F_A is the rate at which colloids approach the collector.

Their research showed that for particles with radius $r_{cd} < 500$ nm, which includes nanoparticles as its smallest components (depending on the definition), diffusion dominates transport. The value of η_T for these very small particles is affected by the shape (spherical to oblate spheroid) of a smooth collector but not by the surface roughness of a spherical collector. For colloidal particles with $r_{cd} > 500$ nm and particularly >1 um, surface roughness affects η_T more strongly. For particles with r_{cd} between 1 and 2 um, the collector shape is relatively unimportant, but for still larger particles, the collector shape is more important. Thus, η_T may respond differently to the shape and roughness of the collector for nanoparticles versus larger particles. This is mostly because the small particles transport primarily by diffusion and not by gravity-related sedimentation.

Most of DLVO theory for modeling particle attachment and detachment to/from a collector and for calculating *sticking efficiency* α is based on the mobile particle radius of curvature being greater than the thickness of the double layer (the *Debye length).* This is often not the case for nanoparticles (See problems 9 and 10 at the end of this chapter). Many of the forces at surfaces, such as electrostatic and van der Waals, can extend over distances that are large in comparison with the size of nanoparticles. Forces that are not considered in classic DLVO theory (Chapter 6), such as hydration/solvation and Born repulsion, are likely to be very important for nanoparticle attachment. Various extensions to DLVO theory are being considered. In some instances, empirical data are used to estimate α for a given nanoparticle and collector surface, and then that value is applied to the same nanoparticle with a different collector—but with varying results. Bresme and Oettel (2007) further described some unique considerations that must be given to nanoparticle attachment at solution interfaces, which are the topics of ongoing research.

Another important factor in nanoparticle mobility through porous media is the fact that nanoparticles in solution often form aggregates, particularly at pH values near the p.z.c. (for charged nanoparticles). Hydrophobic nanoparticles tend to aggregate to minimize the contact area with water. The interactions of nanoparticle aggregates with collector surfaces do not always agree with predictions based on electrostatic considerations and current theories of colloid transport. The stabilities of nanoparticle aggregates in response to solution conditions and flow parameters, particularly shear, are not yet well understood.

Additional research is addressing questions such as how the presence of natural organic matter and of various dispersing agents applied to manufactured nanoparticles may affect nanoparticle deposition, along with the effects of changes to collector surfaces over time. For example, Lecoanet et al. (2004) compared the mobilities of a range of natural and synthetic nanoparticles in columns packed with glass beads. Hydrophobic fullerene particles tended to be far less mobile than more hydrophilic fullerol particles treated to enhance dispersion. Espinasse et al. (2007) showed that deposition of colloidal aggregates of C 60 fullerene (see below) on porous media in saturated conditions increased with (1) increasing ionic strength, (2) the presence of polysaccharide-type organic matter, and (3) lower Darcy velocies, but it decreased in the presence of humic-like organic compounds. Chen et al. (2008) observed that TiO_2 nanoparticle mobility was affected by the saturated/unsaturated ratio or by the ratio of pore space filled by solution versus gas in porous media and suggested that the nanoparticles may adsorb to the air–water interface. This behavior could alter vadose zone transport in a manner similar to that observed for bacteria (Chapter 4).

THE EMERGENCE OF NANOTECHNOLOGY

With realization of the importance of "nano," the field of *nanotechnology* has exploded, as scientists and engineers work to take advantage of the unique properties of *nanomaterials*. One of the most high profile of the engineered nanomaterials is the *fullerenes*, which are

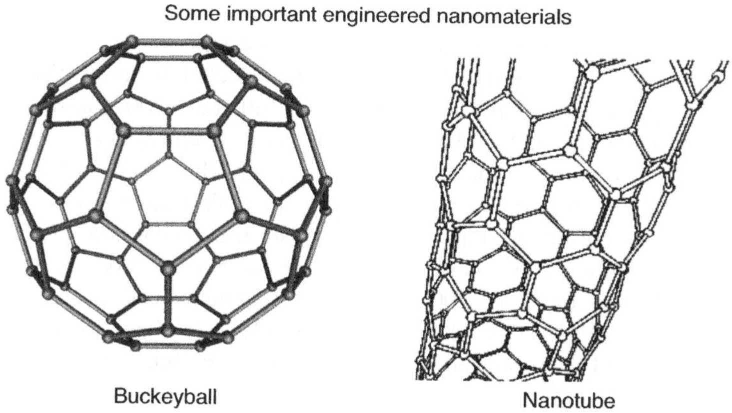

FIGURE 13.11 Buckminsterfullerene (C_{60}) (or "Buckeyball") and a carbon nanotube. Both from http://en.wikipedia.org/wiki/Fullerene accessed, Nov. 1, 2007.

carbon molecules that consist of 60 or 70 C atoms (Figure 13.11). Kroto, Curl, and Smalley were awarded the 1996 Nobel Prize in Chemistry for their discovery of fullerenes, and this and subsequent research was one of the keys to igniting widespread interest in nanoscale structures and materials. The most famous form of fullerene, the buckeyball, is about 1 nm in diameter. *Nanotubes* (Figure 13.11) have become one of the hot items in nanoscale research; they are cylindrical structures similar to fullerenes and about 1.3 nm in diameter with unique properties, including high tensile strength, high electrical conductivity, high ductility, high heat resistance, and very little chemical activity.

The EPA (2005) lists four different types of intentionally manufactured nanoparticles: (1) *carbon-based materials* such as the fullerenes and carbon nanotubes described above; (2) *metal-based materials* such as *quantum dots*, nanogold and nanosilver, and reactive metal oxides such as ZnO and TiO_2; (3) *dendrimers*, which are nanopolymers constructed of branched units; and (4) *nanocomposites*, which are combinations of different nanoparticles or of nanoparticles with larger particles, for example clay-containing nanocomposites used in manufacturing of automobiles. In addition, nanoparticles may be manufactured unintentionally as a result of grinding and other processes. Quantum dots, which are dot-shaped nanoscale semiconductors with fluorescent or light absorption properties, are a particularly "hot" area of manufacturing and research.

POTENTIAL ENVIRONMENTAL EFFECTS OF ENGINEERED NANOPARTICLES

Nanoparticle engineering represents one of the world's fastest growing new industries. This explosion of nanoparticle research, development, and incorporation into technologies raises many questions about potential environmental and human health hazards of nanoparticles themselves as well as about production pathways (Oberdorster et al., 2005; Wiesner et al., 2006). Engineered nanoparticles are finding their way into the environment in increasing numbers, because of either intentional (for example, as tracers in ground water) or unintentional (for example, from degradation of rubber tires or automobile emissions) release.

Many nanotoxicity studies focus on mammalian, including human, lung exposure (Obserdorster et al. 2005), which is of concern because silicosis and asbestosis are often debilitating or lethal lung diseases that result from particle inhalation. Yet other studies have focused on aquatic ecosystems. For example, Griffitt et al. (2007) showed that copper nanoparticles are acutely toxic to zebrafish (*Danio rerio*), exerting a damaging effect on the zebrafish gills different from that of dissolved copper. Zhu et al. (2006b) showed that fullerenes can be toxic to the aquatic species *Daphnia* and flathead minnow.

ZnO nanoparticles are being manufactured for use in personal care products such as sunscreens and as pigments in paints, and they can be acutely toxic to some bacteria (Adams et al., 2006). Lin and Xing (2008) found that ZnO nanoparticles had a (phyto)toxic effect on Ryegrass (*Lolium perenne*). The ZnO particles adsorbed into root tissues and cells (Figure 13.12) and damaged the root tissues, which inhibited seedling growth. The

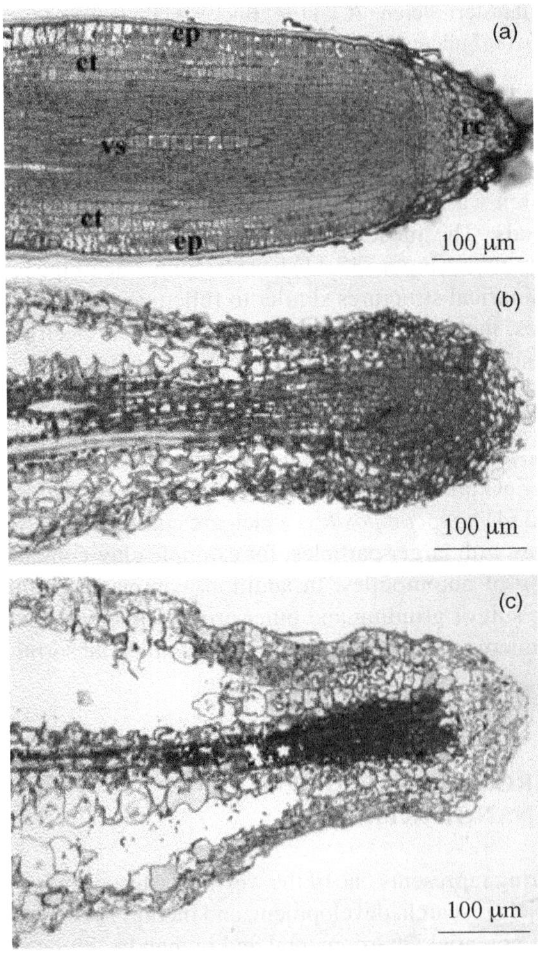

FIGURE 13.12 ZnO nanoparticles may damage plant roots, as shown in these light microscope images of sections of ryegrass root tips under conditions of (a) control without nanoparticles, (b) 100 mg L^{-1} ZnO nanoparticles, and (c) 1000 mg L^{-1} Zn^{2+}. rc = rootcap, ep = epidermis, ct = cortex, vs = vascular cylinder. From Lin and Xing, 2008 (with permission).

ZnO nanoparticles were translocated from root to shoot in only very limited amounts. The phytotoxicity could not be explained by dissolution of ZnO particles, but rather it seemed to be caused by the particles themselves. Although the mechanism of toxicity of nanoparticles often is not fully understood, a "warning flag" has been raised regarding the need to better understand potential toxic effects of nanoparticles on humans as well as on the various components of ecosystems.

At the same time, nanoparticles are being explored for their potential to provide new technologies for water treatment and other environmental uses. Mashino et al. (1999) found that fullerenes modified to be cationic suppressed the growth of the bacterium *Escherichia coli*, which is important because some *E. coli* strains are pathogens.

UV light is commonly used for disinfection, but pathogenic viruses tend to be resistant. Badireddy et al. (2007) investigated the production of reactive oxygen species (ROS; singlet oxygen and superoxide) through the UV photosensitization of polydrydroxylated fullerene. They showed that the ROS produced by different pathways could inactivate viruses. UV light can also activate photocatalytic TiO_2 to produce ROS. Sunlight may also be capable of this activation, which would open the door for broad use in the developing world (Shannon et al., 2008). Finally, carbon nanotubes are being investigated for potential use in desalination, as they offer the potential for high throughput of water at relatively low energy consumption, and ion selectivity through functionalization (Shannon et al., 2008). These observations suggest that nanoparticles may provide new technologies for water and wastewater treatment. However, a good deal of research needs to be completed on the behavior of fullerenes, nanotubes, TiO_2, and other nanoparticles relative to the many constituents of complex environmental systems before we can consider field-scale implementation. Nanoparticles have tremendous potential for environmental applications, but we need to understand them at a fundamental and mechanistic level in order to use them wisely.

QUESTIONS FOR FURTHER THOUGHT

1. (a) Compare the order of magnitude difference between a hydrous ferric oxide (HFO) nanoparticle 10 nm in diameter and the diameter of the Earth.
   (b) Compare the difference in scale between the Earth and your fist versus your fist and a 1-nm diameter nanoparticle.
   (c) Compare the size of a nanoparticle versus the diameter of a pore on human skin.
   (d) What are the potential implications for the use of nanoparticles in cosmetics and sunscreens?
2. Lét's say a natural organic matter (NOM) molecule has a molecular weight of 1000 Da and composition ~50% C, ~45% O, and the remainder H, S, and N (by weight). Is NOM in the nanoparticle size range?
3. Kroto, Curl, and Smalley were awarded the 1996 Nobel Prize in Chemistry for their discovery of fullerenes. Was the discovery of fullerenes worthy of the Nobel Prize? Explain your reasoning.
4. Comment on how different modes of sorption to smectite may provide insight into the uniqueness of nanoscale processes in the environment.

5. Comment on how nanotechnology may potentially benefit or harm the environment. How can we best ensure that nanotechnology provides the maximum benefit with the minimum harm?
6. (a) Plot the surface area-to-volume ratio (S:V) for a perfect (and smooth) spherical particle versus a cubic particle with size (radius r, or side length *l*) ranging from 1 nm to 1 mm.
   (b) Next, plot S:V of a cubic particle as a function of side length (*l*) and a spherical particle as a function of diameter (2r) from 1 nm to 1 mm.
   (c) Compare S:V of spheres versus cubes of comparable sizes (2r versus *l*).
   (d) What are the potential implications for structure and reactivity of nanoparticles?
7. What are some challenges that nanoparticles might pose for understanding and modeling the mobilities of radionuclides in the subsurface?
8. A question for extended library/web-based research: Many biologists believe that the lower size limit for bacteria is ~200 nm long. Yet, there have been reports of potential considerably smaller "nanobacteria." One of the factors that propelled this question into controversy was reports of potential fossilized "nanobacteria" associated with Martian meteorite ALH84001. Write an essay addressing the arguments for an against the existence of nanobacteria. Do you think that they can and do exist? Explain your reasoning.
9. Some research has suggested that the unit cell of ferrihydrite is hexagonal with dimensions of a ~5.95 Å and c ~ 9.06 Å for a 6-nm particle and a ~5.92 Å and c~8.90 Å for a 2-nm particle. Assume particle shape is roughly spherical. If this is so, then
   (a) approximately how many unit cells would be in a 2-nm particle of ferrihydrite?
   (b) A 6-nm particle? For an hexagonal structure, the unit cell dimension a = 2r, and the unit cell volume $V = 8(2)^{1/2}\ r^3$.
10. (a) Plot how the thickness of the double layer (in nm) changes (when all other parameters are kept constant) with NaCl concentration from 0.0001 to 0.1 M, and then as temperature changes from 275 to 325 K for 0.1-M NaCl.
    (b) What insight does this provide for our understanding of colloid/nanoparticle mobility in porous media?
11. The definition of nanoparticle is sometimes based on a change of properties rather than a rigid size per se. How particles interact with surfaces and in porous media is known to change when the size of the particles becomes less than the Debye length $1/\kappa$. Based on your answer to problem (10), what would be the maximum size of a "nanoparticle" for a quartz sand aquifer containing low-ionic-strength (0.001 M) water?

FURTHER READING

Banfield, J.F., and Navrotsky, A. (eds). 2001. Nanoparticles and the Environment. Reviews in Mineralogy and Geochemistry vol. 44. Mineralogical Society of America, Chantilly, VA, 349 p.

Hanson, G.W. 2008. Fundamentals of Nanoelectronics. Pearson/Prentice Hall, Upper Saddle River, NJ, 385 p.

Hochella, M.F., Jr., Lower, S.K., Maurice, P.A., Penn, R.L., Sahai, N., Sparks, D.L., and Twining, B.S. 2008 Nanominerals, mineral nanoparticles, and earth chemistry. *Science* 21, 1631–1635.

Kelsall, R., Hamley, I., and Geoghegan, M. 2005. Nanoscale Science and Technology. John Wiley and Sons, Inc., Chichester, West Sussex, England, 456 p.

Waychunas, G.A., Kim, C.S., and Banfield, J.A. 2005. Nanoparticulate iron oxide minerals in soils and sediments: Unique properties and contaminant scavenging mechanisms. *Journal of Nanoparticle Research* 7, 409–433.

14

THE BIG PICTURE: INTERFACE PROCESSES AND THE ENVIRONMENT

Environmental surfaces and interfaces are key components of many environmental issues of local to international concern. Here, we discuss four topics of environmental importance: reactive transport, acid rain, acid mine drainage, and global climate change. The chapter ends by discussing scaling phenomena, which include new embedded sensor network technologies that are helping scientists to scale from fine-scale environmental processes to understand local, regional, and global phenomena.

REACTIVE TRANSPORT MODELS FOR METALS AND RADIONUCLIDES IN POROUS MEDIA

Contamination of subsurface environments with heavy metals and radionuclides is a pressing environmental problem at locations such as the U.S. Department of Energy's (DOE's) Hanford Site in Washington State, the Oak Ridge Reservation in Tennessee, and the Savannah River Site in Georgia. Incorporation of sorption into hydrologic transport models is essential for understanding and predicting the subsurface fate and transport of pollutants, and for assessing risk. Yet, it can be difficult to come up with sorption equations that work well in transport models, especially over the wide range of conditions such as pH, ionic strength, temperature, and surface properties of geologic materials, which can be encountered at contaminated sites.

The K_d Approach Encounters Difficulties for Metals and Radionuclides

The linear distribution coefficient or K_d approach as described in Chapter 8 has been applied successfully to many cases of hydrophobic organic compound (HOC) mobility

---

Environmental Surfaces and Interfaces from the Nanoscale to the Global Scale, by Patricia A. Maurice

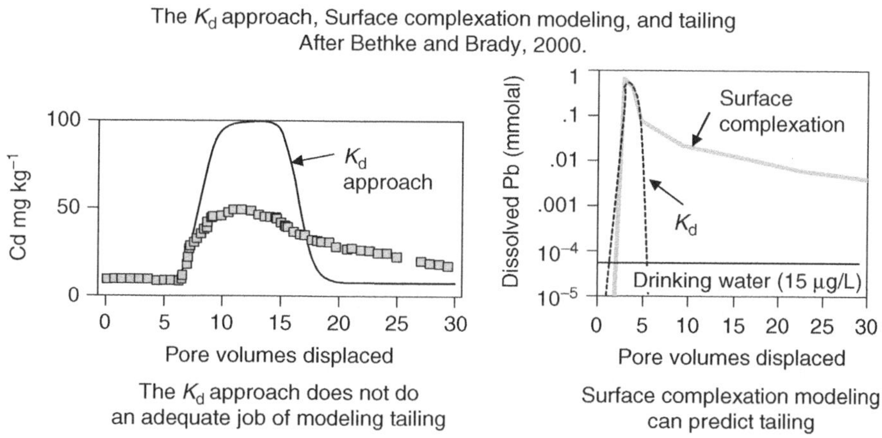

FIGURE 14.1 The K_d versus SCM approach and its behavior relative to prediction of tailing. On the left, the K_d approach fails to predict observed tailing in a laboratory-scale experiment on Cd adsorption. On the right, the SCM approach does a much better job of predicting tailing of Pb in column effluent. Based on and using data from Bethke and Brady, 2000.

in porous media. This approach works well for HOCs because their sorption is driven primarily by physical rather than by chemical interactions. The K_d approach generally does not work well for ion adsorption because it is based on a set of empirical observations under a given set of conditions of solution and mineral surface properties, and there is no ability to expand to other conditions, such as have been considered in surface complexation models (SCMs; Chapter 6). The gain in simplicity is often offset by the constraints of oversimplicity. The K_d approach (1) does not allow for mass balance on sorption sites and permits unlimited sorption when in reality limited sites are available, (2) cannot account for effects of competition between different ions for adsorption sites, (3) does not account for electrostatic effects that develop from changes in charge that occur as metals accumulate at a surface, and (4) cannot accommodate spatial and temporal variability in solution conditions and sorbent properties as would commonly be encountered in a heterogeneous aquifer (Bethke and Brady, 2000; Curtis et al. 2006).

One of the problems with applying the K_d approach to ions is the failure to predict "tailing" accurately. Tailing is a phenomenon in which after the flushing of a contaminated aquifer to remove a metal or radionuclide, the concentrations remain above targets set by regulation (Bethke and Brady, 2000). Example's of the tailing phenomenon and the failure of the K_d approach to predict tailing are shown in Figure 14.1.

Comparison of the K_d versus Surface Complexation Modeling Approaches

In contrast to the K_d approach, modeling based on the surface complexation theory (Chapter 6) takes into account complexation reactions that occur at mineral surfaces, such as for a metal M

$$>SOH + H^+ \rightleftarrows >SOH_2{}^+ \quad (14.1)$$

$$>SOH \rightleftarrows >SO^- + H^+ \quad (14.2)$$

$$>SOH + M^{2+} \rightleftarrows >SOM^+ + H^+. \quad (14.3)$$

The surface complexation modeling approach accounts for electrostatic effects in the mass-action equations, such as for (14.3)

$$Ke^{-\psi F/RT} = \frac{m_{>SOM^+} a_{H^+}}{m_{>XOH} a_{M^{2+}}}, \tag{14.4}$$

where the concentration of surface complexes is used as an approximation for activity. The surface potential ψ is calculated based on the surface charge density and the solution ionic strength (Bethke and Brady, 2000).

Bethke and Brady (2000) compared K_d versus surface complexation approaches for modeling Pb mobility in column experiments. Both approaches were coupled with the Xt model of reactive transport (Bethke, 1997), which takes into account diffusion, dispersion, and advection. The K_d approach failed to describe correctly many important factors that the surface complexation model proved better at handling, including (1) the distribution of Pb in the column, (2) the concentration of Pb in the water passing through the column, and (3) the response of the column to flushing. As shown in Figure 14.1, the K_d approach did not predict tailing, whereas the surface complexation approach did. This result has implications for transport and cleanup strategies relative to the drinking water standard for Pb.

The shortcomings of the K_d approach are appreciated by hydrologists, geochemists, and environmental engineers, who are working to replace K_d modeling with SCMs. However, SCMs are computationally more costly, often require additional geochemical research and parameterization, and do not always correctly model mobility of contaminants in the field. Applying the SCM approach to reactive transport involves combining surface complexation models developed from laboratory experiments on well-characterized surfaces with detailed characterization of the mineralogical content of a study site. As a result, the SCM approach has been applied only occasionally to field studies.

Davis et al. (2004) listed the following four major sources of uncertainty in applying SCMs to field-scale reactive transport: (1) types and abundances of surface sites, (2) effects of solutes on mineral surface charge and potentials, (3) competitive adsorption between different adsorptives, and (4) inconsistencies between different published SCM approaches.

Curtis et al. (2006) applied a modified SCM approach to reactive transport of U(VI) in groundwater at a former uranium mill near Naturita, CO. In oxic environments at circumneutral pH, the mobility of U(VI) can be controlled at least in part by adsorption reactions. The modeling was simplified by using three "generic" surface sites representing average properties of sediment surfaces with very strong, strong, and weak binding of U(VI). Fitting procedures were applied to adsorption data as a function of pH to get around problems associated with modeling the electrical double layer in field situations. The SCM modified model fit field data well enough to suggest that a simplified approach might make SCMs more useful and widely applicable to field studies of reactive transport; however, such modifications need to be widely tested.

ACID RAIN EFFECTS ON CHEMICAL WEATHERING

An excellent example of surface interactions involves the effects of acid rain on ecosystems and engineered materials such as bridges and building stones.

What Makes Rainfall Acidic?

Natural rainfall, along with other forms of *wet deposition* such as snow, sleet, and fog, should have a pH ~5.4. It is naturally acidic because of carbonic acid formed by CO_2 and H_2O

$$CO_2(g) + H_2O \rightleftarrows H_2CO_3 \tag{14.5}$$

and

$$H_2CO_3 \rightleftarrows H^+ + HCO_3^-, \tag{14.6}$$

and because of sulfur-containing compounds (primarily dimethyl sulfide) from plants. However, in many parts of the planet, rainfall is considerably more acidic because of natural sources such as volcanic emissions and/or acidity from human-made sources. Primary sources of acidity include emissions of sulfur dioxide (SO_2) and nitrogen oxides (NO_x) that result from fossil fuel combustion, which turn into sulfuric and nitric acid during reaction with water and oxygen in the atmosphere (Figure 14.2).

The sequence of reactions by which SO_2 is converted to sulfuric acid in the atmosphere involves a free radical such as $OH^{\cdot}$ and the following series of reactions (Masters and Ela, 2008):

$$SO_2 + OH^{\cdot} \rightarrow HOSO_2^{\cdot} \tag{14.7}$$

$$HOSO_2^{\cdot} + O_2 \rightarrow SO_3 + HO_2^{\cdot} \tag{14.8}$$

$$SO_3 + H_2O \rightarrow H_2SO_4. \tag{14.9}$$

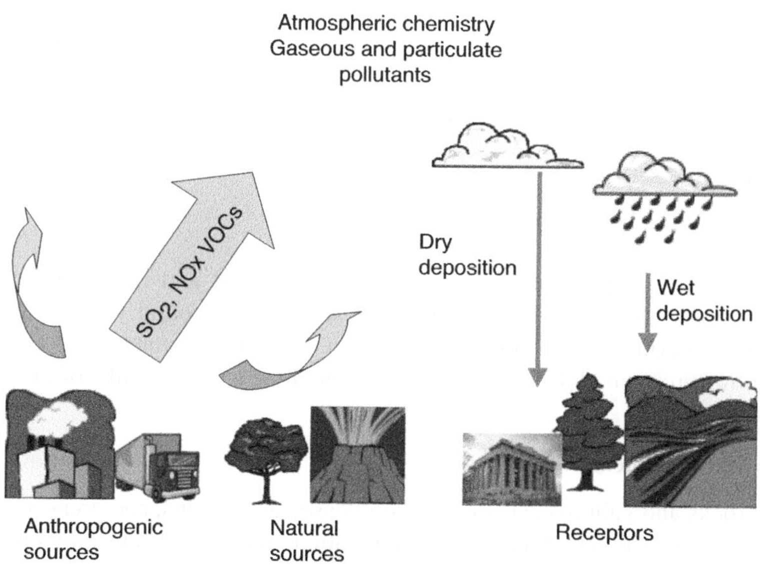

FIGURE 14.2 Wet and dry deposition sources and receptors, both natural and anthropogenic. VOC = volatile organic compound.

FIGURE 14.3 A wet/dry collector for wet and dry deposition sampling and collection. When rainfall hits a sensor, the roof moves from covering the "wet" collection bucket to the "dry" collection bucket. Photo taken of Aerochem/Metrics® sampler in the New Jersey Pinelands (by P. Maurice).

The H_2SO_4 quickly forms micron-scale particulate aerosol particles with water vapor or other atmospheric particles (Masters and Ela, 2008). The deposition of acidic particles to ecosystems can increase acid loading (Figures 14.2, 14.3). NH_3 and NH_4^+ are also significant components of N in atmospheric deposition.

Effects of Acid Rain

Acidic deposition was first identified in the nineteenth century in England by R.A. Smith (Driscoll et al., 2001). In the 1960s and 1970s, reports of acidic precipitation and acidification of surface water began to appear for Scandinavia and the United States, and then other parts of the world. Figure 14.2 shows the processes involved in acidic deposition, and Figure 14.4 shows the average pH of rainfall to the United States in 2006. The more eastern portions of the country tend to receive the most acidic rainfall and are often underlain by rock types that are poorly capable of buffering the acidity.

Limestone and marble building and monument stones can be particularly sensitive to weathering damage by acid rain (Figure 14.5). In many instances, this enhanced weathering of limestone buildings includes gympsum formation

$$\underset{\text{calcite}}{CaCO_3(s)} + H_2SO_4(aq) + H_2O \rightarrow \underset{\text{gypsum}}{CaSO_4 \cdot 2H_2O(s)} + CO_2(g). \quad (14.10)$$

The gypsum tends to spall off, leaving buildings damaged and monuments appearing mottled. This same type of reaction can occur when underground waters pass through sulfide-mineral-bearing rock formations and then enter a limestone cave. Gypsum can grow along fractures, which leads to destabilization of the roof and walls of the cave.

Acid Rain and Chemical Weathering

Calcareous rocks and gabbros can weather to release base cations, and areas underlain by such bedrocks are relatively resistant to acidification. When acid rain falls on land with rocks that are resistant to weathering, such as granitic rocks and other rock types that have predominantly aluminosilicate minerals, the acidity is not substantially neutralized and can pass through soils and into streams and lakes. Acidic deposition has been shown

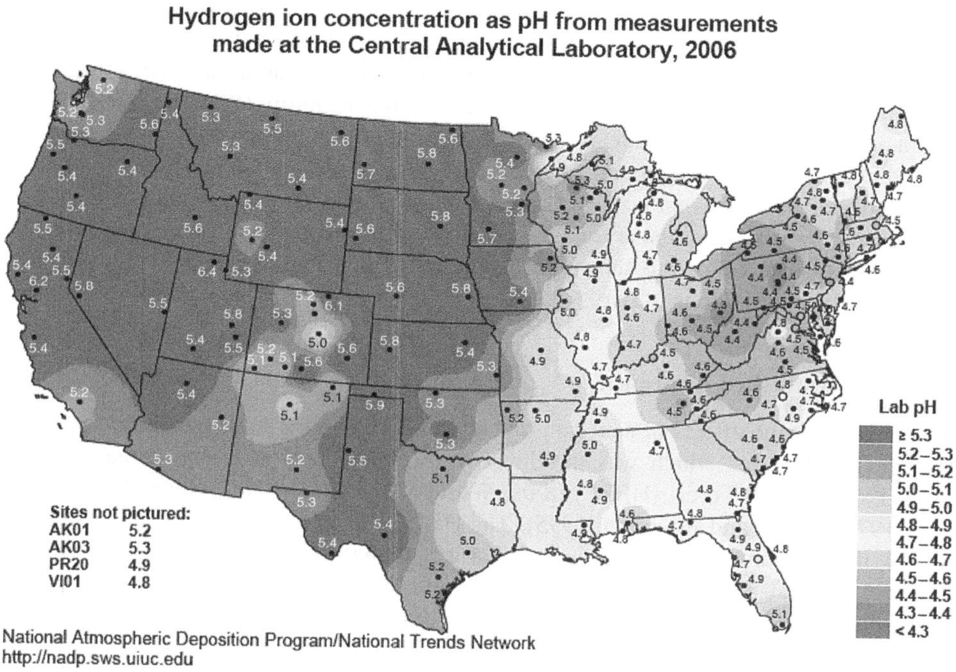

FIGURE 14.4 The pH of rainfall in the United States in 2006. From the National Acidic deposition program. http://nadp.sws.uiuc.edu/isopleths/maps2006/phlab.gif.

to increase the concentrations of protons and strong acid anions (SO_4^{2-}, NO_3^-) and to enhance leaching of base cations (Na^+, K^+, Ca^{2+}, Mg^{2+}) in soils in northeastern portions of the United States, such as the Adirondack Mountains (NY) and the White Mountains (NH). In soils without a considerable acid-neutralizing capacity, Al can be leached and mobilized. Sulfur can accumulate in soils, and the S storage can delay the response of

FIGURE 14.5 Acid rain weathers carbonate to gypsum on a marble baluster at the Organization of American States Building in Washington, D.C. Courtesy of U.S. Geological Survey.

watersheds to decreases in rainfall acidity that result from clean air legislation. Changes to soils can have complex effects on plants; for example, tree diebacks have been observed in parts of North America and Europe. Sometimes, finely disseminated carbonate minerals in predominantly granitic areas can contribute some buffering ability (White et al., 2005).

Not only rock composition but also the pathways that water takes within a watershed can affect the level of acidification and Al toxicity of surface waters. Certain aqueous species of Al (primarily inorganic rather than organic-bound) can be toxic to aquatic and plant life. Some surface waters have undergone long-term acidification; others experience episodic acid events such as following snowmelt. Both acidity and Al can result in fish kills and changes in the distributions of fish.

The Small Watershed Approach

Many studies of the effects of acid rain on chemical weathering have relied on the "small watershed approach," which was developed by a researchers working at the Hubbard Brook Experimental Forest (HBEF) in the White Mountains of New Hampshire (Likens et al., 1977; Likens and Bormann, 1995). The HBEF was chosen for study at least in part because the watershed was underlain by what was thought to be an impermeable granitic basement, so that groundwater inputs and outputs would be minimal. Chemical inputs and outputs were determined by monitoring precipitation to the basin and stream outflow. As the importance of groundwater was recognized, well monitoring also was included. Measurements of groundwater inflow and outflow both in terms of volume and chemical flux can be difficult to obtain, but they are now considered in most watersheds studies. A mass balance can be used to infer geochemical processes, as described below.

NETPATH and PHREEQC

To determine the effects of mineral weathering on solutes in a stream, the following mass balance equation can be used:

$$\text{Watershed inputs} - \text{Watershed outputs} = \text{Weathering inputs} + \Delta\text{Exchange pool} + \Delta\text{Biomass}. \quad (14.11)$$

The model NETPATH (Plummer et al., 1994; downloadable from USGS website as of Nov. 2007) is often used to identify a "plausible" set of chemical reactions based on input to the model of data on water chemical composition. NETPATH (an interactive code for modeling net geochemical reactions along a flow path) is an inverse model that is thermodynamically based. NETPATH takes a starting or input concentration and an output concentration, subtracts the input from the output, and provides appropriate geochemical reactions that could potentially explain the difference. The solutions are not unique but can be constrained with mineralogic information. NETPATH works according to the generalized equation

$$\text{Initial water} + \textit{Reactants} = \text{Final water} + \textit{Products}, \quad (14.12)$$

based on constraints such as the need for reactions to be balanced. In different applications, NETPATH accounts for conservation of mass, electrons, and selected isotopes, depending on how the model is run, to find plausible sets of mineral reactions and gas input/output that fit the changes in water chemistry. Mineral precipitation and dissolution generally are considered, along with cation and anion exchange.

NETPATH is often combined with the model PHREEQC, which was developed by David Parkhurst and colleagues at the U.S. Geological Survey (1997, with many subsequent revisions), for speciation, batch-reaction, one-dimensional transport, and inverse geochemical calculations.

Sr isotopes are often used as an independent check on results of weathering studies using the mass-balance approach or in conjunction with the mass-balance approach. Sr is often thought to be a good surrogate for Ca, although this is not always the case. For example, in a study of weathering at Panola Mountain, Georgia, White et al. (2001) found that Ca and Sr were distributed differently in feldspar phases and that their behaviors in soils were different (White et al. 2001).

The Clean Air Act and Acid Rain Over Time

SO_2 and NO_x emissions from power plants have been decreasing in recent years in the United States, because of amendments to the Clean Air Act in 1970. According to the U.S. Enviornmental Protection Agency (EPA) (www.epa.org), from 1990 to 2006, SO_2 emissions from the power sector fell 40%. NO_x emissions have also fallen since 1990, although NO_x emissions are not capped as are SO_2 emissions. One of the principle ways in which SO_2 emissions have been decreased in the United States is through installation of wet scrubbers that react SO_2 from flue gases in coal-fired power plants with crushed limestone ($CaCO_3$)

$$CaCO_3(s) + SO_2(g) + 2H_2O \rightarrow CaSO_3{\cdot}2H_2O\,(slurry) + CO_2\,(g) \tag{14.13}$$

or lime (CaO)

$$CaO(s) + SO_2(g) + 2H_2O \rightarrow CaSO_3{\cdot}2H_2O\,(slurry) \tag{14.14}$$

to produce a sludge that is sometimes oxidized to gypsum ($CaSO_4$). The "recaptured" gypsum can be used as gypsum wallboards and underflooring in construction.

Driscoll et al. (2003) studied the chemical responses of lakes in the Adirondack Mountains region of New York State to declines in acid deposition since the late 1970s. Marked decreases in $SO_4{}^{2-}$ concentrations were observed for the lakes since 1982, with a mean rate of decline of 2.06 μeq L^{-1} y^{-1} (Figure 14.6). The lakes also showed decreased base cation (C_b; Ca, Mg, Na, K) concentrations accompanying the decrease in strong acid anion (C_a; SO_4 and NO_3) concentrations. Most lakes showed increases in acid-neutralizing capacity (ANC) and decreases in H^+ concentrations (hence, increase in pH). The changes in the concentrations of Al in lakes were small, but some lakes underwent a shift from more toxic inorganic dissolved Al species to less toxic organically bound Al. Nonetheless, in 2000, many of the lakes still contained inorganic Al >2 uM, which is the value considered to be toxic to aquatic organisms (Driscoll, 2003).

ANC is generally calculated from the negative of the sum of base cations minus the sum of strong acid anions

$$ANC = -([Na^+] + [K^+] + 2[Ca^{2+}] + 2[Mg^{2+}] - [Cl^-] - 2[SO_4{}^{2-}] - [NO_3{}^-]), \tag{14.15}$$

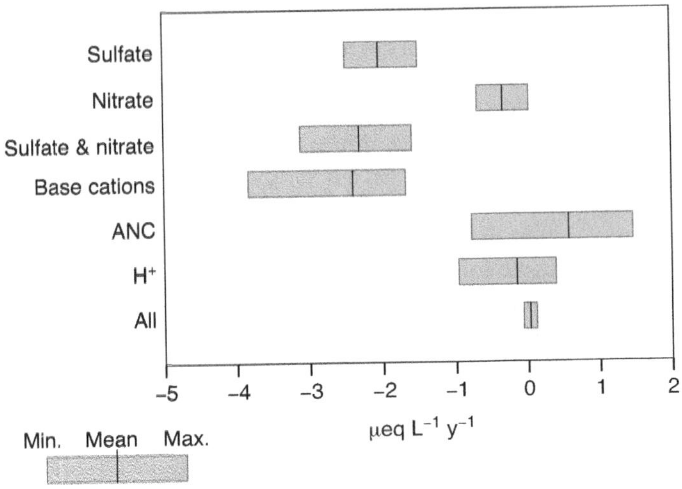

FIGURE 14.6 Mean change in solute concentrations from 1982 to 2000 in 16 lakes in the Adirondack Mountains. In this figure, the range of rate is indicated as a horizontal bar, and the mean as a vertical line within the bar; positive indicates an increase and negative a decline. Based on Driscoll et al., 2004.

where the concentrations are multiplied by charge because a divalent cation can neutralize twice as many OH^- groups as a monovalent cation. A divalent anion can neutralize twice as many H^+ as a monovalent anion. For some waters, organic acids, Al, and other components can be important to ANC and need to be considered.

Although many nations of the world are addressing acid rain issues, much still needs to be done. Significant challenges still exist that range from the generation of acid rain from automobiles and industry in nations such as China and Russia. There is concern about burning of high S coal as petroleum becomes increasingly expensive. Not all watersheds that experience decreases in acidic deposition have shown positive responses in terms of soils and surface waters. Acid rain remains a major environmental problem in many portions of the globe.

ACID MINE DRAINAGE

Acid mine drainage (AMD), which is a form of acid rock drainage (ARD), refers to the outflow of acidic water from metal or coal mines. In many areas, this drainage derives from abandoned mines. However, as the demand for metals such as Cu, Ni, Cd, Au, and Ag grows, mining activities are increasing. Other areas where earth has been disturbed, such as in road construction zones, can also contribute ARD to the environment. The liquid that drains from coal stocks, coal handling facilities, coal washeries, coal waste tips, and even coal piles stored for powerplant use can be highly acidic. ARD occurs naturally in some environments because of rock weathering processes. But it is exacerbated by large-scale earth disturbances that occur from mining and other large construction activities. The problem is particularly severe when the rocks contain sulfide minerals.

The Environmental Problem

Although many different chemical reactions and processes can contribute to ARD, pyrite oxidation is by far the greatest contributor.

First, pyrite is oxidized by ferric iron (Fe $^{3+}$)

Pyrite

$$FeS_2(s) + 14Fe^{3+}(aq) + 8H_2O \rightarrow 15Fe^{2+}(aq) + 2SO_4{}^{2-}(aq) + 16H^+(aq), \quad (14.16)$$

then the Fe^{3+} is "regenerated" via microbially mediated oxidation of Fe^{2+}

$$4Fe^{2+}(aq) + O_2 + 4H^+(aq) \rightarrow 4Fe^{3+}(aq) + 2H_2O \quad (14.17)$$

(Drever, 1997).

When ores have been processed for the desired metals, the large amount of waste rock and tails materials (rocks that have been broken up through processing and milling of the ore into higher surface area rock fragments) are deposited in the enviroment. Once the rocks are broken up and new surfaces are exposed to water and oxygen, pyrite and other reduced sulfidic minerals, such as pyrrhotite, chalcopyrite, and aresenopyrite, oxidatively dissolve to form iron(II) ions, sulfate ions, and hydrogen ions. The sulfate ions become sulfuric acid. In some ARD systems, temperatures can be elevated because of exothermic pyrrhotite ($Fe_{(1\text{-}x)}S$) oxidation, with low-to-negative pH. A negative pH can occur when water evaporates from acidic pools, increasing the hydrogen ion concentration. A variety of extremophile microorganisms can survive in highly acidic waters, and the activity of microbial biofilms (Chapter 12) is an active area of research.

In addition to acidification, another harmful effect of acid mine drainage is contamination by trace and semimetals, such as Ni, Cu, Pb, As, Al, Zn, and Mn, which can be toxic. These pollutants can damage ecosystems and make water unusable for fishing, recreation, consumption, and other uses.

Nanoparticles and AMD

Fe and Mn (hydr)oxides, which are generally present as nanoparticles or coatings on other grains (Figure 14.7), play an important role in biogeochemical processes in AMD streams despite their low overall abundance.

Hochella et al. (2004, 2005) investigated nanoparticles in streams and sediments at the acid-impacted Clarks Fork Superfund Site. They reported nanoparticle (hydr)oxides in the form of two- and six-line ferrihydrite and Mn hydrous oxide. The particles sorbed As, Pb, Cu, and Zn and played an important role in the transport of these metals over long distances in the Clarks Fork river. Other important minerals included jarosite (a basic hydrous sulfate of potassium and iron), which takes up Pb; clays, which carry Cu and Zn; and secondary sulfides, including nanoclusters of chalcopyrite and sphalerite (Hochella et al., 2005).

Hydrobiogeochemical and Photoreductive Processes

McKnight and Duren (2004) described a complex array of hydrobiogeochemical and photoreductive processes that control Fe cycling (Figure 14.8) and diel (daily) fluctuations in Fe(II) concentrations in three Rocky Mountain ARD streams. During the morning,

FIGURE 14.7 Scanning electron microscope (SEM) image of Fe oxide nanoparticle-coated silicate grain from the Clarks Fork Superfund Site. From Hochella et al., 2005 (with permission).

increased light intensity results in increased photoreduction of dissolved Fe(III)

$$\text{(i) } Fe(III)OH^{2+} \underset{h\nu}{\rightarrow} Fe^{2+} + OH^{\cdot} \tag{14.18}$$

and of Fe(III) associated with particulate Fe oxides. Fe(II) is consumed by the *Fenton reaction* to produce Fe(III) according to the overall reaction

$$\text{(ii) } H_2O_2 + 2Fe^{2+} \rightarrow 2Fe^{3+} + 2OH^{-}. \tag{14.19}$$

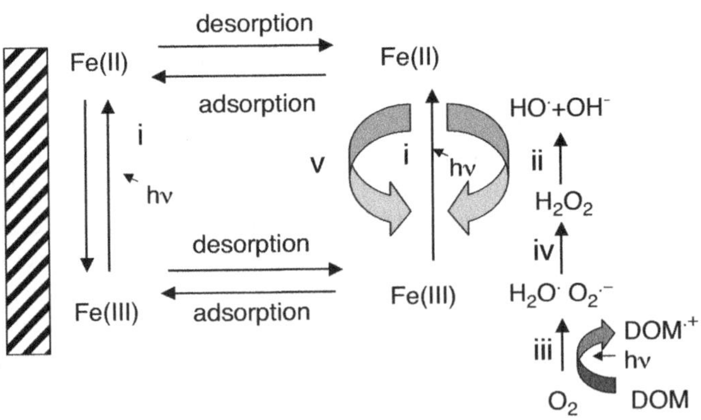

FIGURE 14.8 Schematic diagram showing the processes influencing the chemical speciation and cycling of Fe in the presence of Fe oxide solid phases and photoproducts of DOM. The roman numerals refer to reactions discussed in the text. Based on Voelker et al., 1997, and McKnight and Duren, 2004.

Incident radiation also photolyzes dissolved organic matter (DOM) to produce superoxide radical ($O_2^{\cdot -}$) and H_2O_2

$$\text{(iii) } DOM + O_2 \underset{h\nu}{\rightarrow} DOM^{\cdot} + O_2^{\cdot -} \tag{14.20}$$

ARD streams have low DOM concentrations (McKnight and Duren, 2004), which limits this reaction. At low pH, the superoxide becomes protonated to peroxyl radical, which rapidly converts to hydrogen peroxide (H_2O_2)

$$\text{(iv) } HO_2^{\cdot} + HO_2^{\cdot} \rightarrow H_2O_2 + O_2. \tag{14.21}$$

Fe-oxidizing bacteria such as *Acidithiobacillus ferrooxidans* oxidize Fe(II) to Fe(III) according to

$$\text{(v) } Fe^{2+} - e^{-} \rightarrow Fe^{3+} \tag{14.22}$$

(McKnight and Duren, 2004).

Abiotic oxidation can also occur but is considerably slower in the ARD streams reflecting the typically low pH values (<3). The Roman numerals in Equations 14.18–14.22 correspond to the numbers in the schematic Figure 14.8.

For the three AMD streams investigated by McKnight and Duren (2004), the average photoreduction rates of Fe(III) to Fe(II) ranged from 5.56×10^{-4} to 1.39×10^{-3} uM s^{-1} during the midday when Fe(II) stream concentration was highest.

Fe(III)oxide particles, many of which are nanoparticles, that are present in the stream or as coatings on rocks in the streambed undergo photochemical dissolution as well as biomineralization. These processes help to control the light-related fluctuations in Fe(II) concentrations in the stream. Sullivan et al. (1998) suggested that photocatalyzed dissolution of Fe oxides in ARD drainage streams could release adsorbed phosphate, making phosphate available for biota during the daylight hours.

Biofilms and AMD

Haack and Warren (2003) investigated biofilm processes in an AMD drainage stream in Ontario, Canada, combining field measurements and laboratory molecular-scale investigations. Substantial metal accumulation was observed within biofilms from May through September when temperatures were warm enough to support significant bacterial activity. Metals could be taken up both by the organic components of the biofilm (Ni, Co) and by amorphous Mn oxyhydroxides (Ni, Co, and Cr), which often occurred as microbial bioprecipitates. Mn oxyhydroxides were observed to be more important metal sorbents than were Fe (hydr)oxides within the biofilms.

Profiles of pH and dissolved oxygen were recorded within the biofilm using a microelectrode array (Figure 14.9) that could measure pH and dissolved oxygen at the micrometer scale. This information was coupled with sampling for metals analysis over high-resolution temporal scales and imaging of the biofilm for microorganisms and associated minerals. The biofilm-associated Mn oxyhydroxides dissolved during the day when the oxic–anoxic boundary within the biofilm was deepest and precipitated at night when the biofilm was largely anoxic. This was accompanied by clear diel trends in Cr, Ni, and Co metal dynamics

FIGURE 14.9 A biofilm sampler in an acid mine drainage stream in Ontario, Canada. Courtesy of L. Warren, McMaster University.

in the biofilms (Haack and Warren, 2003). Haack and Warren (2003) suggested that Mn(II) in the water could reduce highly mobile Cr(VI) to less mobile Cr(III) according to the overall reaction:

$$Cr(VI) + 3Mn(II) \rightleftarrows Cr(III) + 3Mn(III) \tag{14.23}$$

during dark hours. The Cr(III) would sorb to Mn hydroxides. The observed rates of formation of biofilm Mn oxyhydroxides at pH values less than 4 could only be explained if the biofilm microorganisms were driving the reaction. During daylight hours, biotic or abiotic dissolution of manganese hydroxide could lead to oxidation of Cr(III) to Cr(VI), coupled with reduction of Mn(VI) to Mn(III). Phylogenetic characterization to the biofilm (Warren et al., 2004) revealed a complex array of microbial associations and biogeochemical processes.

Potential Remediation Strategies

Acid Neutralization A typical treatment to prevent AMD is neutralization of the acidic and often metal-laden discharge by addition of limestone in a tailings treatment pond. The limestone raises the pH sufficiently to precipitate the metals as oxides within the tails pond, which prevents their discharge offsite. The tails water is typically treated secondarily by bubbling CO_2 to lower the pH to that of the receiving water offsite of the mine. This method can fail as Fe and aluminum oxide precipitates may form on the limestone and prevent its dissolution, inactivating its pH buffering capacity. Often, fine limestone powders are added to promote rapid dissolution and effective pH buffering. In either case, sulfate can remain mobile through the system.

Reactive Barriers Another potential strategy for AMD remediation is the use of organic reactive barriers promoting respiration of SO_4^{2-} to hydrogen sulfide by sulfate-reducing bacteria, with subsequent sulfide mineral precipitation, according to (Scherer et al., 2000)

$$2SO_4^{2-} + 4CH_2O + 4H^+ \rightarrow 2H_2S + 4CO_2 + 4H_2O \quad (14.24)$$

$$Fe^{2+} + H_2S \rightarrow FeS(s) + 2H^+. \quad (14.25)$$

The consumption of protons in the respiration step helps to attenuate the acidity of the AMD. The organic components of the reactive barriers help to maintain reducing conditions. Two issues with this approach are how long the reducing conditions and microbial activity can be maintained and the potential for clogging the barrier as sulfide minerals precipitate.

For many historical mine systems, where operations began before an understanding of how acid mine drainage occurred, the creation of acid mine drainage has continued long past the active metal extraction lifespan of the mine. Thus, many mining companies, or governments in the cases of abandoned mines, must continue treatment of acid mine drainage even when active metal extraction activities have stopped. New mining operations in most developed countries (e.g., United States and Canada) and many developing countries must now typically have both an environmental plan to deal with their wastes more effectively as well as a closure plan in place that explicitly outlines how they will minimize and prevent any deleterious effects from mining activities "from cradle to grave."

ENVIRONMENTAL PARTICLES AND CLIMATE CHANGE

One of the most profound environmental issues facing the world today is global climate change. According to the Intergovernmental Panel on Climate Change (IPCC, 2007),

> Warming of the climate system is unequivocal, as is now evident from observations of increases in global average air and ocean temperatures, widespread melting of snow and ice, and rising global average sea level.

Increase in greenhouse gas emissions, primarily CO_2, is believed to be causing an increase in global temperatures, with alteration of precipitation patterns, increased melting of polar glaciers and sea level rise, and many other environmental changes. Figure 14.10 shows how the CO_2 concentration of the atmosphere has changed over time since the

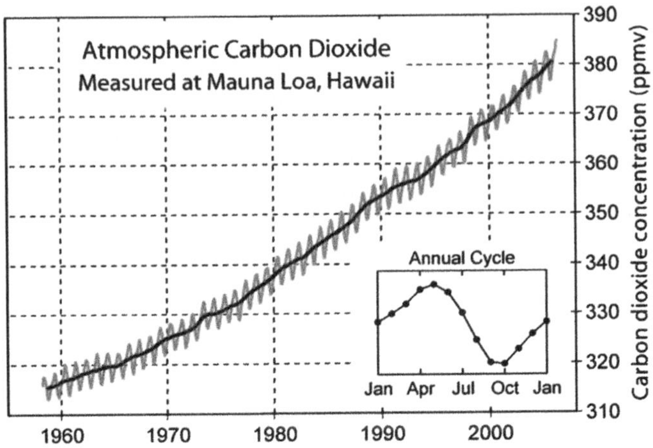

FIGURE 14.10 The change in atmospheric CO_2 as measured at the Mauna Loa Observatory, Hawaii. Courtesy of Global Warming Art; http://www.globalwarmingart.com/wiki/Image: Mauna_Loa_Carbon_Dioxide_png.

mid-1950s. Annual oscillations are caused by seasonal variations in the rates of photosynthesis and respiration, driven primarily by the Northern Hemisphere, which contains the bulk of the continents. During the spring and summer, photosynthesis dominates and removes CO_2 from the atmosphere

$$6CO_2 + 6H_2O \underset{h\nu}{\rightarrow} C_6H_{12}O_6 + 6O_2, \tag{14.26}$$

whereas during the fall and winter respiration dominates and releases CO_2 to the atmosphere

$$6C_6H_{12}O_6 + 6O_2 \rightarrow 6CO_2 + 6H_2O + \text{energy}. \tag{14.27}$$

CO_2 and other greenhouse gases, such as water vapor, nitrous oxide (N_2O), methane (CH_4), halocarbons, and ozone (O_3), act to absorb infrared energy as it leaves the Earth's surface, thus trapping the energy and causing global warming.

Many factors contribute to climate and climate change, and the interactions lead to additional unknowns. The Earth receives a continuous influx of energy from the sun equal to 342 W m^{-2} on average. Approximately 31% of the incoming solar energy is reflected back by clouds and aerosols in the atmosphere or by the Earth's surface. The rest is absorbed at the Earth's suface, which results in an approximate balance, although, in reality, there is always a slight imbalance.

Climate Forcing and Feedbacks

Climate change is driven by climate forcing and climate feedbacks (Figures 14.11, 14.12). A *climate forcing* is an energy imbalance imposed on the climate system either externally by a volcanic eruption or a change in solar influx or by human activities, such as anthropogenic

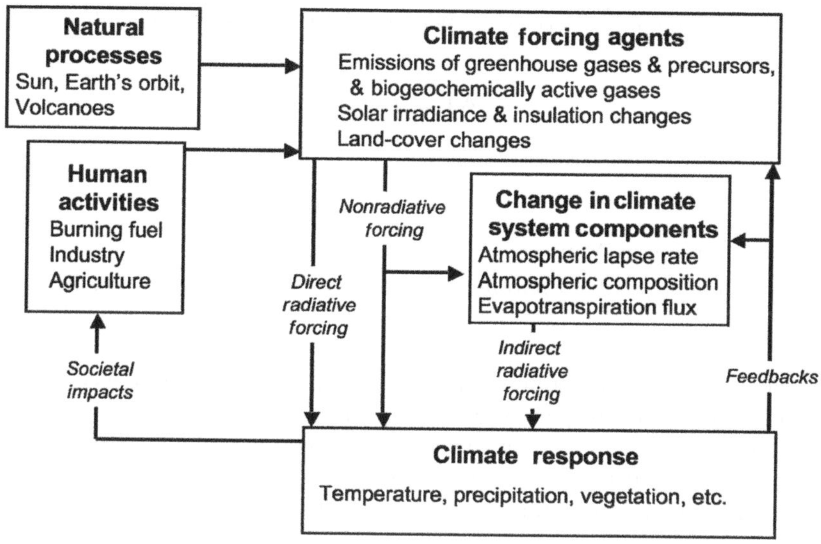

FIGURE 14.11 A schematic showing the conceptual framework for climate forcing, response, and feedbacks in the present Earth climate conditions. The boxes include examples of components and variables that are involved. Based on NRC, 2005.

emissions of greenhouse gases. A *climate feedback* is a process within the climate system that acts to either increase (positive feedback) or dampen (negative feedback) the climate reponse to the specific climate forcing. An example of a positive feedback is warming, which results in melting of snow and ice, decreasing the Earth's reflectivity (or albedo), and causing the Earth to warm even more.

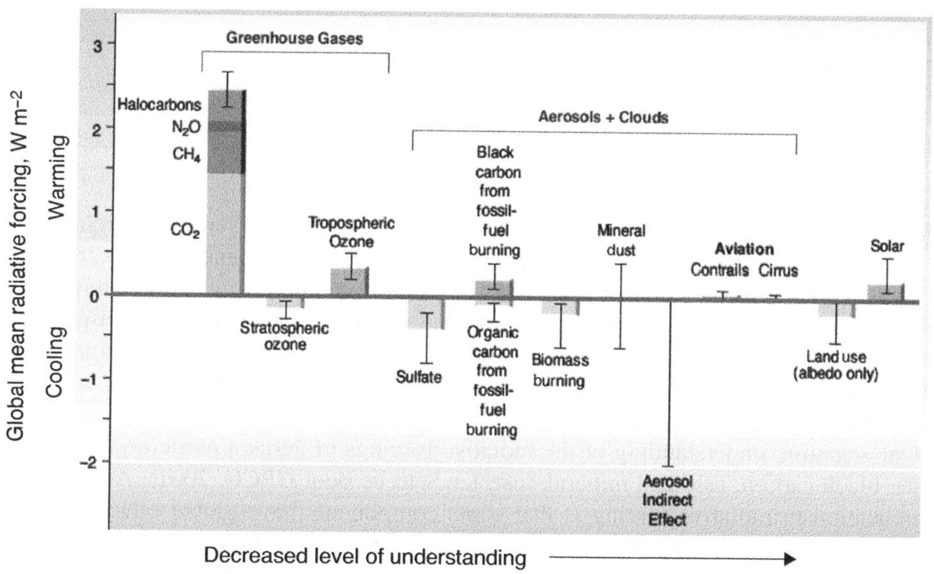

FIGURE 14.12 Components of global mean radiative forcing, indicating degree of understanding by the scientific community (from high on left to very low on right). From IPCC, 2001.

Change in the global mean surface temperature ΔT (in °C) can be calculated from

$$\Delta T = f\lambda, \quad (14.28)$$

where

f = radiative forcing in W m^{-2}
λ = *climate change feedback parameter or climate sensitivity parameter* in °CW^{-1} m^2 (NRC 2005).

The parameter λ accounts for how various interactions and feedback mechanisms may modify surface temperature responses to radiative forcing, and it can be difficult to calculate.

Radiative forcing for individual greenhouse gases that are relatively long lived in the atmosphere can be calculated from formulas that relate atmospheric concentrations at the present-day versus preindustrial times (1750). For CO_2, for example

$$f(t) = 5.35\ \ln[CO_2(t)/CO_2(1750)], \quad (14.29)$$

where t is a given year (NRC, 2001; 2005). The assumption of a linear relationship between ΔT and f seems to work well for the main greenhouse gases, but it breaks down for absorbing aerosols. For aerosols, another parameter known as "efficacy" (E) is the ratio of λ_i for a given forcing agent (i) to 1 for a doubling of atmospheric CO_2 concentration

$$E = \lambda_i/\lambda_{CO_2}. \quad (14.30)$$

From this relationship, an effective forcing f_e is defined by

$$f_e = fE \quad (14.31)$$

(NRC, 2005). For direct radiative forcing of sulfate aerosols, the National Research Council (NRC, 2005) reports $f = -0.34$ Wm^{-2}, $\Delta T = -0.24$°C, $E = 0.83$, and $f_e = -0.28$ Wm^{-2}.

Table 14.1 compares concentrations of well-mixed greenhouse gases in preindustrial times versus 1998. The atmospheric lifetime and radiative forcing of these gases is also summarized. Figure 14.12 shows radiative forcing for the different atmospheric components since preindustrial times (1750). A positive forcing tends to warm the system, whereas a negative forcing tends to cool it.

Our scientific understanding of the radiative forcings of aerosol particles, such as sulphate, black carbon, ash, and mineral dust, tends to be poor (IPCC, 2001). An additional complication to radiative forcing is that some components have global effects, whereas others are more regional to continental in scale. For example, Chung and Seinfeld (2005) calculated that black carbon radiative forcing is likely to result in annual surface temperature increases in the northern hemisphere of ~0.1 to 0.3 K greater than in the Southern Hemisphere.

TABLE 14.1 Concentrations of well-mixed greenhouse gases in Preindustrial (ca 1750) versus in 1998.

Greenhouse gas	Preindustrial conc.	1998 conc.	Rate of conc. change	Atmospheric lifetime
CO_2	~280 ppm	365 ppm	1.5 ppm/year	5–200 years
CH_4	~700 ppb	1745 ppb	7.0 ppb/year	12 years
N_2O	~270 ppb	314 ppb	0.8 ppb/year	114 years
CFC-11	0	268 ppt	~1.4 ppt/year	45 years
HFC-23	0	14 ppt	0.55 ppt/year	260 years
CF_4	40 ppt	80 ppt	1 ppt/year	>50,000 years

CFC-11 is chlorofluorocarbon-11; HFC-23 is hydrofluorocarbon-23.
Data from NRC, 2005.

Volcanoes and Climate

The interplay between environmental particles and climate is dramatically illustrated by explosive volcanic events. Atmospheric particles ejected from individual volcanic eruptions have played major roles in short-term regional to global climate change. For example, the 1815 eruption of Tambora Volcano on the island of Sumbawa, Indonesia, resulted in "the year without a summer" in 1816 North America and Europe. The violent 1991 eruption of Mount Pinatubo in the Philippines ejected huge quantities of ash and SO_2 into the atmosphere (Figure 14.13) and led to a significant average global temperature drop in 1992–1993. SO_2 released from volcanoes forms sulfate in the atmosphere, which also contributes to negative radiative forcing. The eruption of Mount Pinatubo provided an

FIGURE 14.13 Ash column generated by the eruption of Mount Pinatubo on June 12, 1991. Courtesy of U.S. Geological Survey.

estimated -3 W m^{-2} global mean surface radiative forcing and resulted in short-term surface cooling of ~0.4–0.5°C (NRC, 2005).

Ash and sulfate from volcanoes represent examples of how atmospheric particles can play a role in radiative forcing. In addition to direct radiative forcing, volcanic eruptions can have more indirect effects, such as altering precipitation patterns and evapotranspiration rates. Volcanoes also represent major earth-moving and earth-sculpting events, which offer new highly reactive surfaces for weathering and soil formation, sometimes altering streamflow and other hydrologic parameters. Volcanoes can even have implications for decisions on where and how to bury radioactive waste, which ultimately impacts the use of nuclear energy.

CO_2 and Weathering

CO_2 is one of the main greenhouse gases that absorb outgoing infrared radiation from the Earth's surface, and it also affects and is affected by weathering. The atmospheric concentration of CO_2 may influence mineral weathering rates directly or indirectly. Mineral weathering may affect atmospheric concentrations of CO_2, particularly at geological time scales. Mineral weathering is further connected to global climate change because changes in precipitation and temperature may affect weathering rates (Chapter 10).

Navarre-Stitchler and Thyne (2007) investigated the effects of variations in P_{CO_2} (partial pressure of CO_2 in the atmosphere) from geothermal sources on rates of plagioclase weathering in Sierra Nevada watersheds, using data from springs and streams. They found that dissolution rates are proportional to $P_{CO_2}{}^{0.45}$, when the observed variations are attributed solely to changes in P_{CO_2} (Figure 14.14). Other researchers have come up with different measurements or estimates of this proportionality, and models are adjusted accordingly.

The effects of mineral weathering on CO_2 over geologic time scales (10^4 to 10^9 years) were addressed by Robert Berner (e.g., 1995; 2004), who is one of the founders of

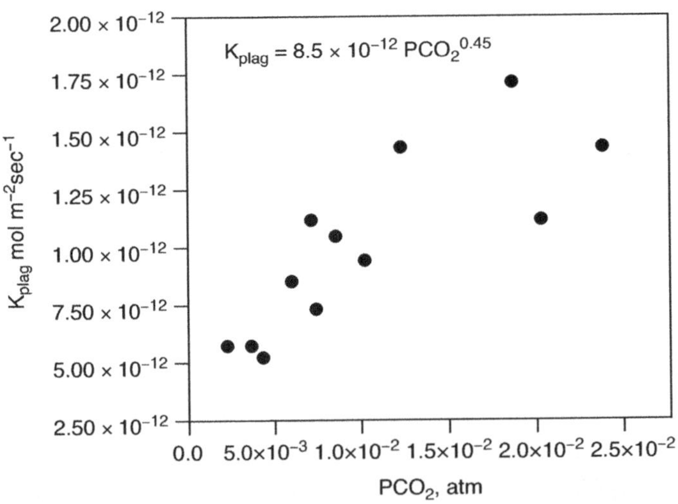

FIGURE 14.14 Correlation between plagioclase dissolution rates and P_{CO_2}. Revised from Navarre-Stitchler and Thyne, 2007.

mineral–water interface geochemistry. He developed a detailed set of equations to describe the effects of weathering of Mg and Ca carbonates versus Mg and Ca silicates on CO_2 cycling. Weathering of carbonate minerals over time scales of 10^5 years or longer does not impact CO_2 concentration in the atmosphere because the carbonate that is taken up balances that released. On shorter time scales, carbonate from weathering may accumulate in the oceans. On any time scale, the weathering of Ca and Mg silicate minerals can affect the CO_2 concentration of the atmosphere.

Carbon dioxide in soil waters can enhance weathering of carbonate and silicate minerals, according to the following reactions (Berner, 2004): for carbonates

$$CO_2 + H_2O + CaCO_3(s) \rightarrow Ca^{2+} + 2HCO_3{}^- \quad (14.32)$$

$$2CO_2 + 2H_2O + CaMg(CO_3)_2(s) \rightarrow Ca^{2+} + Mg^{2+} + 4HCO_3{}^- \quad (14.33)$$

for calcium and magnesium silicates

$$2CO_2 + 3H_2O + CaSiO_3(s) \rightarrow Ca^{2+} + 2HCO_3{}^- + H_4SiO_4 \quad (14.34)$$

$$2CO_2 + 3H_2O + MgSiO_3(s) \rightarrow Mg^{2+} + 2HCO_3{}^- + H_4SiO_4. \quad (14.35)$$

In the oceans, dissolved Ca (and Mg) and carbonate are taken up, and release CO_2 according to

$$2HCO_3{}^- + Ca^{2+} \rightarrow CaCO_3(s) + CO_2 + H_2O. \quad (14.36)$$

Organic matter is buried, which can be illustrated by the generalized equation

$$CO_2 + H_2O \rightarrow CH_2O + O_2. \quad (14.37)$$

If we consider the weathering of carbonates versus the formation of carbonate minerals in the ocean, each mole of CO_2 produced is balanced by a mole of CO_2 consumed.

However, manipulation of Ca and Mg silicate weathering equations with the equation for carbonate formation can be summarized by the following:

$$CaSiO_3(s) + CO_2(g) \rightarrow CaCO_3(s) + SiO_2(s) \quad (14.38)$$

$$MgSiO_3(s) + CO_2(g) \rightarrow MgCO_3(s) + SiO_2(s), \quad (14.39)$$

which reveals that the weathering of Ca and Mg silicate minerals results in the net removal of one CO_2 molecule from the atmosphere (see Berner, 2004). Weathering of Ca and Mg silicates is important to the CO_2 content of the atmosphere because one mole of CO_2 is consumed per mole of Ca or Mg silicate dissolved. Thus, over geologic time, silicate weathering and atmospheric CO_2 have been linked.

As noted, volcanic eruptions demonstrate a link between environmental particles and climate. Volcanic activity has other effects on climate over geologic time. CO_2 released by volcanic activity enhances weathering rates. Volcanic activity not only is a direct source of atmospheric CO_2 but also creates CO_2 sinks. Seafloor volcanic activity creates basaltic oceanic crust that consumes CO_2 on weathering (Wallmann, 2001). Because fresh volcanic rocks expose highly reactive surface area, they tend to dissolve rapidly and begin to consume

FIGURE 14.15 Arenal Volcano in Costa Rica, erupting with volcanic gases. July 2008.

CO_2 quickly when exposed. These sorts of processes are included in models of atmospheric P_{CO2} over geologic time.

The GEOCARB model described below separates basalt weathering from that of other silicates, and it assumes that (1) all subaerial basal weathering occurs in subduction zones and (2) the rate of basalt formation over time can be quantified by rates of seafloor spreading. Berner (2007, personal communication) currently uses an estimate of 30–35% of silicate weathering associated with basalt, and the remainder from other silicate rock types. Thus, the link between basalt weathering and volcanic activity is another excellent example of the link between environmental surfaces and climate change.

The release of CO_2 by volcanic outgassing has been the topic of much debate (see, for example, Berner, 2004), but it is clearly much less than the amount currently released by anthropogenic activity. According to the U.S. Geological Survey (www.usgs.gov), based on studies at Kilauea Volcano, Hawaii, a volcanic eruption can release 8,000 to 30,000 (metric) tonnes of CO_2 into the atmosphere per day. Volcanoes on land (Figure 14.15) and under the ocean release ~200 million metric tonnes of CO_2 annually. Anthropogenic CO_2 emissions total ~28 billion tonnes per year (Energy Information Administration website), so that volcanic sources equal less than 1% of the CO_2 annually released by anthropogenic activity.

Modeling the C Cycle Over Geologic Time

If we are to predict the future, then we must be able to understand the past. To provide broad insight into critical zone interactions over geologic time, Berner (1991, 1994, 2004) and Berner and Kothavala (2001) developed the GEOCARB (to version GEOCARB III) models for the evolution of the carbon cycle and of atmospheric CO_2 over Phanerozoic time. These models are based on geological, geochemical, biological, and climatological data and on equations describing a wide variety of processes. GEOCARB modeling is only applicable to data at geological time spans, but it has proven helpful for organizing the scientific community's thinking on the processes and complex interactions involved in climate change, especially the sensitivity of CO_2 to different critical zone interactions. Like all good models, GEOCARB has raised the level of debate, and it has evolved over time as various validation and verification (V&V) processes are applied (Chapter 6) and as the scientific community points out gaps and produces new data and approaches.

GEOCARB modeling is based on equations that describe C and C-isotope mass balance along with consideration of the rates of weathering and degassing and how they have changed over geologic time through the Phanerozoic (e.g., Berner and Kothavala, 2001; Berner, 2004). Development of the relevant equations is discussed in detail by Berner (2004). The processes involved in weathering and considered in the model are as follows (Berner, 2004): temperature, which is affected by the greenhouse effect, solar evolution, and processes of continental drift; physical erosion, which is affected by the topographical relief of continents; precipitation and runoff, which are affected by the greenhouse effect and continental drift; land vegetation, which has evolved through the Phanerozoic; and lithology as it relates to exposure of different rock types to weathering. Other processes are considered, such as CO_2 outgassing, along with that of CH_4.

Typically, the results of GEOCARB modeling are presented as plots of RCO_2 (the ratio of the mass of CO_2 in the atmosphere at some time t/mass of CO_2 in the atmosphere today) on the y-axis and time before present in millions of years on the x-axis. The model can be run using parameters calculated by different investigators or with different perturbations to the system, such as various schemes for the evolution of land plants or for how different types of plants affect weathering. An example from Berner and Kothavala (2001) that shows how different estimates of activation energies for Ca-Mg silicate weathering may affect models of RCO_2 over the Phanerozoic is shown in Figure 14.16.

Given the complexities of social, scientific, and engineering aspects of climate change, it is essential that we develop interdisciplinary approaches. Environmental surfaces and interfaces are important components of the climate system, and they must be considered in developing strategies for the future of our planet.

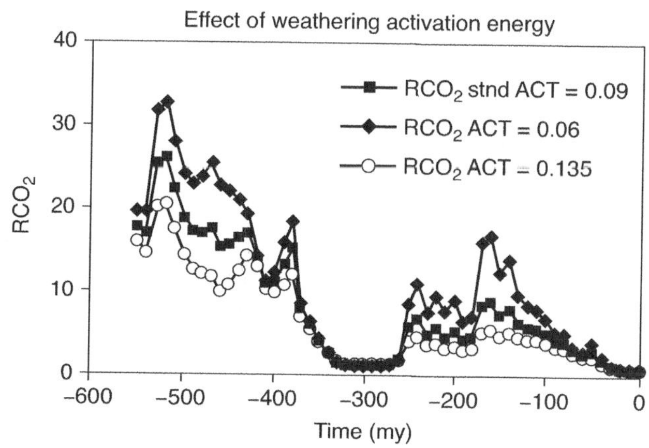

FIGURE 14.16 An example of GEOCARB model output showing how different estimates of activation energies (ACT) for Ca-Mg silicate weathering may affect calculations of RCO_2 over the Phanerozoic. From Berner and Kothavala, 2001 (with permission).

SCALING PHENOMENA: INTEGRATING OBSERVATIONS FROM THE ATOMIC TO THE WATERSHED TO THE GLOBAL SCALE

The Concept of the Macroscope

Most scientists are famliar with the idea of a microscope, which can be used to view "the infinitely small," and a telescope, which can be used to view "the infinitely great." In 1977, Joël de Rosnay proposed that understanding complex systems, such as organisms, economies, and the workings of the world, require a macroscope approach merging information from different scales to investigate and understand "the infinitely complex." According to de Rosnay (1979):

> The macroscope is unlike other tools. It is a symbolic instrument made of a number of methods and techniques borrowed from very different disciplines. It would be useless to search for it in laboratories and research centers, yet countless people use it today in the most varied fields. The macroscope can be considered the symbol of a new way of seeing, understanding, and acting.

The concept of the macroscope is illustrated in the cartoon in Figure 14.17.

To understand the phenomena of elemental cycling, pollutant transport, climate change, and other complex environmental phenomena, it is essential to apply a broad range of techniques and approaches from many different disciplines and at many different scales, in essence, to use a "macroscope" approach. The ultimate goal of investigators who study environmental surface and interface processes is to integrate data and understanding from the molecular scale to much larger scales, in many cases even attaining the global scale. This process is shown schematically in Figure 14.18. Generally, multiscale research requires collaborations by experts in a variety of disciplines.

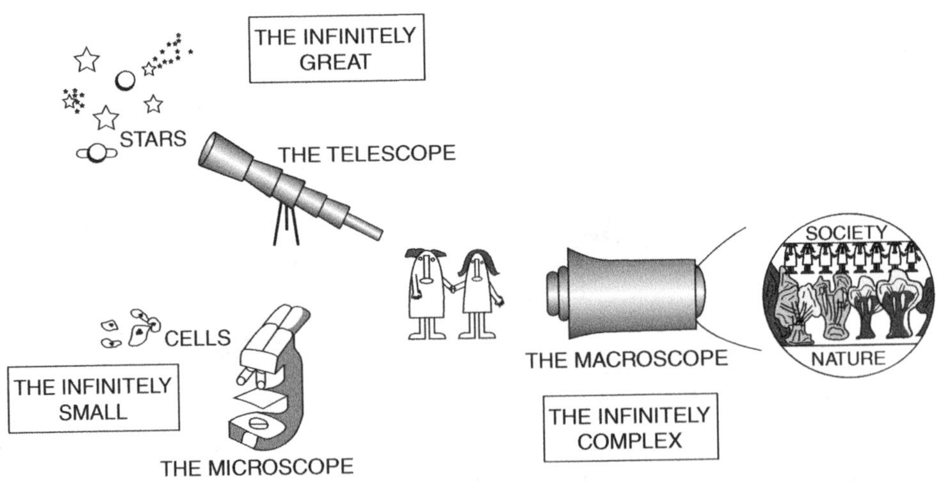

FIGURE 14.17 From "The Macroscope" by Joël de Rosnay. Originally published in 1979 by Harper & Row, (New York). Downloaded from http://pespmc1.vub.ac.be/MACRBOOK.html.

FIGURE 14.18 A schematic representation of the importance of environmental surface observations at different scales, from use of atomic scale techniques such as atomic force microscopy (AFM) (right) to global and even space-bound investigations (left). AFM imaging schematic courtesy of S. Lower (Ohio State University; biofilm sampler courtesy L. Warren (McMaster University); photo of Earth courtesy of NASA.

Embedded Sensor Network Systems

One innovation for scaling phenomena is embedded network sensor systems (e.g., EMNET). *Embedded networks* are an emerging technology that takes advantage of modern embedded processors' networking capabilities to link together state-of-the-art sensors and to embed a computational intelligence in the environment, actually allowing for both adaptive monitoring and control of that environment (Seders et al., 2007). Such sensor systems for the environment are in active development and operation (e.g., Maurice and Harmon, 2007). For example, Seders et al. (2007) developed an embedded sensor network for a lake, as shown in Figure 14.19. The sensor-containing pods interact and communicate electronically with one another, while also sending data to and receiving communication from a laptop computer.

One vision for the future of environmental research is a Global Earth Observation System that would link embedded sensors to data at many scales, even remotely sensed data, through a satellite link system (Figure 14.20). The key to the ultimate success of sensor systems is the development of robust and field-deployable sensors for components that range from pH (although probes exist, they are expensive and subject to problems such

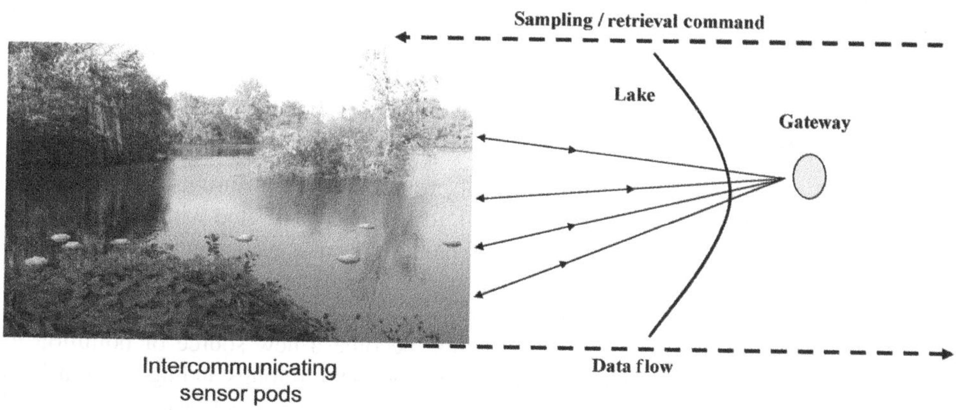

FIGURE 14.19 Schematic of an embedded sensor network for a lake. Courtesy of L. Seders, P. Maurice, J. Talley, and M. Lemmon, University of Notre Dame.

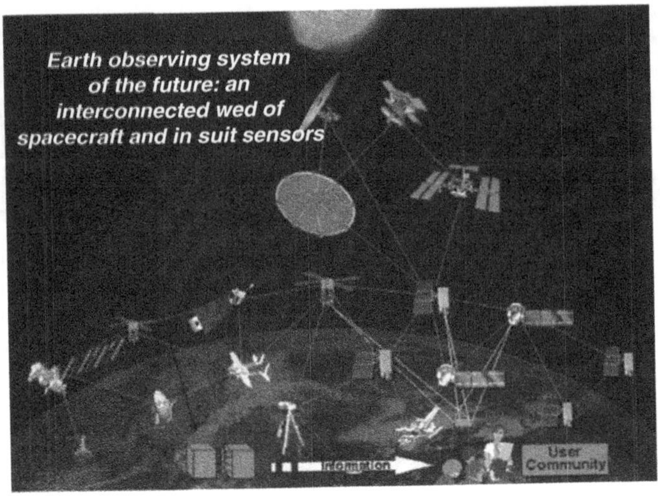

FIGURE 14.20 An earth observing system. From IGOS, 2004. (Credited in the IGOS report to George Mason University).

as clogging and drift) to dissolved oxygen concentration to concentrations of individual metals of interest.

Sensors for Surface and Interface Phenomena

Increasingly, sensors are being developed specifically to address surface and interface phenomena. For example, flexible, miniature, and inexpensive nitrate sensors are being fabricated by electropolymerizing pyrrole onto carbon fiber substrates, using nitrate as a dopant (Bendikov et al., 2005). The resulting sensor (length 1 cm or less; diameter 7 to 10 um; Figure 14.21) is a size and shape that is ideal for deployment in the pore space of soils and sediments. The sensors system shown in Figure 14.21 can quantify nitrate concentrations to just below 10^{-4} M (3 ppm). Although many challenges remain to be worked out, the fact that such sensors are under development and that sensor networking capabilities are growing rapidly means that the future of investigations of environmental surfaces and interfaces is bright.

New Opportunities: New Challenges

Environmental embedded sensor networks will provide new opportunities and pose new challenges. How will new sensors be calibrated, tested, and integrated with existing measurement devices? How will large data sets be handled from a computer networking and computer science standpoint? How will scientists use rapidly expanding data bases effectively to build new understanding, including predictive modeling capabilities? Will widely distributed sensors, likely including nanosensors, become a new source of pollution to the environment? Will environmental scientists be able to build longstanding relationships with chemical engineers, electrical engineers, computer scientists, surface chemists, and researchers in other disciplines so that they can keep complex new systems running and at state-of-the-art well into the future? Can sensors and sensor networks be made marketable

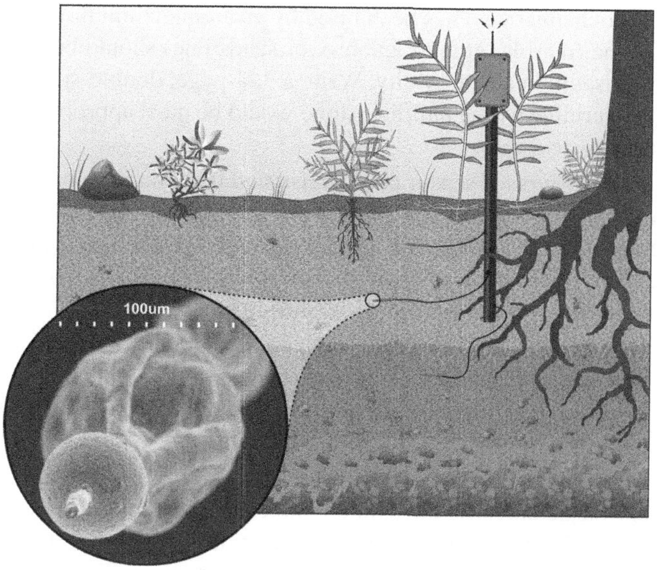

FIGURE 14.21 Illustration of an in situ nitrate microsensor, including an SEM image of a sensor. Courtesy of Jason Fisher, University of California, Merced; and Tatya Bendikov, University of California at Los Angeles.

so that industry will take over the burden of developing new technologies for wide distribution without at the same time making those technologies too expensive for use in scientific research? Will governments embrace new technologies, while at the same time making sure that safeguards are in place to ensure that they are use wisely and in an ecologically sound fashion?

What these questions demonstrate, as a fitting end to this text, is that investigations of environmental surfaces and interfaces from the nanoscale to the global scale are not only complex scientifically but also in terms of interdisciplinary and societal interactions. Investigators in this field need to think always in terms of building bridges and collaborations, while remembering that as scientists and engineers, our ultimate responsibility is to serve society and to enjoy our explorations of the complex world around us simultaneously.

QUESTIONS FOR FURTHER THOUGHT

1. Consider a soil mineral particle in a tropical soil, a temperate soil, and an Arctic soil.
   (a) How might an increase in temperature of 2° and either an increase or a decrease in precipitation affect that particle over the next 10,000 years?
   (b) What factors would we need to know to help determine potential effects on a single soil particle of changing climate over a 10,000-year time interval?
   (c) What other factors besides temperature and precipitation, might need to be considered?

2. You are a geochemist who has been hired by an architectural firm to determine what building stone (marble, granite, gabbro, or sandstone) should be used as facing for a new art museum in a nearby city. Write a 3–4-page, double-spaced proposal for a project to determine which building stone would be most appropriate to use in terms of resistance to acid rain.
3. Imagine that any type of sensor could exist, and that you have been given the job of designing an environmental sensor network to monitor feldspar weathering rates in a forested ecosystem such as the HBEF. Write a 3–5-page essay, including at least one schematic illustration, of your optimum network design, including the factors that would need to be considered in the design strategy.
4. Consider McDonalds Branch basin, which is the small watershed shown schematically at the end of Chapter 2. What are some ways in which a small coast-plain watershed such as this might be impacted by climate change? Consider the processes and reactions shown.
5. CO_2 impacts on weathering have been receiving a good deal of attention by researchers particularly in the United States because of the interest of the U.S. Department of Energy in using various strategies of injecting CO_2 into wells for deep burial, which is known as "carbon sequestration." The question of how injected CO_2 may impact rock formations, including biogeochemical processes, is therefore of great significance. According to the Department of Energy's carbon sequestration website (Dec. 2007), "Geologic formations such as oil and gas reservoirs, unmineable coal seams, and underground saline formations are potential options for storing CO_2. Storage in basalt formations and organic rich shales is also being investigated." Moreover, according to the DOE, "Carbon sequestration is one of the most promising ways for reducing the buildup of greenhouse gases in the atmosphere." Based on your understanding of hydrobiogeochemistry, what are some challenges that might occur related to geologic CO_2 sequestration, and what factors will need to be considered?
6. A sample of surface (stream) water from McDonalds Branch basin had pH 4.5, dissolved organic carbon concentration 20 mgC L^{-1}, and Fe concentration 5 ug L^{-1}. After the analysis showing the above concentrations, the sample was set out in the sun for 1 hour in a constant-temperature chamber open to the sunlight. When it was brought back in, the DOC had dropped to only 2 mgC L^{-1}, the pH had increased slightly, and the Fe concentration was very low. The sample was then allowed to sit in the dark for several days under refrigeration to prevent microbial activity. After coming back to room temperature, most of the DOC and Fe concentration had returned to the original values. Explain chemically what might have happened, using chemical equations for processes where you can.
7. In 2007, former U.S. Vice President Al Gore and the Intergovernmental Panel on Climate Change were joint recipients of the Nobel Peace Prize for their international efforts on climate change. Is this an appropriate topic/theme for the Nobel Peace Prize? Explain.
8. Why is weathering of anorthite investigated more often in climate change studies than weathering of potassium feldspar or albite?
9. What are some factors that might contribute to observations of diel (over the course of the day) fluctuations in metal concentrations in acid rock drainage streams?

10. (a) Why is it important to take into consideration possible cation exchange in studies of acid rain effects on mineral weathering in a terrestrial watershed?
    (b) Why might it be important to consider the effects of plants?
11. Discuss the work of Bethke and Brady (2000) in terms of model verification and validation. You may choose to find and read the original manuscript.
12. (a) Plot the change in global mean surface temperature ΔT that would be predicted if the atmospheric concentration of CO_2 were to increase by ~0.001 or 0.005 ppm per year since 1998 for 100 years and with climate feedback parameters $\lambda =$ 0.20, 0.40, and 0.80°C w^{-1} m^2.
    (b) What implications do increases in atmospheric CO_2 concentrations and estimates of λ have for our understanding of global warming phenomena?

FURTHER READINGS

Berner, R.A. 2004. The Phanerozoic Carbon Cycle: CO_2 and O_2. Oxford University Press, Oxford, UK, 158 p.

Bethke, C.M., and Brady, P.V. 2000. How the K_d approach undermines groundwater cleanup. *Ground Water* 38, 435–443.

Blowes, D.W., Ptacek, C.J., Jambor, J.L., and Weisener, C.G. 2003. The geochemistry of acid mine drainage. In Lollar, B.S. (ed), Treatise on Geochemistry vol. 9. Elsevier, New York, pp. 149–204.

Driscoll, C.T., Lawrence, G.B., Bulger, A.J., Butler, T.J., Cronan, C.S., Eagar, C., Lambert, K.F., Likens, G.E., Stoddard, J.L., and Weathers, K.C. 2001. Acidic deposition in the Northeastern United States: Sources and inputs, ecosystem effects, and management strategies. *BioScience* 51, 180–198.

GLOSSARY OF TERMS

absorption—a sorption process wherein the sorbate enters internal pore space within the sorbent.

acidic deposition—commonly termed "acid rain." Acidic deposition may include both wet (e.g., rain, sleet, snow, and fog) and dry forms. Most natural precipitation is acidic; therefore, the term is generally reserved for cases in which the acidity is increased by anthropogenic activities such as burning of fossil fuels.

acid mine drainage (AMD)—the outflow of acidic water from metal or coal mines. A form of acid rock drainage.

acid neutralizing capacity (ANC)—a measure of the overall buffering ability of a solution against acidification. ANC is generally calculated from the negative of the sum of base cations minus the sum of strong acid anions

$$ANC = -([Na^+] + [K^+] + 2[Ca^{2+}] + 2[Mg^{2+}] - [Cl^-] - 2[SO_4^{2-}] - [NO_3^-]);$$

however, in some waters, other components must be considered.

acid rock drainage (ARD)—the outflow of acidic water, generally from earth that has been disturbed, such as at mines or construction sites.

activated carbon—a form of carbon that is usually formed by processing charcoal to make it extremely porous, with a high internal surface area for sorption.

activation energy—an energy barrier to a reaction occurring; activation energy is the minimum energy necessary for a specific chemical reaction to occur. The activation energy of a reaction is usually denoted by E_a and has units of kilojoules per mole.

activity—"effective" concentration, which takes into consideration nonideal behavior such as charge shielding. For a dissolved species i, activity $a_i = \gamma_i m_i$, where γ_i is the activity coefficient and m_i is the molality.

adatom—from "adsorbed atom," which is a single atom that lies on a crystal surface; it is essentially the opposite of a single-atom vacancy.

admicelle—a term that refers to a micelle-like structure adsorbed to a surface; the surface analog of a micelle.

(ad)sorbate—a substance that is accumulating or (ad)sorbing from air or solution onto a solid surface.

(ad)sorbent—a solid substance on which an (ad)sorbate such as an inorganic ion or an organic compound is accumulating.

adsorption—two-dimensional accumulation at the external surface of an adsorbent.

(ad)sorptive—as used in this text, the substance that is (ad)sorbing when it is still in solution but has the ability to (ad)sorb. This is not a standard definition.

adsorption edge—a plot with concentration of adsorbate per unit adsorbent on the y-axis and pH on the x-axis is called an adsorption edge if the plot is an s-shaped curve (as commonly encountered for many cations).

adsorption envelope—a plot with concentration of adsorbate per unit adsorbent on the y-axis and pH on the x-axis is called an adsorption envelope if the plot is the reverse of an s-shape curve (as commonly encountered for many anions).

adsorption isotherm—a plot of amount adsorbed per unit surface area, generally at a given pH, ionic strength, and temperature. Units are generally g m^{-2} or moles m^{-2}.

advection—the process whereby solutes that are contained within groundwater are transported by the bulk motion of the fluid.

adventitious carbon—a term referring to carbon that is present at surfaces exposed to air for even short periods of time. After exposure to air, surfaces generally contain small amounts of hydrocarbons and oxidized carbon that can be detected when the sample is analyzed by surface techniques such as X-ray photoelectron spectroscopy (XPS).

Aeolian—wind-blown.

Aerobic respiration—the process whereby simple sugar is converted to water and carbon dioxide in the presence of oxygen; it is part of ATP production.

albedo—the albedo of a given surface is the percent of incident sunlight that is reflected back. Snow and ice have very high albedo values; ocean and land surfaces have lower albedo values.

aliphatic—a simple straight- or branched-chain organic compound.

Alkaline earth metals—a series of elements in Group II of the periodic table, including Be, Mg, Ca, Sr, Ba, and Ra. These elements tend to lose two electrons to form doubly charged cations.

Alkali metals—a series of elements in Group I of the periodic table, including Li, Na, K, Rb, Cs, and Fr. Although H is part of this group, its behavior is generally different. These elements tend to lose one electron to form singly charged cations.

allotropy—the property of some elements to take on more than one structural arrangement of bonds; the different forms of which are known as allotropes.

amorphous—used to refer to a solid that does not have a defined crystal structure, not crystalline.

amphiphilic—refers to a chemical compound that has both hydrophobic and hydrophilic properties, = amphipathic.

amphoteric—in chemistry, an amphoteric substance can react as either an acid or a base.

angiosperm—one of the flowering plants, the largest group of land plants on the Earth today.

Ångstrom (Å)—a unit of length equal to 1×10^{-10} meter or 0.1 nm.

anion—a negatively charged ion, such as Cl^-.

anode—an electrode through which (positive) electric current flows into a polarized electrical device. In a galvanic cell, the anode is the negative terminal, whereas in an electrolytic cell, the anode is the positive terminal.

Antarctic Dry Valleys—the McMurdo Dry Valleys are a group of valleys in Antarctica located in Victoria Land west of McMurdo Sound. They have an extremely cold and dry climate.

aquiclude—a solid, impermeable layer beneath an aquifer.

aquifer—a groundwater unit that contains a volume of water that can be supplied at reasonable cost to humans.

Archaea—a group of single-celled microorganisms that, like the bacteria, are prokaryotes. In the Woese system of classification, the three main branches of evolutionary descent are the bacteria, the eukaryota, and the Archaea.

aromatic—an organic compound that has a structure with conjugated double bonds.

Arrhenius equation—a formula describing the temperature dependence of a rate constant (k): $k = Ae^{-E_a/RT}$, where E_a is the activation energy, R is the ideal gas constant, and T is the temperature (in Kelvin). A is known as the "preexponential factor." In many but certainly not all instances, the reaction rate doubles for every $\sim$10°C increase in temperature.

asbestosis—an occupational lung disease caused by exposure to airborne asbestos particles.

assimilatory—a type of metabolism in which the chemical component, such as sulfur, is assimilated into organic compounds.

(ATP) adenosune triphosphate—adenosine triphosphate; an adenosine-derived nucleotide that contains high-energy phosphate bonds and that serves as a source of energy by a cell for many metabolic processes.

atrazine—1-chloro-3-ethlyamino-5-isopropylamino-2,4,6-triazine; a widely used herbicide of environmental concern.

autotroph—referring to an organism that is capable of synthesizing its own "food" (organic compounds) from inorganic substances, using light or chemical energy.

bandgap—a region between the electron valence band and the conduction band devoid of allowed energy states. There is virtually no bandgap in metals, a large bandgap in insulators, and an intermediate and generally small bandgap in semiconductors. The bandgap of a semiconductor such as hematite may vary considerably with the concentration of impurities.

Banded iron formations (BIF)—laminated chemical sediments that consist of millimeter-thick layers of hematite or magnetite interbedded with quartz or chert and that formed between 3.8 and 0.8 Ga (GA = thousand million years ago). The potential role(s) of microbial processes in BIF formation is the subject of much ongoing research.

base cations—positively charged ions such as Ca^{2+}, Mg^{2+}, Na^+, and K^+ that increase pH of water, making it less acidic. In areas impacted by acid rain, the release of base cations through mineral weathering and exchange reactions can help to decrease the acidity.

batch reactor—a type of reaction vessel in which there is no inflow or outflow. Commonly, a beaker, flask, (capped) centrifuge tube, or bottle. Frequently used for sorption experiments; sometimes it is used for mineral growth and dissolution experiments.

bauxite—a rock or soil rich in aluminum hydroxide.

BCF theory (Burton, Cabrera, Frank)—a theory that describes how crystal growth by birth and spread or at steps and dislocations varies as a function of saturation state. Other factors such as temperature and impurities also contribute to the mode of growth.

bedload—part of the stream load consisting of the heavier, coarser grained earth materials that travel along the bottom of the stream by sliding, rolling, or taking short skips or jumps in a process known as saltation.

Bemidji site—a location in Minnesota where a crude oil spill contaminated an aquifer; the site of many hydrobiogeochemical investigations.

Bentonite—a swelling clay formed from volcanic ash and used in a wide variety of applications, such as adhesives, cements, and drilling muds.

BET surface area—surface area of a particle as measured by the method of Brunauer, Emmett, and Teller (1938). The method works by measuring gas adsorption/desorption to/from a surface, and by applying a formula based on theoretical considerations to calculate surface area from these measurements and the weight of material.

bidentate—when ligands bind to a metal atom in solution or at a surface through only one bond, the ligand is said to be binding in a monodentate fashion; when through two bonds, bidentate.

bilayer—a structure, such as a film or membrane, that consists of two molecular layers. A phospholipid bilayer is the basic structural unit of a biological membrane.

binuclear—when a ligand binds to two metal sites on a surface, the binding is said to be binuclear; when to one site, it is said to be mononuclear.

biofilm—a microbial community held together by extracellular attachment features such as exopolysaccharides (EPS) in a distinct architecture. A biofilm may also include abiotic components, such as mineral particles or tooth or bone structures that serve as support and sometimes as sources of nutrients.

biological weathering—refers to rock breakdown by biological processes that may be physical and/or chemical.

biologically controlled mineralization—biomineralization in which the cellular activities of the organism play a direct role in controlling nucleation and growth, along with the morphology and depositional location of the mineral formed. Biologically controlled mineralization may be extracellular, intercellular, or intracellular.

biologically induced mineralization—biomineralizaton in which the cell promotes mineralization by altering solution conditions and/or providing surfaces for heterogeneous nucleation, but it does not directly control the mineral type or habit.

biomineralization—the process whereby living organisms catalyze or otherwise mediate the formation of minerals.

black carbon—carbon particles that form by incomplete combustion of fossil fuel and wood (forming soot) or biomass (forming charcoals).

Born repulsion forces—short-range intermolecular forces that fall off exponentially with distance; develop when neighboring atoms are close enough that their electron clouds begin to overlap. The strong repulsive force is a result of the Pauli exclusion principle.

Brownian motion—the random movement of particles suspended in a liquid or gas.

bryophytes—the nonvascular plants.

Burger's vector (b)—a vector that represents the magnitude and direction of lattice distortion of a dislocation in a crystal lattice.

capacitance (C)—a measure of the amount of electric charge stored for a given electric potential. A common form of capacitor for charge storage is a set of parallel plates. For plates with charge +Q and −Q, and a voltage difference V between the plates, $C = Q/V$.

capillary fringe—the portion of the unsaturated zone at the upper boundary of the water table, where water is pulled upward against the force of gravity by capillary forces.

carbon black—industrially produced and used in many rubber compounds, carbon black is one of the most widely released anthropogenic nanoparticles.

catalyst—a substance that increase the rate of a reaction without modifying the overall Gibbs energy change in the reaction. By definition, a catalyst is not consumed over the course of the reaction.

cathode—an electrode through which (positive) electric current flows out of a polarized electrical device.

cation—a positively charged ion.

cation exchange capacity (CEC)—a term used used to describe the ability of minerals to adsorb and exchange cations. It is important to define CEC in terms not only of excess of cations but also as deficit (negative adsorption) of anions.

chemical kinetics—the study of rates and mechanisms of reactions.

chemical potential—in thermodynamics, the rate of change of the Gibbs energy of a system with respect to the change in the number of moles of a particular constituent.

chemical thermodynamics—the study of energy and its transformations. Chemical thermodyamics deals with the overall energy changes that occur when going from reactants to products.

chemical weathering—the process in soils and sediments whereby minerals are dissolved and sometimes new minerals are formed through chemical reactions at or near the Earth's surface.

chemolithotrophs—organisms that obtain their energy from inorganic molecules, such as iron, sulfur, and hydrogen sulfide, and that assimilate carbon as CO_2, bicarbonate, or carbonate. The chemolithotrophs can be especially effective in enhancing dissolution of minerals, such as Fe and Mn oxides.

chemoorganotrophs—organisms that obtain energy from organic molecules.

chemostat—(the chemical environment is static). A type of continuously stirred tank reactor (CSTR) used in geomicrobiology. A chemostat consists of a nutrient reservoir and a growth chamber. By controlling the concentration of a limiting nutrient, the bacterial culture can be maintained under carefully controlled conditions over long periods of time at a specific growth phase.

chemotrophs—organisms that access energy that is stored in chemical components; most bacteria are chemotrophs.

chlorosis—in botany, the yellowing or whitening of what should be green plant tissue resulting from a decrease in chlorophyll, often as a result of a deficiency of one or more nutrients such as Fe.

clay—a sedimentary material with particle size ≤ 2 um in diameter.

clay mineral—fine-grained, hydrous minerals that are members of the phyllosilicates. Some examples include kaolinite and montmorillonite.

colloid—colloidal particles have at least one dimension that is between 1 nm and 1 μm. A colloidal suspension consists of colloidal-sized particles.

complex—a complex is a dissolved species that exists because of the association of a cation and an anion or a neutral molecule. Surface complexes may also occur.

conditional stability constant—also known as an apparent constant. A conditional stability constant is a concentration quotient that is not a true equilibrium constant. A common instance where a conditional stability constant would be used would be where pH is fixed at a particular value in determining the constant; in this case, the conditional stability constant would vary with pH. Conditional stability constants are often used for surface complexation reactions.

confined aquifer—an aquifer that is confined between two impermeable units known as aquitards.

congruent dissolution—mineral dissolution in which the mineral dissolves stoichiometrically.

coprecipitation—occurs when a precipitate includes chemical species from the aqueous solution along with species released by dissolution of the mineral sorbent.

covalent bond—a chemical bond that forms by the sharing of one more electrons, generally an electron pair.

critical radius—in mineral nucleation theory, the smallest number of atoms or ions that must come together to form a stable nucleus for mineral growth. This term can also be applied to other types of systems, such as bubble nucleation and formation.

critical zone—the heterogeneous, near-surface Earth environment where the hydrosphere, the biosphere, the geosphere, and the atmosphere interact with one another and with the human sphere in a complex manner that is of critical importance to life on Earth.

crystalline—used to refer to a solid with a defined crystal structure.

continuously stirred tank reactor (CSTR)—a type of chemical reaction vessel that has inlet and outlet flow and is stirred. Most commonly run at steady state, but it can also be run at transient state(s).

cyanobacteria—a form of bacteria that obtain their energy through photosynthesis. Formerly known as blue-green algae.

cystolith—small carbonate concretions that form in some plant leaves.

Darcy's Law—in hydrology, the empirical relationship, $Q = -K(dh/l)A$ where Q is flow (L^3/T; T is time), K is hydraulic conductivity, dh/dl is change in head over a given distance, A is the cross-sectional area (L^2). Darcy's Law is a fundamental tenant of hydrology that water always flows from locations of greater total head to locations of lesser total head.

Debye length—the thickness of the double layer, designated $1/\kappa$.

denitrification—the process whereby microorganisms reduce nitrate or nitrite to N_2 gas.

desferrioxamine—a chelating agent that is a siderophore (Fe(III)-binding ligands produced by some aerobic microorganisms and graminaceous plants). It is characterized by hydroxamine functional groups and occurs in various chain or cyclic structures (DFO-B, D, etc.). It is also known as desferoxamine (IUPAC).

diffuse ion swarm—if a solute near a surface does not form a surface complex, it is adsorbed in the diffuse ion swarm. It can sometimes be difficult to distinguish between ions that form outer-sphere complexes and those that are in the diffuse-ion swarm.

diffuse layer—the region in which non-specifically adsorbed ions are accumulated and distributed through the combined influences of electric field and thermal motion. Counter ions in direct contact with the surface are said to be located in the Stern layer, whereas ions further away from the surface are in the diffuse layer or Gouy layer.

diffusion—the spontaneous mingling of particles of two or more different substances caused by random thermal motion.

dipole—in an electric dipole, there is a separation of positive and negative charge. Magnetic dipoles may also occur.

dislocation—dislocations are types of crystallographic defects that occur within a crystal structure. In an *edge dislocation*, an extra partial plane of atoms is introduced midway through the crystal, distorting nearby planes of atoms. *Screw dislocations* form when one portion of crystal is out of alignment with another in a way that can be visualized by imaging slicing a stack of paper perpendicular to the stack halfway through and then lifting one side relative to the other.

dispersion—in groundwater, a solute spreading phenomenon brought about by mechanical mixing and molecular diffusion.

dissimilatory—a type of metabolism in which the chemical component, such as S or Fe, is not incorporated into organic compounds.

dissolved load—the components of the overall load of a stream that would by definition pass through a 0.45-μm filter. As such, the *dissolved load* includes ions, molecules, as well as many small colloidal particles, including nanoparticles.

DLVO theory—named after two groups of researchers, Derjaguin and Landau, and Verwey and Overbeek who independently developed a theory for interactions between charged surfaces and colloid stability. Describes the force between charged surfaces that interact through a liquid medium. The DLVO theory considers van der Waals attraction and electrostatic repulsion as the sole factors operating between surfaces, by which the surfaces interact. Extended DLVO theory considers additional interactions such as acid-base behaviors, hydrophobic interactions, and so on.

driving force—the driving force associated with crystal growth is related to the change in chemical potential $\Delta\mu$, and hence to supersaturation state (σ), according to driving force $= \Delta\mu/k_B\mathrm{T} = \sigma$, where k_B is the Boltzmann constant and T is absolute temperature.

ecohydrology—a new scientific discipline that links hydrology with ecological processes that are involved in the hydrologic cycle.

efflux—a mechanism responsible for extrusion of toxic substances from a cell; for example, bacteria may remove some toxins such as Cd by efflux.

electrical potential (ψ)—refers to the potential energy per unit of charge; the difference in electrical potential between two points is the voltage (V).

electron tunneling—Occurs in quantum mechanics because electrons behave as waves of energy. In scanning tunneling microscopy (STM), the tip and sample wave functions overlap such that when under an applied bias, there is some finite probability to find the electron on the other side of the energy barrier. Thus, an electron may "tunnel" across the barrier from tip to sample or sample to tip, depending upon the direction of the bias voltage.

electronic (or electrical) double layer (EDL)—refers to a model describing the interface between a charged surface and the solution in which that surface is immersed. The EDL consists of the charged surface and a diffuse layer of counterions near the surface, on the solution side of the interface.

electroosmosis—the movement of liquid relative to a stationary charged surface in an applied electric field.

electrophoresis—the motion of charged colloidal particles relative to stationary liquid in an applied constant electric field.

electrophoretic mobility (u)—migration of a component in an applied electric field. $u = U_E/E$, where U_E is the steady-state velocity of a charged particle moving in response to an electric field of magnitude E.

elementary reaction—an individual defined step in a reaction; this term is used for a chemical reaction in which one or more chemical species react directly to form products in a single reaction step and with a single transition state.

Elovich equation—an equation commonly used to describe the kinetics of chemisorption.

embedded network—an emerging technology that embeds a computational intelligence in the environment that allows adaptive monitoring and control of that environment.

embryo—in crystal nucleation, a small (metastable) particle whose size is beneath the critical radius.

enthalpy—in chemical thermodynamics, the heat content of a system.

entropy—in chemical thermodynamics, a measure of the randomness of atoms or molecules in a system.

enzyme—a biomolecule that catalyzes a chemical reaction.

epitaxy—an ordered crystalline growth on a monocrystalline substrate, wherein the overgrowth takes on the lattice structure and orientation of the substrate.

equilibrium—in chemistry, the state in which the rates of forward and reverse reactions are equal. At equilibrium, $\Delta G = 0$.

etch pit—a hole or pit in a surface that forms during dissolution, often at a dislocation; small etch pits may nucleate at a point defect or impurity.

eukaryote—a single-cell or multicellular organism with a membrane-bound, structurally discrete nucleus and other well-developed subcellular structures.

eutrophication—a reduction in the concentration of dissolved oxygen in water, often resulting from an oversupply of nutrients such as phosphate. Can lead to algal blooms and changes to food webs.

evapotranspiration—a term referring to the combination of evaporation from open bodies and surfaces and plant transpiration.

exopolysaccharides (EPS)—polymers composed of sugar residues and secreted by a microorganism into the surrounding environment; the "glue" that holds biofilms together.

extensive property—a property that depends on the amount of a sample

extremophile—an organism that lives in an extreme environment, such as in highly acidic (e.g., pH 2) and/or hot conditions (>60°C).

facultative anaerobic—a microorganism that makes ATP by aerobic respiration if oxygen is present but can switch to fermentation.

Faraday constant—the magnitude of electric charge per mole of electrons; F = 96,486 coulombs $mole^{-1}$.

fatty acid—(hydrophilic) carboxylic acid, often with a long, unbranched (hydrophobic) aliphatic tail that is either saturated or unsaturated.

fermentation—the process whereby energy is derived from the oxidation of organic compounds, such as carbohydrates, using an endogenous electron acceptor; usually another organic compound, which is reduced. Fermentation can be used by some bacteria, algae, fungi, and a few protozoa along with higher organisms including humans under special circumstances, such as during intense exercise.

ferric iron—Fe(III), oxidized form.

ferromagnetism—the most commonly encountered form of magnetism in everyday life. It is defined as the phenomenon whereby materials such as iron become magnetized in an external magnetic field and remain magnetized for a period of time after the material is no longer in the field.

ferromanganese nodules—(often called manganese nodules) are microscopic to centimeter-scale concretions found in different parts of the ocean floor. They consist of concentric layers of Fe and Mn hydroxides around a core, and they can contain potentially economically viable concentrations of Mn, Ni, Cu, and Co.

ferrous iron—Fe(II), reduced form.

Firmicutes—a group of bacteria, most but not all of which are gram positive.

Freundlich equation—for sorption, $q = \Gamma = K_f C^n$, where q is the sorbate sorption density in mg m^{-2} (or sometimes mg g^{-1}), C is the equilibrium concentration of sorptive remaining in solution during sorption in mg L^{-1}, n is a constant expressing isotherm curvature, and K_f is the Freundlich isotherm coefficient.

Frumkin equation—when organic molecule sorption is dominated by hydrophobic interactions, the sorption isotherm data are often fit to a Frumkin equation. The Frumkin equation is $\Theta/(1\text{-}\Theta) = BC\, e^{2\alpha\Theta}$, where Θ is the fraction of adsorbent surface covered by adsorbate, C is the concentration of the adsorptive in aqueous solution, B is the adsorption constant, and α is the interaction coefficient.

fullerene—as originally defined, "a compound composed solely of an even number of carbon atoms which form a cage-like fused-ring polycyclic system with twelve five-membered rings and the rest six-membered rings; the archetypal example is the sixty atom structure, where the atoms and bonds delineate a truncated icosahedron." More broadly, "any closed-cage structure having more than twenty carbon atoms consisting entirely of three coordinate carbon atoms." (IUPAC recommendation; http://www.chem.qmul.ac.uk/iupac/fullerene/).

fulvic acid—soluble form of humic substance that is soluble under the full range of environmental pH conditions.

fungus—a eukaryotic organism of the kingdom Fungi; heterotrophic with a chitonous cell wall. May be multicellular or occur as single cells.

Geobacter—a proteobacteria genus that undergoes aerobic respiration and that has been the subject (particularly by D. Lovley and colleagues) of many biogeochemical investigations from probing mechanisms of dissimilatory metal reduction to use in microbial fuel cells.

GEOCARB—model by R. Berner (Yale University) for the evolution of the carbon cycle and of atmospheric CO_2 over Phanerozoic time.

Gibbs energy—(*G*). According to IUPAC, "Enthalpy minus the product of thermodynamic temperature and entropy." Although the IUPAC recommended name is Gibbs energy, the former term Gibbs free energy is still widely used.

Gibbs plot—a manner of plotting chemical composition data, usually for rivers, in which the ratio $[Na^+]/([Na^+] + [Ca^{2+}]$ is plotted on the x-axis of the diagram and the total dissolved salts in ppm on the y-axis.

Gouy Chapman theory—the theory describing charge distribution near a flat surface, in its most simple form for a symmetrical electrolyte. The charge in the diffuse layer σ_d in C m^{-2} (Coulombs per meter squared) is related to the potential at the surface, ψ (the Stern potential) in volts through the following equation, which is valid for a symmetrical electrolyte $\sigma_d = (8RT\varepsilon\varepsilon_o\, c \times 10^3)^{1/2} \sinh(Z\psi F/2RT)$, where R is the molar gas constant (8.314 J $mol^{-1}K^{-1}$), T is absolute temperature (K), ε is dielectric constant of water (= 78.5 at 25°C), ε_o the permitivity of free space (8.854×10^{-12} $C^2\ J^{-1}\ m^{-1}$), c is the molar electrolyte concentration, Z is the ionic charge on the symmetrical electrolyte (e.g., 1 for Na, Cl), and F is the Faraday's constant

gram negative/gram positive—the eubacteria have been categorized into two broad groups, gram positive and gram negative, depending on the characteristics of their cell walls. This designation is based on positive or negative reaction to the gram stain. The differences in staining is related to differences in cell wall characteristics.

graminaceous plants—grasses.

gymnosperms—"naked seed" plants; cone-bearing trees are gymnosperms.

gypsum—a hydrated calcium sulfate mineral.

half-life, $t_{1/2}$—the quantity of time it takes for something to decrease to half of its original value; commonly used in radioactive decay to describe the time it takes for the number of moles of a radioactive element to decrease to half its original value.

Hammett constant (σ)—used for substituted benzene molecules to represent the ability of a group to withdraw or donate electronic density as a result of its inductive and resonance effect.

Hanford Site—a decommissioned complex in Southeastern Washington State that was involved in nuclear production beginning with the Manhattan Project and during the Cold War. The Hanford Site is overseen by the U. S. Department of Energy. It is of considerable environmental concern because of radioactive waste and other contaminants.

heterogeneous nucleation—in crystal nucleation in solution, formation of one solid phase on another.

heterogeneous reaction—a reaction that involves more than one phase.

heterotroph—an organism that uses organic carbon from other sources and cannot create its own organic carbon "food".

homogeneous nucleation—in crystal nucleation in solution, formation of a solid phase directly in solution rather than on the surface of another solid phase.

homogeneous reaction—a reaction involving only one phase.

humic acid—soluble form of humic substances that are not soluble in water at below pH 2 but are soluble (to varying degrees) under less acidic conditions.

humic substances—"naturally occurring, biogenic, heterogeneous organic substances that can be characterized as being yellow to black in color, of high molecular weight, and refractory" (Aiken et al., 1985 definition). Humic and fulvic acids are amphiphilic, containing hydrophobic moieties in the form of a carbon skeleton and hydrophilic functional groups.

HYDRAQL—a computer model for thermodynamic speciation calculations in aqueous solution, including surface complexation modeling (Papelis et al., 1988).

hydraulic conductivity—a parameter in groundwater flow. In analogy with electrical conductivity, hydraulic conductivity depends on the flow properties of the geologic materials and of the water. The value of hydraulic conductivity tends to be high for sands and low for clays.

hydrogen bonding—refers to a type of dipole–dipole interaction force that exists between an electronegative atom and a hydrogen atom bonded to another electronegative atom. The energy of the attraction is close to that of a very weak covalent bond.

hydrologic cycle—the complex system whereby water circulates at and near the Earth's surface, along flow paths connecting reservoirs.

hydrologic head—the potential energy of the mass of groundwater at a given point relative to a specified base level, most often sea level.

hydrolysis—a chemical reaction in which a molecule or component is broken down by reaction with water. Hydrolysis may also refer to breaking down of the water molecule into hydroxides and protons, as occurs in surface hydrolysis to form protonated/deprotonated surface sites.

hydrophilic—"water loving" solutes that are polarizable and tend to dissolve more readily in water than in oil. Surfaces may also be hydrophilic.

hydrophobic—"water fearing" substances that are nonpolar and tend to associate more readily with nonpolar solvents than with water. Surfaces may also be hydrophobic.

hyperaccumulation—in phytoremediation, a plant is termed a hyperaccumulator if it has the ability to concentrate pollutants to some minimum percentage varying according to the particular pollutant (e.g., Pb, Cr) involved.

hyperthermophiles—organisms that are adapted to growing at temperatures above about 60°C. Some hyperthermophilic bacteria can grow at temperatures near the boiling point of water. Hyperthermophiles are extremophiles.

hyporheic zone—the area of subsurface sediment and porous space adjacent to (and underlying) a stream through which the stream water readily exchanges.

igneous rocks—are formed by solidification of molten rock by cooling, either below the land surface as intrusive rocks or on the surface as extrusive rocks. Examples are granite and basalt.

incongruent dissolution—mineral dissolution wherby the release of constituents to solution does not reflect the original stoichiometric relationships in the mineral phase.

inner-sphere complex—a surface complex that forms by direct binding of an adsorbate to a surface, with no intervening water molecules.

in situ microcosm—an approach used in geochemistry and geomicrobiology in which a reaction vessel with a (mineral) sample of interest is embedded at a field site. An *in situ* microcosm can be a simple holey bottle buried in soil or sediment or a more complex vessel that pumps solution from a given source such as a river or a hydrothermal lake through a packed column.

insulator—a material that resists the flow of electrical current.

intensive property—a property that does not depend on the amount of material in a sample.

interface—formally, the boundary between two condensed phases.

Intergovernmental Panel on Climate Change (IPCC)—a scientific body that has been given the task of evaluating the risk of climate change cause by humic activity, established by the United Nations. The IPCC shared the 2007 Nobel Peace Prize with former U. S. Vice President Albert Gore for its efforts related to climate change.

International Union of Pure and Applied Chemistry (IUPAC)—IUPAC is the world authority on critically evaluated data such as atomic weights and chemical nomenclature, terminology, and standardized methods for measurement.

intrinsic stability constant—a "true" stability constant that is not conditional on specific conditions. The intrinsic stability constant is determined from measured conditional stability constants. The term "intrinsic" refers to the concept that the stability constant is an intrinsic physical chemical parameter that relates concentrations of the components of a reaction at equilibrium.

intrinsic surface charge density—(σ_{in}). For a surface in solution, the sum of permanent and net proton surface charge densities.

ionic potential, I_p—the ratio of charge to radius of an ion, the ionic potential $I_P = z/r$.

isoelectric point—according to IUPAC: "The pH value at which the net electric charge of an elementary entity is zero." In the context of this book, the elementary entity would be a particle such as a colloidal mineral particle or a bacterium.

isomorphic substitution—in mineralogy, substitution of one ion such as Si^{4+} by another such as Al^{3+} without a change in crystal structure.

IUPAC "Gold Book"—a compendium of chemical terminology by the International Union of Pure and Applied Chemistry, informally known as the "Gold Book." For this text, the Gold Book was accessed at http://goldbook.iupac.org.

kinetics—in chemistry, the subdiscipline that focuses on reaction rates and mechanisms.

Langmuir equation—applied for adsorption

$$q = \Gamma = \frac{K_q q_m C}{1 + K_q C},$$

where q is the adsorbed density (mg m^{-2}) on the surface, q_m is the maximum adsorbed density (mg m^{-2}), K_q is the adsorption equilibrium constant ($L(mg)^{-1}$), and C is the equilibrium concentration (more properly, activity, although concentration is often used instead) in solution (of the same component that is adsorbing).

latent heat of vaporization—the energy required to transform a given quantity of a substance into a gas (or vapor).

lattice misfit (d)—in heterogeneous mineral nucleation, the term "misfit" describes the mismatch between arrays of atoms at the interface between the substrate crystal and the overgrowth crystal. The lattice misfit d is calculated from $d = D_a/a$, where a is the lattice parameter of the stress-free substrate crystal and D_a is the difference in lattice parameters between the substrate and the overgrowth.

Lennard-Jones potential—describes the interaction forces that occur between a pair of neutral atoms or molecules as they approach one another. At long ranges, there is van der Waals attraction, whereas at shorter separation distances, there is strong repulsion as a result of the Pauli exclusion principle.

lichen—a fungus living symbiotically with photosynthetic algae.

ligand—an anion or a neutral molecule that can combine with a cation (or a cation center at a surface) to form a complex.

ligand exchange reaction—the chemical change that occurs when one ligand is replaced by another, for example in solution on a metal or at a surface site.

linear distribution coefficient, K_d—is a simple approach that has been used widely to incorporate adsorption into transport models. This approach is based on an equation for "partitioning" of a pollutant between dissolved and sorbed phases: $C_{i(s)} = K_d\, C_{i(aq)}$, where $C_{i(s)}$ is the molar concentration of the adsorbate on the surface, $C_{i(aq)}$ is the molar concentration of the adsorptive in solution, and K_d is the distribution coefficient in units of volume solvent over weight adsorptive (Lg^{-1} or cm^3g^{-1}).

liquid crystal—"a thermodynamic stable phase characterized by anisotropy of properties without the existence of a three-dimensional crystal lattice, generally lying in the temperature range between the solid and isotropic liquid phase, hence the term mesophase." (http://plc.cwru.edu/tutorial/enhanced/files/glossary/glossary.htm).

macroscope—"a symbolic instrument made of a number of methods and techniques borrowed from very different disciplines" (de Rosnay, 1979).

magnetotactic bacteria—a class of bacteria that have the ability to orient themselves along the magnetic field lines of Earth's magnetic field, because of intracellular magnetic particles known as magnetosomes.

mean residence time—the amount of time that the average drop of water spends between entering and leaving a given reservoir of the hydrologic cycle.

meniscus—a curve in the surface of a liquid in a narrow container that is produced in response to the interactions with the surrounding surface and through capillary action. It can be either concave or convex depending on the nature of the liquid–surface interactions.

mesophiles—organisms that grow within the temperature range from 25 to ~45°C.

metamorphic rock—a type of rock that forms by transformation caused by heat and/or temperature of an igneous, sedimentary, or other metamorphic rock. Some examples of typical metamorphic rocks are schist and gneiss.

methanogens—obligate anaerobic archaea that carry out *methanogenesis,* or the production of methane.

micelle—an aggregate of surfactant molecules dispersed as a liquid colloid. Typically in aqueous solutions, a micelle forms with the hydrophobic tails oriented inward and the hydrophilic heads facing outward to solution.

Michaelis-Menten kinetics equation—an equation that describes the kinetics of many enzyme-catalyzed reactions.

mineral—a naturally occurring solid substance that has a characteristic chemical composition, a highly ordered atomic structure, and specific physical properties.

mixotrophs—organisms that may "derive energy simultaneously from the oxidation of reduced carbon compounds and oxidizable inorganic compounds, or they may derive their carbon simultaneously form organic carbon and CO_2, or they may derive their energy totally from the oxidation of an inorganic compound but their carbon from inorganic compounds." (Ehrlich, 2002).

model V&V—verification and validation. Verification is about getting the "simulation right" whereas validation is about getting the "right simulation."

molecular dynamics (MD) modeling—in this modeling approach, the forces on individual atoms are evaluated using an energy forcefield. Initial positions and velocities of atoms and/or molecules are assigned by the modeler. Following a short time step, new atomic positions are calculated based on Newton's classic equation of motion. By modeling over a large number of time steps, a time-dependent trajectory can be calculated of all the atomic motions.

molecularity—the *molecularity* of an elementary reaction is the number of molecules (or atoms or ions) that come together to react. A unimolecular reaction involves a single atom, molecule, or ion, as in radioactive decay. A bimolecular reaction involves two molecules (or atoms or ions) that come together to react.

Monte Carlo simulation—for surface processes, the Monte Carlo approach models surface structure and dynamic processes on surfaces based on statistical sampling with random numbers. Molecular structures and configurations are optimized based on analysis of a large number of randomly generated trial configurations.

nanometer (nm)—a unit of length equal to 1×10^{-9} meter or 10 Ångstroms.

nanoparticle—no single consistent definition of a nanoparticle is available. Generally, a particle with at least one dimension that is as small ~1–100 nm in scale. However, sometimes the term nanoparticle is used for particles in about this size range whose behavior deviates substantially from that of larger particles.

Nernst equation—the Nernst equation is used to determine the equilibrium reduction potential of a half-cell in an electrochemical cell. There is also another form of the Nernst equation for the total cell potential.

$$E_{cell} = E^0_{cell} - \frac{RT}{zF} \ln \ Q$$

$$E_{red} = E^0_{red} - \frac{RT}{zF} \ln \ \frac{^a\text{Red}}{^a\text{Ox}},$$

where E_{red} is the half-cell reduction potential, E^0_{red} is the standard half-cell reduction potential, E_{cell} is the cell potential or electromotive force, E^0_{cell} is the standard cell potential, R is the ideal gas constant, T is the absolute temperature, F is the Faraday constant, a_{Red} is the activity of the reduced species, a_{Ox} is the activity of the oxidized species, Q is the reaction quotient, and z is the number of electrons transferred in the half-cell or full cell reaction.

NETPATH—a computer model that is an "Interactive Code For Modeling Net Geochemical Reactions Along a Flow Path" (Plummer et al., 1994). NETPATH is an inverse geochemical model used to identify potential geochemical reactions that might explain a change from initial to final solution conditions along a flow path. Modeling is often constrained with mineralogical data.

nontronite—an Fe(III)-rich smectite mineral.

octanol-water partition coefficient, K_{ow}—for a given chemical compound, the ratio of the concentration of the compound in the organic solvent octanol to that in water at equilibrium and a specified temperature. Related to solubility in water.

Ostwald Ripening—a process whereby many small particles of a mineral in solution with high surface area eventually evolve to fewer larger particles, to minimize the specific surface area.

OTIS model—a model for one-dimensional transport with inflow and storage (Runkel, 1998).

outer-sphere surface complex—a surface complex that forms by relatively weak interaction with a surface; no chemical bond is formed with the surface, and water molecule(s) occur between the sorbate and the mineral surface.

ozone—O_3; naturally occurring ozone in the stratosphere helps to block ultraviolet (UV) radiation. A pollutant in the lower atmosphere.

paramagnetism—a form of magnetism that occurs only in the presence of an externally applied magnetic field.

Pauli exclusion principle—as applies to fermions, including electrons, protons, and neutrons. For electrons in a single atom, no two electrons can have exactly the same four quantum numbers. The Pauli exclusion principle forbids electrons from occupying the same quantum state. The Pauli exclusion principle results in a strong repulsive force between two atoms or molecules at close approach when electron orbitals begin to overlap.

pedogenesis—the process of formation or genesis of soils.

peristerite—a feldspar with alternating layers of albite and orthoclase, such that the feldspar has an irridescent appearance.

permeability—for porous media, a measure of the interconnectedness of pores.

Phanerozoic—in geologic time, the Eon from ~540 million years ago to the present time.

phenocryst—a conspicuous, usually large, and often (but not always) euhedral crystal embedded in the groundmass of a porphyritic igneous rock.

phloem—the food-conducting tissue of a vascular plant; the phloem transports organic nutrients, especially sugar.

phospholipid—a class of lipids; occur as a component of all biological membranes. Amphiphilic means amphipatic molecules that have both hydrophobic and hydrophilic components.

photoheterotrophs—organisms that derive all or part of their energy from sunlight but assimilate organic carbon as the carbon source.

photolithotrophs—organisms that obtain energy directly from sunlight and assimilate carbon as CO_2, bicarbonate, or carbonate.

PHREEQC—a computer program for speciation, batch-reaction, one-dimensional transport, and inverse geochemical calculations. (Parkhurst and Appelo, 1999).

physical weathering—refers to the breakdown of rocks through primarily nonchemical processes such as wind, ice wedging, or abrasion caused by to currents or wave action.

phytolith—small precipitate of Si or Ca formed in many plants; a "plant stone".

phytoremediation—the use of plants in remediation of one or more contaminants.

piezoelectric scanner—in the piezoelectric effect, a material generates a voltage when deformed; however, in the converse piezoelectric effect, the material deforms or moves in response to an applied electric field. In atomic force and scanning tunneling microscopies, the piezoelectric elements undergo submicron-scale accurate movements in response to an applied electric field, allowing potential atomic-scale resolution.

pili—thread-like structures that are present on some bacteria. Some pili may act as "nanowires" to transfer electrons and play a role in the bacterial metal reduction. Nanowires have been observed on some *Geobacter*, *Shewanella*, *Pseudomonas*, and other bacteria.

Piper diagram—a commonly used diagram for representing the chemical composition of groundwater (or sometimes surface water) and for distinguishing different hydrogeochemical facies.

plug flow reactor—most commonly, a column reactor used to represent a continuous, flowing system for determination of reaction kinetics.

polycyclic aromatic hydrocarbons (PAHs)—are organic compounds that consist of fused aromatic rings and that do not contain heteroatoms or carry substituents. They can occur in oil and coal and as byproducts of the burning of fossil fuels.

polydisperse—for polymers, referring to a mixture (such as humic and fulvic acids) that consists of a variety of molecules with a distribution of molecular weights.

polymer—an organic *polymer* is a macromolecule composed of repeating units bound by covalent bonds.

porosity—the fraction of the bulk volume of rock or sediment that is composed of voids that may be filled with water and/or air or soil gases.

prokaryote—a cell or an organism that lacks a membrane-bound, structurally discrete nucleus and other subcellular compartments, such as bacteria and cyanobacteria.

proteobacteria—a group of gram-negative bacteria that have been divided into five classes or subdivisions based on 16S rRNA analysis: α, β, γ, δ, and ε.

psychrophiles—organisms that can grow at low temperatures as associated with the polar regions

p.z.c.—the pH of the point of zero charge on a particle, where charge on the particle $\sigma_p = 0$.

p.z.n.p.c.—the pH of the point of zero net proton charge on a particle, where $\sigma_H = 0$.

quantum yield—of a radiation-induced process is the number of times that a defined event, such as a specific reaction step, occurs per photon absorbed by the system. The quantum yield therefore is a measure of the efficiency with which absorbed radiation produces some effect. For fluorescence, the ratio of photons emitted to photons absorbed.

quorum sensing—the communication between cells. Many bacteria use quorum sensing to regulate their gene expression according to cell density.

radiative forcing—a term used to refer to agents that change the energy inputs and outputs through a part of the atmosphere known as the tropopause. A positive forcing tends to warm the system, whereas a negative forcing tends to cool it.

reaction order—an empirically observed quantity that describes how a reaction rate is proportional with respect to a given species.

regolith—unweathered bedrock underlying a soil, which is generally lower in organic matter and microbial activity than the overlying soil zones.

rhizosome—the area of a soil directly surrounding roots that contains associated microorganisms

saturated zone—the region of underground water in which pore spaces are filled with water completely.

saturation index—a measure of how far from equilibrium a solution is with respect to a given mineral. SI = Q/K_{eq}, where Q is the reaction quotient and K_{eq} is the equilibrium constant. Often given as $\log_{10}(Q/K_{eq})$.

Savannah River Site—a nuclear processing center in South Carolina that is of environmental concern because of radionuclide and other contaminants. Today, most activity at the site is related to environmental remediation.

secondary growth—a term sometimes applied to nanoparticles, referring to the growth by assembly of aggregates; it is also known as oriented attachment.

secondary mineral—a mineral that forms as the result of decomposition of a primary mineral or from reprecipitation of decomposition products of a primary mineral.

sedimentary rock—a type of rock that forms by transport and deposition or through chemical precipitation. Examples are sandstone and limestone.

sedimentation potential—the electric field that is created when charged particles move relative to a stationary liquid.

semiconductor—a solid material whose conductivity is intermediate between that of an insulator and a conductor.

serpentinite—a metamorphic rock that contains one or more serpentine minerals.

Shewanella—a genus of bacteria that has been used in many biogeochemical experiments, such as on dissimilatory metal reduction.

siderophore—a low-molecular-weight organic Fe(III) chelating compound produced by some aerobic microorganisms and graminaceous plants.

silicosis—an occupational lung disease caused by exposure to airborne crystalline silica dust.

singlet oxygen—$^1O_2$ an energized but uncharged form of molecular oxygen that can be toxic to cells. Singlet oxygen can be generated chemically or photochemically.

SNBs—acronym for substituted nitrobenzenes.

sorption—a general term that refers to accumulation at a surface when the mechanism is not known or when referring to several mechanisms (e.g., absorption, adsorption) simultaneously.

sorption isotherm—a plot at constant temperature of concentration of sorbate sorbed on the y-axis and sorptive equilibrium concentration on the x-axis.

specific heat capacity—a measure of the heat energy that is needed to increase the temperature of a unit quantity of a substance by a certain interval of temperature. Also known as heat capacity.

speleothem—a secondary mineral deposit in a cave, such as a stalactite or a stalagmite.

spodosol—also known as podzol. A type of soil characterized by thick, sandy horizons composed of residual, resistant primary minerals such as quartz, and only scarce amounts of secondary minerals. Spodosols have distinct layering, including an uppermost organic horizon where leaf litter and other organic debris are broken down by microbial activity, an underlying leached layer, and beneath that a zone of accumulation of Fe and Al hydroxides.

stalactite—a mineral formation that hangs from the ceiling of a cave; it is often made of calcite.

stalagmite—a mineral formation that grows upward from the floor of a cave; it is often made of calcite.

Stern layer—a layer of ions directly bound to a surface, as an addition to the double layer. Suggested by and later named after Stern.

stomata—or stoma; a tiny opening or pore on a plant used for gas exchange, which is usually on the underside of a leaf.

streaming potential—the electric field that is created when liquid is caused to flow along a stationary charged surface.

structural surface charge—permanent surface charge on a mineral caused by isomorphic substitution.

superoxide—a highly reactive form of oxygen that forms when molecular oxygen gains a single electron to form O_2^-. Superoxide is a free radical.

surface—formally, the exterior boundary of a condensed phase in a vacuum or gaseous environment. However, we often speak of surfaces in aqueous solution (more formally, interfaces).

surface complexation models (SCMs)—models that consider the formation of surface complexes, which are innerspherically bound at the surface along with electrostatic interactions in modeling adsorption at surfaces.

surface-controlled dissolution—in surface-controlled dissolution, the rate-determining step is a reaction at the surface, probably most often the detachment from the surface of a metal or metal-ligand "activated complex."

surface excess—the positive or negative excess concentration of a component at or near the surface relative to the concentration in the bulk. Used commonly for the mineral–water interface in discussion of adsorption.

surface free energy—the free energy associated with the formation of a surface. The surface free energy γ is defined as $\gamma \equiv (\delta G/\delta A)_{P,T}$ (pressure, temperature, remain constant) where G = Gibbs free energy and A = surface area.

surface precipitation—three-dimensional accumulation of an adsorbate on an adsorbent, in which a new surface phase is formed. It can be difficult to distinguish sometimes whether precipitation in the presence of a surface is through surface precipitation or homogeneous nucleation in solution.

surface tension—surface free energy, which is a term applied most commonly to liquids.

surface reconstruction—the process by which atoms at the surface of a crystal assume a different structure than that of the bulk to make up for lack of full bonding that would occur in the bulk solid. Surface reconstruction may result in a different unit cell at the surface than in the bulk solid.

surface relaxation—a primarily vertical movement of atoms at a surface in vacuum in response to dangling bonds. Relaxation may also occur in air or solution.

surfactant—a "surface active" agent molecule that has a hydrophobic "head" and a hydrophilic tail. Surfactants are amphiphilic.

suspended load—the fine-grained sediment that is transported by the stream, such as silt, clay, and fine sand, the smallest particles of which overlap with the operationally defined dissolved load.

synchrotron—in physics, a subatomic particle accelerator. The synchrotron is a particular type of cyclic particle accelerator in which the electric field, which is used to accelerate the particles, and the magnetic field, which is used to turn the particles, are carefully synchronized with the traveling particle beam. Used for certain surface analytical applications such as X-ray absorption spectroscopy (XAS) because the synchrotron permits high brightness and high intensity, as well as highly collimated and tunable radiation.

tailing—desorption of a contaminant may result in significant "tailing" when an adsorption isotherm is nonlinear. When porous media are flushed, removal of an adsorbed contaminant may show significant "tailing" such that relatively high concentrations of the contaminant remain in the effluent for long periods of time after the peak concentration is achieved and passed.

ternary surface complex—a surface complex that contains both a ligand and a metal. The ligand may be either organic or inorganic. Two types of ternary surface complex are Type A: >S—O—M—L and Type B: >S—L—M, where >S is the surface site, M a metal, O oxygen, and L a ligand.

thermodynamics—a branch of physics that studies the effects of changes in temperature, pressure, and volume on physical systems at the macroscopic scale by analyzing the collective motion of their particles. In the context of environmental surfaces, thermodynamics is the subdiscipline of chemistry that focuses on the energy associated with different states and the energy changes involved in reactions. Important for determining equilibrium conditions and directions of spontaneous change.

thermophiles—organisms that are adapted to life at a high temperature (higher than 45°C).

throughfall—water that results as precipitation passes through a canopy.

transition metal—loosely stated, this term refers to any element in the d-block of the periodic table. A stricter definition refers to an element whose atom has an incomplete d subshell, or which can give rise to cations with an incomplete d subshell. This stricter definition excludes certain elements, such as Zn, Cd, and Hg, which are allowed by the looser definition.

Transition State Theory—states that an *activated complex* occurs when there is an energetically unstable complex present at the surface. This transitional state represents an energy barrier that must be overcome in processes such as mineral growth or dissolution.

transport-controlled dissolution—in transport-controlled dissolution, the rate-determining step is a transport step, such as diffusion from bulk solution to the surface. Dissolution tends to be sensitive to flow (or stir) rate except at fairly high flow rates, where transport is fast.

tunneling current—(I_T) In STM, the tunneling current occurs when a small bias voltage is applied to the sample relative to the tip, which causes electrons to tunnel across the gap. This tunneling current is a function of the electronic properties of the tip and the sample. It is extremely sensitive to the distance between the tip and the sample.

ultrahigh vacuum (UHV)—a vacuum environment at 1×10^{-9} to 1×10^{-12} Torr (100 nPa to 100 pPa), which is commonly needed for many surface chemical analyses, such as X-ray photoelectron spectroscopy (XPS). Creating and maintaining a UHV environment requires special instrumentation and procedures such as baking a vacuum chamber.

unconfined aquifer—an aquifer whose surface is the water table.

underground water—water beneath the earth surface, which includes groundwater, soil water, and water in the capillary fringe.

unit cell—the smallest group of atoms, ions, or molecules whose arrangement defines a crystal's characteristic symmetry and whose repetition at regular intervals in three dimensions produces the crystal lattice.

unsaturated zone (vadose zone)—the region beneath the land surface in which the pore spaces are not completely filled with water; rather, at least some pores contain gases.

vadose zone—a term for the unsaturated zone beneath the land surface but above the saturated zone.

van der Waals forces—in environmental surface chemistry, this term is used for a variety of forces, most commonly for relatively weak intermolecular forces that are not caused by covalent bonds or by purely electrostatic interactions. The Van der Waals forces may be either attractive or repulsive. They result primarily from interactions between the dipole moments of adjacent molecules. However, if no dipole moment is present, then they may result from mutual electrostatic interactions of the electrons and the nuclei, resulting in induced dipole and quadrupole interactions.

VOCs—acronym for volatile organic compounds.

volume diffusion—when a crystal grows, reactants are consumed in the local area, leading to a dilute region. Solutes will tend to diffuse from the bulk solution toward the growing surface in a process known as volume or bulk diffusion.

water budget—a calculation that equates water coming into a basin with that leaving a basin, taking into consideration changes in water storage within the basin.

water table—the top of an unconfined aquifer where groundwater pressure is equal to atmospheric pressure.

Wulff rule—when the surface free energy of a crystal is minimized, that is, an equilibrium situation, the perpendicular distance from a given crystal face to the center of the crystal divided by the surface free energy of that face is constant for all faces of the crystal. This is known as the Wulff rule or the Gibbs-Curie-Wulff theorem.

Xt model—a model for reactive transport (Bethke, 1997).

xylem—one of the two types of tissue transport systems in vascular plants. Its main function is to transport water.

Young-Laplace equation—shows the relationship among the pressure difference (ΔP), surface tension (γ), and radii of curvature (R_x, R_y) of the surface, $\Delta P = \gamma\ (1/R_x + 1/R_y)$, where ΔP is the pressure difference on either side of the surface, γ is the surface tension, and R_x and R_y are the radii of curvature of the axes that are parallel to the surface.

zeta potential—electrochemical potential in a colloidal system.

REFERENCES

Adams, L.K., Lyon, D.K., and Alvarez, P.J.J. 2006. Comparative eco-toxicity of nanoscale TiO_2, SiO_2, and ZnO water suspensions. *Water Research* 40, 3527–3532.

Adamson, A.W. 1982. Physical Chemistry of Surfaces 4th ed., John Wiley and Sons, New York, 664 p.

Adamson, A.W., and Gast, A.P. 1997. Physical Chemistry of Surfaces 6th ed., Wiley-Interscience, New York, 808 p.

Aiken, G.R. 1985. Isolation and concentration techniques for aquatic humic substances. In Aiken, G.R., McKnight, D.M., Wershaw, R.L., and MacCarthy, P. (eds), Humic Substances in Soil, Sediment, and Water, John Wiley and Sons, New York, pp. 363–385.

Aiken, G.R., McKnight, D.M., Thorn, K.A., and Thurman, E.M. 1992. Isolation of hydrophilic organic acids from water using nonionic macroporous resins. *Organic Geochemistry* 18, 567–573.

Albrecht-Gary, A.-M., and Crumbliss, A.L. 1998. The coordination chemistry of siderophores: Thermodynamics and kinetics of iron chelation and release. In Sigel, A., and Sigel, H. (eds), Iron Transport and Storage in Microorganisms, Plants and Animals, Vol. 35 of *Metal Ions in Biological Systems*, M. Dekker, Inc., New York, pp. 239–327.

Alexandre, A., Meunier, J.D., Colin, F., and Koud, J.-M. 1997. Plant impact on the biogeochemical cycle of silicon and related weathering processes. *Geochimica et Cosmochimica Acta* 61, 677–682.

Allen, P.A. 1997. Earth Surface Processes, Wiley-Blackwell, New York, 416 p.

Allpike, B.P., Heitz, A., Joll, C.A., Kagi, R.I., Abbt-Brown, G., Frimmel, F.H., Brinkmann, T., Her, N., and Amy, G. 2005. Size exclusion chromatography to characterize DOC removal in drinking water treatment. *Environmental Science and Technology* 39, 2334–2342.

Allwood, A.C., Walter, M.R., Kamber, B.S., Marshall, C.P., and Burch, I.W. 2006. Stromatolite reef from the Early Archaean era of Australia. *Nature* 441, 714–718.

Amacher, M.C. 1991. Methods for obtaining and analyzing kinetic data. In D.L. Sparks and D.L. Suarez (eds), Rates of Soil Chemical Processes, Soil Science Society of America, Madison, WS, pp. 714–725.

Amal, R., Raper, J.A., and Waite, T.D. 1992. Effect of fulvic acid adsorption on the aggregation kinetics and structure of hematite particles. *Journal of Colloid and Interface Science* 151, 244–257.

Amirbahman A., Kent, D.B., Curtis, G.P., and Davis, J.A. 2006. Kinetics of sorption and abiotic oxidation of arsenic(III) by aquifer materials. *Geochimica et Cosmochimica Acta* 70, 533–547.

Amores, D.R., and Warren, L.A. 2007. Identifying when microbes biosilicify: The interconnected requirements of acidic pH, colloidal SiO_2 and exposed microbial surface. *Chemical Geology* 240, 298–312.

Ams, D.A., Fein, J.B., Dong, H., and Maurice, P.A. 2004. Experimental measurements of the adsorption of *Bacillus subtilis* and *Pseudomonas mendocina* onto Fe-oxyhydroxide-coated and uncoated quartz grains. *Geomicrobiology Journal* 21, 511–519.

Ams, D.A., Maurice, P.A., Hersman, L.E., and Forsythe, J.H. 2002. Siderophore production by *P. mendocina* in the presence of kaolinite: Evidence for bacterial responses to Fe stress. *Chemical Geology* 188, 161–170.

Anastasio, C., and Martin, S.T. 2001. Atmospheric nanoparticles. In Banfield, J.F., and Navrotsky, A. (eds), Nanoparticles and the Environment, Reviews in Mineralogy and Geochemistry 44, pp. 293–349.

Anderson, S.J., and Sposito, G. 1991. Cesium adsorption methods for measuring accessible structural surface charge. Soil Science Society of Amerrica J 55, 1569–1576.

Angove, M.J., Fernandes, M.B., and Ikhsan, J. 2002. The sorption of anthracene onto goethite and kaolinite in the presence of some benzene carboxylic acids. Journal of Colloid and Interface Science 247, 282–289.

Arai, Y., and Sparks, D.L. 2001. ATR–FTIR Spectroscopic investigation on phosphate adsorption mechanisms at the ferrihydrite–water interface. Journal of Colloid and Interface Science 241, 317–326.

Armelao, L., Barreca, D., Bottaro, G., Gasparotto, A., Maccato, C., Maragno, C., Tondello, E., Stangar, U.L., Bergant, M., and Mahne, D. 2007. Photocatalytic and antibacterial activity of TiO_2 and Au/TiO_2 nanosystems. Nanotechnology 18, Art. No. 375709.

ASTM. 2008. American Society for Testing and Materials, www.astm.org.

Atlas, R.M. 1997. Principles of Microbiology. 2nd ed. Wm. C. Brown, Dubuque, IA.

Back, W. 1966. Hydrochemical facies and ground-water flow patterns in northern part of Atlantic Coastal Plain: U.S. Geological Survey Professional Paper 498-A, 42 p.

Badireddy, A.R., Hotze, E.M., Chellam, S., Alvarez, P., and Wiesner, M.R. 2007. Inactivation of bacteriophages via photosensitization of fullerol nanoparticles. Environmental Science and Technology 41, 6627–6632.

Banfield, J.F., and Zhang, H. 2001. Nanoparticles in the environment. In Banfield, J.F., and Navrotsky, A. (eds), Nanoparticles and the Environment, Mineralogical Society of America, Washington, DC, pp. 1–58.

Barr, D.P., and Aust, S.D. 1994. Mechanisms white rot fungi use to degrade pollution. Environmental Science and Technology 28, 78A–87A.

Barton, T.T., and Ogle, K.M. 2002. Water quality and environmental isotopic analyses of groundwater samples collected from the Wasatch and Fort Union Formations in areas of coalbed methane development—implications to recharge and ground-water flow, Eastern Powder River Basin, WY. Water-Resources Investigations Report 02-4045. Cheyenne, WY, 2002.

Bazylinski, D.A., and Frankel, R.B. 2004. Magnetosome formation in prokaryotes. Nature Reviews 2, 217–230.

Beller, H.R., Chain, P.S.G., Letain, T.E., Chakicherla, A., Larimer, F.W., Richardson, P.M., Coleman, M., Wood, A.P., and Kelly, D.P. 2006. The genome sequence of the obligately chemolithoautotrophic, facultatively anaerobic bacterium *Thiobacillus denitrificans*. Journal of Bacteriology 188, 1473–1488.

Bendikov, A. Kim, J. and Harmon, T.C. 2005. Development and environmental application of a nitrate selective microsensor based on doped polypyrrole films. Sensors and Actuators B-Chemical. 106, 512–517

Benedict, W.S., Gailar, M., and Plyler, E.K. 1956. Rotation-vibration spectra of deuterated water vapor. Journal of Chemical Physics 24, 1139–1165.

Bendikov et al. 2005.

Bennett, P.C., Rogers, J.R., and Choi, W.J. 2001. Silicates, silicate weathering, and microbial geology. Geomicrobiology Journal 18, 3–19.

Bennett, P.C., Engel, A.S., Roberts, J.A. 2006. Counting and imaging bacteria on mineral surfaces. In P. A. Maurice and L. A. Warren, eds. Methods of Investigating Microbe-Mineral Interactions. CMS Workshop Lectures Series No. 14, The Clay Minerals Society, Chantilly, Virginia. 128–165.

Bergendahl, J., and Grasso, D. 1999. Prediction of colloid detachment in a model porous media: Thermodynamics. AIChE Journal 45, 475–484.

Bergquist, B.A., Wu, J., and Boyle, E.A. 2007. Variability in oceanic dissolved iron is dominated by the colloidal fraction. Geochimica et Cosmochimica Acta 71, 2960–2974.

Berner, E.K., and Berner, R.A. 1987. The Global Water Cycle: Geochemistry and Environment, Prentice-Hall, Englewood Cliffs, NJ, 397 p.

Berner, E.K., and Berner, R.A. 1996. Global Environment: Water, Air, and Geochemical Cycles, Prentice Hall, Upper Saddle River, NJ.

Berner, E.K., Berner, R.A., and Moulton, K.L. 2003. Plants and mineral weathering: Present and past. In Treatise on Geochemistry Vol. 5, Elsevier, LTD, New York, pp. 169–188.

Berner, R.A. 1980. Early Diagenesis: A Theoretical Approach. In Princeton Series in Geochemistry. Princeton University Press, Princeton, NJ, 241 p.

Berner, R.A. 2004. The Phanerozoic Carbon Cycle: CO_2 and O_2, Oxford University Press, Oxford, UK, 160 p.

Berner, R.A. 1991. A model for atmospheric CO_2 over Phanerozoic time. American Journal of Science. 291, 339–376.

Berner, R.A. 1994. GEOCARBII: A revised model model for atmospheric CO_2 over Phanerozoic time. American Journal of Science 294, 56–91.

Berner, R.A. and Kothvala, Z. 2001. GEOCARB III: a revised model of atmospheric CO_2 over Phanerozoic time. American Journal of Science 291, 339–376.

Bethke, C.M. 1997. Modelling transport in reacting geochemical systems. Comptes Rendus de l-Académie des Sciences 324, 513–528.

Bethke, C.M., and Brady, P.V. 2000. How the K_d approach undermines ground water cleanup. Ground Water 38, 435–443.

Beveridge, T.J. 1981. Ultrastructure, chemistry, and function of the bacterial wall. International Review of Cytology 72, 229–317.

Beveridge, T.J. 1988. The bacterial surface: general considerations towards design and function. Canadian Journal of Microbiology 34, 363–372.

Beveridge, T.J., and Doyle, R.J. (eds), 1989. Metal Ions and Bateria, John Wiley and Sons, New York.

Beveridge, T.J., and Koval, S.F. 1981. Binding of metals to cell envelopes of *Escherichia coli* K-12. Applied and Environmental Microbiology 42, 325–331.

Bickel, H., Bosshardt, R., Gaeumann, E., Reusser, P., Vischer, E., Voser, W., Wettstien, A., and Zaehner, H. 1960. Ueber die Isolierung und Charakterisierung der Ferrioxamine A-F, neuer Wuchsstoffe der Sideramin-Gruppe. Helvetica Chimica Acta 43, 2118–2128.

Blum, A.E., and Lasaga, A.C. 1987. Monte Carlo simulations of surface reaction rate laws. In Stumm, W. (ed), Aquatic Surface Chemistry, John Wiley and Sons, New York, pp. 255–292.

Blum, A.E., and Stillings, L.L. 1995. Feldspar dissolution kinetics. In White, A.F., and Brantley, S.L. (eds), Chemical Weathering Rates of Silicate Minerals. Reviews in Mineralogy Vol. 31, *Mineralogical Society of America*, Washington, DC, pp. 291–351.

Blum, A.E., Yund, R.A., and Lasaga, A.C. 1990. The effect of dislocation density on the dissolution rate of quartz. Geochimica et Cosmochimica Acta 54, 283–298.

Boily, J.-F., Sjöberg, S., and Persson, P. 2005. Structures and stabilities of Cd(II) and Cd(II)-phthalate complexes at the goethite/water interface. Geochimica et Cosmochimica Acta 69, 3219–3235.

Borrok, D., Fein, J.B., and Kulpa, C.F. 2004. Proton and Cd adsorption onto natural bacterial consortia: Testing universal adsorption behavior. Geochimica et Cosmochimica Acta 68, 3231–3238.

Boyanov, M.I., Kelly, S.D., Kemner, K.M., Bunker, B.A., Fein, J.B., and Fowle, D.A. 2003. Adsorption of cadmium to *Bacillus subtilis* bacterial cell walls: A pH-dependent X-ray absorption fine structure spectroscopy study. Geochimica et Cosmochimica Acta 67, 3299–3311.

Boyd, E.S., Cummings, D.E., and Geesey, G.G. 2007. Mineralogy influences structure and diversity of ba communities associated with geologic substrata in a pristine aquifer. Microbial Ecology 54, 170–182.

Boyd, S.A., Sheng, G., Teppen, B.J., and Johnston, C.T. 2001. Mechanisms for the adsorption of substituted nitrobenzenes by smectite clays. Environmental Science and Technology 35, 4227–4234.

Brady, P.V., Cygan, R.T., and Nagy, K.L. 1996. Molecular controls on kaolinite surface charge. Journal of Colloid and Interface Science 183, 356–364.

Brantley, S.L. 2003. Reaction kinetics of primary rock-forming minerals under ambient conditions. In Drever, J.I. (ed), Treatise on Geochemistry Vol. 5, Surface and Groundwater, Weathering and Soils, Elsevier, LTD, New York, pp. 73–117.

Brantley, S.L., and Chen, Y. 1995. Chemical weathering rates of pyroxenes and amphiboles. In White, A.F., and Brantley, S.L. (eds), Chemical Weathering Rates of Silicate Minerals. Reviews in Mineralogy Vol. 31, Washington, DC, pp. 119–172.

Brantley, S.L., Chesley, J.T., and Stillings, L.S. 1998. Isotopic ratios and release rates of strontium measured from weathering feldspars. Geochimica et Cosmochimica Acta 62, 1493–1500.

Brantley, S.L., Crane, S.R., Crerar, D.A., Hellmann, R., and Stallard, R. 1986. Dissolution at dislocation etch pits in quartz. Geochimica et Cosmochimica Acta 50, 2349–2361.

Brantley, S.L., Ruebush, S., Jang, J.-H., and Tien, M. 2006. Analysis of (bio)geochemical kinetics of Fe(III) oxides. In Maurice, P.A., and Warren, L.A. (eds), Methods for Study of Microbe-Mineral Interactions Vol. 14, CMS Workshops Lectures, pp. 79–116.

Brantley, S.L., White, A.F., and Hodson, M.F. 1999. Surface area of primarily silicate minerals. In Jamveit, B., and Meakin, P. (eds), Growth, Dissolution, and Pattern Formation in Geosystems, Springer, New York, pp. 291–326.

Bresme, F., and Oettel, M. 2007. Nanoparticles at fluid interfaces. Journal of Physics—Condensed Matter 19, 413101.

Brimhall, G.H., Lewis, C.J., Ague, J.J., Dietrich, W.E., Hampel, J., Tague, T., and Rix, P. 1988. Metal enrichment in bauxites by deposition of chemically mature Aeolian dust. Nature 333, 819–824.

Brock, T.D. 1985. Life at high temperatures. Science 230, 132–138.

Brock, T.D. 1994. Life at High Temperatures, Yellowstone Association for Natural Science, History & Education, Inc., Yellowstone National Park, WY.

Brown, G.E., Henrich, V.E., Casey, W.H., Clark, D.L., Eggleston, C., Felmy, A., Goodman, D.W., Gratzel, M., Maciel, G., McCarthy, M.I., Nealson, K.H., Sverjensky, D.A., Toney, M.F., and Zachara, J.M. 1999. Metal oxide surfaces and their interactions with aqueous solutions and microbial organisms. *Chemical Reviews* 99, 77–174.

Brown, G.E., and Parks, G.A. 2001. Sorption of trace elements on mineral surfaces: Modern perspectives from spectroscopic studies, and comments on sorption in the marine environment. International Geology Review 43, 963–1073.

Brown, G.E., Parks, G.A., and O'Day, P.A. 1995. Sorption at mineral-water interfaces: Macroscopic and microscopic perspectives. In Vaughan, D.J., and Pattrick, R.A.D. (eds), Mineral Surfaces Vol. 5, The Mineralogical Society Series, Chapman & Hall, London, 370 p.

Browning, R., and Hochella, M.F. Jr. 1999. Auger Electron Spectroscopy. In Perry, D.L. (ed), Instrumental Surface Analysis of Geologic Materials, VCH, New York, pp. 87–120.

Brunauer, S., Emmett, P.H., and Teller, E. 1938. Adsorption of gases inmultimolecular layers. Journal of Physical Chemistry 60, 309–319.

Brutsaert, W. 2005. Hydrology: An Introduction, Cambridge University Press, Cambridge, UK, 605 p.

Bryant, D.A., Costas, A.M.G., Maresca, J.A., Chew, A.G.M., Klatt, C.G., Bateson, M.M., Tallon, L.J., Hostetler, J., Nelson, W.C., Heidelberg, J.F., and Ward, D.M. 2007. *Candidatus Chloracidobacterium thermophilum*: An aerobic phototrophic acidobacterium. Science 317, 523–526.

Brydson, R.M.D., and Hammond, C. 2005. Generic methodologies for nanotechnology: Characterization. In Kelsall, R., Hamley, I.W., and Geoghegan, M. (eds), Nanoscale Science and Technology, John Wiley and Sons, New York, pp. 56–129.

Buffle, J. 2006. The key role of environmental colloids/nanoparticles for the sustainability of life. Environmental Chemistry 3, 155–158.

Burda, C., Chen, X.B., Narayanan, R., and El-Sayed, M.A. 2005. Chemistry and properties of nanocrystals of different shapes. Chemical Reviews 105, 1025–1102.

Burken, J.G., and Schnoor, J.J. 1997. Uptake and metabolism of atrazine by poplar trees. Environmental Science and Technology 31, 1399–1406.

Burns, P.C., Kubatko, K.A., Sigmon, G., Fryer, B.J., Gagnon, J.E., Antonio, M.R., and Soderholm, L. 2005. Actinyl peroxide nanospheres. Chemie International Edition 44, 2135–2139.

Burton, W.K., Cabrera, N., and Frank, F.C. 1951. The growth of crystals and the equilibrium structure of their surfaces. Philosophical Transactions of the Royal Society of London Series A-Mathematical and Physical Sciences 243, 299–358.

Buss, H.L., Lüttge, A., and Brantley, S.L. 2007. Etch pit formation on iron silicate surfaces during siderophore-promoted dissolution. Chemical Geology 240, 326–342.

Cabaniss, S.E., Zhou, Q., Maurice, P.A., Chin, Y.-P., and Aiken, G.R. 2000. A log-normal distribution model for the molecular weight of aquatic fulvic acids. Environmental Science and Technology 24, 1103–1109.

Cabrera, N., and Levine, M.M. 1956. On the dislocation theory of evaporation of crystals. Philos. *Magazine* Series 8, 1, 450–458.

Cabrera, N., and Vermilyea, D.A. 1958. The growth of crystals from solution. In Growth and Perfection of Crystals, John Wiley and Sons, New York, pp. 393–410.

Camesano, T.A., and Logan, B.E. 2000. Probing bacterial electrosteric interactions using atomic force microscopy. Environmental Science and Technology 34, 3354–3362.

Camesano, T.A., Natan, M.J., and Logan, B.E. 2000. Imaging modified bacterial cells using atomic force microscopy. Langmuir 16, 4563–4572.

Cardenas, E., Wu, W.-M., Leigh, M.B., Carley, J., Carroll, S., Gentry, T., Luo, J., Watson, D., Gu, B., Grinder-Vogel, M., Kitanidis, P.K., Jardine, P.M., Zhou, J., Criddle, C.S., Marsh, T.L., and Tiedje, J.M. 2008. Microbial communities in contaminated sediments, associated with bioremediation of uranium to submolar levels. Applied and Environmental Microbiology 74, 3718–3729.

Carrasco, N., Kretzschmar, R., Pesch, M-L., and Kraemer, S.M. 2008. Effects of anionic surfactants on ligand-promoted dissolution of iron and aluminum hydroxides. Journal of Colloid and Interface Science. 321, 279–287.

Casey, W.H. 2006. Large aqueous aluminum hydroxide molecules. Chemical Reviews 106, 1–16.

Casey, W.H., and Ludwig, C. 1995. Silicate mineral dissolution as a ligand-exchange reaction. In White, A.F., and Brantley, S.L. (eds), Chemical Weathering Rates of Silicate Minerals, Reviews in Mineralogy Vol. 31, Washington, DC, pp. 87–117.

Catalano, J.G., Park, C., Fenter, P., and Zhang, Z. 2008. Simultaneous inner- and outer-sphere arsenate adsorption on corundum and hematite. Geochimica et Cosmochimica Acta 72, 1986–2004.

Chadwick, O.A., Derry, L.A., Vitousek, P.M., Huebert, B.J., and Hedin, L.O. 1999. Changing sources of nutrients during four million years of ecosystem development. Nature 397, 491–497.

Chapelle, F.H. 2001. Ground-Water Microbiology and Geochemistry 2nd ed., John Wiley and Sons, New York, 477 p.

Chapelle, F.H. 2004. Geochemistry of Groundwater. In Drever, J.I. (ed), Treatise on Geochemistry Vol. 5., Surface and Groundwater, Weathering and Soils, Elsevier, New York, pp. 425–449.

Chappell, M.A., Laird, D.A., Thompson, M.L., Li, H., Teppen, B.J., Aggarwal, V., Johnston, C.T., and Boyd, S.A. 2005. Influence of smectite hydration and swelling on atrazine sorption behavior. Environmental Science and Technology 39, 3150–3156.

Charlet, L., Dise, N., and Stumm, W. 1993. Sulfate adsorption on variable charge soil and reference materials. Agriculture, Ecosystems & Environmental 47, 87.

Chattoraj, D.K., and Birdi, K.S. 1984. Adsorption and the Gibbs Surface Excess, Plenum Press, New York, 471 p.

Chaudhuri, S.K., and Lovley, D.R. 2003. Electricity generation by direct oxidation of glucose in mediator less microbial fuel cells. Nature Biotechnology 21, 1229.

Cheah, S.-F., Brown, G.E., and Parks, G.A. 1999. Structure and composition of copper(II)-2,2'-bipyridine sorption complexes on amorphous SiO_2. Geochimica et Cosmochimica Acta 63, 3229–3246.

Chen, H.W., Su, S.F., Chien, C.T., Lin, W.H., Yu, S.L., Chou, C.C., Chen, J.J., and Yang, P.C. 2006. Titanium dioxide nanoparticles induce emphysema-like lung injury in mice. FASEB Journal 20, 2393–2395.

Chen, L.X., Sabatini, D.A., and Kibbey, T.C.G. 2008. Role of the air-water interface in the retention of TiO_2 nanoparticles in porous media during primary drainage. Environmental Science and Technology 42, 1916–1921.

Chernyshova, I.V., Hochella, M.F. Jr., and Madden, A.S. 2007. Size-dependent structural transformations of hematite nanoparticles. 1. Phase transition. Physical Chemistry Chemical Physics 9, 1736–1750.

Chi, F.-H., and Amy, G.L. 2004. Transport of anthracene and benz(a)anthracene through iron-quartz and three aquifer materials in laboratory columns. Chemosphere 55, 512–524.

Chin, T.J. 1993. Physical hydrology of the dry valley lakes. In Green, W.J., and Friedman, E.I. (eds), Physical and Biogeochemical Processes in Antarctic Lakes, American Geophysical Union, Washington, DC, pp. 1–51.

Chiou, C.T., Kile, D.E., Rutherford, D.W., Sheng, G.Y., and Boyd, S.A. 2000. sorption of selected organic compounds from water to a peat soil and its humic-acid and humin fractions: Potential sources of the sorption nonlinearity Environmental Science and Technology 34, 1254–1258.

Chiou, C.T., Peters, L.J., and Freed, V.H. 1979. Sorption of selected organic compounds from water to a peat soil and its humic-acid and humin fractions: Potential sources of the sorption nonlinearity. Science 206, 831 p.

Chorover, J., and Sposito, G. 1993. Measurement of surface charge components, Technical Report, NSF Grant EAR 9221258, University of California at Berkeley, 48 p.

Chorover, J., and Sposito, G. 1995. Surface charge characteristics of kaolinitic tropical soils. Geochimica et Cosmochimica Acta 95, 875–994.

Chung, S.H., and Seinfeld, J.H. 2005. Climate response of direct radiative forcing of anthropogenic black carbon. Journal of Geophysical Research 110, D11102.

Claessens, J., van Lith, Y., Laverman, A.M., and van Capellen, P. 2005. Acid-base activity of live bacteria: Implications for quantifying cell wall charge. Geochimica et Cosmochimica Acta 70, 267–276.

Clapp, C.E., Harrison, R., and Hayes, M.H.B. 1991. Interactions between organic macromolecules and soil inorganic colloids and soils. In Bolt, G.H., DeBoodt, M.F., Hayes, M.H.B., and McBride, M.B. (eds), Interactions at the Soil Colloid-Soil Solution Interface, Kluwer Academic Press, Germany Dordrecht, pp. 409–468.

Cloud, P.E. 1965. The significance of Gunflint (Precambrian) microflora—photosynthetic oxygen may have had important local effects before becoming a major atmospheric gas. Science 148, 27–35.

Cohen, A.L., and McConnaughey, T.A. 2003. Geochemical perspectives on coral mineralization. In Dove, P.M., DeYoreo, J.J., and Weiner, S. (eds), Biomineralization. Reviews in Mineralogy and Geochemistry Vol. 54, Mineralogical Society of America and Geochemical Society, Washington, DC, pp. 151–187.

Conley, D.J. 2002. Terrestrial ecosystems and the global biogeochemical silica cycle. Global Biogeochemical Cycles 16, 1121 68-1-68-8.

Cooney, C.M. 1996. Sunflowers remove radionuclides from water in ongoing phytoremediation field tests. Environmental Science and Technology 30, 194A.

Cornell, R.M., and Schwertmann, U. 2003. The Iron Oxides: Structure, Properties, Reactions, Occurrences, and Uses 2nd ed., Wiley-VCH, Weinheim, Germany 664 p.

Costerton, J.W., Lewandowski, Z., Caldwell, D.E., Korber D.R., and Lappin-Scott, H.M. 1995. Microbial biofilms. Annual Review of Microbiology 49, 711–745.

CRC Handbook of Chemistry and Physics. Accessed online June 24, 2008.

Curie, P. 1885. Sur les questions des ordre. Bulletin de la Société minéralogique de France 8, 145–150.

Curtis, G.P., Davis, J.A., and Niftz, D.L. 2006. Simulation of reactive transport of uranium (VI) in groundwater with variable chemical conditions. Water Resources Research 42, W04404.

Cygan, R.T. 2001. Molecular modeling in mineralogy and geochemistry. In Cygan, R.T., and Kubicki, J.D. (eds), Reviews in Mineralogy and Geochemistry Vol. 42, Molecular Modeling Theory: Applications in the Geosciences. Geochemical Society and Mineralogical Society of America, Washington, DC, pp. 1–35.

Daniel, D.E., Applegate, J.S., Anspaugh, L., Croff, A.G., Ewing, R.C., Locke, P.A., Maurice, P.A., Rogers, R., Smith, A.E., Theofanous, T.G., Wong, J., and Lowenthal, M.D. 2005. Risk and decisions about disposition of transuranic and high-level radioactive waste, Final Report of the Committee on Risk-Based Approaches for Disposition of Transuranic and High-Level Radioactive Waste, National Academy/National Research Council Division of Earth and Life Sciences.

Davis, J.A., and Kent, D.B. 1990. Surface complexation modeling in aqueous geochemistry. In Hochella, M.F. Jr., and White, A.F. (eds), Mineral Water Interface Geochemistry, Reviews in Mineralogy Vol. 23, The Mineralogical Society of America, Washington, DC, pp. 177–279.

Davis, J.A., Meece, D.E., Kohler, M., and Curtis, G.P. 2004. Approaches to surface complexation modeling of uranium (VI) adsorption on aquifer sediments. Geochimica et Cosmochimica Acta 68, 3621–3641.

Davis, K.J., Nealson, K.H., and Lüttge, A. 2007. Calcite and dolomite dissolution rates in the context of microbe-mineral surface interactions. Geobiology 5, 191–205.

Delin, G.N., Essaid, H.H., Cozzarelli, I.M., Lahvis, M.H., and Berkins, B.A. 1998. Ground water contamination by crude oil near Bemidji, MN USGS Fact Sheet 084-98. U.S. Geological Survey, Mounds View, MN.

Derjaguin, B.V., and Landau, L. 1941. Theory of the stability of strongly charged lyophobic sols and of the adhesion of strongly charged particles in solution of electrolytes. Acta Physicochimica URSS 14, 633–662.

De Rosnay, J., 1979. The Macroscope: A Systematic Approach to Complexity, Harper and Row, New York.

DeYoreo, J.J., and Dove, P.M. 2004. Shaping crystals with biomolecules. Science 306, 1301–1302.

DeYoreo, J.J., and Vekilov, P.G. 2003. Principles of crystal nucleation and growth. In Dove, P.M., DeYoreo, J.J., and Weiner, S. (eds), Biomineralization, Reviews in Mineralogy and Geochemistry Vol. 54, pp. 56–93.

Dhungana, S., Anthony, C.R., and Hersman, L.E. 2007. Effect of exogenous reductant on growth and iron mobilization from ferrihydrite by the Pseudomonas mendocina ymp strain. Applied and Environmental Microbiology 73, 3428–3430.

DiChristina, T.J., Moore, C.M., and Haller, C.A. 2002. Dissimilatory Fe(III) and Mn(IV) reduction by *Shewanella putrefaciens* requires *ferE*, a homolog of the *pulE* (*gspE*) type II protein secretion gene. Journal of Bacteriology 184, 142–151.

Dietz, A.C., and Schnoor, J.L., 2001. Advances in phytoremediation. Environmental Health Perspectives 109, 163–168.

Dixit, S., and Hering, J.G. 2003. Comparison of arsenic(V) and arsenic(III) sorption onto iron oxide minerals: Implications for arsenic mobility. Environmental Science and Technology 37, 4182–4189.

Drever, J.I. 1997. The Geochemistry of Natural Waters: Surface and Groundwater Environments 3rd ed., Prentice Hall, New Jersey, 436 p.

Driscoll, C.T., Driscoll, K.M., Roy, K.M., and Mitchell, M.J. 2003. Chemical responses of lakes in the Adirondack region of New York to declines in acidic deposition. Environmental Science and Technology 37, 2036–2042.

Driscoll, C.T., Lawrence, G.B., Bulger, A.J., Butler, T.J., Cronan, C.S., Eagar, C., Lambert, K.F., Likens, G.E., Stoddard, J.L., and Weathers, K.C. 2001. Acidic deposition in the northeastern United States: Sources and inputs, ecosystem effects, and management strategies. BioScience 51, 180–198.

Du, Z., Li, H., and Gu, T., 2007. A state of the art review on microbial fuel cells: A promising technology for wastewater treatment and bioenergy. Biotechnology Advances 25, 464–482.

Duc, M., Lefevre, G., and Fedoroff, M. 2006. Sorption of selenite ions on hematite. Journal of Colloid and Interface Science 298, 556–563.

Duckworth, O.W., and Sposito, G. 2005. Siderophore-manganese (III) interactions. I. Air-oxidation of manganese (II) promoted by desferrioxamine B. Environmental Science and Technology 39, 6037–6044.

Dyer, A. 1995. Zeolite surfaces and reactivity. In Vaughan, D.J., and Pattrick, R.A.D. (eds), Mineral Surfaces, Chapman and Hall, London, pp. 333–353.

Dzombak, D.A., and Morel, F.M.M. 1990. Surface Complexation Modeling: Hydrous Ferric Oxide, John Wiley and Sons, New York, 393 p.

Eggleston, C.M. 1999. The surface structure of alpha-Fe_2O_3 (001) by scanning tunneling microscopy: Implications for interfacial electron transfer reactions. American Mineralogist 84, 1061–1070.

Eggleston, C.M. 1994. High-resolution scanning probe microscopy: Tip-surface interaction, artifacts, and applications in mineralogy and geochemistry. In Nagy, K.L., and Blum, A.E. (eds), Scanning Probe Microscopy of Clay Minerals, 7, Clay Minerals Society, Boulder, CO, pp. 1–90.

Eggleston, C.M., and Hochella, M.F. Jr. 1992. The structure of hematite (001) surfaces by scanning tunneling microscopy: Image interpretation, surface relaxation, and step structure. American Mineralogist 77, 911–922.

Ehrlich, H.L. 2002. Geomicrobiology, 4th ed., Marcel Dekker, Inc., New York, 768 p.

Eisenberg, D., and Kauzmann, W. 1969. The Structure and Properties of Water, Oxford University Press, Oxford, UK, 296 p.

Eisenlohr, L., Meteva, K., Gabrovsek, F., and Dreybrodt, W. 1999. The inhibiting action of intrinsic impurities in natural calcium carbonate minerals to their dissolution kinetics in aqueous H_2O-CO_2 solutions. Geochimica et Cosmochimica Acta 63, 989–1001.

Elzinga, E.J., and Sparks, D.L. 2007. Phosphate adsorption onto hematite: An in situ ATR-FTIR investigation of the effects of pH and loading level on the mode of phosphate surface complexation. Journal of Colloid and Interface Science 308, 53–70.

Encyclopædia Britannica Online. 2008. Chemical element. Retrieved June 22, 2008, from Encyclopeadia Britannica Online: http://search.eb.com/eb/article-81236.

Eng, P.J., Trainor, T.P., Brown, G.E. Jr., Waychunas, G.A., Newville, M., Sutton, S.R., and Rivers, M.L. 2000. Structure of the hydrated α-Al_2O_3 (0001) surface. Science 288, 1029–1033.

Engel, A.S., Stern, L.A., and Bennet, P.C. 2004. Microbial contributions to cave formation: New insights into sulfuric acid epleogeneis. Geology 32, 369–372.

EPA. 2005. Nanotechnology white paper. The U.S. Environmental Protection Agency, Science Policy Council, Washington, DC (http://www.epa.gov/osa/pdfs/EPA_nanotechnology_white_paper_external_review_draft_12-02-2005.pdf).

Erel, Y., Blum, J.D., Roueff, E., and Ganor, J. 2004. Lead and strontium isotopes as monitors of experimental granitoid mineral dissolution. Geochimica et Cosmochimica Acta 68, 4649–4663.

Espinasse, B., Hotze, E.M., and Wiesner, M.R. 2007. Transport and retention of colloidal aggregates of C-60 in porous media: Effects of organic macromolecules, ionic composition, and preparation method. Environmental Science and Technology 41, 7396–7402.

Ewers, W.E. 1983 Chemical factors in the depositon and diagenesis of banded iron-formation. In Trandall, A.F., and Morris, R.C. (eds), Iron Formation: Facts and Problems, Elsevier, New York, pp. 491–512.

Fein, J.B. 2002. The effects of ternary surface complexes on the adsorption of metal cations and organic acids onto mineral surfaces. In Water-Rock Interactions, Ore Deposits, and Environmental Geochemistry: A Tribute to David A. Crerar, The Geochemical Society Special Publications No. 7, St. Louis, MO, pp. 365–378.

Fendorf, S.E., and Sparks, D.L. 1996. X-ray absorption fine structure spectroscopy. In Sparks, D. Methods of Soil Analysis, Part 3. Chemical Methods, Soil Science Society of America, Madison, WI, 377–416.

Fenter, P. 2002. X-ray reflectivity as a probe of mineral-fluid interfaces: A user guide. Applications of Synchrotron Radiation in Low-temperature Geochemistry and Environmental Sciences 49, 149–220.

Fenter, P., Rivers, M., Sturchio, N., and Sutton, S. (eds). 2002. Applications of Synchrotron Radiation in low Temperature Geochemistry and Environmental Science. Reviews in Mineralogy and Geochemistry Vol. 49. The Geochemical Society and the Mineralogical Society of America, Washington, DC, 579 p.

Fenter, P., Park, C., Nagy, K.L., and Sturchio, N.C. 2007. Resonant anomalous X-ray reflectivity as a probe of ion adsorption at solid–liquid interfaces. Thin Solid Films 515, 5654–5659.

Ferris, F.G., 2005. Biogeochemical properties of bacteriogenic iron oxides. Geomicrobiology Journal 22, 79–85.

Field, C.B., and Chatfield, N. 2000. Global change and the terrestrial carbon cycle: The Jasper Ridge CO_2 experiment. In Ernst, W.G. (ed), Earth Systems Science, Cambridge University Press, Cambridge, UK, pp. 297–324.

Findlay, S., Carlough, L., Crocker, M.T., Gill, H.K., Meyer, J.L., and Smith, P.J. 1986. Bacterial growth on macrophyte leachate and fate of bacterial production. Limnology and Oceanography 31, 1335–1341.

Finnegan, M.P., Hengzhong, Z., and Banfield, J.F. 2007. Phase stability and transformation in titania nanoparticles in aqueous solutions determined by surface energy. Journal of Physical Chemistry C 111, 1962–1968.

Fischer, L., Brummer, G.W., and Barrow, N.J. 2007. Observations and modelling of the reactions of 10 metals with goethite: adsorption and diffusion processes. European Journal of Soil Science 58, 1304–1315.

Fitts, J.P., Brown, G.E. Jr., and Parks, G.A. 2000. Structural evolution of Cr(III) polymeric species at the γ-Al_2O_3/water interface. Environmental Science and Technology 34, 5122–5128.

Forbes, T.Z., McAlpin, G., Murphy, R., and Burns, P.C. 2008. Metal-oxygen isopolyhedra assembled into fullerene topologies. Angewandte Chemie International Edition, 47, 2824–2827.

Forsythe, J.H., Maurice, P.A., and Hersman, L.E. 1998. Attachment of a *Pseudomonas* sp. to Fe(III)-(hydr)oxide surfaces. Geomicrobiology 15, 293–308.

Fortin, D., and Langley, S. 2005. Formation and occurrence of biogenic iron-rich minerals. Earth-Science Reviews 72, 1–19.

Fortin, D., Ferris, F.G., and Beveridge, T.J. 1997. Surface-mediated mineral development by bacteria. In Banfield, J.F., and Nealson, K.H. (eds), Reviews in Mineralogy, Vol. 35. MSA, Washington, DC, pp. 161–180.

Freeze, R.A., and Cherry, J.A. 1979. Groundwater, Prentice-Hall, Englewood Cliffs, NJ, 604 p.

Fripiat, J., Cases, J., François, M., and Letellier, M. 1982. Thermodynamic and microdynamic behavior of water in clay suspensions and gels. Journal of Colloid Interface Science 89, 378–400.

Fukuski, K., and Sato, T. 2005. Using a surface complexation model to predict the nature and stability of nanoparticles. Environmental Science and Technology 39, 1250–1256.

Fukushi, K., and Sverjensky, D.A. 2007. A predictive model (ETLM) for arsenate adsorption and surface speciation on oxides consistent with spectroscopic and theoretical molecular evidence. Geochimica et Cosmochimica Acta 71, 3717–3745.

Furrer, G., and Stumm, W. 1986. The coordination chemistry of weathering. I. Dissolution kinetics of δ-Al_2O_3 and BeO. Geochimica et Cosmochimica Acta 50, 1847–1860.

Garrels, R.M., and Christ, C.L. 1965. Solutions Minerals, and Equilibria, Freeman-Cooper, San Francisco, CA, 450 p.

Garrels, R.M., and Mackenzie, F.T. 1967. Origin of e chemical compositions of some springs and lakes. In Gould, R.F. (ed), Equlibrium Concepts in Natural Water Chemistry, Advances in Chemistry Series 67, American Chemical Society, Washington, DC, pp. 222–242.

Garrity, D.P., Vidal, E.T., and O'Toole, J.C. 1984. Genotypic variation in the thickness of silica deposition on flowering rice spikelets. Ann. Bot. 54, 413–421.

Geesey, G.G., and Mitchell, A.C. 2008. Need for direct measurements of coupled microbiological and hydrological processes at different scales in porous media systems. Journal of Hydrologic Engineering 13, 28–36.

Gephardt, R.E., and Lundgren, R.E. 1998. Hanford tank cleanup: A guide to understanding the technical issues, Battelle Press, Columbus, OH.

Germer, S., Neill, C., Krusche, A.V., Neto, S.C.G., and Elsenbeer, H. 2007. Seasonal and within-event dynamics of rainfall and throughfall chemistry in an open tropical rainforest in Rondonia, Brazil. Biogeochemistry 86, 155–174.

Gibbs, J.W. 1878. On the equilibrium of heterogenous substances. Trans. Conn. Acad. 3, 343–524. Reprinted in The Scientific Papers of J. Willard Gibbs Vol. 1, Longmans, Green, and Co., Toronto, 1906 (Dover, New York, 1961).

Gibbs, J.W. 1906a. On the equilibrium of heterogeneous substances, in The Scientific Papers of J.W. Gibbs, 1, Longman Green & Co., London.

Gibbs, J.W. 1906b. Certain suggestions on geophysical research. *American Journal of Science* 21, 461–462.

Gibbs, J.W. 1931. The Collected Works of J.W. Gibbs Vol. 1. Longmans, Green, New York.

Gibbs, R.J. 1970. Mechanisms controlling world water chemistry. Science 170, 1088–1090.

Gilbert, B., Huang, F., Zhang, H.Z., Waychunas, G.A., and Banfield, J.F. 2004. Nanoparticles: Strained and stiff. Science 305, 651–654.

Gilmer, G.H. 1976, Growth on imperfect crystal faces. I. Monte Carlo growth rates. Journal of Crystal Growth 35, 15–28.

Gilmer, G.H. 1980. Computer models of crystal growth. Science 208, 355–363.

Gislason, S.R., Arnorsson, S., and Armannsson, H. 1996. Chemical weathering of basalt in southwest Iceland: Effects of runoff, age of rocks and vegetative/glacial cover. Americal Journal of Science 296, 837–907.

Glasauer, S., Langley, S., and Beveridge, T.J. 2002. Intracellular iron minerals in a dissimilatory iron-reducing bacterium. Science 295, 117–119.

Goldberg, S. 2002. Competitive adsorption of arsenate and arsenite on oxides and clay minerals. Soil Science Society of America Journal 66, 413–421.

Goldich, S.S. 1938. A study in rock-weathering. Journal of Geology 46, 17–58.

Gooseff, M.N., McKnight, D.M., Lyons, W.B., and Blum, A.E. 2002. Weathering reactions and hyporheic exchange controls on stream water chemistry in a glacial meltwater stream in the McMurdo Dr Valleys. Water Resources Research 38, 1279–1296.

Gorby, Y.A., Beveridge, T.J., and Blakemore, R.P. 1988. Characterization of the bacterial magnetosome membrane. Journal of Bacteriology 170, 834–841.

Gorby, Y.A., Yanina, S., McLean, J.S., Rosso, K.M., Moyles, D., Dohnalkova, A., Beveridge, T.J., Chang, I.S., Kim, B.H., Kim, K.S., Culley, D.E., Reed, S., Romine, M.F., Saffarini, D.A., Hill, E.A., Shi, L., Elias, D.A., Kennedy, D.W., Pinchuk, G., Watanabe, K., Ishii, S., Logan, B., Nealson, K.H., and Fredrickson, J.K. 2006. Electrically conductive bacterial nanowires produced by Shewanella oneidensis strain MR-1 and other microorganisms. Proceedings National Academy of Sciences 103, 11358–11363.

Gorman-Lewis, D., Fein, J.B., Soderholm, L, Jensen, M.P., and Chiang, M.-H. 2005. Experimental study of neptunyl adsorption onto *Bacillus subtilis*. Geochimica et Cosmochimica Acta 69, 5837–4844.

Grahame, DC 1947. The electrical double layer and the theory of electrocapillarity. Chemical Review 41, 441–501.

Grasso, D., Subramanian, K., Butkus, M., Strevett, K., and Bergendahl, J. 2002. A review of non-DLVO interactions in environmental colloidal systems. Reviews in Environmental Science and Bio/Technology 1, 17–38.

Gratz, A.J., and Hillner, P.E. 1993. Poisoning of calcite growth viewed in the atomic force microscope (AFM). Journal of Crystal Growth 129, 789–793.

Gratz, A.J., Hillner, P.E., and Hansma, P.K. 1993. Step dynamics and spiral growth on calcite. Geochimica et Cosmochimica Acta 57, 491–495.

Griffitt, R.J., Weil, R., Hyndman, K.A., Denslow, N.C., Powers, K., Taylor, D., and Barber, D.S. 2007. Exposure to copper nanoparticles causes gill injury and acute lethality in Zebrafish (*Danio Rerio*). Environmental Science and Technology 41, 8178–8186.

Grimm, V. 2002. Visual debugging: A way of analyzing, understanding and communicating bottom-up simulation models in ecology. Natural Resource Modeling 15, 23–38.

Gu B., Schmitt J., Chen Z., Liang L., and McCarthy J.F. 1995. Adsorption and desorption of different organic matter fractions on iron oxide. Geochimica et Cosmochimica Acta 59, 219–229.

Haack, E.A., Johnston, C., and Maurice, P.A. 2008. Siderophore sorption to montmorillonite. Geochimica et Cosmochimica Acta 72, 3381–3397.

Haack, E.A., and Warren, L.A. 2003. Biofilm hydrous manganese oxyhydroxides and metal dynamics in acid rock drainage. Environmental Science and Technology 37, 4138–4147.

Hamley, I.W. 2005. Self-assembling nanostructured molecular materials and devices. In Kelsall, R., Hamley, I., and Geoghegan, M. (eds), Nanoscale Science and Techology, John Wiley & Sons, Chichester, England, pp. 343–376.

Hammett, L.P. 1937. The effect of structure upon the reactions of organic compounds. Benzene derivatives. Journal of American Chemical Society 59, 96–103.

Hammett, L.P. 1940. Physical Organic Chemistry, McGraw-Hill, New York.

Hansch, C., Leo, A., and Taft, R.W. 1991. A survey of Hammett substituent constants and resonance and field parameters. Chemical Reviews 91, 165–195.

Hanson, G.W. 2008. Fundamentals of Nanoelectronics, Pearson/Prentice Hall, Upper Saddle River, NJ, 385 p.

Hayes, K.F., Roe, A.L., Brown, G.E. Jr., Hodgson, K.O., Leckie, J.O., and Parks, G.A. 1987. In-situ X-ray absorption study of surface complexes: Selenium oxyanions on a-FeOOH. Science 238, 783–786.

Hayes, K.F., and Leckie, J.O. 1988. Modeling ionic strength effects on cation adsorption at hydrous oxide/solution interaces. Journal of Colloid Interface Science 115, 564–572.

Healy, T.W., and Fuerstenau, D.W. 2007. The isoelectric point/point—of zero-charge of interfaces formed by aqueous solutions and nonpolar solids, liquids, and gases. Journal of Colloid and Interface Science 309, 183–188.

Hedges, J.I., Eglinton, G., Hatcher, P.G., Kirchman, D.L., Arnosti, C., Derenne, S., Evershed, R.P., Kogel-Knabner, I., de Leeuw, J.W., Littke, R., Michaelis, W., and Rullkotter, J. 2000. The molecularly-uncharacterized component of nonliving organic matter in natural environments. Organic Geochemistry 31, 945–958.

Hepinstall, S., Turner, B., and Maurice, P.A. 2005. Effects of siderophores on Cd and Pb adsorption to kaolinite. Clays and Clay Minerals 53, 557–563.

Herrmann, A.M., Ritz, K., Nunan, N., Clode, P.L., Pett-Ridge, J., Kilburn, M.R., Murphy, D.V., O'Donnell, A.G., and Stockdale, E.A. 2007. Nano-scale secondary ion mass spectrometry—a new analytical tool in biogeochemistry and soil ecology: a review article. Soil Biology and Biochemistry 39, 1835–1850.

Hersman, L.E., Forsythe, J.H., Ticknor, L.O., and Maurice, P.A, 2001. Growth of *Pseudomonas mendocina* on Fe(III)(hydr)oxides. Applied Environmental Microbiology 67, 4448–4453.

Hiebert, F.C., and Bennett, P.C., 1992. Microbial control of silicate weathering in organic-rich ground water. Science 258, 278–281.

Hiemstra, T., VanRiemsdjik, W.H., and Bolt, G.H. 1989. Multisite proton adsorption modeling at the solid-solutio interface of (hydr)oxides—A new approach. 1. Model description and evaluation of intrinsic reaction constants. Journal of Colloid and Interface Science 133, 91–104.

Hiemstra, T., and VanRiemsdijk, W.H. 1996. A surface structural approach to ion adsorption: The charge distribution (CD) model. Journal of Colloid and Interface Science 179, 488–508.

Hochella, M.F. Jr. 1990. Atomic structure, microtopography, composition, and reactivity of mineral surfaces. In Hochella, M.F., and White, A.F. (eds), Reviews in Mineralogy, Vol. 23, Mineral-Water Interface Geochemistry, Mineralogical Society of America, Washington, DC, pp. 87–132.

Hochella, M.F. Jr. 1995. Mineral surfaces: Their characterization and their chemical, physical, and reactive nature. In Vaughan, D.J., and Pattrick, R.A.D. (eds), Mineral Surfaces. The Mineralogical Society Series Vol. 5, Chapman and Hall, London, pp. 17–60.

Hochella, M.F. Jr., Lower, S.K., Maurice, P.A., Penn, R.L., Sahai, N., Sparks, D.L., and Twining, B.S. 2008. Nanominerals, mineral nanoparticles, and Earth chemistry. Science 21, 1631–1635.

Hochella, M.F. Jr., and Madden, A.S. 2005. Earth's nano-compartment for toxic metals. Elements 1, 199–203.

Hochella, M.F. Jr., Madden, A.S., Moore, J.N., Kasama, T., Putnis, A., and Putnis, C.V. 2004. Environmentally important nanoparticles from a massive acid mine drainage site. Geochimica et Cosmochimica Acta 68, A103–A103.

Hochella, M.F. Jr., Moore, J.N., Putnis, C.V., Putnis, A., Kasama, T., and Eberl, D.D. 2005. Direct observation of heavy metal-mineral association from the Clark Fork River Superfund Complex: Implications for metal transport and bioavailability. Geochimica et Cosmochimica Acta 69, 1651–1663.

Hodson, M.J., White, P.J., Mead, A., and Broadley, M.R. 2005. Phylogenetic variation in the silicon composition of plants. *Annals of Botany* 96, 1027–1046.

Hogg, R., Healy, T.W., and Fuerstenau, D.W. 1966. Mutual coagulation of colloidal dispersions. Transactions of the Faraday Society 62, 1638–1651.

Holdren, G.R. Jr., and Berner, R.A. 1979. Mechanism of feldspar weathering: I. Experimental studies. Geochimica et Cosmochimica Acta 43, 1161–1171.

Holmén, B.A., and Casey, W.H. 1996. Hydroxamate ligands, surface chemistry, and the mechanism of ligand promoted dissolution of goethite [α-FeOOH(s)]. Geochimica et Cosmochimica Acta 60, 4403–4416.

Holmén, B.A., Tejedor-Tejedor, M.I., and Casey, W.H. 1997. Hydroxamate complexes in solution at the goethite-water interface: A cylindrical internal reflection Fourier transform infrared spectroscopy study. Langmuir 13, 2197–2206.

Hornberger, G.M., Scanlon, T.M., and Raffensperger, J.P. 2001. Modelling transport of dissolved silica in a forested headwater catchment: The effect of hydrological and chemical time scales on hysteresis in the concentration-discharge relationship. Hydrological Processes 15, 2029–2038.

Huang, P.M., and Bollag, J.-M. 1998. Minerals-Organics-Microorganisms Interactions in the Soil Environment. In Huang, P.M., Senesi, N., and Buffle, J. (eds), Structure and Reactions of Soil Particles, John Wiley and Sons, Chichester, England, pp. 1–105.

Hurlbut, C.S., and Klein, C. 1985. Manual of Mineralogy, after J.D. Davis. 20th ed. John Wiley and Sons, New York.

IPCC, 2007: Intergovernmental Panel on Climate Change, Climate Change 2007: The Physical Science Basis, Summary for Policymakers (Feb. 2007).

International Union of Pure and Applied Chemistry (IUPAC) 2005–2007. Compendium of Chemical Terminology—the Gold Book, Accessed at http://goldbook.iupac.org.

Israelachvili, J.N. 1985. Intermolecular and Surface Forces, Academic Press, New York.

Israelachvili, J.N. 1992. Intermolecular and Surface Forces 2nd ed., Academic Press, San Diego, CA, 450 p.

Israelachvili, J.N., and Adams, E.G. 1978. Measurement of forces between two mica surfaces in aqueous electrolyte solutions in the range 0-100 nm. Journal of Chemical Society, Faraday Transactions I. 74, 975 p.

ITRC, Interstate Technology and Regulatory Cooperation Work Group Phytotechnologies Work Team. 2001. Phytotechnology, white paper report.

Izbicki, J.A., Danskin, W.R., and Mendez, G.O. 1998. Chemistry and isotopic composition of ground water along a section near the Newmark Area, San Bernardino County, CA. U.S. Geological Survey Water Resources Investigation Report 97-4179.

Jaillard, B. 1987. Les Structures Rhizomorphes Calcaires: Modèles de Réorganisation des Minéraux du Sol par les Racines. Montpellier, France: Institut National Recerche Agronomie Laboratoire Sciences Sol. 221 p.

Jaillard, B., Guyon, A., and Maurin, A.S. 1991. Structure and composition of calcified roots, and their identification in calcareous soils. Geoderma 50, 197–210.

Jamtveit, B., and Meakin, P. (eds). 1999. Growth, Dissolution and Pattern Formation in Geosystems, Kluwer Academic Publishers, Dordrecht, The Netherlands.

Johnsson, P.A., and Barringer, J.L. 1993. Water quality and hydrogeochemical processes in McDonalds Branch Basin, New Jersey Pinelands, 1984-1988. U.S. Geological Survey Water-Resources Investigation Report 91-4081, U.S. Geological Survey, West Trenton, NJ, 111 p.

Johnston, C.T. 1996. Sorption of organic compounds on clay minerals: A surface functional group approach. In Sawhney, B. (ed), Organic Pollutants in the Environment, Clay Minerals Society, Boulder, CO, pp. 1–44.

Johnston, C.T. 1999. Fourier transform infrared and raman spectroscopy. In Perry, D.L. (ed), Instrumental Surface Analysis of Geologic Materials, VCH, New York, pp. 121–155.

Johnston, C.T., Boyd, S.A., Teppen, B.J., and Sheng, G. 2004. Sorption of nitroaromatic compounds on clay surfaces. In Auerbach, S.M., Carrado, K.A., and Dutta, P.A. (eds), Handbook of Layered Materials, Marcel Dekker, New York, pp. 155–189.

Johnston, C.T., DeOliveira, M.F., Teppen, B.J., Sheng, G.Y., and Boyd, S.A. 2001. Spectroscopic study of nitroaromatic-smectite sorption mechanisms. Environmental Science and Technology 24, 4767–4772.

Johnston, C.T., Farmer, W., and Aochi, Y. 1996. Fourier transform infrared and Raman spectroscopy. In Sparks, D. (ed), Methods of Soil Analysis, Part 3. Chemical Methods, Soil Science Society of America, Madison, WI, pp. 269–321.

Johnston, C.T., Sheng, G., Teppen, B.J., Boyd, S.A., and De Oliveira, M.F. 2002. Spectroscopic study of dinitrophenol herbicide sorption on smectite. Environmental Science and Technology 36, 5067–5074.

Johnston, C.T., Sposito, G., and Erickson, C. 1992. Vibrational probe studies of water interactions with montmorillonite. Clays Clay Minerals 40, 722–730.

Johnston, C.T., and Tombacz, E. 2002. Surface chemistry of soil minerals. In Dixon, J.B. and Schulze, D.G. (eds), Soil Mineralogy with Environmental Applications, Soil Science Society of America, Madison, WI, pp. 37–67.

Jones, B., Renaut, R.W., and Rosen, M.R. 2001. Microbial construction of siliceous stalactites at geysers and hot springs: Examples from the Whakarewarewa Geothermal Area, North Island, New Zealand. Palaois 16, 73–94.

Jones, B., Renaut, R.W., Rosen, M.R., and Ansdell, K.M. 2002. Coniform stromatolites from geothermal systems, North Island, New Zealand. Palaois 17, 84–103.

Jones, L.H.P., Milne, A.A., and Sanders, J.V. 1966. Tabashir: An opal of plant origin. Science 151, 464–466.

Junta, J., and Hochella, M.F. Jr. 1994. Manganese (II) oxidation at mineral surfaces: A microscopic and spectroscopic study. Geochimica et Cosmochimica Acta 58, 4985–4999.

Kalinowski, B.E., Liermann, L.J., Brantley, S.L., Barnes, A., and Pantano, C.G. 2000. X-ray photoelectron evidence for bacteria-enhanced dissolution of hornblende. Geochimica et Cosmochimica Acta 64, 1331–1343.

Kalinowski, B.E., and Schweda, P. 1996. Kinetics of muscovite, phlogopite, and biotite dissolution and alteration at pH 1–4, room temperature. Geochimica et Cosmochimica Acta 60, 367–385.

Kamath, R., Rentz, J.A., Schnoor, J.L., and Alvarez, P.J.J., 2004. Phytoremediation of hydrocarbon-contaminated soils: principles and applications. Petroleum Biotechnology: Developments and Perspectives Studies in Surface Science and Catalysis 151, 447–478.

Kaplan, I.R., and Bartley, J.K. 2000. Global biogeochemical cycles: Carbon, sulfur, and nitrogen. In Ernst, W.G. (ed), Earth Systems Science, Cambridge, UK, pp. 278–296.

Kappler, A., Pasquero, C., Konhauser, K.O., and Newman, D.K. 2005. Deposition of banded iron formations by anoxygenic phototrophic Fe(II)-oxidizing bacteria. Geology 33, 865–868.

Kashefi, K., and Lovley, D.R. 2003. Extending the upper temperature limit for life. Science 301, 934.

Kelly, D.P., and Wood, A.P. 2000. Reclassification of some species of *Thiobacillus* to the newly designated genera *Acidithiobacillus* gen. nov., *Halothiobacillus* gen. nov. and *Thermithiobacillus* gen. nov. International Journal of Systematic and Evolutionary Microbiology 50, 489–500.

Kelly, S.D., Kemner, K.M., Carley, J., Criddle, C., Jardine, P.M., Marsh, T.L., Phillips, D., Watson, D., and Wu, W.-M. 2008. Speciation of uranium in sediments before and after in situ biostimulation. Environmental Science and Technology 42, 1558–1564.

Kelly, S.D., Kemner, K.M., Fein, J.B., Fowle, D.A., Boyanov, M.I., Bunker, B.A., and Yee, N. 2002. X-ray absorption fine strcture determination of pH-dependent U-bacterial cell wall interactions. Geochimica et Cosmochimica Acta 66, 3855–3871.

Kelsall, R.W., Hamley, I.W., and Geoghegan, M. (eds) 2005. Nanoscale Science and Technology. John Wiley and Sons. Chichester, England, 456 p.

Kemner, K.M., Kelly, S.D., Lai, B., Maser, J., O'Loughlin, E.J., Sholto-Douglas, D., Cai, Z.H., Schneegurt, M.A., Kulpa, C.F., and Nealson, K.H. 2004. Elemental and redox analysis of single bacterial cells by X-ray microbeam analysis. Science 306, 686–687.

Kendall, T.A., Hochella, M.F., and Becker, U. 2005. Computational modeling of azotobactin—goethite/diaspore interactions: Applications to molecular force measurements and siderophore—mineral reactivity. Chemical Geology 216, 17–35.

Kilham, P. 1990. Mechanisms controlling the chemical composition of lakes and rivers: Data from Africa. Limnology and Oceanography 35, 80–83.

Kirkpatrick, R.J., Kalinichev, A.G., and Wang, J. 2005. Molecular dynamics modelling of hydrated-mineral interlayers and surfaces: Structure and dynamics. Mineralogical Magazine 69, 289–308.

Kitchener, J.A. 1965. Mechanisms of adsorption from aqueous solutions: Some basic problems. The Journal of Photographic Science 13 152–159.

Klein, C. 2001. Manual of Mineral Science, 22nd ed. John Wiley and Sons, New York.

Klein, C., and Beukes, N.J. 1992. Time distribution, stratigraphy, and sedimentologic setting, and geochemistry of Precambrian Iron-Formations. In Schopf, J.W., and Klein, C. (eds), The Proterozoic Biosphere: A Multidisciplinary Study, Cambridge University Press, Cambridge, England.

Klein, C., and Hurlbut, C.S. 1985. Manual of Mineralogy. Revised. After Dana, J.S. John Wiley & Sons, New York.

Konhauser, K.O., Hamade, T., Raiswell, R., Morris, R.C., Ferris, F.G., Southam, G., and Canfield, D.E. 2002. Could bacteria have formed the Precambrian banded iron formations? Geology 30, 1079–1082.

Konhauser, K.O., Phoenix, V.R., Bottrell, S.H., Adams, D.G., and Head, I.M. 2001. Microbial-silica interactions in Icelandic hot springs sinter: Possible analogues for some Precambrian siliceous stromatolites. Sedimentology 48, 415–433.

Kosmulski, M. 2006. pH-dependent surface charging and points of zero charge. III. Update. Journal of Colloid and Interface Science 298, 730–741.

Kosmulski, M. 2001. Chemical Properties of Material Surfaces. Dekker, New York.

Kraemer, S.M. 2004. Iron oxide dissolution and solubility in the presence of siderophores. Aquatic Sciences 66, 3–18.

Kraemer S.M., Cheah, S-F., Zapf, R., Xu, J., Raymond, K.N., and Sposito, G. 1999. Effect of hydroxamate siderophores on Fe release and Pb(II) adsorption by goethite. Geochimica et Cosmochimica Acta 63, 3003–3008.

Krämer, M., von Bohlen, A., Sternemann, C., Paulus, M., and Hergenroder, R. 2007. Synchrotron radiation induced X-ray standing waves. Applied Surface Science 253, 3533–3542.

Krauskopf, K.B., and Bird, D.K. 1995. Introduction to Geochemistry, 3rd ed. WCB McGraw-Hill. Boston, MA, 647 p.

Kuhlbusch, T.A.J., and Crutzen, P.J. 1995. Toward a global estimate of black carbon in residues of vegetation fires representing a sink of atmospheric CO2 and a source of O2. Global Biogeochemical Cycles 9, 491–501.

Langmuir, I. 1916. The evaporation, condensation, and reflection of molecules and the mechanism of adsorption. Physical Review 8, 149–176.

Langmuir, D. 1981. The power exchange function: A general model for metal adsorption onto geological materials. In Tewari, D.H. (ed), Adsorption from Aqueous Solutions, Plenum Press, New York, pp. 1–17.

Langmuir, D. 1997. Aqueous Environmental Chemistry. Prentice Hall, Englewood Cliffs, NJ, 600 p.

Larson, R.A., and Weber, E.J. 1994. Reaction Mechanisms in Environmental Organic Chemistry. Lewis Publishers, Boca Raton, FL, 433 p.

Lasaga, A.C. 1990. Atomic treatment of mineral-water surface reactions. In Hochella, M.F. Jr., and White, A.F. (eds), Mineral-Water Interface Geochemistry. Reviews in Mineralogy Vol. 23. Mineralogical Society of America, Washington, DC, pp. 17–85.

Lasaga, A.C. 1995. Fundamental approaches in describing mineral dissolution and precipitation rates. In White, A.F., and Brantley, S.L. (eds), Chemical Weathering Rates of Silicate Minerals. Reviews in Mineralogy Vol. 31, Mineralogical Society of America, Washington, DC, pp. 23–86.

Lasaga, A.C., and Lüttge, A. 2001. Variation of crystal dissolution rate based on a dissolution stepwave model. Science 291, 2400–2404.

Lasaga, A.C., Soler, J.M., Ganor, J., Burch, T.E., and Nagy, K.L. 1994. Chemical weathering rate laws and global geochemical cycles. Geochimica et Cosmochimica Acta 58, 2361–2368.

Lecoanet, H.F., Bottero, J.Y., Wiesner, M.R. 2005. Laboratory assessment of the mobility of nanomaterials in porous media. Environmental Science and Technology 38, 5164–5169.

Lee, B.D., Sears, S.K., Graham, R.C., Amrhein, C., and Vali, H. 2003. Secondary mineral genesis from chlorite and serpentine in an ultramafic soil toposequence. Soil Science Society of America Journal 67, 1309–1317.

Lee, E.M., and Koopal, L.K. 1996. On protonation of surface with anionic surfactant sorption. Journal of Colloid Interface Science 177, 478–489.

Lee, J.F., Mortland, M.M., Chiou, C.T., Kile, D.E., and Boyd, S.A. 1990. Adsorption of benzene, toluene, and xylene by two tetramethylammonium-smectites having different charge densities. Clays Clay Minerals 38, 113–120.

Leenheer, J.A. 1989. Methods for determination of structural models of fulvic acid from the Suwannee River by convergent independent analyses. In Averett, R.C., Leenheer, J.A., McKnight, D.M. and Thorn, K.A. (eds), Humic Substances in the Suwannee River, Georgia: Interactions, Properties, and Proposed Structures, U.S. Geological Survey Open-File Report 87–557. 377 p.

Lepot, K., Benzerara, K., Brown, G.E., Jr., and Philippot, P. 2008. Microbially influenced formation of 2,724-million-year-old stromatolites. Nature Geoscience 1, 118–121.

Lewandowski, Z., Stoodley, P., and Altobelli, S. 1995. Experimental and conceptual studies on mass transport in biofilms. Water Science and Technology 31, 153–162.

Li, C.-W., and Kingery, W.D. 1984. Solute segregation at grain boundaries in polycrystalline Al_2O_3. In Kingery, W.D. (ed), Structure and Properties of MgO and Al_2O_3 Ceramics, Am. Ceram. Soc, Columbus, OH, pp. 368–378.

Liang, L., and Morgan, J.J. 1990. Coagulation of iron oxide particles in the presence of organic materials. In Melchior, DC, and Bassett, R.L. (eds), Chemical Modeling of Aqueous Systems II. American Chemical Society, Washington, DC, pp. 293–308.

Liermann, L.J., Kalinowski, B.E., Brantley, S.L., and Ferry, J.G. 2000. Role of bacterial siderophores in dissolution of hornblende. Geochimica et Cosmochimica 64, 587–602.

Likens, G. 2004. Some perspectives on long-term biogeochemical research from the Hubbard Brook Ecosystem study. Ecology 85, 2355–2362.

Likens, G.E., and Bormann, F.H. 1995. Biogeochemistry of a Forested Ecosystem 2nd ed. Springer, New York, 159 p.

Likens, G.E., Bormann, F.H., Pierce, R.S., Eaton, J.S., and Johnson, N.M. 1977. Biogeochemistry of a Forested Ecosystem, Springer-Verlag, New York.

Lin, D., and Xing, B. 2008. Root uptake and phytotoxicity of ZnO nanoparticles. Environmental Science and Technology 42, 5580–5585.

Lin, F.-C., and Clemency, C.V. 1981a. The kinetics of dissolution of muscovites at 25°C and 1 atm CO2 partial pressure. Geochimica et Cosmochimica Acta 45, 571–576.

Lin, F.-C., and Clemency, C.V. 1981b. Dissolution kinetics of phlogopite.1. Closed system. Clays Clay Minerals 29, 101–106.

Little, B.J., Wagner, P.A., and Lewandowski, Z. 1997. Spatial relationships between bacteria and mineral surfaces. Reviews in Mineralogy 35, 123–159.

Liu, J., Aruguete, D.M., Jinschek, J.R., Rimstidt, J.D., and Hochella, M.F. Jr. 2009. The non-oxidative dissolution of galena nanocrystals: Insights into mineral dissolution rates as a function of grain size, shape, and aggregation state. Geochimica et Cosmochimica Acta, in press.

Liu, Y., Sethurman, G., Wu, W., Nancollas, G.H., and Grynpas, M. 1996. The crystallization of fluorapatite in the presence of hydroxylapatite seeds and of hydroxylapatite in the presence of fluorapatite seeds. Journal of Colloid and Interface Science 186, 102–109.

Logan, B.E., Hamelers, B, Rozendal, R., Shröder, U, Keller, J., Freguia, S, Aelterman, P., Verstraete, W., and Rabaey, K. 2006. Microbial fuel cells: Methodology and technology. Environmental Science and Technology 40, 5181–5192.

Lord, D.G., Barringer, J.L., Johnsson, P.A., Schuster, P.F., Walker, R.L., Fairchild, J.E., Sroka, B.N., and Jacobsen, E. 1990. Hydrogeochemical data from an acid deposition study at McDonalds Branch basin in the New Jersey Pinelands, 1983–1986. U.S. Gelogical Survey Open-File Report 88-500. 132 p.

Lovern, S.B., and Klaper, R. 2006. *Daphnia magna* mortality when exposed to titanium dioxide and fullerene (C60) nanoparticles. Environmental Toxicology and Chemistry 25, 1132–1137.

Lovley, D.R. 2006. Bug juice: Harvesting electricity with microorganisms. Natural Reviews Microbiology 4, 497–508.

Lovley, D.R., Chapelle, F., and Woodward, J.C. 1994. Use of dissolved H_2 concentrations to determine distribution of microbially catalyzed redox reactions in anoxic groundwater. Environmental Science and Technology 28, 1205–1210.

Lovley D.R., Coates J.D., Blunt-Harris E.L., Phillips E.J.P., and Woodward J.C. 1996. Humic substances as electron acceptors for microbial respiration. Nature 382, 445–448.

Lovley, D.R., Holmes, D.E., and Nevin, K.P., 2004. Dissimilatory Fe(III) and Mn(IV) reduction. Advances in Microbiol Physiology 49, 219–286.

Lovely, D.R., and Phillips, E.J.P. 1988. Novel mode of microbial energy metabolism: Organic carbon oxidation coupled to dissimilatory reduction of iron or manganese. Applied and Environmental Microbiology 54, 1472–1480.

Lovley, D.R., Phillips, E.J.P., Gorby, Y.A., and Landa, E.R. 1991. Microbal reduction of uranium. Nature 350, 413–416.

Lovley, D.R., Stolz, J.F., Nord, G.L., and Phillips, E.J.P. 1987. Anaerobic production of magnetite by a dissimilatory iron-reducing microorganism. Nature 330, 252–254.

Lowenstam, H.L., and Weiner, S. 1989. On Biomineralization, Oxford University Press, New York.

Lower, S.K., Beveridge, T.J., and Hochella, M.F. Jr. 2001. Bacterial recognition of mineral surfaces: Nanoscale interactions between *Shewanella* and α-FeOOH. Science 292, 1360–1363.

Lower, S.K., Maurice, P.A., and Traina, S.J. 1998a. Simultaneous dissolution of hydroxylapatite and precipitation of hydroxylpyromorphite: Direct evidence of homogeneous nucleation. Geochimica et Cosmochimica Acta 62, 1773–1780.

Lower, S.K., Maurice, P.A., Traina, S.J., and Carlson, E.H. 1998b. Aqueous Pb sorption by hydroxylapatite: Applications of atomic force microscopy to dissolution, nucleation, and growth studies. American Mineralogist 83, 147–158.

Lower, S.K., Tadanier, C.J., and Hochella, M.F. 2000. Measuring interfacial and adhesion forces between bacteria and mineral surfaces with biological force microscopy. Geochimica et Cosmochimica Acta 64, 3133–3139.

Lower, S.K., Tadanier, C.J., and Hochella, M.F. Jr. 2001. Dynamics of the mineral-microbe interface: Use of biological force microscopy in biogeochemistry and geomicrobiology. Geomicrobiology Journal 18, 63–76.

Lucas, Y. 2001. The role of plants in controlling rates and products of weathering: Importance of Biological Pumping. Annual Review Earth and Planet Science 29, 135–63.

Luther, G.W., and Rickard, D.T. 2005. Metal sulfide cluster complexes and their biogeochemical importance in the environment. Journal of Nanoparticle Research 7, 389–407.

Lüttge, A., Bolton, E.W., and Lasaga, A.C. 1999. An interferometric study of the dissolution kinetics of anorthite: The role of reactive surface area. American Journal of Science 299, 652–678.

Lüttge, A., and Conrad, P.G. 2004. Direct observation of microbial inhibition of calcite dissolution. Applied and Environmental Microbiology 70, 1627–1632.

Lüttge, A., Zhang, L., and Nealson, K.H. 2005. Mineral surfaces and their implications for microbial attachment: Results from Monte Carlo simulations and direct surface observations. American Journal of Science 305, 766–790.

Lyklema, J. Interfacial tension: Measurement. In Fundamentals of Interface and Colloid Science, Vol. 3. Liquid-Fluid Interfaces. Interfacial tension: Measurement. Academic Press, London, pp. 1.1–1.87.

Lyons, W.B., Welch, K.A., Neumann, K., Toxey, J.K., McArthur, R., Williams, C., McKnight, D.M., and Moorhead, D. 1998. Geochemical linkages among glaciers, streams and lakes within the Taylor Valley, Antarctica. In Priscu, J.C. (ed), Ecosystem Dynamics in a Polar Desert: The McMurdo Dry Valleys, Antarctica, Antarctic Research Series 72. American Geophysical Union, Washington, DC, pp. 77–92.

Lyons W.B., Welch K.A., Nezat C.A., Crick K., Toxey J.K., Mastrine J.A., and McKnight D.M. 1997. Chemical weathering rates and reactions in the Lake Fryxell basin, Taylor Valley: Comparison to temperate river basins. In Lyons, W.B., and Howard-Williams, C. (eds), Ecosystem Processes in Antarctic Ice-Free Landscapes, Balkema, Leiden, The Netherlands, pp. 147–154.

Macinnis, I.N., and Brantley, S.L. 1992. The role of dislocations and surface-morphology in calcite dissolution. Geochimica et Cosmochimica Acta 56, 1113–1126.

Madden, A.S., and Hochella, M.F. Jr. 2005. A test of geochemical reactivity as a function of mineral size: Manganese oxidation promoted by hematite nanoparticles. Geochimica et Cosmochimica Acta 69, 389–398.

Madigan, M.T., Martinko, J.M., and Parker, J. 1997. Brock Biology of Microorganisms 8th ed. Prentice Hall, Upper Saddle River, NJ.

Majors, P.D., McLean, J.S., Pinchuk, G.E., Fredrickson, J.K., Gorby, Y.A., Minard, K.R., and Wind, R.A. 2005. NMR methods for in situ biofilm metabolism studies. Journal of Microbiological Methods 62, 337–344.

Manceau, A., and Charlet, L. 1994. The mechanism of selenate adsorption on goethite and hydrous ferric-oxide. Journal of Colloid Interface Science 168, 87–93.

Mandelbrot, B.B. 1982. The Fractal Geometry of Nature, W.H. Freeman, New York.

Manecki, M., Maurice, P.A., and Traina, S.J. 2000. Uptake of aqueous Pb by Cl-, F-, and OH- apatites: Mineralogic evidence for nucleation mechanisms. American Mineralogist 85, 932–942.

Maniloff, J. 1997. Nanobacteria: Size limits and evidence. Science 276, 1776.

Manning, B.A., Fendorf, S.E., Bostick, B., and Suarez, D.L. 2002. Arsenic(III) oxidation and arsenic(V) adsorption reactions on synthetic birnessite. Environmental Science and Technology 36, 976–981.

Markov, I.V. 1995. Crystal Growth for Beginners, World Scientific, Singapore, 422 p.

Marland, G., Boden, T.A., and Andres, R.J. 2007. Global, regional, and national CO2 emissions. In Trends: A Compendium of Data on Global Change. Carbon Dioxide Information Analysis Center, Oak Ridge National Laboratory, U.S. Department of Energy, Oak Ridge, TN.

Marmier, A., Hoang, P.N.M., Picaud, S., Girardet, C., and Lynden-Bell, R.M. 1998. A molecular dynamics study of the structure of water layers adsorbed on MgO(100). Journal of Chemical Physics 109, 3245–3254.

Marshall, K.C., Cruickshank, H. 1973. Cell surface hydrophobicity and the orientation of certain bacteria at interfaces. Archiv fur Mikrobiogie 91, 29–40.

Marshall, K.C., Stout, R., and Mitchell, R. 1971. Mechanism of the initial events in the sorption of marine bacteria to surfaces. Journal of General Microbiology 68, 337–348.

Martell, A.E., and Smith, R.M. 1976. Critical Stability Constants Vol. 1–4, Plenum Press, New York.

Mashino, T., Okuda, K., Hirota, T., Hirobe, M., Nagano, T., and Mochizuki, M. 1999. Inhibition of *E. coli* growth by fullerene derivatives and inhibition mechanism. Bioorganic & Medicinal Chemistry Letters 9, 2959–2962.

Masters, G.M., and Ela, W.P. 2008. Introduction to Environmental Engineering and Science, 3rd ed., Pearson/Prentice Hall, Upper Saddle River, NJ, 708 p.

Maurice, P.A. 2000. The hydrologic cycle. In Ernst, G. (ed), Earth Systems: Processes and Issues, Stanford University and Cambridge University Press, Cambridge, MA, pp. 135–151.

Maurice, P.A., Anthony, C., Arthurs, L., Giesting, P., Golden, S., Hepinstall, S., Liu, G., Lavarnway, A., Mishra, B., and Tholen, K. 2004. Effects of Cd on natural organic matter adsorption to goethite. Water-Rock Interaction International XI.

Maurice, P.A., and Harmon, T.C. 2007. Environmental embedded sensor networks. Environmental Engineering Science 24, 149–150.

Maurice, P.A., Hochella, M.F. Jr., Sposito, G., and Schwertmann, U. 1995. Hematite dissolution in simple organic acids: Atomic-force microscopy of surface microtopography. Clays and Clay Minerals 43, 29–38.

Maurice, P.A., Lee, Y.-J., and Hersman, L. 2000. Dissolution of Al-substituted goethites by an aerobic *Pseudomonas mendocina var.* bacteria. Geochimica et Cosmochimica Acta 64, 1363–1374.

Maurice, P.A., and Lower, S.L. 2008. Applications of atomic force microscopy in soil analysis. In Ulery, A.L., and Drees, L.R. (eds), Methods of Soil Analysis, Part 5: Mineralogical Methods pp. 299–333.

Maurice, P.A., McKnight, D.M., Leff, L., Fulghum, J.E., and Gooseff, M. 2002. Direct observation of aluminosilicate dissolution in the hyporheic zone of an Antarctic Dry Valley Stream. Geochimica et Cosmochimica Acta 66, 1335–1347.

Maurice, P., and Namjesnik-Dejanovic, K. 1999. Aggregate structures of sorbed humic substances observed in aqueous solution. Environmental Science and Technology 33, 1538–1540.

Maurice, P.A., Vierkorn, M., Hersman, L., Fulghum, J., and Ferryman, A. 2001. Enhancement of kaolinite dissolution by an aerobic *Pseudomonas mendocina* bacteria. Geomicrobiology 18, 21–35.

Maurice, P.A., and Warren, L. 2007. Introduction to geomicrobiology: Microbial interactions with minerals. In Maurice, P.A., and Warren, L. (eds), Methods of Investigating Microbial-Mineral Interactions. Clay Minerals Society Short Course Book,

McAuliffe, C. 1966. Solubility in water of paraffin, cycloparaffin, olefin, acetylene, cycloolefin, and aromatic hydrocarbons, Journal of Physical Chemistry 70, 1267–1275.

McBride, M.B. 1994. Environmental Chemistry of Soils. Oxford University Press, New York, 406 p.

McCarthy J.F., and Zachara, J.M. 1989. Subsurface transport of contaminants—Mobile colloids in the subsurface environment may alter the transport of contaminants. Environmental Science and Technology 23, 496–502.

McConnell, J.R., Edwards, R., Kok, G.L., Flanner, M.G., Zender, C.S., Saltzman, E.S., Banta, J.R., Pasteris, D.R., Carter, M.M., and Kahl, J.D.W. 2007. 20th-Century Industrial Black Carbon Emissions Altered Arctic Climate Forcing, Science 317, 1381–1384.

McKinley, J.P., Zachara, J.M., Heald, S.M., Dohnalkova, A., Newville, M.G., and Suttun, S.R. 2004. Microscale distribution of cesium sorbed to biotite and muscovite. Environmental Science and Technology 38, 1017–1023.

McKenzie, R.M. 1980. The adsorption of lead and other heavy metals on oxides of manganese and iron. Australian Journal of Soil Research 18, 61–73.

McKnight, D.M. 1989. Metal-binding structures. In Averett, R.C., McKnight, D.M., and Thorn, K.A. (eds), Humic Substances in the Suwannee River, Georgia: Interactions, Properties, and Proposed Structures, Denver, CO: U.S. Geological Survey Open-File Report 87-557, pp. 347–350.

McKnight, D.M., Bencala, K.E., Zellweger, G.W., Aiken, G.R., Feder, G.L., and Thorn, K.A. 1992. Sorption of dissolved organic carbon on hydrous aluminum and iron oxides occurring at the confluence of Deer Creek with the Snake River, Summit County, Colorado. Environmental Science and Technology 26, 1388–1396.

McKnight, D.M., and Duren, S.M. 2004. Biogeochemical processes controlling midday ferrous iron maxima in streams affected by acid rock drainage. Applied Geochemistry 19, 1075–1084.

McKnight, D.M., and Thurman, E.M. 1989. Mineral elements and amino acids. In Averett, R.C., Leenheer, J.A., McKnight, D.M., and Thorn, K.A. (eds), Humic Substances in the Suwannee River, Georgia: Interactions, Properties, and Proposed Structures, Denver, CO: U.S. Geological Survey Open-File Report 87-557, pp. 345–347.

Mclintock, I.S. 1967. The Elovich Equation in Chemisorption Kinetics. Nature 216, 1204–1205.

Meier, M., Namjesnik-Dejanovic, K., Maurice, P.A., Chin, Y.-P., and Aiken, G.R. 1999. Fractionation of aquatic natural organic matter upon sorption to kaolinite and goethite. Chemical Geology 157, 275–284.

Metson, J.B. 1999. Secondary Ion Mass Spectrometry. In D.L. Perry, (ed), Instrumental Surface Analysis of Geologic Materials, VCH, New York, pp. 311–345.

Meunier, J.D., Colin, F., and Alarcon, C. 1999. Biogenic silica storage in soils. Geology 27, 835–838, 1999.

Meunier, J.-D. 2003. Le rôle des plantes dans le transfert du silicium á la surface des continents. C.R. Geoscience 335, 1199–1206.

Meybeck, M. 2004. Global occurrence of major elements in rivers. In Drever, J.I. (ed), Treatise on Geochemistry, Vol. 5, Surface and Groundwater, Weathering and Soils, Elsevier, New York, pp. 207–223.

Michaelis, L., and Menten, M. 1913. Die Kinetik der Invertinwirkung, Biochemische Zeitschrift 49, 333–369.

Michel, F.M., Ehm, L., Antao, S.M., Lee, P.L., Chupas, P.J., Liu, G., Strongin, D.R., Schoonen, M.A.A., Phillips, B.L., and Parise, J.B. 2007a. The structure of ferrihydrite, a nanocrystalline material. Science 316, 2726–2729.

Michel, F.M., Ehm, L., Liu, G., Han, W.Q, Antao, S.M., Chupas, P.J., Lee, P.L., Knorr, K., Eulert, H, Kim, J., Grey, C.P., Celestian, A.J., Gillow, J., Schoonen, M.A.A., Strongin, D.R., and Prise, J.B. 2007b. Similarities in 2-and 6-line ferrihydrite based on pair distribution function analysis of X-ray total scattering. Chemistry of Materials 19, 1489–1496.

Michot, L.J., Villiéras, F., François, M., Bihannic, I., Pelletier, M., and Cases, J.-M. 2002. Water organisation at the solid-aqueous solution interface. Comptes Rendus Géoscience 334, 611–631.

Millero, F.J. 2007. The marine inorganic carbon cycle. Chemical Reviews 107, 308–341.

Mishra, B., Haack, E.A., Vasconcelos, I.F., Maurice, P.A., and Bunker, B.A. 2007. XAFS determination of Pb and Cd speciation with siderophores and the metal/siderophore/kaolinite system. AIP Proceedings 882, 196–198.

Molina, M.J., Tso, T., Molina, L.T., and Wang, F.C. 1987. Antarctic stratospheric chemistry of chlorine nitrate, hydrogen-chloride and ice-release of active chlorine. Science 238, 1253–1257.

Moore, D.M., and Reynolds, R.C. 1997. X-Ray Diffraction and the Identification and Analysis of Clay Minerals 2nd ed., Oxford University Press, Oxford, England.

Morse, J.W., Arvidson, R.S., and Lüttge, A. 2007. Calcium carbonate formation and dissolution. Chemical Reviews 107, 342–381.

Mucci, A. 1983. The solubility of calcite and aragonite in seawater at various salinities, temperatures, and one atmosphere total pressure. American Journal of Science 238, 780–799.

Mucci, A. 1987. Influence of temperature on the composition of magnesian calcite overgrowths precipitated from seawater. Geochimica et Cosmochimica Acta 51, 1977–1984.

Muñoz, M.A. 1988. Properties of synthetic goethites and their effect on sulfate adsorption. Ph.D. dissertation, Ohio State University, Columbus OH.

Murrell, J.N., and Jenkins, A.D. 1994. Properties of Liquids and solutions 2nd ed., John Wiley and Sons, Chichester, England.

Nachtegaal, M., Marcus, M.A., Sonke, J.E., Vangronsveld, J., Livi, K.J.T., Van Der Lelie, D., and Sparks, D.L. 2005. Effects of *in situ* remediation on the speciation and bioavailability of zinc in a smelter contaminated soil. Geochimica et Cosmochimica Acta 69, 4649–4664.

Namjesnik-Dejanovic, K., and Maurice, P.A. 2001. Conformations and aggregate structures of sorbed natural organic matter on muscovite and hematite. Geochimica et Cosmochimica Acta 65, 1047–1057.

Navarre-Stitchler, A., and Thyne, G. 2007. Effects of carbon dioxide on mineral weathering rates at earth surface conditions. Chemical Geology 243, 53–63.

Navrotsky, A. 2001. Thermochemistry of nanomaterials. In Banfield, J.F., and Navrotsky, A. (eds), Nanoparticles and the Environment. Reviews in Mineralogy and Geochemistry 44, Mineralogical Society of America and Geochemical Society, Washington, D.C., pp. 73–103.

Navrotsky, A., Mazeina, L., and Majzslan, J. 2008. Size-driven structural and thermodynamic complexity in iron oxides. Science 319, 1635–1638.

Navrotsky, A. 2003. Energetics of nanoparticle oxides: Interplay between surface energy and polymorphism. Geochemical Transactions 4, 34–37.

Nealson, K.H., and Stahl, D.A. 1997. Microorganisms and biogeochemical cycles: What can we learn from layered microbial communities? In Banfield, J.F., and Nealson, K.H. (eds), Geomicrobiology: Interactions Between Microbes and Minerals. Reviews in Mineralogy Vol. 35, Mineralogical Society of America, Washington, DC, pp. 5–34.

Neilands, J.B. 1981. Microbial iron compounds. Annal Review of Biochemistry 50, 715–731.

Neilands, J.B. 1957. Some aspects of iron metabolism. Bacteriology Reviews 21, 101–111.

Nevin, K.P., and Lovley, D.R. 2000. Lack of production of electron-shuttling compounds or solubilization of Fe(III) during reduction of insoluble Fe(III) oxide by Geobacter metallireducens. Applied and Environmental Microbiology 66, 2248–2251.

Newman, D.K., and Kolter, R. 2000. A role for excreted quinones in extracellular electron transfer. Nature 405, 94–97.

Nielsen, A.E. 1964. Kinetics of Precipitation, MacMillan, New York, 151 p.

Nies, D.H. 1999. Microbial heavy-metal resistance. Applied Microbiology and Biotechnology 51, 730–750.

Nordstrom, D.K., and Muñoz, J.L. 1986. Geochemical Thermodynamics, Blackwell Scientific Publications, Palo Alto, CA, 477 p.

Nowack, B., and Bucheli, T.D. 2007. Occurrence, behavior, and effects of nanoparticles in the environment. Environmental Pollution 150, 5–22.

NRC. 2001. Basic Research Opportunities in Earth Science, National Academy Press, Washington, DC, 168 p.

NRC. 2005. Radiative Forcing of Climate Change: Expanding the Concept and Addressing Uncertainties, National Academies Press, Washington, DC, 224 p.

Nugent, M., Maurice, P., and Brantley, S. 1998a. Estimation of feldspar dissolution rates from field-reacted samples. In Arehart, G.B. and Hulston, J.R. (eds) Water-Rock Interaction International IX, Balkema, Rotterdam, p. 387–390.

Nugent, M., Brantley, S. Pantano, C., and Maurice, P. 1998b. The influence of natural mineral coatings on feldspar dissolution. Nature 395, 588–591.

Oberdorster, G., Oberdorster, E., and Oberdorster, J. 2005. Nanotoxicology: An emerging discipline evolving from studies of ultrafine particles. Environmental Health Perspectives 113, 823–839.

Obert, M., Pfeifer, P., and Sernetz, M. 1990. Microbial growth patterns described by fractal geometry. Journal of Bacteriology 172, 1180–1185.

O'Brien, A.K., Rice, K.C., Bricker, O.P., Kennedy, M.M., and Anderson, R.T. 1997. Use of geochemical mass balance modeling to evaluate the role of weathering in determining stream chemistry in five Mid-Atlantic watersheds on different lithologies. Hydrologic Processes 11, 719–744.

O'Brien, R.W., and White, L.R. 1978. Journal of Chemical Society Faraday Transaction 2, 1607.

Ohnesorge, F., and Binnig, G. 1993. True atomic resolution by atomic force microscopy through repulsive and attractive forces. Science 260, 1451–1456.

Okuda, I., and Senesi, N. 1998. Fractal principles and methods applied to the chemistry of sorption onto environmental particles. In Huang, P. M., Senesi, N., and Buffle, J. (eds), Structure and Surface Reactions of Soil Particles, John Wiley and Sons, pp. 77–105.

Oleson, T.A., and Sahai, N. 2008. Oxide-dependent adsorption of a model membrane phospholipids, Dipalmitoylphosphatidylcholine: Bulk adsorption isotherms. Langmuir 24, 4865–4873.

Ostwald, W. 1896. Lehrbruck der Allgemeinen Chemie, Vol. 2, part 1. Leipzig, Germany.

Papelis, C., Hayes, K.F., and Leckie, J.O.. 1988. HYDRAQL: A program for the computation of chemical equilibrium composition of aqueous batch systems including surface-complexation modeling of ion adsorption at the oxide/solution interface. Technical Report no. 306. Department of Civil Eng., Stanford University, Stanford, CA.

Parfitt, R.L., and Atkinson, R.J. 1976. Phosphate adsorption on goethite (alpha-FEOOOH). Nature 264, 740–742.

Park, C., Fenter, P.A., Sturchio, N.C., and Regalbuto, J.R. 2005. Probing outer-sphere adsorption of aqueous metal complexes at the oxide-water interface with resonant anomalous X-ray reflectivity. Physical Review Letters 94, Art. No. 076104.

Parkhurst, D.L., and Appelo, C.A.J. 1999. User's guide to PHREEQC (version 2)- a computer program for speciation, batch-reaction, one-dimensional transport, and inverse geochemical calculations. USGS Water-Resources Investigations Report 99-4259.

Parks, G.A. 1990. Surface energy and adsorption at mineral/water interfacers: An introduction. In Hochella, M.F. Jr., White, A. F. (eds), Reviews in Mineralogy, Vol. 23, Mineral-Water Interface Geochemistry, Mineralogical Society of America. pp. 133–175.

Pasteris, J.D., Wopenka, B., Freeman, J.J., Rogers, K., Valsami-Jones, E., van der Houwen, J.A.M., and Silva, M.J. 2004. Lack of OH in nanocrystalline apatite as a function of degree of atomic order: implications for bone and biomaterials. Biomaterials 25, 229–238.

Pasteris, J.D., Wopenka, B., and Valsami-Jones, E. 2008. Bone and tooth mineralization: Why apatite? Elements 4, 97–104.

Paul, K.W., Kubicki, J.D., and Sparks, D.L. 2006. Quantum chemical calculations of sulfate adsorption at the Al- and Fe-(Hydr)oxide-H2O interface-estimation of Gibbs free energies. Environmental Science and Technology 40, 7717–7724.

Peak, D., Ford, R.G., and Sparks D.L. 1999. An in situ ATR-FTIR investigation of sulfate bonding mechanisms on goethite. Journal of Colloid Interface Science 218, 289–299.

Peak, D., and Sparks, D.L. 2002. Mechanisms of selenate adsorption on iron oxides and hydroxides. Environmental Science and Technology 36, 1460–1466.

Pearsall, D.M., and Piperno, D.R. 1993. Current Research in Phytolith Analysis: Applications in Archaeology and Paleontology. MASCA Research Papers in Science and Archaeology Vol. 10 The University Museum of Archaeology and Anthropology, University of Pennsylvania, Philadelphia, PA, 212 p.

Penn, R.L. and Banfield, J.F. 1998a. Imperfect oriented attachment: Dislocation generation in defect-free nanocrystals. Science 281, 969–971.

Penn, R.L., and Banfield, J.F. 1998b. Oriented attachment and growth, twinning, polytipism, and formation of metastable phases: Insights from nanocrystalline TiO2. American Mineralogist 83, 1077–1082.

Perry, E.C. Jr., Tan, F.C., and Morey, G.B. 1973. Geology and stable isotope geochemistry of the Biwabik iron formation, Northern Minnesota. Economic Geology 68, 1110–1125.

Perdue E.M. 1985. Acidic functional groups of humic substances. In Aiken, G.R., McKnight, D.M., Wershaw, R.L., and Maccarthy, P. (eds), Humic Substances in Soil, Sediment, and Water, John Wiley and Sons, New York pp. 493–526.

Perry, D.L. ed. 1990. Instrumental Surface Analysis of Geologic Materials, VCH Publishers, Inc., New York, 373 p.

Pettijohn, F.J., Potter, P.E., and Siever, R. 1987. Sand and Sandstone, Springer, New York, 553 p.

Pfeifer, P. 1987. Characterization of surface irregularity. In Laszlo, P. (ed), Preparative Chemistry Using Supported Reagends, Academic Press, London, p. 13.

Phillips, D.H., Watson, D.B., Roh, Y., Mehlhorn, T.L., Moon, J.W., and Jardine, P.M. 2006. Distribution of uranium contamination in weathered fractured saprolite/shale and ground water. Journal of Environment Quality 35, 1715–1730.

Philpot M. 2004. Atomic scale modeling of the solid-liquid interface, In Liu, X.Y., and De Yoreo, J.J. (eds), Nanoscale structure and assembly at solid-fluid interfaces Vol. I, Interfacial Structures versus Dynamics, Series on Nanoscience and Technologies, The Netherlands, Kluwer Academic Publisher, 700 p.

Piper, A.M., 1944. A graphic procedure in the geochemical interpretation of water analyses. American Geophysical Union Transactions, 25, 914–923.

Pitchford, A.M., Mazzella, A.T., and Scarborough, K.R. 1988. Soil-gas and geophysical techniques for detection of subsurface organic contamination. EPA Project Summary, EPA/600/S4-88/019. U.S. Environmental Protection Agency, Las Vegas, NV.

Plummer, L.N., Bexfield, L.M., and Anderholm, S.K. 2003. How ground-water chemistry helps us understand the aquifer. In J.R. Bartolino and J.C. Cole (eds) Ground-Water Resources of the Middle Rio Grande Basin, U.S. Geological Survey Circular 1222. pp. 92–93.

Plummer, L.N., Parkhurst, D.L., and Thorstenson, DC 1983. Development of reaction models for groundwater systems. Geochimica et Cosmochimica Acta 4, 665–686.

Plummer, L.N., Prestemon, E.C., and Parkhurst, D.L. 1994. An Interactive Code (NETPATH) for Modeling NET Geochemical Reactions Along a Flow PATH Version 2.0. U.S. Geological Survey Water-Resources Investigations Report 94-4169.

Poirier, N., Derenne, S., Rouzaud, J.N., Largeau, C., Mariotti, A., Balesdent, J., and Maquet, J. 2000. Chemical structure and sources of the macromolecular, resistant, organic fraction isolated from a forest soil (Lacadee, south-west France). Organic Geochemistry 31, 813–827.

Ponthieu, M., Juillot, F., Hiemstra, T., van Riemsdijk, W.H., and Benedetti, M.F. 2006. Metal binding to iron oxides. Geochimica et Cosmochimica Acta 70, 2679–2698.

Porder, S., Hilley G.E., and Chadwick, O.A. 2007. Chemical weathering, mass loss, and dust inputs across a climate by time matrix in the Hawaiian Islands. Earth and Planetary Science Letters 258 414–427.

Post, J.E. 1999. Manganese oxide minerals: Crystal structures and economic and environmental significance. Proceedings of the National Academy of Sciences of the United States of America 96, 3447–3454.

Potter, M. C. 1910. On the difference of potential due to the vital activity of microorganisms. Proceedings University of Durham Philosophical *Society* 3, 245–249.

Potter, M. C. 1911. Electrical effects accompanying the decomposition of organic compounds. Proceedings Royal Society of London B 84, 260–276.

Pulford, I.D., and Watson, C. 2003. Phytoremediation of heavy metal-contaminated land by trees —a review. Environment International 29, 529–540.

Punckt, C., Bolscher, M., Rotermund, H.H., Mikhailov, A.S., Oganic, L., Budiansky, N., Scully, J.R., and Hudson, J.L. 2004. Sudden onset of pitting corrosion on stainless steel as a critical phenomenon. Science 305, 1133–1136.

Rahnemaie, R., Hiemstra, T., and van Riemsdjik, W.H. 2006. Geometry, charge distribution, and surface speciation of phosphate on goethite. Langmuir 23, 3680–3689.

Rancourt, D.G., and Meunier, J.-F. 2008. Constraints on structural models of ferrihydrite as a nanocrystalline material. American Mineralogist 93, 1412–1417.

Ratkowsky, D.A., Olley, J., McMeekin, T.A., and Ball, A. 1982. Relationship between tempearute and growth rate of bacterial cultures. Journal of Bacterio 149, 1–5.

Raymond, K.N., and Dertz, E.A. 2004. Biochemical and physical properties of siderophores. In Cross J.H., Mey, A.R., and Payne S.M. (eds), Iron Transport in Bacteria, ASM Press, Washington, DC, pp. 3–17.

Rea, R.L., and Parks, G.A. 1990. Numerical simulation of coadsorption of ionic surfactants with inorganic ions on quartz. In Melchior, DC, and Basset, R.L. (eds), Chemical Modeling of Aqueous Systems II, American Chemical Society, Washington, DC, pp. 260–271.

Reeder, R.J. 1996. Interaction of divalent cobalt, zinc, cadmium, and barium with the calcite surface during layer growth. Geochimica et Cosmochimica Acta 60, 1543–1552.

Reguera, G., McCarthy, K.D., Mehta, T., Nicoll, J.S., Tuominen, M.T., and Lovley, D.R. 2005. Extracellular electron transfer via microbial nanowires. Nature 435, 1098–1101.

Reguera, G., Nevin, K.P., Nicoll, J.S., Covalla, S.F., Woodard, T.L., and Lovley, D.R. 2006. Biofilm and nanowire production leads to increased current in Geobacter sulfurreducens fuel cells. Applied and Environmental Microbiology 72, 7345–7348.

Reisinger, H.J., Burris, D.R., and Hering, J.G. 2005. Remediating subsurface arsenic contamination with monitored natural attenuation. Environmental Science and Technology, 39, 458A–464A.

Rimstidt J.D., and Dove P.M. 1986. Mineral/solution reaction rates in a mixed flow reactor: Wollastonite hydrolysis. Geochimica et Cosmo-chimica Acta 50, 2509–2516.

Rodriguez-Iturbe, I. 2000. Ecohydrology: A hydrologic perspective of climate-soil-vegetation dynamics. Water Resources Research 36, 3–9.

Rogers, J.R., and Bennett, P.C. 2004. Mineral simulation of subsurface microorganisms: Release of limiting nutrients from silicates. Chemical Geology 203, 91–108.

Roginsky, S.Z., and Zeldovich, J. 1934. An Equation for the Kinetics of Activated Adsorption. Acta Physicochimica USSR 1, 554.

Rosenberg, E., and Ron, E.Z. 1999. High- and low-molecular-mass microbial surfactants Applied Microbiology and Biotechnology 52, 154.

Rosso, K.M., Zachara, J.M., Fredrickson, J.K., Gorby, Y.A., and Smith, S.C. 2003. Nonlocal bacterial electron transfer to hematite surfaces. Geochimica et Cosmochimica Acta 67, 1081–1087.

Rufe, E., and Hochella, M.F., Jr. 1999. Quantitative assessment of reactive surface area of phlogopite during acid dissolution. Science 285, 874–876.

Rowland, F. S., and Cutler, J.C. 2007. Stratospheric ozone depletion by chlorofluorocarbons (Nobel Lecture). In Cutler, J.C. (ed), Encyclopedia of Earth, Environmental Information Coalition, National Council for Science and the Environment, Washington, DC.

Runkel, R.L. 1998. One-dimensional transport with inflow and storage (OTIS)—a solute transport model for streams and rivers, U.S. Geological Survey Water Resources Investigations Report 98–4018.

Runkel, R.L. 1995. Simulation models for conservative and nonconservative solute transport in streams. IAHS Publication 226, 153–159.

Runkel, R.L., Bencala, K.E., Broshears, R.E., and Chapra, S.C. 1996. Reactive solute transport in streams, 1. Development of an equilibrium-based model. Water Resources Research 32, 409–418.

Ryan, J.N., and Gschwend, P.M. 1990. Colloid mobilization in two Atlantic coastal-plain aquifers—Field studies. Water Resources Research 26, 307–322.

Saiers, J.E., and Ryan, J.N. 2005. Colloid deposition on non-ideal porous media: The influences of collector shape and roughness on the single-collector efficiency. Geophysical Research Letters 32, I.21406.

Schindler, P.W. 1990. Co-adsorption of metal ions and organic ligands: Formation of ternary surface complexes. In Hochella, M.F. Jr., and White, A.F. (eds), Mineral-water Interface Geochemistry. Reviews in Mineralogy Vol. 23. Mineralogical Society of America, Washington, DC, pp. 281–307.

Schindler, P.W., and Gamsjager, H. 1972. Acid-base reactions of the TiO_2 (anatase)-water interface and the point of zero charge of TiO_2 suspensions. Kolloid Z.Z. Pollymere 250, 759–763.

Schindler, P.W., and Kamber, H.R. 1968. Die Acidität von silanolgruppen. Helvetica Chimica Acta 51, 1781–1786.

Scherer, M.M., Richter, S., Valentine, R.L., and Alvarez, P.J.J. 2000. Chemistry and microbiology of permeable reactive barriers for in situ groundwater clean up. Critical Reviews in Microbiology 26, 221–264.

Schlautman, M. 1992. Mineral surfaces and humic substances: Partitioning of hydrophobic organic pollutants. Ph.D. dissertation, California Institute of Technology.

Schlich, T. 2002. Molecular Modeling and Simulation, Springer, New York.

Schnitzer M., and Khan S.U. 1978. Soil Organic Matter, Elsevier, New York.

Schnitzer, M. 1978. Humic substances: Chemistry and reactions. In Schnitzer, M., Khan, S.U. (eds), Soil Organic Matter Elsevier Scientific Publishing Company, New York, pp. 1–64.

Schnoor, J.L. 2000. Degradation by Plants—Phytoremediation, Biotechnology, Vol. IIB, Wiley-VCH, Weinheim, Germany.

Schopf, J.W., Hayes, J.M., and Waler, M.R. 1983. Evolution of Earth's earliest ecosystems: Recent progress and unsolved problems. In Schopf, J.W. (ed), Earth's Earliest Biosphere: Its Origin and Evolution, Princeton University Press, Princeton, NJ, pp. 214–239.

Schoonheydt, R.A. 1995. Clay mineral surfaces. In Vaughan, D.J., and Patrick, R.A.D. (eds), In Mineral Surfaces, Chapman and Hall, London, pp. 303–332.

Schroth, B.K., and Sposito, G. 1997, Surface charge properties of kaolinite. Clays and Clay Minerals 45, 85–91.

Schwarzenbach, R.P., Gschwend, P.M., and Imboden, D.M. 1993. Environmental Organic Chemistry. John Wiley & Sons, New York, 881 p.

Schulze, D.G., and Schwertmann, U. 1984. The influence of aluminium on iron oxides. X. Properties of Al-substituted goethites. Clay Minerals 19, 521–539.

Schwarzenbach, R.P., and Westall, J. 1981. Transport of nonpolar organic compounds from surface water to groundwater. Laboratory sorption studies. Environmental Science and Technology 15, 1360–1367.

Schwarzenbach, R.P., Gschwend, P.M., and Imboden, D.M. 2002. Environmental Organic Chemistry 2nd ed., John Wiley and Sons, New York, 1000 p.

Schwertmann, U., and Cornell, R.M. 1991. Iron Oxides in the Laboratory: Preparation and Characterization, VCH, Weinheim, Germany, 137 p.

Scully, N.M., Lean, D.R.S., McQueen, D.J., and Cooper, W.J. 1995. Photochemical formation of hydrogen peroxide in lakes: Effects of dissolved organic carbon and ultraviolet radiation. Canadian Journal of Fisheries and Aquatic Sciences 52, 2675–2681.

Seders, L., Shea, C., Lemmon, M., Talley, J., and Maurice, P.A. 2007. LakeNet: A sensor network for hydrobiogeochemical studies in lakes. Environmental Engineering Science 24, 183–191.

Senesi, N. 1996. Electron spin (or Paramagnetic) resonance spectroscopy. In Sparks, D. (ed), Methods of Soil Analysis, Part 3. Chemical Methods, Soil Science Society of America, Madison, WI, p. 323–356.

Senesi, N., and Boddy, L. 2002. A fractal approach for interactions between soil particles and microorganisms. In Huang, P.M., Bollag, J.-M., and Senesi, N. Interactions Between Soil Particles and Microorganisms John Wiley and Sons, New York, pp. 41–83.

Serne, R.J., Zachara, J.M., and Burke, D.S. 1988. Chemical information on tank supernatants, Cs adsorption from tank liquids into Hanford sediments, and field observations of Cs migration from past tank leaks; PNNL-11495, Pacific Northwest National Laboratory, Richland, WA.

Shannon, M.A., Bohn, P.W., Elimelech, M., Georgiadis, J.G., Mariñas, B.J., and Mayes, A.M. 2008. Science and technology for water purification in coming decades. Nature 452, 301–310.

Shannon, R.D. 1976. Revised effective ionic radii and systematic studies of interatomic distances in halides and chalcogenides. Acta Crystallography A32, 751–767.

Shaw, D.J. 1992. Introduction to Colloid and Surface Chemistry 4th ed., Butterworth Heinemann, Oxford, 306 p.

Shepherd, A. 2001. Activated carbon adsorption for treatment of VOC emissions. *Presented at the 13th Annual EnviroExpo, Boston MA, 2001*. Accessed at www.carbtrol.com/voc.pdf.

Shi, L., Chen, B.W., Wang, Z.M., Elias D.A., Mayer, M.U., Gorby, Y.A., Ni, S., Lower, B.H., Kennedy, D.W., Wunschel, D.S., Mottaz, H.M., Marshall, M.J., Hill, E.A., Beliaev, A.S., Zachara, J.M., Fredrickson, J.K., and Squier, T.C. 2006. Isolation of a high-affinity functional protein complex between OmcA and MtrC: Two outer membrane decaheme c-type cytochromes of Shewanella oneidensis MR0-1. Journal of Bacteriology 188, 4705–4714.

Shieh, W.Y., and Jean W.D. 1998. *Alterococcus agarolyticus*, gen nov., sp. nov., a halophilic thermophilic bacterium capable of agar degradation. Canadian Journal of Microbiology 44, 637–645.

Shrout, J.D., Chopp, D.L., Just, C.L., Hentzer, M., Givskov, M., and Parsek, M.R. 2006. The impact of quorum sensing and swarming motility on *Pseudomonas aeruginosa* biofilm formation is nutritionally conditional. Molecular Microbiology 62, 1264–1277.

Siffert, C., and Sulzberger, B. 1991. Light-induced dissolution of hematite in the presence of oxalate: A case study. Langmuir 19, 1627–1634.

Skinner, H.C.W. 2005. Biominerals. Mineralogical Magazine 69, 621–641.

Skinner, H.C.W., and Jahren, A.H. 2003. Biomineralization. In Schlesinger, W.H. (ed), Biogeochemistry, Treatise on Geochemistry, Vol. 8, Elsevier Science, pp. 117–184.

Smedley, P.L., and Kinniburgh, D.G. 2002. A review of the source, behaviour, and distribution of arsenic in natural waters. Applied Geochemistry 17, 517–568.

Smith, A.H., Lopipero, P.A., Bates, M.N., and Steinmaus, C.M. 2002. Public health-arsenic epidemiology and drinking water standards. Science 296, 2145–2146.

Smith, R.W., and Jenne, E.A. 1991. Recalculation, evaluation, and prediction of surface complexation constants for metal adsorption on iron and manganese oxides. Environmental Science and Technology 25, 525–531.

Solomons, T.W.G. 2008. Organic Chemistry, 9th rev. ed., John Wiley and Sons, New York.

Sparks, D.L.. 1989. Kinetics of Soil Chemical Processes, Academic Press, St. Louis, MO 210 p.

Sparks, D.L. 1998. Kinetics of sorption/release reactions on natural particles. In Huang, P.M., Senesi, N., and Buffle, J. (eds), Structure and Surface Reactions of Soil Particles, Vol. 4, Iupac Series on Analytical and Physical Chemistry of Environmental Systems, John Wiley and Sons, Chichester, England, pp. 413–448.

Sparks, D.L. 2006. Advances in elucidating biogeochemical processes in soils: It is about scale and interfaces. Journal of Geochemical Exploration 88, 243–245.

Sposito G. 1984. The Surface Chemistry of Soils, Oxford University Press, New York, 234 p.

Sposito, G. 1989. The Chemistry of Soils, Oxford University Press, New York, 277 p.

Sposito, G. 2004. The Surface Chemistry of Natural Particles, Oxford University Press, Oxford, New York, 242 p.

Sposito, G., Park, S.H., and Sutton, R. 1999. Monte Carlo simulation of the total radial distribution function for interlayer water in sodium and potassium montmorillonites. Clays and Clay Minerals 47, 192–200.

Sposito, G., and Prost, R., 1982. Structure of water adsorbed on smectites. Chemical Reviews 82, 553–573.

Sprinkle, C.L. 1989. Geochemistry of the Florida aquifer system in Florida and parts of Georgia, South Carolina, and Alabama, USGS Professional Paper 1403-I, Washington, DC.

Stallard, R.F. 1985. River chemistry, geology, geomorphology, and soils in the Amazon and Orinoco basins. In Drever, J.I. (ed), The Chemistry of Weathering, D. Reidel, Dordrecht, Holland, pp. 293–316.

Stallard, R.F., and Edmond, J.M. 1981. Geochemistry of the Amazon.I-Precipitation chemistry and the marine contribution to the dissolved load at the time of peak discharge. Journal of Geophysical Research 86, 9844–9858.

Stern, O. 1924. The theory of the electrolytic double shift. The theory of the electrolytic double shift. Zeitschrift Fur Elektrochemie Und Angewandte Physikalische Chemie 30, 508–516.

Stipp, S.L., Hochella, M.F., Parks, G.A., and Leckie, J.O. 1992. Cd^{2+} uptake by calcite, solid-state diffusion, and the formation of solid-solution: Interface processes observed with near-surface sensitive techniques (XPS, LEED, AND AES). Geochimica et Cosmochimica Acta 56, 1941–1954.

Stipp, S.L.S. 1999. Toward a conceptual model of the calcite surface: Hydration, hydrolysis, and surface potential. Geochimica et Cosmochimica Acta 63, 3121–3131.

Stowell, M.J. 1975. Defects in epitaxial deposits. In Matthew, J.W. (ed), Epitaxial Growth, part B. Academic Press, New York, 680 p.

Stucki, J.W., Lee K., Goodman, B.A., and Kostka, J.E. 2007. Effects of in situ biostimulation on iron mineral speciation in a sub-surface soil. Geochimica et Cosmochimica Acta 71, 835–843.

Stumm, W. 1992. Chemistry of the Solid-Water Interface: Processes at the Mineral-Water and Particle-Water Interface in Natural systems, Wiley Interscience, New York, 428 p.

Stumm, W., and Furrer, G. 1987. The dissolution of oxides and aluminum silicates; examples of surface-coordination-controlled kinetics. In Stumm, W. (ed), Aquatic Surface Chemistry, John Wiley and Sons, New York, pp. 197–219.

Stumm, W., Huang, C.P., and Jenkins, S.R. 1970. Specific chemical interactions affecting the stability of dispersed systems. Croatica et Chemica Acta 42, 223–244.

Stumm, W., and Morgan, J.J. 1996. Aquatic Chemistry, 2nd ed., John Wiley and Sons, New York, 1996.

Stumm, W., and Wieland, E. 1990. Dissolution of oxide and silicate minerals: Rates depend on surface speciation. In Stumm, W. (ed), Aquatic Chemical Kinetics, John Wiley and Sons, New York, pp. 367–400.

Sullivan, A.B., Drever, J.I., and McKnight, D.M. 1998. Diel variation in element concentrations, Peru Creek, Summit County, Colorado. Journal of Geochemical Exploration 64, 141–145.

Sumner, M.E., and Miller, W.P. 1996. Cation exchange capacity and exchange coefficients. In Sparks, D. (ed), Methods of Soil Analysis, Part 3. Chemical Methods. Soil Science Society of America, Madison, WI, pp. 1201–1229.

Sunagawa, I. 2005. Crystals: Growth, Morphology, and Perfection, Cambridge University Press, Cambridge, MA, 295 p.

Suter, D., Banwart, S., and Stumm, W. 1991. Dissolution of hydrous iron(III) oxides by reductive mechanisms. Langmuir 7, 809–813.

Sutheimer, S., Maurice, P., and Zhou, Q. 1999. Dissolution of well and poorly crystallized kaolinites: Al speciation and effects of surface characteristics. American Mineralogist, 84, 620–628.

Sutton, R., and Sposito, G. 2006. Molecular simulation of humic substance-Ca-montmorillonite complexes. Geochimica et Cosmochimica Acta 70, 3566–3581.

Suzuki, Y., Kelly, S.D., Kemner, K.M., and Banfield, J.F. 2002. Nanometre-size products of uranium bioreduction. Nature 419, 134.

Sverjensky, D.A., and Fukushi, K. 2006. Anion adsorption on oxide surfaces: Inclusion of the water dipole in modeling the electrostatics of ligand exchange. Environmental Science and Technology 40, 263–271.

Takahashi, Y., Manceau, A., Geoffroy, N., Marcus, M.A., and Usui, A. 2007. Chemical and structural control of the partitioning of Co, Ce, and Pb in marine ferromanganese oxides. Geochimica et Cosmochimica Acta 71, 984–1008.

Tang, Y.Z., Elzinga, E.J., Lee, Y.J., and Reeder, R.J. 2007. Coprecipitation of chromate with calcite: Batch experiments and X-ray absorption spectroscopy. Geochimica et Cosmochimica Acta 71, 1480–1493.

Tebo, B.M., Ghiorse, W.C., van Waasbergen, L.G., Siering, P.L., and Caspi, R. 1997. Bacterially-mediated mineral formation: Insights into manganese(II) oxidation from molecular genetic and biochemical studies. In Banfield, J.F., and Nealson, K.H. (eds), Geomicrobiology: Interactions Between Microbes and Minerals, Reviews in Mineralogy Vol. 35. Mineralogical Society of America, Washington, DC, pp. 225–266.

Tejedor-Tejedor, M.I., and Anderson, M.A. 1990. Protonation of phosphate on the surface of goethite as studied by CIR-FTIR and electrophoretic mobility. Langmuir 6, 602–611.

Templeton, A.S., Spormann, A.M., and Brown, G.E. 2003. Speciation of Pb(II) sorbed by Burkholderia cepacia/goethite composites. Environmental Science and Technology 37, 2166–2172.

Templeton, A.S., Trainor, T.P., Spormann, A.M., Newville, M., Sutton, S.R., Dohnalkova, A., Gorby, Y., and Brown, G.E. Jr. 2003. Environmental Science and Technology 37, 300.

Templeton, A.S., Trainor, T.P., Traina, S.J., Spormann, A.M., and Brown, G.E. Jr. 2001. Pb(II) distributions at biofilm-metal oxide interfaces. Proceedings of the National Academy of Sciences of the United States 98, 11897–11902.

Teng, H.H. 2004. Controls by saturation state on etch pit formation during calcite dissolution. Geochimica et Cosmochimica Acta 68, 253–262.

Teng, H.H., Dove, P.M., and DeYoreo, J.J. 2000. Kinetics of calcite growth: Surface processes and relationships to macroscopic rate laws. Geochimica et Cosmochimica Acta 64, 2255–2266.

Thompson, H.A., Parks, G.A., and Brown, G.E., 1999. Ambient-temperature synthesis, evolution, and characterization of cobalt-aluminum hydrotalcite-like solids. Clays and Clay Minerals 47, 425–438.

Thompson, J.B., and Ferris, F.G. 1990. Cyanobacterial precipitation of gypsum, calcite, and magnesite from natural alkaline lake water. Geology 18, 995–998.

Tingle, T.N., Hochella, M.F. Jr., Becker, C.H., and Mulhotra, R. 1990. Organic compounds on crack surfaces in olivine from San Carlos, Arizona nd Hualalai Volcano, Hawaii. Geochimica et Cosmochimica Acta 54, 477–485.

Tipping, E. 1981. The adsorption of aquatic humic substances by iron oxides. Geochimica et Cosmochimica Acta 45, 191–199.

Tomaic J., and Zutic V. 1998. Humic material polydispersity in adsorption at hydrous alumina/seawater interface. Journal of Colloid Interface Science. 26, 482–492.

Toner, B., Manceau, A., Marcus, M.A., Millet, D.B., and Sposito, G. 2005. Zinc sorption by a bacterial biofilm. Environmental Science and Technololgy 39, 8288–8294.

Trainor, T.P., Templeton, A.S., and Eng, P.J. 2006. Structure and reactivity of environmental interfaces: Application of grazing angle x-ray spectroscopy and long-period x-ray standing waves. Journal of Electron Spectroscopy and Related Phenomena 150, 66–85.

Trouwborst, R.E., Johnston, A., Koch, G., Luther, G.W. III, and Pierson, B.K. 2007. Biogeochemistry of Fe(II) oxidation in a photosynthetic microbial mat: Implications for Precambrian Fe(II) oxidation. Geochimica et Cosmochimica Acta 71, 4629–4643.

Turner, B.F., and Fein, J.B. 2006. Protofit: A program for determining surface protonation constants from titration data. Computers and Geosciences 32, 1344–1356.

Ueshima, M., and Tazaki, K. 2001, Possible role of microbial polysaccharides in nontronite formation. Clays and Clay Minerals 49, 292–299.

Ulrich, H.-J., Stumm, W., and Cosovic, B. 1988. Adsorption of aliphatic fatty acids on aquatic surfaces. Comparison between two model surfaces: The mercury electrode and γ-Al2O3 colloids. Environmental Science and Technology 22, 37–41.

Um, W., Serne, R.J., Brown, C.F., and Last, G.V. 2007. U(VI) adsorption on aquifer sediments at the Hanford Site. Journal of Contaminant Hydrology 93, 255–269.

Ushakov, S.V., and Navrotsky, A. 2005. Direct measurements of water adsorption enthalpy on hafnia and zirconia. Applied Physics Letters 87, 164103.

Vaccari, D.A., Strom, P.F., and Alleman, J.E. 2006. Environmental Biology for Engineers and Scientists, John Wiley and Sons, Hoboken, NJ, 931 p.

Van der Merve, J.H., and Ball, C.A.B. 1975. Energy of interfaces between crystals. In Matthew, J.W. (ed), Epitaxial Growth, Part B. Academic Press, New York, pp. 493–528.

van der Wal, A., Norde, W., Zenhder, A.J.B., and Lyklema, J. 1997. Determination of the total charge in the cell walls of Gram-positive bactera. Colloids and Surfaces B 9, 81–100.

van Olphen, H., and Fripiat, J.J. 1979. Data handbook for clay minerals and other nonmetallic minerals, Pergamon Press, New York.

Vasconcelos, I.F., Bunker, B.A., and Cygan, R.T. 2007. Molecular dynamics modeling of ion adsorption to the basal surface of kaolinite. Journal of Physical Chemimstry C 111, 6753–6762.

Vasconcelos, I.F., Haack, E.A. Maurice, P.A., and Bunker, B.A. 2008. EXAFS Analysis of Cadmium(II) Adsorption to Kaolinite. Chemical Geology 249, 236–250.

Vaughan, D.J., and Patrick, R.A.D., eds. 1995. Mineral Surfaces, Chapman and Hall, London, 370 p.

Verwey, E.J.W., and Overbeek, J.T.G. 1948. Theory of Stability of Lyophobic Colloids, 1st ed. Elsevier, Amsterdam, The Netherlands, 1 p.

Wall, J.D., and Krumholz, L.R. 2006. Uranium reduction. Annual Review of Microbiology 60, 149–166.

Wallmann, K. 2001. Controls on the Cretaceous and Cenozoic evolution of seawater composition, atmospheric CO2 and climate. Geochimica et Cosmochimica Acta 65, 3005–3025.

Walter, M. R., Bauld, J. & Brock, T. D. 1976. Heliotropism in modern stromatolites. Microbiology and morphogenesis of columnar stromatolites (Conophyton, Vacerrilla) from hot springs in Yellowstone National Park. In Stromatolites, ed. Walter, M. R., Elsevier Science, Amsterdam, pp. 273–310.

Wan, J., Wilson, J.L, and Kieft, T.L. 1994. Influence of the gas-water interface on transport of microorganisms through unsaturated porous media. Applied and Environmental Microbiology 60, 509–516.

Wang, Z., Ainsworth, C.C., Friedrich, D.M., Gassman, P.L., and Joly, A.G. 2000. Kinetics and mechanism of surface reaction of salicylate on alumina in colloidal aqueous suspension. Geochimica et Cosmochimica Acta 64, 1159–1172.

Warmack, R.J., Zheng, X.Y., Thundat, T., and Allison, D.P. 1994. Friction effects in the deflection of atomic-force microscope cantilevers. Review of Scientific Instruments 65, 394–399.

Warren, L.A., and Ferris, F.G. 1998. Continuum between sorption and precipitation of Fe(III) on microbial surfaces. Environmental Science and Technology 32, 2331–2337.

Warren, L.A., Maurice, P.A., Parmar, N., and Ferris, F.G. 2001. Microbially mediated calcium carbonate precipitation: Implications for interpreting calcite precipitation and for solid-phase capture of inorganic contaminants. Geomicrobiology Journal 18, 93–115.

Waychunas, G.A. 2001. Structure, aggregation and characterization of nanoparticles. In Banfield, J.A., and Navrotsky, A. (eds), Nanoparticles and the Environment. Reviews in Mineralogy and Geochemistry Vol. 44, Mineralogical Society of America and Geochemical Society, Washington, DC, pp. 105–166.

Waychunas, G.A., Kim, C.S., and Banfield, J.F. 2005. Nanoparticulate iron oxide minerals in soils and sediments: Unique properties and contaminant scavenging mechanisms. Journal of Nanoparticle Research 7, 409–433.

Waychunas, G.A., Trainor, T.P., Eng, P.J., Catalano, J., Brown, G.E. Jr., Davis, J., Rogers, J., and Bargar, J. 2005. Surface complexation studied via combined grazing-incidence EXAFS and surface diffraction: Arsenate on hematite (0001) and (1012). Analytical and Bioanalytical Chemistry 383, 12–27.

Wazne, M., Korfiatis, G.P., and Meng, X.G. 2003. Carbonate effects on hexavalent uranium adsorption by iron oxyhydroxide. Environmental Science and Technology 37, 3619–3624.

Weaver, M.L., Qui, S.R., Hoyer, J.R., Casey, W.H., Nancollas, G.H., and DeYoreo, J.J. 2007. Inhibition of calcium oxalate monohydrate growth by citrate and the effect of the background electrolyte. Journal of Crystal Growth 306, 135–145.

Weber, K.A., Achenbach, L.A., and Coates, J.D. 2006. Microorganisms pumping iron: anaerobic microbial oxidation and reduction. Nature Reviews Microbiology 4, 752–764.

Weiner, S., and Dove, P.M. 2003. An overview of biomineralization processes and the problem of the vital effect. In Dove, P.M., DeYoreo, J.J., and Weiner, S. (eds), Biomineralization. Reviews in Mineralogy and Geochemistry Vol. 54. Mineralogical Society of America Geochemical Society, Washington, DC, pp. 1–29.

White, A.F. 1995. Chemical Weathering Rates of Silicate Minerals in Soils. In White, A.F., and Brantley, S.L. (eds), Chemical Weathering Rates of Silicate Minerals. Reviews in Mineralogy Vol. 31, Mineralogical Society of America, Washington, DC, pp. 407–461.

White, A.F. 2004. Natural weathering rates of silicate minerals. In Drever, J.I. (ed), Treatise on Geochemistry, Vol. 5, Surface and Groundwater, Weathering and Soils, Elsevier, New York, pp. 133–168.

White, A.F., Blum, A.E., Bullen, T.D., Vivit, D.V., Schulz, M., and Fitzpatrick, J. 1999. The effect of temperature on experimental and natural weathering rates of granitoid rocks. Geochimica et Cosmochimica Acta 63, 3277–3291.

White, A.F., Blum, A.E., Schulz M.S., Bullen, T.D., Harden, J.W., and Peterson, M.L. 1995. Chemical weathering rates of a soil chronosequence on granitic alluvium. 1. Quantification of mineralogical and surface area changes and calculation of primary silicate reaction rates. Geochimica et Cosmochimica Acta 60, 2533–2550.

White, A.F., and Brantley, S.L. 2003. The effect of time on the weathering of silicate minerals: Why to weathering rates differ in the laboratory and the field? Chemical Geology 202, 479–506.

White, A.F., Bullen, T.D., Schulz, M.S., Blum, A.E., Huntington, T.G., and Peters, N.E. 2001. Differential rates of feldspar weathering in granitic regoliths. Geochimica et Cosmochimica Acta 65, 847–869.

White, A.F., Schulz, M.S., Lowenstern, J.B., Vivit, D.V., and Bullen, T.D. 2005. The ubiquitous nature of accessory calcite in granitoid rocks: Implications for weathering, solute evolution, and petrogenesis. Geochimica et Cosmochimica Acta 69, 1455–1471.

White, A.F., Schulz, M.S., Vivit, D.V., Blum, A.E., and Stonestrom, D.A. 2006. Controls on soil pore water solutes: An approach for distinguishing between biogenic and lithogenic processes. Journal of Geochemical Exploration 88, 363–366.

Widdel, F., Schnell, S., Heising, S., Ehrenrich, A., Assmus, B., and Schink, B. 1993. Ferrous iron oxidation by anoxygenic phototrophic bacteria. Nature 362, 834–836.

Wiesner, M.R., Lowry, G.V., Alvarez, P., Dionysiou, D., and Biswas, P. 2006. Assessing the risks of manufactured nanomaterials. Environmental Science and Technology 40, 4336–4345.

Wiggington, N.S., Haus, K.L., and Hochella, M.F. Jr. 2007. Aquatic environmental nanoparticles. Journal of Environmental Monitoring 9, 1–11.

Wijnja, H., and Schulthess, C.P. 2000a. Interaction of carbonate and organic anions with sulfate and selenate adsorption on an aluminum oxide. Soil Science Society of America Journal 64, 898–908.

Wijnja, H., and Schulthess, C.P. 2000b. Vibrational spectroscopy study of selenate and sulfate adsorption mechanisms on Fe and Al (hydr)oxide surfaces. Journal of Colloid and Interface Science 229, 286–297.

Wijnja, H., and Schulthess, C.P. 2002. Effect of carbonate on the adsorption of selenate and sulfate on goethite. Soil Science Society of America Journal 66, 1190–1197.

Wildman, W.E., Jackson, M.L., and Whittig, L.D. 1968. Iron-rich montmorillonite formation in soils derived from serpentinite. Soil Science Society of America Proceedings 32, 787–794.

Wilson, B., Dewers, T., Reches, Z., and Brune, J. 2005. Particle size and energetics of gouge from earthquake rupture zones. Nature 434, 749–752.

Woertz, J.R., and Kinney, K.A. 2002. Nitrogen effects on fungal culture growth and bioreactor performance. Proceedings of the USC-TRG Conference on Biofiltration, Newport Beach, CA.

Woese, C.R., Kandler, O., and Wheelis, M.L. 1990. Towards a natural system of organisms: proposal for the domains Archaea, Bacteria, and Eucarya. Proceedings of the National Academy of Science of the United States of America 87, 4576–4579.

Wood, W.W., and Low, W.H. 1988. Solute geochemistry of the Snake River Plain Regional Aquifer System, Idaho and Eastern Oregon. U.S. Geological Survey Professional Paper 1408-D, 78 p.

Wyllie, P.J. 1971. The Dynamic Earth, John Wiley and Sons, New York.

Xiang, X., Huang, Y., Madey, G., Cabaniss, S., Arthurs, L., and Maurice, P. 2006. Modeling the evolution of natural organic matter in the environment with an agent-based stochastic approach. Natural Resources Modeling Joural 19, 67–90.

Xu, S., and Boyd, S.A. 1995. Alternative model for cationic surfactant adsorption by layer silicates. Environmental Science and Technology 29, 3022–3028.

Xu, W., Johnston, C.T., Parker, P., and Agnew, S.F. 2000. Infrared study of water sorption on Na-, Li-, Ca- and Mg-exchanged (SWy-1 and SAz-1) montmorillonite. Clays Clay Minerals 48, 120–131.

Yang, X.F., Sun, Z.X., Wang, D.S., and Forsling, W. 2007. Surface acid-base properties and hydration/dehydration mechanisms of aluminum (hydr)oxides. *Journal of Colloid Interface Science* 308, 395–404.

Young, J.B., and Henricksen, K. 2003. Biomineralization within vesicles: The calcite of coccoliths. In Dove, P.M., DeYoreo, J.J., and Weiner, S. (eds), Biomineralization, Vol. 54, Reviews in Mineralogy and Geochemistry, The Mineralogical Society of America and the Geochemical Society, Washington, DC, pp. 189–215.

Zachara, J.M., Kukkadapu, R.K., Gassman, P.L., Dohnalkova, A., Fredrickson, J.K., and Anderson, T. 2004. Biogeochemical transformation of Fe minerals in a petroleum-contaminated aquifer. *Geochimica et Cosmochimica Acta* 68, 1791–1805.

Zelazny, L.W., He, L., and Vanwormhoudt, A.M. 1996. Charge analysis of soils and anion exchange. In Sparks, D.L., Page, A.L., Helmke, P.A., Loeppert, R.H., Soltanpour, P.N., Tabatabai, M.A., Johnston, C.T., and Sumner, M.E. (eds), Methods of Soil Analysis 3—Chemical Methods, Soil Science Society of America, Madison, WI, pp. 1231–1254.

Zhou, Q., Cabaniss, S.E., and Maurice, P.A. 2000. Considerations in the use of high-pressure size exclusion chromatography (HPSEC) for determining molecular weights of aquatic humic substances. Water Research 34, 3505–3514.

Zhou, Q., Maurice, P.A., and Cabaniss, S.E. 2001. Size fractionation upon adsorption of fulvic acid on goethite: Equilibrium and kinetic studies. Geochimica et Cosmochimica Acta 65, 803–812.

Zhu, C., Veblen, D.R., Blum, A.E., and Chipera, S.J. 2006a. Naturally weathered feldspar surfaces in the Navajo Sandstone aquifer, Black Mesa, Arizona: Electron microscopic characterization. *Geochimica et Cosmochimica Acta* 70, 4600–4616.

Zhu, S.Q., Oberdorster, E., and Haasch, M.L. 2006b. Toxicity of an engineered nanoparticle (fullerene, C60) in two aquatic species, *Daphnia* and fathead minnow. *Marine Environmental Research* 62, S5–S9.

INDEX

Environmental Surfaces and Interfaces from the Nanoscale to the Global Scale, by Patricia A. Maurice

Printed and bound by CPI Group (UK) Ltd, Croydon, CR0 4YY
16/09/2026
14963415-0006